FUNDAMENTALS OF MICROWAVE AND RADAR ENGINEERING

FUNDAMENTALS OF MICROWAVE AND RADAR ENGINEERING

For B.E./B.Tech. Students of Electronics & Communication
Engineering as per All Indian Technical Universities

Er. K.K. SHARMA
Assistant Professor
Deptt. of Electronics & Communication Engg.
Lingaya's University
Faridabad

S Chand And Company Limited
(ISO 9001 Certified Company)
RAM NAGAR, NEW DELHI - 110 055

S Chand And Company Limited
(ISO 9001 Certified Company)

Head Office: 7361, RAM NAGAR, QUTAB ROAD, NEW DELHI - 110 055
Phone: 23672080-81-82, 66672000 Fax: 91-11-23677446
www.schandpublishing.com; e-mail: info@schandpublishing.com

Branches:

Ahmedabad	:	Ph: 27541965, 27542369, ahmedabad@schandpublishing.com
Bengaluru	:	Ph: 22268048, 22354008, bangalore@schandpublishing.com
Bhopal	:	Ph: 4209587, bhopal@schandpublishing.com
Chandigarh	:	Ph: 2625356, 2625546, 4025418, chandigarh@schandpublishing.com
Chennai	:	Ph: 28410027, 28410058, chennai@schandpublishing.com
Coimbatore	:	Ph: 2323620, 4217136, coimbatore@schandpublishing.com (Marketing Office)
Cuttack	:	Ph: 2332580, 2332581, cuttack@schandpublishing.com
Dehradun	:	Ph: 2711101, 2710861, dehradun@schandpublishing.com
Guwahati	:	Ph: 2738811, 2735640, guwahati@schandpublishing.com
Hyderabad	:	Ph: 27550194, 27550195, hyderabad@schandpublishing.com
Jaipur	:	Ph: 2219175, 2219176, jaipur@schandpublishing.com
Jalandhar	:	Ph: 2401630, jalandhar@schandpublishing.com
Kochi	:	Ph: 2809208, 2808207, cochin@schandpublishing.com
Kolkata	:	Ph: 23353914, 23357458, kolkata@schandpublishing.com
Lucknow	:	Ph: 4065646, lucknow@schandpublishing.com
Mumbai	:	Ph: 22690881, 22610885, 22610886, mumbai@schandpublishing.com
Nagpur	:	Ph: 2720523, 2777666, nagpur@schandpublishing.com
Patna	:	Ph: 2300489, 2260011, patna@schandpublishing.com
Pune	:	Ph: 64017298, pune@schandpublishing.com
Raipur	:	Ph: 2443142, raipur@schandpublishing.com (Marketing Office)
Ranchi	:	Ph: 2361178, ranchi@schandpublishing.com
Sahibabad	:	Ph: 2771235, 2771238, delhibr-sahibabad@schandpublishing.com

© 2011, Er. K.K. Sharma

All rights reserved. No part of this publication may be reproduced or copied in any material form (including photocopying or storing it in any medium in form of graphics, electronic or mechanical means and whether or not transient or incidental to some other use of this publication) without written permission of the copyright owner. Any breach of this will entail legal action and prosecution without further notice.

Jurisdiction: All disputes with respect to this publication shall be subject to the jurisdiction of the Courts, Tribunals and Forums of New Delhi, India only.

First Edition 2011
Reprint 2016
Reprint 2018

ISBN : 978-81-219-3537-1   **Code** : 1010 448

PRINTED IN INDIA

By Vikas Publishing House Pvt. Ltd., Plot 20/4, Site-IV, Industrial Area Sahibabad, Ghaziabad-201010 and Published by S Chand And Company Limited, 7361, Ram Nagar, New Delhi-110 055.

*This book is dedicated
to
my GURU JI*

Ananth Sri Vibhushit Indraprasth
and Haryana Peethadheeshwar
Srimad Jagadguru Ramanujacharya
Swami Sri Sudarshanacharya Ji
Maharaj

PREFACE

In the recent years microwaves are being used in radar system, satellite communication, broadcasting of Television signals, astronomic research and in many applications in industry. This book **Fundamentals of Microwave and Radar Engineering** is intended to be used as a textbook that relatively covers all topics to the subject at undergraduate and postgraduate engineering levels. The fundamental concepts of microwave engineering are explained in a simple manner. Although the subject of Microwave and Radar Engineering is of particular interest to the specialists in that field, it is also of interest to persons, especially the military and civilian users of radar, satellite and navigational systems.

The author has 20 yrs of vast exposure on different types of radar system used in Indian Air Force as well as at the civil Airports. The author also has teaching experience at Air Force Technical College, Bangalore and has experience of installation and commissioning of radars.

The book has been written in a very simple language keeping in focus the syllabi prescribed for the subject in different institutions and universities. All the topics of this book has been illustrated with the photograph and solved problems. The students preparing for the examinations can understand the concept in a better way by doing the solved and unsolved problems at the end of each chapter. Although utmost care has been taken to minimize the errors, author expect from the readers to provide feedback so that in future edition this book will be more useful.

FARIDABAD   **Er. K.K. SHARMA**
E- mail- krishan_ksharma@yahoo.com

ACKNOWLEDGEMENT

I take this opportunity to express my gratitude to the respected Prof G.V.K Sinha, Chancellor of Lingaya's University, Dr. K.K. Aggarwal, Chief Patron Lingaya's University, Dr. P. Gadde, Pro Chancellor, Lingaya's University and Dr. S. Mukhopadhyay, Pro Vice-Chancellor, Lingaya's University, Faridabad for providing innovative and encouraging environment to work on this project.

I would like also to express my thanks to Dr. Pratap Singh, Director, LIMAT, Faridabad, for constant inspiration and encouragement. .

I would like to express my sincere thank to Dr. SVAV Prasad (HOD E&CE), Lingaya's University, Faridabad, for their valuable technical suggestions and constant encouragement, without, which this book would not have, come into existence.

I would like to express my sincere gratitude to Dr. K.K. Gupta, Professor in ECE Department (Manav Rachna International University, Faridabad) and Cdr Rajendera Kumar, Professor (HOD E&CE), Echelon Institute of Technology, Faridabad, for constant inspiration and motivation.

I would also like to thank to the Management, teachers, students and supporting staff of Lingaya's University.

I am grateful to my wife Sunita, my daughters Aparna and Vaishali, and son Aman for supporting me to complete this project.

I would like to express my gratitude to all the authors whose work has been freely consulted in the preparation of this text.

I deeply express my thanks to Mr. Pranav Debnath for supporting me to type the manuscript in time.

A special thanks to the publishers S. Chand & Company Limited for publishing this book in such a beautiful get-up and well in time.

Lastly, I am grateful to my father and mother without their blessing this book would have not come into existence.

Er. K.K. Sharma

4. MICROWAVE CAVITY RESONATORS 109–125
- Introduction ... 109
- Advantages of Cavity Resonators ... 109
- Types of Cavity Resonators ... 110
- Operation of Microwave Cavity Resonator ... 110
- Field Expressions ... 113
- Resonating Condition ... 113
- Advantages and Features ... 113
- Applications of Cavity Resonators ... 114
- Field Expression for Rectangular Cavity Resonator ... 115
- Quality Factor (Q) of Cavity Resonator ... 120
- Transmission Line Resonator ... 122
- Summary ... 124
- Review Questions ... 125

5. MICROWAVE COMPONENTS AND DEVICES 126–186
- Introduction ... 126
- Scattering Parameters ... 127
- Waveguide Plumbing ... 129
- Microwave Junctions or Waveguide Junctions ... 129
- Hybrid Ring ... 144
- Waveguide Impedance Matching ... 148
- Waveguide Terminations ... 148
- Waveguide Bends ... 151
- Waveguide Twist ... 153
- Flexible Waveguide ... 153
- Waveguide Joints (Rotating and Permanent Joints) ... 153
- Directional Coupler ... 155
- Hybrid Coupler ... 163
- Waveguide and Microwave Attenuators ... 166
- Ferrite Devices (Ferrite, Isolator, Circulator, Insulator, Gyrator) ... 170
- Waveguide Phaseshifter ... 178
- Microwave filters ... 180
- Summary ... 182
- Review Questions ... 186

6. MICROWAVE TUBES 187–244
- Introduction to Microwave Tubes ... 187
- Frequency Limitations of conventional tubes ... 188
- Microwave Tubes ... 192

CONTENTS

1. INTRODUCTION TO MICROWAVE   1 – 6
- Introduction   1
- Brief History of Microwave Technology   1
- Microwave Frequencies   2
- Characteristic Features of Microwaves   3
- Advantages of Microwaves Over Low frequences   3
- Applications of Microwave   4
- Summary   6
- Review Questions   6

2. TRANSMISSION LINES THEORY   7 – 29
- Introduction   7
- Basic Transmission Line Equations   8
- Characteristic Impedance   13
- Transmission Line Parameters   14
- Matched Load   15
- Transmission Line Discontinuities and Load Impedances   16
- Input Impedance   16
- Directions of Travel on a Transmission Line   17
- Transformation of Impedance   18
- Impedance Matching   19
- Stub Matching   21
- Transmission Line Sections as Circuit Elements   22
- Some Practicalities   23
- Smith Chart   24
- Insertion Loss   28
- Summary   28
- Review Questions   29

3. MICROWAVE TRANSMISSION LINES   30 – 108
- Introduction   30
- Types of Transmission Lines   31
- Losses in Transmission Lines   42
- Length of a Transmission Lines   43
- Waveguides   44
- Circular Waveguide   74
- Planar Transmission Lines   83
- Summary   91
- Solved Problems   98
- Review Questions   107

4. MICROWAVE CAVITY RESONATORS   109 – 125
- Introduction   109
- Advantages of Cavity Resonators   109
- Types of Cavity Resonators   110
- Operation of Microwave Cavity Resonator   110
- Field Expressions   113
- Coupling to Cavity   113
- Tuning of Cavities   113
- Applications of Cavity Resonators   114
- Field Expression for Rectangular Cavity Resonator   115
- Quality Factor (Q) of Cavity Resonators   120
- Transmission Line Resonator   122
- Summary   124
- Review Questions   125

5. MICROWAVE COMPONENTS AND DEVICES   126 – 186
- Introduction   126
- Scattering Parameters   127
- Waveguide Plumbing   129
- Microwave Junctions or Waveguide Junctions   129
- Hybrid Ring   144
- Waveguide Impedance Matching   145
- Waveguide Terminations   148
- Waveguide Bends   151
- Waveguide Twist   153
- Flexible Waveguide   153
- Waveguide Joints(Rotating and Permanent joints)   153
- Directional Coupler   155
- Hybrid Coupler   163
- Waveguide and Microwave Attenuators   166
- Ferrites Devices (Ferrite Attenuator, Circulators, Isolator, Gyrator)   170
- Waveguide Transition   178
- Microwave Filters   180
- Summary   182
- Review Questions   186

6. MICROWAVE TUBES   187 – 244
- Introduction to Microwave Tubes   187
- Frequency Limitations of conventional tubes   188
- Microwave Tubes   192

	• Klystrons	192
	• The Traveling Wave Tube (TWT)	212
	• Backward Wave Oscillators	219
	• Microwave Cross Field Tubes	223
	• Magnetron Oscillators	224
	• Summary	235
	• Solved Problems	237
	• Review Questions	242
7.	**MICROWAVE SEMICONDUCTOR DEVICES**	**245 – 320**
	• Introduction	245
	• Microwave Semiconductor Diodes	246
	• Tunnel Diode Devices	246
	• Varactor Diode	254
	• Step-Recovery Diode	259
	• Bulk-Effect Semiconductors	268
	• The Point-Contact Diode	279
	• Crystal Diodes	282
	• Schottky Barrier Diode	283
	• PIN Diodes	284
	• Stimulated Emission and Associated Devices	288
	• Avalanche Transit Time Devices	291
	• Microwave Transistors	297
	• Microwave Field Effect Transistors	303
	• MOS Field-Effect Transistors	308
	• CMOS	315
	• MOSFET Memory	316
	• Summary	317
	• Review Questions	319
8.	**MICROWAVE MEASUREMENTS**	**321 – 344**
	• Introduction	321
	• Detection of Microwave Signal	321
	• Frequency and Wavelength Measurements	324
	• Measurement of VSWR	327
	• Impedance Measurement	329
	• Insertion Loss and Attenuation Measurements	332
	• Power Measurement	334
	• Radiation Pattern Measurements	340
	• Network Analyzer	341

- Summary — 342
- Review Questions — 343

9. MICROWAVES ANTENNAS — 345 – 360
- Introduction — 345
- Types of Microwaves Antennas — 345
- Parabolic Reflector — 345
- Cassegrain Antenna — 350
- Horn Antennas — 351
- Slot Antennas — 356
- Microstrip Antennas — 357
- Summary — 359
- Review Questions — 360

10. RADIO WAVE PROPAGATION – AN OVERVIEW — 361 – 374
- Introduction — 361
- Radio Links — 361
- EM Wave Propagation : Introduction — 362
- Mechanism of Propagation — 367
- Vertical and Oblique Incidence – Critical Frequency and Critical Angle — 370
- Skip Distance, Skip Zone and Multiple Hop Transmission — 371
- MUF, LUF and OUF (Usable Frequencies) — 371
- Fading — 372
- Summary — 373
- Important Formulas — 374
- Review Questions — 374

11. GROUND WAVE PROPAGATION — 375 – 400
- Introduction — 375
- Friss Power Transmission Equation and Free Space Path Loss — 375
- Space Wave Propagation — 376
- Surface Wave Propagation — 389
- Summary — 393
- Important Formulas — 394
- Solved Problems — 395
- Review Questions — 399

12. SKY WAVE PROPAGATION – THE IONOSPHERE WAVES — 401 – 425
- Introduction: Ionosphere — 401
- Characteristics of Different Ionized Regions — 402
- Ionospheric Variations — 403
- Effect of Ionosphere on Radio Waves — 405

- Summary — 416
- Important Formulas — 417
- Solved Problems — 417
- Review Questions — 422

13. RADAR — 426 – 461

- Introduction — 426
- Principle of RADAR — 427
- Radar Block Diagram — 428
- Range Determination — 429
- Common Parameters of Radar Pulse — 429
- Applications of Radar — 430
- Radar Frequencies — 431
- Classification of Radar — 432
- Basic Pulse Radar System — 433
- Simple form of Radar Equation — 435
- Effect of Pulse Width (PW) on Range — 437
- Pulse Repetition Frequency (PRF) and Range Ambiguities — 438
- Minimum Detectable Signal — 441
- Radar Receiver Noise and Signal to Noise Ratio — 441
- Continuous Wave (CW) Radar — 442
- Frequency Modulated CW Radar — 444
- Moving Target Indicator (MTI) and Doppler Radar — 445
- Receiver Noise — 448
- Low Noise Front Ends — 449
- Duplexer — 450
- Radar Display — 451
- Radar Antennas — 452
- Historial Overview — 454
- Solved Problems — 455
- Unsolved Problems — 459
- Review Questions — 460

14. MICROWAVE COMMUNICATION SYSTEMS — 462 – 488

- Introduction — 462
- Propagation of Microwave Frequencies — 462
- Applications of Microwave Frequencies in Communication — 463
- Summary — 486
- Review Questions — 488

CHAPTER 1

INTRODUCTION TO MICROWAVE

OBJECTIVES
- Introduction
- Brief History of Microwave Technology
- Microwave Frequencies
- Characteristic Features of Microwave
- Advantages of Microwaves Over Low Frequencies
- Applications of Microwave

1.1. INTRODUCTION

Microwave is the electromagnetic (EM) wave spectrum in the frequency range from 1 GHz to 100 GHz. In other words, the wavelength of EM waves at microwave frequencies is very small; typically from 30 cm to 0.3 mm. Sometimes higher frequencies (extending up to 600 GHz) are also called 'microwaves'. Before discussing the principles and applications of microwave frequencies, the meaning of the term microwave as it is used in this book must be established. On the surface, the definition of a microwave would appear to be simple because, in electronics, the prefix "micro" normally means a millionth part of a unit. Micro also means small, which is a relative term, and it is used in this sense in the book. Microwave is a term loosely applied to identify electromagnetic waves above 1000 megahertz in frequency because of the short physical wavelengths of these frequencies because of the short physical wavelengths of these frequencies. Microwaves present several interesting and unusual features not found in other portions of the electromagnetic frequency spectrum. These features make 'microwaves' uniquely suitable for various useful applications. Small antennas and other small components are made possible by microwave frequency applications.

1.2. BRIEF HISTORY OF MICROWAVE TECHNOLOGY

The scientists of nineteenth century laid the foundation of communication using RF wave and wire less technology, which has affected modern environment. In 1845 Michel Faraday observed the effect of magnetic field on the light propagation. In the year 1864, James. C. Maxwell developed the four basic equations of electromagnetic theory of light. He put forth fundamental relation of electro magnetic field and also summed up the research observations of Poisson, Laplace, Faraday and Gauss. He also predicted about the propagation of electrical signals in free space. In 1893, Heinrich Hertz subsequently verified (by conducting an experiment) to show that a parabolic antenna fed by an element i.e. dipole on excitation by a spark discharge, sends a signal to an identical receiving arrangement at a distance. G. Marconi transmitted wireless signals across the Atlantic Ocean successfully. William Thompson developed the wave-guide theory and the mode properties of propagation through hollow metallic wave-guide.

The development in this field still continues and interested readers may find more information in the IEEE transaction on microwave theory and techniques (vol-MTT-32, SEPTEMBER 1984).

1.3. MICROWAVE FREQUENCIES

Table 1.1 and 1.2 shows the IEEE frequency based designation and microwave frequency band designation respectively.

Figure 1.1 shows the EM spectrum with the main applications.

Table 1.1. IEEE frequency based designation

Band Designation	Frequency Range	Wavelength Range (in Free Space)
LVF	3 – 30 kHz	10 – 100 km
LF	30 – 300 kHz	1 – 10 km
MF	300 – 3000 kHz	100 m – 1 km
FH	3 – 30 MHz	10 – 100 m
VHF	30 – 300 MHz	1 – 10 m
UHF	300 – 3000 MHz	10 cm – 1 m
SHF	3 – 30 GHz	1 – 10 cm
EHF	30 – 300 GHz	0.1 – 1 cm

Table 1.2. Microwave Frequency based designation

Frequency Bands	Old	New
500 – 1000 MHz	L	C
1 – 2 GHz	S	D
2 – 4 GHz	S	E
3 – 4 GHz	S	F
4 – 6 GHz	C	G
6 – 8 GHz	C	H
8 – 10 GHz	X	I
10 – 12.4 GHz	X	J
12.4 – 18 GHz	Ku	J
18 – 20 GHz	K	J
20 – 26.5 GHz	K	K
26.5 – 40 GHz	Ka	K

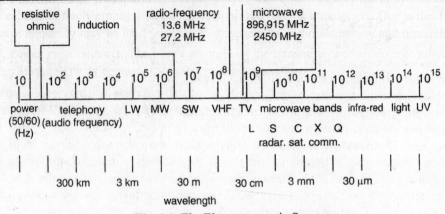

Fig. 1.1. The Electromagnetic Spectrum.

INTRODUCTION TO MICROWAVE

Microwave frequencies present special problems in transmission, generation, and circuit design that are not encountered at lower frequencies. Conventional circuit theory is based on voltages and current while microwave theory is based on electromagnetic fields. The concept of electromagnetic field interaction is not entirely new, since electromagnetic fields the basis of antenna theory. However, many students of electronics find electromagnetic field theory in the simplest terms possible but many of the concepts are still somewhat difficult to thoroughly understand. Therefore, you must realize that this book will require very careful study for you to properly understand microwave theory.

1.4. CHARACTERISTIC FEATURES OF MICROWAVES

The main characteristic features of microwave come from the small of wavelengths (30 cm to 0.3 mm) in relation to the sizes of components or devices normally used. Short wavelength energy offers distinct advantages in many applications. For instance, excellent directivity can be obtained using relatively small antenna and low-power transmitters. These features are ideal for use in both military and civilian radar and communication applications.

Because of small wavelength, the phase varies rapidly with distance; consequently the techniques of circuit analysis and design, of measurements and power generation, and amplification at these frequencies distinct from those at lower frequencies. For dealing with these small wavelengths, methods of circuit analysis and representation need to be modified. Analysis of circuits based on Kirchhoff's laws and voltage-current concepts are sufficient to describe the circuit behavior at microwave frequencies. It is necessary to analyze the components or circuits in terms of electric and magnetic fields associated with it.

It's not only the analytic techniques, even the methods of measurements are also become specialized at these frequencies. Measurements are carried out in respects of phase differences, field amplitudes and power carried by the waves.

Because of small wavelengths, use of lumped elements at microwave frequencies becomes difficult. For realizing the lumped behavior of a resistor, a capacitor or an inductor size must be much smaller than the wavelength and due to this reason microwave systems employ distributed circuit elements, which are made up of small sections of transmission lines and waveguides. For examples, a half wavelength section of transmission line constitutes a resonant circuit to be used in place of an L-C resonator and a quarter wavelength sections, on the other hand, is used as an impedance transformer.

The generation of microwaves has a variety of devices- both in vacuum tube and in semiconductor device fields. It was attempted to generate microwave frequencies by using a low frequency source (triode or transistor) but the operation is limited by the fact that transit time of the carriers through the devices becomes comparable to the time period of the wave. This problem has been solved by technological innovations and by totally new ideas.

1.5. ADVANTAGES OF MICROWAVES OVER LOW FREQUENCIES

Following are the various advantages of microwaves over low frequencies:

1. Directivity: The first main characteristic of microwave is the directivity. We know that is the frequency increased the directivity increases and beam width decreases.

Beam width 'θ_B' of the parabolic reflector is given by $\theta_B = 140° \times \lambda/D$

where,

θ_B = Beam width (Degree), λ = Wavelength (cm) and D = Diameter of antenna (cm).

For example at 60 GHz (wavelength = 0.5 cm) for 1-degree beam width, the diameter of the parabolic antenna is given by

$$D = 140 \times \lambda / \theta_B$$

$$D = 140 \times 0.5 = 70 \text{ cm}$$

If the frequency is 300 MHz ($\lambda = 100$ cm) for 1-degree beam width, the diameter of the parabolic antenna will be

$$D = 140 \times 100 = 140 \text{ m}$$

In second case where frequency is 300 MHz, for 1° beam width, the size of the antenna increases up to 140m, whereas at higher frequency i.e., 60 GHz, the size of the antenna is only 70cm which is very small than the 140m.

Microwave frequencies are said to posse's quasi-optical properties because it has extremely directed beam just as an optical lens focuses the light rays.

2. More radiated power or gain: As the frequency increases the λ is decreases and the relation between power radiated and λ is given by

$$P_r = \mu_0 r^2 . I_0^2 (l/\lambda)^2$$

where,

Pr = Power radiated, I_0 = IAC current, l = length of the antenna, λ = Wavelength of the frequency used.

By analysis of the above relation, it is clear that as λ (Wavelength of the frequency) is decreasing the gain or power radiated is increasing. In other works, the power is inversely proportional to λ^2, and high gain can be designed which is highly impracticable at lower frequencies.

3. More bandwidth at higher frequency: A large number of communication channels or Television programs can be sent through a single microwave carrier over one communication link. For example, to transmit 100 channels of TV programs, the bandwidth required is approximately 600 MHz which is just a 1% of 6 MHz.

4. Penetration or transparency property: The microwave signals travel by line of sight and are capable of freely propagation through the ionized layer in atmosphere. Due to this feature microwave frequency may be used in space communication, radio astronomy and research.

5. Power requirements: The power required by the transmitter and receiver at microwave frequency is quite low as compared to low frequency operation.

1.6. APPLICATIONS OF MICROWAVE

Most of the applications of microwave arise from the properties and the advantageous features of the microwave frequency. In modern technology the microwaves have a wide range of applications as shown in Fig. (1.2).

INTRODUCTION TO MICROWAVE

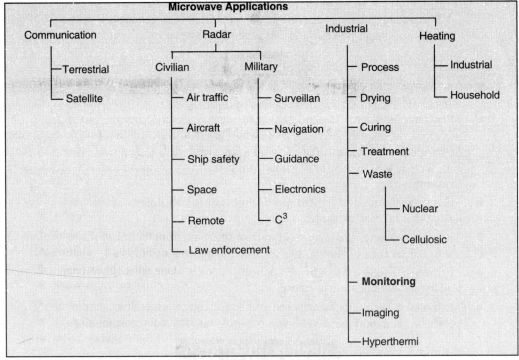

Fig. 1.2. Microwave applications.

The application may be divided broadly in areas as appended below:

1. Communication: In the field of communications microwave frequency has many applications such as Television, Telephone and Satellite communication.

2. Radars: In military, radars are used for the following purposes:
  (a) Surveillance
  (b) Navigations
  (c) Guidance of weapons
  (d) Electronic warfare (ECM and ECCM) etc.

The civilian applications of radars are as follows:
  (a) Air traffic control (ATC)
  (b) Aircraft navigations
  (c) Ship safety
  (d) Space vehicles
  (e) Remote sensing
  (f) Law enforcement

3. Industrial applications:
  (a) Processing control (in food industry)
  (b) Drying machines used in food, paper, clothes and textile industries.
  (c) Waste treatments
  (d) Rubber, plastics, chemical and forest product industries
  (e) Biomedical applications (Monitoring of heart beat, lung water detection and imaging of human body)

(f) Microwave oven for industry as well as household application.

(g) Microwave techniques are now being introduced in extremely fast operations. Pulses with very small pulse widths are used in high speed logic circuits.

SUMMARY

- Microwave is the electromagnetic (EM) wave spectrum in the frequency range from 1 GHz to 100 GHz.
- Microwave is a term loosely applied to identify electromagnetic waves above 1000 megahertz in frequency because of the short physical wavelengths of these frequencies. Short wavelength energy offers distinct advantages in many applications. For instance, excellent directivity can be obtained using relatively small antennas and low-power transmitters.
- Microwave frequencies present special problems in transmission, generation, and circuit design that are not encountered at lower frequencies.
- The main characteristic features of microwave come from the small of wavelengths (30 cm to 0.3 mm) in relation to the sizes of components or devices normally used.
- Microwave frequency has many applications such as in the field of communication, Radars, industrial and many more.
- Microwaves has many advantages over low frequencies such as its directivity, more power or gain, more bandwidth and requires less power for transmissions.

REVIEW QUESTIONS

1. What do you understand by microwave frequencies?
2. Discuss the advantages of microwave frequencies.
3. Explain the various applications of the microwave.
4. Discuss the characteristics of microwave.
5. Explain the reasons for using microwave frequencies.
6. What is the microwave frequency classifications?

CHAPTER 2

TRANSMISSION LINES THEORY

OBJECTIVES
- Introduction
- Basic Transmission Line Equations
- Reflection Coefficient and Standing Wave
- Characteristic Impedance
- Transmission Line Parameters
- Matched Load
- Discontinuities and Load Impedances
- Transmission Line
- Input Impedance
- Directions of Travel on a Transmission Line
- Transformation of Impedance
- Impedance Matching
- Stub Matching
- Transmission Line Sections as Circuit Elements
- Some Practicalities
- Smith Chart
- Components of a Smith Chart
- Insertion Loss

2.1. INTRODUCTION

A transmission line is a structure used to guide the flow of electromagnetic energy from one point to another point.

It is used, for example, to transfer the output RF energy of a transmitter to an antenna. This energy will not travel through normal electrical wire without great losses. Although the antenna can be connected directly to the transmitter, the antenna is usually located some distance wave from the transmitter. A transmission line is used to connect the transmitter and the antenna. The transmission line has a signal purpose for both the transmitter and the antenna. This purpose is to transfer the energy output of the transmitter to the antenna with the least possible power loss. How well this is done depends on the special physical and electrical characteristics (impedance and resistance) of the transmission line.

In addition to two wire and coaxial transmission lines commonly employed at lower frequencies, a special type of transmission structures known as waveguides is widely used at microwave frequencies. Waveguides are single conductor in the form of tubular structure, rectangular or circular, capable of guiding high frequency electromagnetic waves through them.

This transmission line may be of any physical structure; that is, it may be made of two parallel wires or two parallel plates or coaxial conductors, or it may be of hollow conductor variety (waveguides). The general characteristics of electromagnetic wave propagation in these lines are the same. The preference depends only on the frequency of wave propagation and the use to which these lines are put.

In addition to their use for transmission of microwaves, short sections of transmission lines and waveguides are also used as circuit elements (equivalent to capacitors and inductors) at these frequencies. This chapter is a brief review of transmission lines and waveguides.

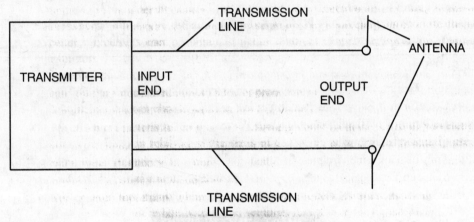

Fig. 2.1. Basic transmission line

All transmission lines have two ends refer fig. (2.1). The end of a two-wire transmission line connected to a source is ordinarily called the input end or the generator end. Other names given to this end are transmitter end, sending end, and source. The other end of the line is called the Output end or receiving end. Other names given to the output end are load end and sink.

We can describe a transmission line in terms of its impedance. The ratio of voltage to current (V_{in}/I_{in}) at the input end is known as the **input impedance** (Z_{in}). This is the impedance presented to the transmitter by the transmission line and is known as the **output impedance** (Z_{out}). This is the impedance presented to the load by the transmission line and its source. If an infinitely long transmission line could be used, the ratio of voltage to current at any point on that transmission line would be some particular value of impedance. This impedance is known as the **characteristic impedance.**

2.2. BASIC TRANSMISSION LINE EQUATIONS

Since standard circuit theory cannot be employed on an electrical network at microwave frequencies, an alternative analysis must be applied to the system. Transmission line theory is a tool which bridges the gap between circuit theory and a complete field analysis. Transmission lines lie between a fraction of a wavelength and many wavelengths in size. Conversely, in circuit analysis, the physical dimensions of the network are much smaller than the wavelength. A transmission line is considered a distributed parameter network as opposed to a circuit which consists of lumped elements. Consequently, the voltages and currents associated with a propagating wave in a t-line can vary in both phase and magnitude over the length of the line.

In general, if we examine a transmission line, we will find four parameters, i.e., series resistance (R), series inductance (L), shunt capacitance (C) and shunt conductance (G), distributed along the whole length of the line. If R, L, C and G be these primary constants per unit length of the line, then the unit length of the line may be represented by an equivalent circuit of the type shown in Fig. 2.2. Naturally, a relatively long piece of line would contain several such identical sections as shown in Fig. 2.3.

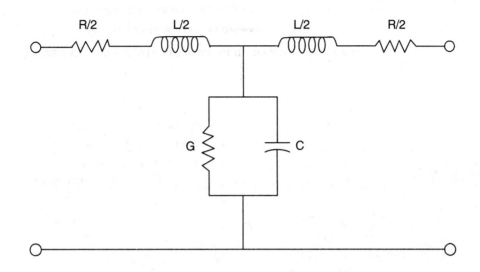

Fig. 2.2. Equivalent circuit of unit length of transmission line.

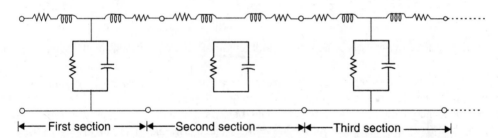

Fig. 2.3. A long piece of transmission line as a multi T- section line

The series impedance and shunt admittance per unit length of the line are given by:

$$Z = R + j\omega L \qquad \qquad ...2.1$$
$$Y = G + j\omega C. \qquad \qquad ...2.2$$

The expressions for voltage and current per unit length are, respectively,

$$dV/dz = (R + j\omega L)\ I \qquad \qquad ...2.3$$
$$dI/dz = (G + j\omega C)\ V \qquad \qquad ...2.4$$

Where negative sign indicates decrease in voltage and current as z increases. The current and voltage are measured from the receiving end; i.e., at receiving end, z = 0 and line extends in negative z-direction.

Differentiating Eqs. (2.3) and (2.4) and combining them,

$$\frac{d^2V}{dz^2} = \gamma^2 V$$

and $\quad \dfrac{d^2 I}{dz^2} = \gamma^2 I$ ..... 2.5 and 2.6

These are wave equations of voltage and current respectively propagating on the line; where

$$\gamma = \sqrt{ZY} = \sqrt{(R + j\omega L)(G + j\omega C)} \qquad ...2.7$$

is called the **propagation constant** which is in general a complex quantity and so may be defined as

$$\gamma = \alpha + j\beta$$

$$\gamma = \sqrt{\alpha^2 + \beta^2} \qquad ...2.8$$

α, called the **attenuation constant**, is the real part of Eq. (2.7), and β, the phase constant is the imaginary part. Thus, propagation constant γ is a measure of the phase shift and attenuation per unit length along the line. Separating γ into real and imaginary parts, we have,

$$\alpha = \left[\dfrac{\sqrt{(R^2 + \omega^2 L^2)(G^2 + \omega^2 C^2)} + (RG - \omega^2 LC)}{2} \right]^{1/2}$$

$$\beta = \left[\dfrac{\sqrt{(R^2 + \omega^2 L^2)(G^2 + \omega^2 C^2)} - (RG - \omega^2 LC)}{2} \right]^{1/2} \qquad ...2.9 \text{ and } 2.10$$

α is measured in decibels or nepers per unit length of the transmission line (1 neper = 8.686 decibels).

β is the phase shift per unit length of transmission line and is measured in radians per unit length of this line. The phase constant β is the imaginary part of equation 1.8 which shows the phase dependence with z of both forward and backward waves. If z changes from z_1 by a wavelength λ_1 to $z_1 + \lambda_1$, the phase of the wave must change by 2π.

Now

$$\beta(z_1 + \lambda_1) - \beta(z_1) = 2\pi$$

or $\quad \beta = 2\pi/\lambda_1$, or $\lambda_1 = 2\pi/\beta$ ...2.11

where λ_1 is the distance along the line corresponding to a phase change of 2π radians. The phase velocity $V_p = f\lambda_1$, where f is the signal frequency.

The solutions of voltage and current wave Eqs. (2.5) and (2.6) may be written as

$$V = V_1 e^{-\gamma z} + V_2 e^{+\gamma z}$$

$\rightarrow + z \quad - z \leftarrow$

$$I = I_1 e^{-\gamma z} + I_2 e^{+\gamma z}$$

$\rightarrow + z \quad - z \leftarrow$ ...2.12 and 2.13

These solutions are shown as the sum of two waves; the first term indicates the wave travelling in positive z-direction, i.e., incident wave, and the second term indicates the wave travelling in the negative z-direction, i.e., reflected wave. Where,

TRANSMISSION LINES THEORY

V_1 = Sending voltage amplitude
V_2 = Reflected voltage amplitude
I_1 = Sending current amplitude
I_2 = Reflected current amplitude

2.2.1 REFLECTION COEFFICIENT AND STANDING WAVE

The ratio of the reflected wave to that of the incident wave is known as reflection coefficient (Γ),

$$\Gamma = V_- / V_+ \qquad \ldots 2.14$$

For a lossless transmission line $\alpha = 0$;

$$V(z) = V_+ \, e^{-j\beta z} (1+\Gamma \, e^{2j\beta z}) \qquad \ldots 2.15$$
$$I(z) = Y_0 \, V_+ \, e^{-j\beta z} (1-\Gamma \, e^{2j\beta z}) \qquad \ldots 2.16$$

Since z may be measured from any point, let us consider that z to zero at the load end as shown in Fig. 2.4.

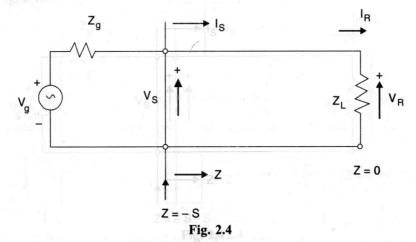

Fig. 2.4

At, $z = 0$ (Load end), we have
$$V_R = I_R Z_L$$
Equations 2.15 & 2.16, may be combined to yield
$$Z_L = Z_0 (1+\Gamma)/(1-\Gamma) \text{ or}$$
$$\Gamma = Z_L - Z_0 / Z_L + Z_0 \qquad \ldots 2.17$$

We may define a generalized reflection coefficient, $\Gamma(z)$ as the ratio of reflected and incident components inclusive of propagation phases, that is

$$\Gamma(z) = V_-(z) / V_+(z) = \Gamma \, e^{2j\beta z} \qquad \ldots 2.18$$

Equations 2.15 & 2.16, may be again written as

$$V(z) \, V_+ \, e^{-j\beta z} \{1+\Gamma(z)\} \qquad \ldots 2.19$$
$$I(z) = Y_0 \, V_+ \, e^{-j\beta z} \{1-\Gamma(z)\} \qquad \ldots 2.20$$

Vector $V(z) / V_+$ may be shown as in figure 2.5

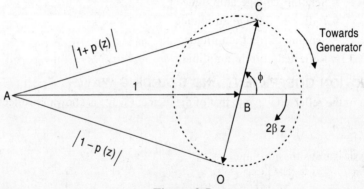

Figure 2.5

As one moves towards the generator, Γ (z) rotates in clockwise direction. The maximum value of the magnitude of V (z) / V$_+$ is 1 +| Γ| and the minimum value is 1- | Γ|.

Same way, the variation of I (z) / (Y$_0$ V$_+$), may also be illustrated by a rotating vector. It is seen that the position of the current maximum and vice versa.

Fig. 2.6 (a) shows the normalized voltage and current plotted along the line length. Fig. 2.6 (b,c,& d) shows some simple cases of standing wave pattern for voltage and current.

Case-1, when Z_L = 0 than Γ =-1, as shown in figure 2.6(b).

Case-2, when Z_L = ∞ than Γ = 1, as shown in figure 2.6 (c).

Case-3, when Z_L = Z_o, Γ = 0, there are no standing waves and the resulting pattern shown in Fig. 2.6(d), is a straight line representing a uniform distribution of the voltage and current along the line.

The ratio of the maximum voltage to the minimum voltage is reflected as voltage standing wave ratio (VSWR) and may be written as

$$\text{VSWR} = V_{max} / V_{min} = 1 +| \Gamma | / 1 - | \Gamma | \qquad \ldots 2.21$$

Fig. 2.5, the vector diagram, shows that the position of the first voltage minimum from the load end towards the generator is defined by the relation,

$$2\beta\, d_{min} = \Phi + \pi \qquad \ldots 2.22$$

where Φ, is the phase angle of Γ at the load end. If we know the position of the first voltage minimum, the phase angle of Γ can be find easily. The magnitude of Γ may be found by measuring VSWR and using equation 2.22.

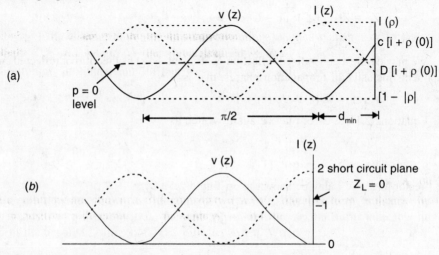

TRANSMISSION LINES THEORY

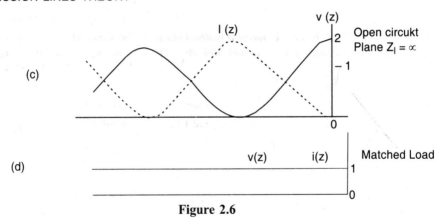

Figure 2.6

2.3. CHARACTERISTIC IMPEDANCE

When a generator is suddenly connected to a length of transmission line, there will be a time delay until the signal travels to the other end of the line. During this time, the generator has no knowledge of how long the line might be; to all intents and purposes there is no difference (during this time) between a short line and a very long line. The generator has to supply energy to the line, to establish the electromagnetic fields around the conductors. Energy is stored in the capacitance between the conductors, proportional to the square of the voltage between the wires, and in the magnetic field around the conductors which represents series inductance, proportional to the square of the current on the conductors. Even though the line may be open circuit at the other end, the generator does not (yet) know this and supplies current; the product of current and voltage is the rate at which energy is supplied to the line; this is obtained by multiplying the stored electric plus magnetic energy per unit length of line, by the velocity at which the signals travel.

The ratio of voltage (between the wires) to current (along one wire and back along the other) has dimensions of impedance or resistance. At a single frequency, on a lossless line, the current is in phase with the voltage and the impedance is real. It is called the "Characteristic Impedance". [The usual algebraic symbol for the Characteristic Impedance is Zo.] It does not depend on what is connected to the ends of the line, but only on the line geometry and material construction.

The Characteristic Impedance, although real and looking like a resistance, is actually lossless, non-dissipative" impedance, nothing gets hot as a result of supplying energy to this resistance. All that happens is that energy is transferred from the generator and stored temporarily in the transmission line. At some later time, possibly a great many transit times later, it can be extracted and returned to the generator, or used to make a real resistive dissipative load get hot.

A RF voltage applied across the conductors of an infinite line causes a current I to flow. By this observation, the line looks like an impedance which is denoted by Z_0 and is known as **characteristic impedance Z_0.**

It may also be defined as the impedance measured at the input of the transmission line when its length is infinite.

$$Z_0 = \frac{V_1}{I_1}.$$

The expression for current I, using Eqs. (2.16) and (2.17), is given by

$$I = \frac{1}{(R+j\omega L)} \cdot \frac{\partial V}{\partial z} = \frac{-1}{(R+j\omega L)}(-\gamma)(V_1 e^{-\gamma z} - V_2 e^{+\gamma z})$$

FUNDAMENTALS OF MICROWAVE AND RADAR ENGINEERING

$$I = I_1 e^{-\gamma z} - I_2 e^{\gamma z} \quad \text{(Phase reversed due to reflection).} \quad ...2.23$$

For infinite line there are no reflections, that is, V_2 and I_2 are zero. So we have

$$I_1 e^{-\gamma z} = \frac{\gamma V_1 e^{-\gamma z}}{(R + j\omega L)}$$

or

$$\frac{V_1}{I_1} = \frac{R + j\omega L}{\sqrt{(R + j\omega L)(G + j\omega C)}}$$

$$Z_0 = \sqrt{\frac{R + j\omega L}{G + j\omega C}} = R_0 + jX_0 \quad ...2.24$$

Where R_0 and X_0 are the real and imaginary parts of Z_0. R_0 should not be mistaken for R while R is in ohms per meter; R_0 is in ohms.

For loss-less line (R and G being zero)

$$Z_0 = \sqrt{L/C} \text{ and } \beta = \omega\sqrt{LC} \quad ...2.25$$

$$\gamma = \sqrt{(R + jwL)(G + j\omega C)}$$

$$\gamma = \sqrt{j\omega L \cdot j\omega C}$$

$$\gamma = j\omega\sqrt{LC}$$

$$\gamma = j\beta$$

$$\alpha = 0$$

$$\beta = \omega\sqrt{LC}$$

$$\frac{\omega}{\beta} = 1/\sqrt{LC}$$

The ratio $\frac{\omega}{\beta}$ is known as the **phase velocity** of the wave.

The characteristic impedance may be summed up as follows:
1. The impedance seen from the sending end of an infinitely long line.
2. The value which impedance of the load must have to match the load to the line.
3. The impedance seen looking towards the load at any point on a matched line-moving along the line produces, there is no changes in impedance towards the load.

2.4. TRANSMISSION LINE PARAMETERS

If we consider an infinite lossless transmission line, we can determine the inductance L and capacitance C per unit length from geometric field considerations. The three physical embodiments that are of interest are the two-wire transmission line, the coaxial transmission and the microstrip transmission line (a simple parallel-plate approximation).

Parameter	Two-wire	Coaxial	Microstrip
L	$\frac{\mu}{p} \text{ in } (D/a)$	$\frac{\mu}{2p} \text{ in } (b/a)$	$\mu T/W$
C	$\frac{p\varepsilon}{\ln(D/a)}$	$\frac{2p\varepsilon}{\ln(b/a)}$	$\varepsilon W/T$

TRANSMISSION LINES THEORY

In this table, **D** and **a** are the center-to-center spacing and wire radius of the two-wire line, **b** and **a** are the outer and inner radius of the coaxial line and **T** and **W** are the dielectric thickness and conductor width of the microstrip line. For two-wire line, the expressions include the approximation

$$\cosh^{-1}(D/2a) \sim \ln(D/a) \text{ for } D/2a \gg 1.$$

If we solve for Zo of coaxial and microstrip line, we have

1. For coaxial line (note use of ln and $\log_{10}$ in different references),

$$Z_0 = \frac{377}{2p\sqrt{\varepsilon_r}} \ln(b/a)$$

2. For microstrip line, ignoring fringing fields.

$$Z_0 =\sim \frac{377}{\sqrt{\varepsilon_r}} T/W$$

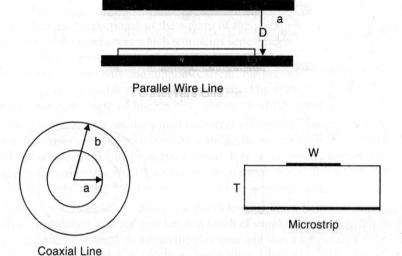

Fig. 2.7

2.5. MATCHED LOAD

If the transmission is uniform and infinite, the wave in the +z direction will continue indefinitely and never return in the -z direction.

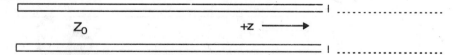

Fig. 2.8. Infinite Transmission Line.

If the uniform transmission line is truncated and connected instead to a lumped resistive load $R_L = Z_0$, the entire +z wave is dissipated in the load, which has the same effect as if an infinite line of characteristic impedance Zo were attached at the same point. This matched impedance condition is a unique situation in which all the power of the +z wave is delivered to the load just as if it were an infinite transmission line, with no reflected waves generated in the -z direction. Boundary conditions at a matched load are the same as for the infinite transmission line.

Fig. 2.9. Matched Termination Same as Infinite Line.

2.6. TRANSMISSION LINE DISCONTINUITIES AND LOAD IMPEDANCES

If the wave on a transmission line of characteristic impedance Zo arrives at a boundary with different Zo, or at a discontinuity, lumped load or termination of Z. Zo, the single wave moving in the +z direction cannot simultaneously satisfy the boundary conditions relating V(z) to I(z) on both sides of the boundary. On one side of the boundary $V(z)/I(z) = V+/I+ = Zo$ and on the other side $V(z)/I(z) = (V++V-)/(I+-I-) = Z_L$. As in the case for a plane wave reflecting from a dielectric or conducting boundary, transmitted and reflected waves are required to satisfy all the boundary conditions.

Waves can exist traveling independently in either direction on a linear transmission line. If a wave in the -z direction is formed by a complete or partial reflection of the +z wave by some discontinuity such as a lumped load of Z. Zo, the two waves are by definition coherent and an interference pattern will exist.

Even though the waves are traveling in opposite directions, the interference pattern will be stationary with respect to the point of reflection, and will thus be a standing wave such as may be found on the strings of musical instrument (of course, these are also defined by a wave equation). The standing wave interference pattern is present both in the resulting V(z) and I(z).

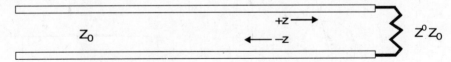

Fig. 2.10. Mismatched Load Creates Reflected Wave.

2.7. INPUT IMPEDANCE

We can see that the finite transmission line terminated by its Z_0 [Fig. 2.11(a)] has input impedance also equal to Z_0, that is, the finite line of characteristic impedance Z_0 has an input impedance Z_0 when it is terminated in Z0. A line terminated in its characteristic impedance will absorb all the power and there will be no reflection and hence it behaves as an infinite line.

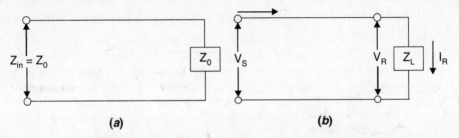

Fig. 2.11. (a) Finite transmission line terminated by its Z_0;
(b) Finite transmission line terminated by its Z_L.

TRANSMISSION LINES THEORY

One can obtain the expression for input impedance of line when it is terminated in an impedance Z_L [Fig. 2.11(b)] located at $z = 0$ as

$$Z_{in} = \frac{V_s}{I_s} = \frac{V_R \cosh \gamma z + Z_0 I_R \sinh \gamma z}{I_R \cosh \gamma z + \frac{V_R}{Z_0} \sinh \gamma z} \qquad ...2.26$$

$$Z_{in} = Z_0 \left[\frac{Z_L + Z_0 \tanh \gamma z}{Z_0 + Z_L \tanh \gamma z} \right]$$

where Vs = Voltage at the sending end
Is = Current at the sending end
z = Length of the line
Z_0 = Characteristic impedance
$\left. \begin{array}{l} V_R = \text{Voltage at the receiving end} \\ I_R = \text{Current at the receiving end} \end{array} \right\}$ at $z = 0$

If the line is short-circuited ($Z_L = 0$), we have short-circuited input impedance, Zsc, given by ($V_R = 0$),

$$Z_{sc} = Z_0 \tanh \gamma z \qquad ...2.27$$

The open-circuited input impedance ($Z_L = \infty$), Z_{oc}, can be found by putting $z_L = \infty$ and $I_R = 0$ in Eq. (2.26),

$$Z_{oc} = Z_0 \coth \gamma z \qquad ...2.28$$

The product of Eqs. (2.27) and (2.28) gives

$$Z_0 = \sqrt{Z_{oc} + Z_{sc}} \qquad ...2.29$$

2.8. DIRECTIONS OF TRAVEL ON A TRANSMISSION LINE

There are only two directions of travel on a transmission line, namely from generator to load, and from load to generator. These are termed "forward" and "backward" waves respectively. There is no physical difference you can measure between the ends of a line, so Zo must be the same if either end is driven. But the current in the backward wave is in the negative direction compared to the forward wave, for a generator connected to the "load" end pushes its current into the transmission line's hot conductor in the same way as the generator connected to the "generator" end does. However, the generators are facing in opposite directions and therefore so are the currents they supply to the transmission line also in opposite directions. The voltages, of course, are in the same sense in the forward and backward waves, being between the "hot" and "return" conductors. If we take the convention for current flow to be from "generator" end to "load" end, the backward wave has mathematically characteristic impedance -Zo. The sign reversal has physical meaning; the currents are directed oppositely as we have seen, although the voltages have the same sense.

Considering the product of current and voltage, power is delivered by the generator to the forward wave, but absorbed by the generator from the backward wave. A forward wave absorbs power from the generator end but delivers it to the load end; real impedance that sources power is negative impedance. Looking in to the load end of the cable, connected at the other end of the cable to a generator having series internal resistance Zo, we see matched impedance -Zo because power is being "forced down our throat".

As we have seen, the transmission line conveys energy from one place to another. If **we short circuit the line,** there is no energy dissipation in the short; the voltage times current across the short is zero; the energy is all reflected as it has to go somewhere; and a return wave is set up having the same size or amplitude as the incident wave. At the short circuit the voltage has to be zero for all time; so the sum of forward wave voltage and backward wave voltage must be zero. This can only happen for non-zero forward wave amplitude if the reflected wave has opposite polarity to the forward wave. There is no constraint on the current however; so the current in the short is not zero, being the sum of the currents in forward and backward wave.

If we **open-circuit the line** (cut the end off), the current is zero, but there is no constraint on the voltage. The sum of forward wave current and backward wave current is zero for all time; the backward wave voltage is equal to the forward wave voltage amplitude.

Mathematically, we say if the forward wave voltage amplitude is written V+, the backward wave amplitude is written V−, and similarly for the currents I+ and I−, then because of the −Zo property of backward wave impedance we have

$$V+ = Zo\ I+$$

and

$$V- = -Zo\ I-.$$

The total voltage at the load is (V+ + V−) and the total current at the load is (I+ + I−) because the currents are measured in the same direction along the line, namely in the direction of travel of the forward wave. You should show that if V+ = V− then the total current (I+ + I−) is zero; and that if I+ = I− then the total voltage (V+ + V−) is zero.

For other **conditions of load than short circuit or open circuit**, the load impedance ZL sets the ratio of total voltage to total current at the load attachment point.

Thus, $Z_L = (V+ + V-)/(I+ + I-)$.

Taken with the equations above

$$V+ = Zo\ I+ \text{ and } V- = -Zo\ I-$$

The ratio of backward wave voltage amplitude V− to forward wave voltage amplitude V+ is given by

$$(V-/V+) = (ZL - Zo)/(ZL + Zo).$$

This ratio is called the **"complex reflection coefficient"**.

In general the wave amplitudes V+, V−, I+ and I− are complex amplitudes in respect of waves which are sinusoidal or single frequency waves. The response of the transmission line to a complex signal is got by making a Fourier superposition of different frequencies in the usual manner. Fourier superposition's can be made in the spatial domain as well as the time domain; the spatial frequency is the reciprocal of the wavelength in the same way that the temporal frequency is the reciprocal of the period.

2.8. TRANSFORMATION OF IMPEDANCE

Suppose we take an instantaneous picture of a wave travelling from generator to load. The voltage and current will vary along the transmission line. If the signal is complex the variation in space may be similar to the shape of the time dependent wave when observed on a one-shot oscilloscope trace. If the signal is sinusoidal, only a single frequency is present. That means that only a single wavelength is observed. The load end of the picture shows the voltage which set out from the generator at an earlier time; the generator end of the picture shows the voltage setting out now. We are observing a "time history" of the waveform of the generator, spread out along the line.

For waves travelling from generator to load, the load end shows earlier history and the generator end comparatively later. The situation is reversed for waves travelling from load to generator, possibly produced by a reflection at the load.

TRANSMISSION LINES THEORY

In what follows, the term "forward wave" is taken to include the sum of all the disturbances travelling in the forward direction, no matter how many times they may have been reflected from the ends of the line. Similarly, "reverse wave" includes the sum of all disturbances travelling in the backward direction.

Considering the forward waves, the phase angle of the sinusoidal waves becomes less as we move towards the load. This is because the phase of a sinusoidal oscillation is continually advancing in time as time progresses [$\exp(j \omega t)$] with phase angle (ωt) and so at earlier times, ie towards the load, the phase is less.

Considering the backward waves, the situation is reversed, and towards the load the phase is more.

Now we add the forward and backward wave voltages, and similarly the forward and backward wave currents, to get the total voltage and current on the line. It is the ratio of this total voltage to this total current which is the impedance we would measure if we cut the line and put it on an impedance bridge.

Because $V+ = Z_o\, I+$ but $V- = -Z_o\, I-$, a little mathematical experimentation shows that the measured impedance must depend on position along the line unless $V- = 0$ everywhere.

Open circuit or short circuit? At an open circuit, the total current is zero

$$I+ \;+\; I- \;=\; 0$$

so $\quad I+ = -I-$ and $V+ = V-$

If we travel a quarter of a wavelength towards the generator, the phase angle of $V+$ has advanced by 90 degrees, whereas that of $V-$ has retarded by 90 degrees, and so the total voltage is now zero (they are the same size and there is 180 degrees phase difference between them). The currents, which were 180 degrees out of phase, are now in phase. The total current is large and the total voltage zero, so looking into the "generator" end of a 1/4 wavelength of transmission line, with an open circuit at the "load" end, we see a perfect short circuit.

Another way of looking at this surprising result is that at the load end there is a real open circuit: As we move towards the generator, along the line, we add the shunt capacitance between the line conductors, and the series inductance of the conductor loop. When we have moved 1/4 wavelength towards the generator these components form a series tuned circuit at the generator frequency across the generator terminals, which looks like a perfect short circuit. There is danger in this point of view, as the inductance and capacitance of the line are "distributed" along the line rather than being concentrated at a single point. It is necessary to solve the transmission line equations to get a proper understanding of the phenomenon.

2.9. IMPEDANCE MATCHING

Suppose we think about a very long length of transmission line, and somewhere along it we make an imaginary join between the left half and the right half. The right half presents real lossless impedance (equal to the characteristic impedance) to waves approaching the junction from the left half. But there is no actual discontinuity at the imaginary join, so there is nothing to give rise to a reflection here. If we replace the right half transmission line with a resistor having resistance equal to the real lossless characteristic impedance of the line, there is no way the waves arriving at the junction can tell the difference between the resistor and the long right-half transmission line it has replaced. Thus there can be no reflection. We have "matched" the impedance of the line with the same value resistance. All the power delivered by the generator is dissipated in the resistance; there is no reflected wave amplitude or power, and there is no backward travelling wave. As far as the generator is concerned, there is no way it can know how long the length of line is, or even that there is a line there at all; it cannot tell the difference between a resistive load Zo and a very long matched line having characteristic impedance Zo.

Over a wide range of frequencies, in lossless line, the impedance Zo is substantially independent of frequency. Thus if a line is matched at one frequency, it will also be matched at another frequency, under these conditions. There are other conditions on impedance matching under which transmission line is matched only at a single frequency, or over a small range of frequencies.

A number of techniques can be used to eliminate reflections when the characteristic impedance of the line and the load impedance are mismatched. Impedance matching techniques can be designed to be effective for a specific frequency of operation (narrow band techniques) or for a given frequency spectrum (broadband techniques).

A common method of impedance matching involves the insertion of an impedance transformer between line and load.

An impedance transformer may be realized by inserting a section of a different transmission line with appropriate characteristic impedance. A widely used approach realizes the transformer with a line of length $\lambda/4$.

The quarter-wavelength transformer provides narrow-band impedance matching. The design goal is to obtain zero reflection coefficients exactly at the frequency of operation.

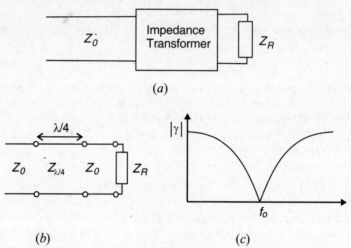

Fig. 2.12. (a, b & c) An impedance transformer

The length of the transformer is fixed at $\lambda/4$ for design convenience, but is also possible to realize generalized transformer lines for which the length of the transformer is a design outcome. A broadband design may be obtained by a cascade of $\lambda/4$ line sections of gradually varying characteristic impedance.

It is not possible to obtain exactly zero reflection coefficient for all frequencies in the desired band. Therefore, available design approaches specify a maximum reflection coefficient (or maximum VSWR) which can be tolerated in the frequency band of operation.

Another broadband matching approach may use a tapered line transformer with continuously varying characteristic impedance along its length. In this case, the design obtains reflection coefficients lower than a specified tolerance at frequencies exceeding a minimum value.

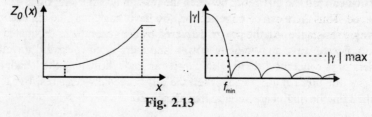

Fig. 2.13

TRANSMISSION LINES THEORY

Another narrow-band approach involves the insertion of a shunt imaginary admittance on the line. Often, the admittance is realized with a section (or stub) of transmission line and the technique is commonly known as stub matching. The end of the stub line is short-circuited or open-circuited, in order to realize an imaginary admittance. Designs are also available for two or three shunt admittances placed at specified locations on the line.

Other narrow-band examples involve the insertion of series impedance (stub) along the line, and the insertion of a series and a shunt element in L-configuration.

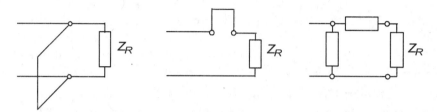

Fig. 2.14

2.10. STUB MATCHING
Why match?

Matching the impedance of a network to the impedance of a transmission line has two principal advantages.

1. All the incident power is delivered to the network.
2. Second, the generator is usually designed to work into impedance close to common transmission line impedances.

If it does so it is better behave, the load impedance has no reactive part which can pull the generator frequency, and the VSWR on the line is unity or close to unity so the line length is immaterial and the line connecting the generator to the load is non-resonant.

Single stub matching

If you look at the SMITH chart you will find a circle of constant real impedance r=1 which goes through the open circuit point and the centre of the chart. If you plot any arbitrary impedance on the SMITH chart and follow round at constant radius towards the generator, you must cross the $r = 1$ circle somewhere. This transformation at constant radius represents motion along the transmission line towards the generator. One complete circuit of the SMITH chart represents a travel of one half wavelengths towards the generator. At this intersection point your generalized arbitrary load impedance r + jx has transformed to 1 + jx', so at least the real part of the impedance equals the characteristic impedance of the line. Note x' is different from x in general.

At this point you cut the line and add a pure reactance –jx'. The total impedance looking into the sum of the line impedance and –jx' is therefore 1 + jx' – jx' = 1 and the line is matched.

Why stubs?

Stubs are shorted or open circuit lengths of transmission line intended to produce a pure reactance at the attachment point, for the line frequency of interest. Any value of reactance can be made, as the stub length is varied from zero to half a wavelength.

Again, look at the SMITH chart and find the outer circle where the modulus of the reflection coefficient is one. On this circle are the SHORT and OPEN points, and all values of positive and negative reactance. The resistance is zero everywhere. To generate a specified reactance, start at a short circuit (or maybe an open) and follow around towards the generator until the desired reactance is obtained. Cut the stub this number of wavelengths long.

It is important to keep the total stub length as short as possible, if wider bandwidths are required. Every time you add a half wavelength to the stub length the reactance of the stub comes back to the same value. It is good design practice to make stubs in the range 0 to half wavelengths long. However, this may require an impractically short stub, so then one can make the stub just a little over half wavelengths.

Short or open stubs?

If one is allowed to use either short or open stubs at will, one can always keep the total stub length in the range 0 to quarter wavelengths. A length of transmission line of quarter wavelengths takes us half way round the SMITH chart and transforms an open into a short, or vice versa. On microstrip it is usually easier to leave stubs open circuit, for constructional reasons. On coax line or parallel wire line, a short circuit stub has less radiation from the ends: it is difficult to make a perfect non-radiating open circuit as there are always some end effects on the line.

2.10. TRANSMISSION LINE SECTIONS AS CIRCUIT ELEMENTS

At microwave frequencies, short sections of transmission lines are widely used as basic circuit elements. These sections may be used either as two-port elements or as single port elements (the other end of the transmission line section is either kept open or shorted).

Two-Port Elements

Transmission line sections which are half wavelength ($\lambda/2$), quarter wavelength ($\lambda/4$) or one-eighth wavelength ($\lambda/8$) long, have some specific interesting characteristics, are discussed as follows:

1. Half Wavelength ($\lambda/2$) sections:

In case when $l = \lambda/2$, $Z_{in} = Z_L$ ...2.27

Input impedance is equal to the load impedance and $\lambda/2$ sections may thus consider as one to one impedance transformer. $\lambda/2$ sections are inserted in transmission line circuits, whenever physical spacing between two ports of a circuit is required without changing the electrical behavior at a particular frequency.

2. Quarter Wavelength ($\lambda/4$) sections:

In case when $l = \lambda/4$,

$$Z_{in} = Z_0^2 / Z_L \qquad 2.28$$

This relation shows that if Z_L is resistive Z_{in} is also resistive but has a different value. Thus a quarter wavelength section of suitable impedance Z_0 may be used to transform a resistive impedance Z_L to another impedance value Z_{in}.

When

$Z_L = 0$, Z_{in} tends to infinity. This property may be utilized to provide high impedance stub supports for RF structure and as well as RF choke arrangements for DC bias in microwave circuits. Quarter Wavelength ($\lambda/4$) are widely used for impedance transformation.

3. One-eighth wavelength sections:

When $L = \lambda/8$ and $\tan(\beta l)$ equals unity

Then $Z_{in} = Z_0 \{(Z_L + j Z_0) / (Z_0 + j Z_L)\}$ ...2.29

i.e. If Z_L is real then $|Z_{in}|$ equals Z_0.

When Z_L is complex and $Z_0 = |Z_L|$, Z_{in} becomes real and is given by

$$Z_{in} = |Z_L| \sqrt{\{(R^2/X + X + |Z_L|) / (R^2/X + X - |Z_L|)\}} \qquad 2.30$$

Where X and R are resistive and z is reactive part.

It is to be noted that a $\lambda/8$ section can be used to transform a complex impendence Z_L in to a real impedance Z_{in} in when $Z_0 = |Z_L|$.

TRANSMISSION LINES THEORY

Single Port Element

Open circuited transmission line sections:

When a section of a transmission line is open circuited at the load end, Z_L tends to infinity and Z_{in} may be given as

$$Z_{in} = -j Z_0 \cot(\beta \ell) \qquad ...2.31$$

Z_{in} is pure reactance. It is capacitive for $(\beta \ell) < \pi/2$ and inductive for $\pi/2 < (\beta \ell) < \pi$.

Thus an open circuited section may also be used as an inductive and capacitive reactance.

Short circuited transmission line sections:

When a section of transmission is short circuited at load end, $Z_L = 0$ and Z_{in} may be expressed as

$$Z_{in} = j Z_0 \tan(\beta l) \qquad ...2.32$$

which shows that Zin is purely reactive. It is capacitive for $\pi/2 < (\beta \ell) < \pi$ (i.e. $\lambda/4 < (\ell) < \lambda/2$) and inductive for $(\beta l < \pi/2$ (i.e. $< \lambda/4$). Thus a short circuited transmission line section may be used as capacitive reactance by suitable choice of its length.

Since it is often more convenient to have ideal shot circuit termination rather than an ideal open circuit termination (particularly in case of coaxial line), short circuited section of transmission lines find more frequent use.

2.11. SOME PRACTICALITIES

Some of the factors which determine whether (or not) a practical transmission line behaves as intended include, but are not limited to, the following:-

1. Dimensional uniformity including
   (*a*) twists
   (*b*) regions which have been stretched
   (*c*) fastenings (clouts, staples, attached wires)
   (*d*) kinks
   (*e*) bends
2. Dielectric properties including
   (*a*) effects of water
   (*b*) effects of ageing
   (*c*) loss tangent
   (*d*) effects of temperature
   (*e*) material dispersion (frequency dependence)
3. Behavior in undesigned-for modes, eg.
   (*a*) currents on outside of coaxial sheath
   (*b*) common mode transmission on twin feeder
   (*c*) waveguide modes (non TEM) on TEM lines
   (*d*) TEM transmission down waveguide and parallel structures
   (*e*) dielectric waveguide modes on insulated lines

- Typically, twin feeder is very much more subject to variability in the wet than is coaxial cable. However, a common cause of coaxial cable failure is the wicking of water up the semi-air-spaced dielectric separating inner conductor from sheath.

- All outdoor cables, and some indoor cables, suffer from the effects of light. The insulators harden and crack and can let in water.
- Cables which are overstressed, electrically, can develop carbonised conducting paths between the conductors of the cable. Failure is then progressive.
- Transmission line with exposed conductor surfaces can experience oxidation and an increase in losses due to skin effect heating.
- Different cable technologies have their own range of practical characteristic impedance values.

2.11. SMITH CHART

For complex transmission line problems, the use of the formulae becomes increasingly difficult and inconvenient. An indispensable graphical method of solution is the use of Smith Chart.

Smith Chart is a *polar plot* of the complex reflection coefficient (called γ *gamma* herein), or also known as the 1-port scattering parameter s or s_{11}, for reflections from a normalized complex load impedance $z = r + jx$; the normalized impedance is a complex dimensionless quantity obtained by dividing the actual load impedance Z_L in ohms by the characteristic impedance Zo (also in ohms, and a real quantity for a lossless line) of the transmission line.

The contours of $z = r + jx$ (dimensionless) are plotted on top of this polar reflection coefficient (complex γ) and form two orthogonal sets of intersecting circles. The centre of the SMITH chart is at $\gamma = 0$ which is where the transmission line is "matched", and where the normalized load impedance $z=1+j0$; that is, the resistive part of the load impedance equals the transmission line impedance, and the reactive part of the load impedance is zero.

The complex variable $z = r + jx$ is related to the complex variable γ by the formula

$$z = r + jx = \frac{1+\gamma}{1-\gamma} \qquad \ldots 2.33$$

and of course, the inverse of this relationship is

$$\gamma = \frac{z-1}{z+1} = \frac{(r-1)+jx}{(r+1)+jx} \qquad \ldots 2.34$$

From this chart we can read off the value of γ for a given z, or the value of z for a given γ. The modulus of γ, which is written $|\gamma|$, is the distance out from the centre of the chart, and the phase angle of γ, written $\arg(\gamma)$, is the angle around the chart from the positive x axis. There is an angle scale at the perimeter of the chart.

On a lossless transmission line the waves propagate along the line without change of amplitude. Thus the size of γ, or the modulus of γ, $|\gamma|$, doesn't depend on the position along the line. Thus the impedance "transforms" as we move along the line by starting from the load impedance $z = ZL/Zo$ and plotting a circle of constant radius $|\gamma|$ travelling towards the generator. The scale on the perimeter of the SMITH chart has major divisions of $1/100$ of a wavelength; by this means we can find the input impedance of the loaded transmission line if we know its length in terms of the wavelength of waves travelling along it.

TRANSMISSION LINES THEORY

- The plot you usually see is the inside of the region bounded by the circle $|\gamma|=1$. Outside this region there is reflection gain; in this outside region, the reflected signal is larger than the incident signal and this can only happen for $r < 0$ (negative values of the real part of the load impedance). Thus the perimeter of the SMITH chart as usually plotted is the $r=0$ circle, which is coincident with the $|\gamma|=1$ circle.

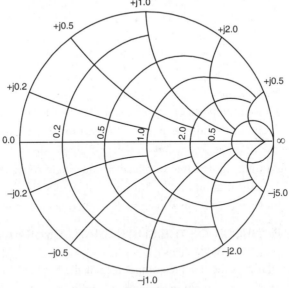

Fig. 2.15. Smith Chart.

- The $r=1$ circle passes through the centre of the SMITH chart. The point $\gamma = 1$ angle 0 is a singular point at which r and x are multi-valued.
- The SMITH chart represents both impedance and admittance plots. To use it as an admittance plot, turn it through 180 degrees about the centre point. The directions "towards the generator" and "towards the load" remain in the same sense. The contours of constant resistance and constant reactance are now to be interpreted as constant (normalised) conductance g, and (normalised) susceptance s respectively.
- To see this property of the SMITH chart we note first that the admittance y is the reciprocal of the impedance z (both being normalised). Thus inverting the equation above we see that

$$y = g + js = \frac{1-\gamma}{1+\gamma} \qquad \ldots 2.35$$

and this is the same formula that we had above if we make the substitution $\gamma \longrightarrow (-\gamma)$. Of course, inverting the SMITH chart is the same as rotating it though 180 degrees or *pi* radians, since $(-\gamma) = (\gamma)(exp\{j\ pi\})$.

- Admittance plots are useful for shunt connected elements; that is, for elements in parallel with the line and the load.

2.11.1 COMPONENTS OF A SMITH CHART

1. **Horizontal line:** The horizontal line running through the center of the Smith chart represents either the resistive or the conductive component. Zero resistance or conductance is located on the left end and infinite resistance or conductance is located on the right end of the line.
2. **Circles of constant resistance and conductance:** Circles of constant resistance are drawn on the Smith chart tangent to the right-hand side of the chart and its intersection with the

centerline. These circles of constant resistance are used to locate complex impedances and to assist in obtaining solutions to problems involving the Smith chart.

3. **Lines of constant reactance:** Lines of constant reactance are shown on the Smith chart with curves that start from a given reactance value on the outer circle and end at the right-hand side of the center line.

2.11.2 SOLUTIONS TO MICROWAVE PROBLEMS USING SMITH CHART

The types of problems for which Smith charts are used include the following:
1. Plotting a complex impedance on a Smith chart
2. Finding VSWR for a given load
3. Finding the admittance for a given impedance
4. Finding the input impedance of a transmission line terminated in a short or open.
5. Finding the input impedance at any distance from a load Z_L.
6. Locating the first maximum and minimum from any load
7. Matching a transmission line to a load with a single series stub.
8. Matching a transmission line with a single parallel stub
9. Matching a transmission line to a load with two parallel stubs.

2.11.3 WHY IS ONE CIRCUIT OF THE SMITH CHART ONLY HALF A WAVELENGTH?

We remember that the SMITH chart is a polar plot of the complex reflection coefficient, which represents the ratio of the complex amplitudes of the backward and forward waves.

Imagine the forward wave going past you to a load or reflector, then travelling back again to you as a reflected wave. The total phase shift in going there and coming back is twice the phase shift in just going there. Therefore, there is a full 360 degrees or 2π radians of phase shift for reflections from a load HALF a wavelength away. If you now move the reference plane a further HALF wavelength away from the load, there is an additional 360 degrees or 2π radians of phase shift, representing a further complete circuit of the complex reflection (SMITH) chart. Thus for a load a whole wavelength away there is a phase shift of 720 degrees or 4π radians, as the round trip is 2 whole wavelengths. Thus in moving back ONE whole wavelength from the load, the round trip distance is actually increasing by TWO whole wavelengths, so the SMITH chart is circumnavigated twice.

2.11.4 PRECISENESS OF THE SMITH CHART

It might be thought that the SMITH chart is only a rough and ready calculator since points can only be determined and plotted on it to within a certain tolerance depending on the size of the print copy of the chart. However, the angular scale at the edge has divisions of 1/500 of a wavelength (0.72 degrees) and the reflection coefficient scale can be read to a precision of 0.02. A little thought shows that this is quite sufficient for most purposes.

For example, if the wavelength in coaxial cable at 1 GHz is 20 cm, the SMITH chart locates the position along the cable to 20/500 cm or 0.4 mm and it is clear to anyone who has handled cable at 1GHz that it cannot be cut to this precision.

Normally 3 significant figures in the reflection coefficient is more than ample; angles can be quoted to the nearest degree and normalised impedances and admittances to about 1%. For, it is going to be very difficult to construct a real circuit which is accurately described by more precision than this.

Since many people now rely on computer modeling of transmission lines, they have lost sight of the precision limits of the descriptions of their physical circuit implementations. If your matching

TRANSMISSION LINES THEORY

circuit requires parameters to be chosen more closely than about a percent in order to work, you probably won't be able to make it physically.

We notice that at low frequencies (approaching zero) the SMITH chart plot starts from the open circuit point (large resistance and reactance), because the dipole is open-circuit to DC. If we look, on the other hand, at the equivalent plot for a 500mm circumference loop antenna, we see that it starts from the short circuit point on the SMITH chart, as the loop is short circuit at DC.

2.11.5 ADVANTAGES OF THE SMITH CHART

The SMITH chart is particularly elegant for the following reasons.

1. It is a direct graphical representation, in the complex plane, of the complex reflection coefficient.
2. It is a Riemann surface, in that it is cyclical in numbers of half-wavelengths along the line. As the standing wave pattern repeats every half wavelength, this is entirely appropriate. The number of half wavelengths may be represented by the winding number.
3. It may be used either as an impedance or admittance calculator, merely by turning it through 180 degrees.
4. The inside of the unity *gamma* circular region represents the passive reflection case, which is most often the region of interest.
5. Transformation along the line (if lossless) results in a change of the angle, and not the modulus or radius of *gamma*. Thus, plots may be made quickly and simply.
6. Many of the more advanced properties of microwave circuits, such as noise figure and stability regions, map onto the SMITH chart as circles.
7. The "point at infinity" represents the limit of very large reflection gain, and so therefore need never be considered for practical circuits.
8. The real axis maps to the Standing Wave Ratio (SWR) variable. A simple transfer of the plot locus to the real axis at constant radius gives a direct reading of the SWR.

2.11.6 PLOTTING A COMPLEX IMPEDANCE ON A SMITH CHART

To locate a complex impedance, $Z = R+\text{-}jX$ or admittance $Y = G +\text{-} jB$ on a Smith chart, normalize the real and imaginary part of the complex impedance. Locating the value of the normalized real term on the horizontal line scale locates the resistance circle. Locating the normalized value of the imaginary term on the outer circle locates the curve of constant reactance. The intersection of the circle and the curve locates the complex impedance on the Smith chart.

2.11.7 FINDING THE VSWR FOR A GIVEN LOAD

1. Normalize the load and plot its location on the Smith chart.
2. Draw a circle with a radius equal to the distance between the 1.0 point and the location of the normalized load and the center of the Smith chart as the center.
3. The intersection of the right-hand side of the circle with the horizontal resistance line locates the value of the VSWR.

2.11.8 FINDING THE INPUT IMPEDANCE AT ANY DISTANCE FROM THE LOAD

1. The load impedance is first normalized and is located on the Smith chart.
2. The VSWR circle is drawn for the load.
3. A line is drawn from the 1.0 point through the load to the outer wavelength scale.
4. To locate the input impedance on a Smith chart of the transmission line at any given distance from the load, advance in clockwise direction from the located point, a distance in wavelength equal to the distance to the new location on the transmission line.

2.12. INSERTION LOSS

The reduction in power level is observed when a network is introduced between a source and load for impedance matching. This power loss is known as insertion loss.

Insertion loss may be further described as,

Insertion Loss = Input Power – Output power

If the receiving end current without the inserted network be I_1 and the receiving end current with the inserted network be I_2, then

Insertion Loss = $\log_e | I_1 / I_2 | = 20 \log_{10} I_1 / I_2$ Nep

The insertion loss is governed by the value of the source and load impedance and is not a fixed quantity. Sometime, insertion of a network may cause a phase shift called insertion shift between I_1 and I_2.

SUMMARY

- A transmission line is a structure used to guide the flow of electromagnetic energy from one point to another point.
- The ratio of voltage to current (V_{in}/I_{in}) at the input end is known as the **input impedance** (Z_{in}).
- This is the impedance presented to the transmitter by the transmission line and is known as the **output impedance** (Z_{out}).
- The ratio of voltage to current at any point on that transmission line would be some particular value of impedance. This impedance is known as the **characteristic impedance.**
- **Propagation constant**

$$\gamma = \alpha + j\beta$$

$$\gamma = \sqrt{\alpha^2 + \beta^2}$$

- α, called the **attenuation constant**, α is measured in decibels or nepers per unit length of the transmission line (1 neper = 8.686 decibels).
- β is the phase shift per unit length of transmission line and is measured in radians per unit length of this line.
- A RF voltage applied across the conductors of an infinite line causes a current I to flow. By this observation, the line looks like an impedance which is denoted by Z_0 and is known as **characteristic impedance Z_0.**
- It may also be defined as the impedance measured at the input of the transmission line when its length is infinite.

- $Z_0 = \dfrac{V_1}{I_1}.$

- The ratio $\dfrac{\omega}{\beta}$ is known as the phase velocity of the wave

$$Z_{in} = \frac{V_1}{I_s} = \frac{V_R \cosh \gamma z + Z_0 I_R \sinh \gamma z}{I_R \cosh \gamma z + \dfrac{V_R}{Z_0} \sinh \gamma z}$$

$$Z_{in} = Z_0 \left[\frac{Z_L + Z_0 \tanh \gamma z}{Z_0 + Z_L \tanh \gamma z} \right]$$

TRANSMISSION LINES THEORY

- A common method of impedance matching involves the insertion of an impedance transformer between line and load.
- An impedance transformer may be realized by inserting a section of a different transmission line with appropriate characteristic impedance. A widely used approach realizes the transformer with a line of length $\lambda/4$.
- At microwave frequencies, short sections of transmission lines are widely used as basic circuit elements. These sections may be used either as two-port elements or as single port elements (the other end of the transmission line section is either kept open or shorted).
- For complex transmission line problems, the use of the formulae becomes increasingly difficult and inconvenient. An indispensable graphical method of solution is the use of Smith Chart.
- **Smith Chart** is a *polar plot* of the complex reflection coefficient.
- Insertion Loss = $\log_e | I_1 / I_2 | = 20 \log_{10} I_1 / I_2$ Nep

REVIEW QUESTIONS

1. Define the characteristic impedance of transmission lines.
2. What do you mean by the term "impedance matching"?
3. What is a stub? Why short circuited stub is always preferred?
4. What are the drawbacks of a single stub?
5. Explain single-stub impedance matching in transmission systems. Derive the relevant expressions.
6. Explain impedance matching in waveguides and coaxial lines.
7. What are
    (i) $\lambda/8$- stub lines?
    (ii) $\lambda/4$- stub lines?
    (iii) $\lambda/2$- stub lines?
8. Find out the condition for minimum attenuation in a transmission line.
9. What is quarter wave transformer? Mention its use.
10. Compare single and double-stub matching.
11. What is the utility of "Smith chart"?
12. What do you mean by voltage standing wave ratio (VSWR)? What is its significance?
13. What do you mean by Insertion loss?
14. Write shorts notes on
    (i) Matched load
    (ii) Transmission line discontinuities
    (iii) Input impedance
    (iv) Transformation of impedance

CHAPTER 3

MICROWAVE TRANSMISSION LINES

OBJECTIVES
- Introduction
- Types of Transmission Lines
- Losses in Transmission Lines
- Length of a Transmission Line
- Microwave Transmission Lines
- Waveguides
- Rectangular Waveguide
- Circular waveguide
- Planar transmission lines

3.1. INTRODUCTION

The conventional open wire transmission lines are not suitable for microwave frequency transmissions due to high radiation loss. Many different types of microwave transmission lines have been developed over the years. In an evolutionary sequence from rigid rectangular and circular waveguide, to flexible coaxial cable, to planar strip line to micro strip line, microwave transmission lines have been reduced in size and complexity. The Transmission media used are mainly classified according to the frequency used as following (Refer Fig. 3.1):

Frequencies for communication

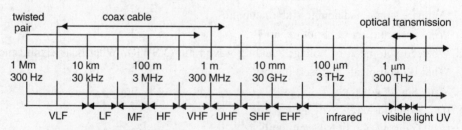

VLF = Very Low Frequency
LF = Low Frequency
MF = Medium Frequency
HF = High Frequency
VHF = Very High Frequency
UHF = Ultra High Frequency
SHF = Super High Frequency
EHF = Extra High Frequency
UV = Ultraviolet Light

Frequency and wave length:
$$\lambda = c/f$$
wave length λ, speed of light $c \cong 3 \times 10^8$ m/s, frequency f

Fig. 3.1. Frequency Band used in Communication.

MICROWAVE TRANSMISSION LINES

3.2. TYPES OF TRANSMISSION LINES

Generally there are six types of transmission mediums, which are as:

1. Parallel-line
2. Twisted pair
3. Shielded pair
4. Coaxial line
5. Strip type substrate transmission lines and
6. Wave guides.

The use of a particular line depends, among other things, on the applied frequency, the power-handling capabilities, and the type of installation.

At microwave frequencies following transmission lines are generally employed.

1. Multi conductor lines:
   - (a) Co-axial lines
   - (b) Coplanar lines
   - (c) Strip lines
   - (d) Micro strip lines
2. Waveguides:
   - (a) Rectangular wave guides
   - (b) Circular wave guides
   - (c) Elliptical waveguides
   - (d) Ridge wave guides
3. Open-boundary structures:
   - (a) Open wave-guides
   - (b) Dielectric rods

Microwave frequencies present special problems in transmission, generation, and circuit design that are not encountered at lower frequencies. Conventional circuit theory is based on voltages and current while microwave theory is based on electromagnetic fields.

3.2.1 PARALLEL LINE

One type of parallel line is the **two-wire open line** shown in figure 3.2. This line consists of two wires that are generally spaced from 2 to 6 inches apart by insulating spacers. This type of line is most often used for power lines, rural telephone lines, and telegraph lines. It is sometimes used as a transmission line between a transmitter and an antenna or between an antenna and receiver. An advantage of this type of line is its simple construction. The principal disadvantages of this type of line are the high radiation losses and electrical noise pickup because of the lack of shielding. The changing fields created by the changing current in each conductor are producing radiation losses.

Open Wire is traditionally used to describe the electrical wire strung along a telephone poles. There is a single wire strung between poles. No shielding or protection from noise interference is used. We are going to extend the traditional definition of Open Wire to include any data signal path without shielding or protection from noise interference. This can include multiconductor cables or single wires. This media is susceptible to a large degree of noise and interference and consequently not acceptable for data transmission except for short distances less than 8meter.

A transmission line may be a pair of conducting wires held apart by an insulator or dielectric. They come in a variety of construction geometries. The simplest and least expensive form is **two-wire (ribbon) cable**. This type of transmission line is commonly used to connect a Television antenna to television receiver. This line is essentially the same as the two wire open line except that uniform spacing is assured by embedding the two wires in a low loss dielectric, usually polyethylene. Since the wires are embedded in the thin ribbon of polyethylene space is partly air and partly polythene. Fig. 3.2 shows the open wire cable.

32   FUNDAMENTALS OF MICROWAVE AND RADAR ENGINEERING

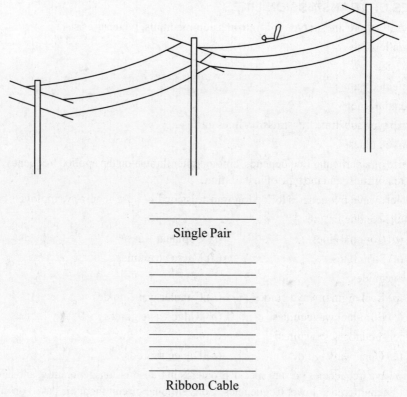

Fig. 3.2. Parallel/ Open wire.

3.2.2 SHIELDED PAIR

The **shielded pair**, shown in figure (3.3), consists of parallel conductors separated from each other and surrounded by a solid dielectric. The conductors are contained within braided copper tubing that acts as an electrical shield. The assembly is covered with a rubber or flexible composition coating that protects the line from moisture and mechanical damage. Outwardly, it looks much like the power cord of a washing machine or refrigerator.

The principal advantage of the shielded pair is that the conductors are balanced to ground; that is, the capacitance between the wires is uniform throughout the length of the line. This balance is due to the uniform spacing of the grounded shield that surrounds the wires along their entire length. The braided copper shield isolates the conductor form stray magnetic fields.

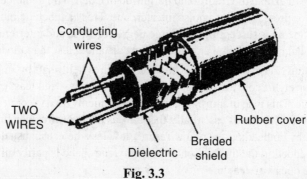

Fig. 3.3

MICROWAVE TRANSMISSION LINES

3.2.3 TWISTED PAIR CABLES

Twisted pair cable consists of two insulated copper wires, typically 1mm thick, twisted together. The wires are twisted in a helical form, just like DNA molecule as shown in fig. 6. This twisting reduces the sensitivity of the cable to EMI (Electromagnetic interference) and also reduces the tendency of the cable to radiate radio frequency noise that interferes with nearby cables and electronic components. Each pair would consist of a wire used for the positive data signal and a wire used for the negative data signal. Any noise that appears on one wire of the pair would occur on the other wire. Because the wires are opposite polarities, they are 180 degrees out of phase (180 degrees - phasor definition of opposite polarity). When the noise appears on both wires, it cancels or nulls itself out at the receiving end.

This is because the transmitted signals from the twisted cables tend to cancel each other out. Twisted pairs are described as differential transmission lines, or balanced. The common application of the twisted pair is the telephone system. Mostly nearly all telephones are connected to the local telephone exchange.

The degree of reduction in noise interference is determined specifically by the number of turns per foot. Increasing the number of turns per foot reduces the noise interference. To further improve noise rejection, a foil or wire braid shield is woven around the twisted pairs. This "shield" can be woven around individual pairs or around a multi-pair conductor (several pairs).

Twisted Pair cables are most effectively used in systems that use a balanced line method of transmission: polar line coding (Manchester Encoding) as opposed to unipolar line coding (TTL logic).

It is not used for transmitting high frequency because of the high dielectric losses that occur in the rubber insulation. When the line is wet, the losses increase greatly.

Unshielded Twisted Pair

Fig. 3.4. Twisted Pair cable.

Twisted pairs can be used several kilometers without amplification but for the distance more than approximately 5 km, repeaters are needed. Two types of twisted pair cable are used in LANs are;

(*i*) Unshielded twisted pair(UTP)

(*ii*) Shielded twisted pair (STP)

Unshielded Twisted Pair

Cables without a shield are called **Unshielded Twisted Pair** or UTP. Twisting the wires together results change of characteristic impedance of the cable. Typical impedance for UTP is 100 ohm for Ethernet 10BaseT cable.

UTP or Unshielded Twisted Pair cable is used on Ethernet 10BaseT and can also be used with Token Ring. It uses the RJ line of connectors (RJ45, RJ11, etc.)

Unshielded Twisted Pair Cable (UTP) consists of two conductors and each surrounded by some insulating material. This is the most common type of cable used in house wiring and in telecommunication.

Shielded Twisted Pair

Cables with a shield are called **Shielded Twisted Pair** and commonly abbreviated STP. Shielded pair cable contains two wires surrounded and separated by a solid dielectric. The dielectric is contained within a copper braid that shields the conductors from external noise sources. The entire construction is housed in a flexible, waterproof cover. Twisted pairs may be used for transmitting either digital or analog signals. The bandwidth depends on the thickness of the wire and the distance traveled.

STP or Shielded Twisted Pair is used with the traditional Token Ring cabling or ICS - IBM Cabling System. It requires a custom connector. IBM STP (Shielded Twisted Pair) has a characteristic impedance of 150 ohms.

Figure 3.5 (a, b & c) shows the (a)Two-wire twisted-pair Unshielded cable (b) Two-wire twisted-pair shielded cable and (c) Four pair Unshielded twisted cable CAT-5 type.

The use of this kind of cable is limited by two factors:

(*a*) **Attenuation**   (*b*) **cross-talk**.

There are three principle sources of attenuation.

Resistance (or impedance) **losses** are simply the loss resulting from the resistance of the wires. This loss is minimized by the choice of a metal with low resistivity. Copper is chosen for this reason. (Gold is even better, and is in fact used on satellites to reduce losses.)

Dielectric losses are caused by the heating effects when a varying electric field passes through a dielectric (insulator).

Radiation losses occur because the cable acts as an antenna. All these losses increase with frequency.

(*a*) Two-wire twisted-pair Unshielded cable   (*b*) Two-wire twisted-pair shielded cable

(*c*) Four pair Unshielded twisted cable CAT-5 type

Fig. 3.5. (a, b and c) Types of Twisted-pair cables.

When a transmission line can act as an antenna, it can also act as a receiver. Lines prone to radiation loss are also susceptible to **pick-up**, or **cross-talk**. The shielded pair is designed to reduce this pick-up.

All these lines have strong attenuation at frequencies above 1MHz. They are generally used for low bit-rate communication. Two-wire ribbon cable is standard for the connection of individual telephone receivers. Twisted pair(s) is the normal method of connection for computer terminals and short high bit-rate connections.

MICROWAVE TRANSMISSION LINES

Attenuation increases with both frequency and length. It is usually specified in dB/m at a particular frequency. Because of this fact, it is not possible to give hard-and-fast rules concerning the bandwidth availability of transmission lines. A twisted pair can support rates of several Mbps over short distances (meters), but over long distances (kilometers) will be completely unsuitable at these data-rates.

Categories of Unshielded Twisted –Pair (UTP) Cable

The Electronic Industries Association (EIA) has developed standards to classify UTP cable in to five categories. Categories are specified by cable quality, with 5 as the highest and 1 as the lowest.

CAT-1: This cable is used for voice communication in telephone system.

CAT-2: This cable is suitable for voice as well as for data transmission with the data transfer rate is up to 5 Mbps.

CAT-3: This type of cable requires minimum 3 to 4 twists per feet of length and capable of data transferring rate up to 10 Mbps.

CAT-4: Capable of transferring all types of data with the speed of 66 Mbps.

CAT-5: This has very high speed data rate up to 100 Mbps and mostly used in internet networking and Local Area Network (LAN).

Connectors for STP and UTP

The common connector mostly used is RJ-45 (Registered Jack) for STP and UTP cables. It has 8 conductors and this connector is a keyed connector, means connector can be inserted in only in one direction.

Applications of STP and UTP

a. These cables are used for Local area network.
b. Used for local loop in telephone lines
c. Used to provide high data rate connection used in DSL lines
d. Used in telephone lines to provide voice and data channel.

3.2.4 COAXIAL LINES

Coaxial cable gets its name because two conductors share a common axis; Coax is shorthand for Co-axial. For long distances, or data-rates in excess of several Mbps, coaxial cable is used. Generally the coaxial cable is used for higher frequency ranging from 10 kHz to 500MHz.

Components of **coaxial cables** are as follows (refer Fig. 3.6 a):

- A center (inner) conductor: Usually solid copper wire and sometimes is made of stranded wire.
- An outer conductor: Forms a tube surrounding the center conductor. This conductor can consist of braided wires, metallic foils or both. The outer conductor, often called shield, serves as a ground and also protects the inner conductor from electromagnetic interference.
- An insulation layer: This layer keeps the outer conductor spaced evenly from the inner conductor.
- A plastic encasement (jacket) protects the cable from damage.

Coaxial cable has a central wire, surrounded by a dielectric, in turn concentrically sheathed in a braided conductor. The cable is finally surrounded in a water-proof, flexible sheath. Coaxial cable is familiar is the cable used to connect our television to the Ariel or to the cable operator. The main advantage of this method of construction is its resistance to radiation losses. The outer conductor acts to shield out any external fields, whist preventing any internal fields escaping. The outer shield protects the inner conductor from outside electrical signals.

The essential function of coaxial cables is to transmit high frequency energy and signals with low loss, with reflection less matching and without Phase and Amplitude distortion. For such transmission, Characteristic Impedance and Attenuation are important concepts. Characteristic Impedance (Z_o) is determined by the ratio of dielectric OD to conductor diameter; hence a large cable and a very small cable can have the same Z_o but the smaller cable will have higher losses.

3.2.5 Ideal coaxial line

An ideal coaxial line consists of two cylindrical perfect conductors coaxial to each other as shown in Fig. 3.6 (a). The space between the two conductors is filled with a homogeneous lossless dielectric having dielectric constant ε_r. both the conductors are at different potentials. The outer conductor is normally grounded. Since we have the inner conductors, the natural mode of propagation will be TEM modes. Fig. 3.6(b) shows the electric and magnetic field distributions.

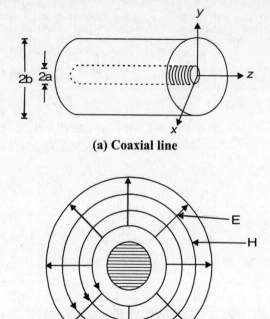

(a) Coaxial line

(b) Field distribution

Fig. 3.6

As there is no cut off frequency for coaxial line, it may be referred as broad band line.

For propagation along position 'z' directions, the electric field in the dielectric space is given as

$$E = E_\rho = E_0(\rho, \phi) \, e^{-j\beta z} = V.a_\rho \, e^{-j\beta z}/\rho \ln(b/a) \qquad \ldots 3.1$$

And
$$H = H_\phi = E_\rho/\eta = V.\, a\phi e^{-j\beta z}/\eta\rho \ln(b/a) \qquad \ldots 3.2$$

Where, β = phase constant = $\omega\sqrt{\mu\varepsilon}$

V = potential of the inner conductor w.r.t. outer conductor

η = intrinsic impedance = $\sqrt{\mu/\varepsilon}$

a = radius of inner conductor

MICROWAVE TRANSMISSION LINES

b = radius of outer conductor

α = 0 (attenuation constant is 0 for ideal line)

The density of current on the surface of the inner conductor is given as

$$J_s = a_\rho \times H = V.a_z e^{-j\beta z} / \eta a \ln(b/a) \qquad \ldots 3.3$$

The power flow through the coaxial line may be given as

$$P = \frac{1}{2} \text{Re} \int_a^b \int_0^{2\pi} (E \times H) \, ds$$

Therefore $P = \pi v^2 / \eta \ln(b/a)$ ...3.4

Where $ds = \rho \cdot d\rho \cdot d\phi$

The characteristic impedance of the symmetric coaxial line for the non-magnetic dielectric is written as

$$Z_0 = \left[\sqrt{\mu/\varepsilon} \ln(b/a)\right] 2\pi$$

$$= \left[\sqrt{\mu_0 \mu_r / \varepsilon_0 \varepsilon_r} \ln(b/a)\right] / 2\pi$$

$= [\eta_0/\varepsilon_r \ln(b/a)] 2\pi$ (since $\mu_r = 1$ for non-magnetic dielectric)

Where η_0 = intrinsic impedance of free space i.e., $120\pi \, \Omega$

Therefore $Z_0 = 120\pi / 2\pi \sqrt{\varepsilon_r} [\ln(b/a)] \, \Omega$

$$Z_0 = [60/\sqrt{\varepsilon_r} \ln(b/a)] \, \Omega \qquad \ldots 3.5$$

Practical coaxial line with small losses

An ideal coaxial line is impractical an impossible to realize. Any practical coaxial line will have some losses such as dielectric and power losses due to ferrite conductivity of the conductors.

If the conductivity of the dielectric is σ_d, then the attenuation and phase constants are expressed as,

$$\alpha_d = \sigma_d / 2 \sqrt{\mu/\varepsilon} \qquad \ldots 3.6$$

and phase constant $\beta_d = \omega \sqrt{\mu\varepsilon}$ ...3.7

Because of finite conductivity σ_c of the conductors, the surface impedance is given by

$$\eta_c = (1+j) \sqrt{\omega\mu/2\sigma_c} = 1 + j/\sigma_c \delta \qquad \ldots 3.8$$

where $\delta = 1/\sqrt{\pi f \mu \sigma_c}$ = skin depth ... 3.9

The distributed line parameters of coaxial lines are given by

$$R = R_s / 2\pi (1/a + 1/b) \qquad \ldots 3.10$$

$$G = \omega \varepsilon'' C/\varepsilon' \qquad \ldots 3.11$$

$$L = \sqrt{(\mu_0 \varepsilon'} \, Z_0)$$

$$= [\mu_0 / 2\pi \ln(b/a)] \qquad \ldots 3.12$$

$$C = 2\pi\varepsilon' / \ln(b/a) \qquad \ldots 3.13$$

Where $R_s = \sqrt{\pi f \mu / \sigma_c}$, ε = complex dielectric constant of the lossy dielectric ($\varepsilon = \varepsilon' - j\varepsilon''$)

For low-loss lines,

$\alpha_c = R/2Z_0$ = attenuation constant due to conductor losses ...3.14

$\alpha_d = GZ_0/2$ = attenuation constant due to dielectric losses ...3.15

Total attenuation constant,

$$\alpha = \alpha_c + \alpha_d$$
$$\alpha = R/2Z_0 + GZ_0/2 \qquad \text{...3.16}$$

Characteristics impedance $Z_0 = \sqrt{L/C}$

Substituting for Z_0 in equation (16), we get

$$\alpha = \tfrac{1}{2}[R\sqrt{C/L} + G\sqrt{L/C}] \qquad \text{...3.17}$$

Modes of operation in coaxial lines

Generally coaxial lines are used up to 3GHz for power transmission and impedance is 50Ω! to have minimum attenuation. Whenever coaxial lines are operated at high frequencies, generation of higher order modes takes place. The cut-off frequencies of these modes are high; as a result, mode interference will not occur at normal operating frequencies.

The following higher order modes are possible in coaxial lines:

(i) TE_{11} mode   (ii) TM_{01} mode

The electric and magnetic field distributions in the dielectric space of the coaxial line for TE_{11} and TM_{01} modes are shown in Fig. 3.7.

The cut-off wavelength (λ) of these modes are given as

$$\lambda_{c11} H \approx \pi(a+b) \qquad \text{...3.18}$$

for TE_{11} mode and

$$\lambda_{c01} H \approx 2(b-a) \qquad \text{..3.19}$$

for TM_{01} mode.

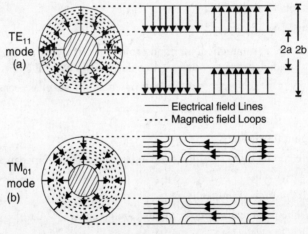

Fig. 3.7(a and b)

To avoid interference with higher modes, the average circumference $\pi(a+b)$ should be less than the operating wavelength of the coaxial line.

Power handling capability of microwave coaxial lines

The power that can be transmitted through a microwave coaxial line may not be increased beyond a certain limit. This limit is due to the electric breakdown at the region where the electric field intensity is maximum. In case of breakdown, normally a spark discharge with a large sound intensity at usual atmospheric pressures and a glow discharge occur at low pressures. Due to the discontinuity at the breakdown point, large reflection of the signal occurs and this reflected signal may cause damage to the transmitting equipment. Since the temperature at the breakdown point is high, it causes oxidation of the conductor and burn out of the cable may occur.

MICROWAVE TRANSMISSION LINES

The breakdown electric field depends on the breakdown peak voltage and the dimensions of the coaxial cable.

Breakdown electric field is given by

$$E_{bd} = V_{peak}/a[\ln(b/a)] \qquad \ldots 3.20$$

Where

E_{bd} = breakdown electric field in coaxial line.
V_{peak} = breakdown peak voltage.
a = radius of inner conductor.
b = radius of outer conductor.

The breakdown power may be determined by

$$P_{bd} = V^2_{peak}/2Z_0 \qquad \ldots 3.21$$

Putting the value of V_{peak} from equations 3.20 and Z_0 from 3.5, we get for air dielectric

$$P_{bd} = a^2 E^2_{bd} [\ln(b/a)]^2/2 \times (60/\sqrt{1}) \ln(b/a) \quad (\varepsilon_r = 1)$$
$$P_{bd} = a^2 E^2_{bd} \ln(b/a)/120 \qquad \ldots 3.22$$

From the equation 3.22, the result shows that the power capacity can be increased by using a larger coaxial cable with larger value of 'a' and 'b', keeping the ratio of (b/a) constant for same value of Z_0.

But propagation of higher modes limits the maximum operating frequency for a given cable size. Thus, it is always, there is an upper limit on the power capacity of coaxial line for a given maximum frequency f_{max}.

The maximum power for the maximum frequency is expressed as

$$P_{bd} = 5.8 \times 10^{12} [E_{bd}/f_{max}]^2 \qquad \ldots 3.23$$

High energy transmission losses

The problem in high energy transmission is the losses because of

  (*i*) cable dielectric
  (*ii*) cable construction
  (*iii*) radiation, and
  (*iv*) interference with adjoining circuits.

The coaxial cable construction has been invented to overcome these problems, in which signals and energy are contained within an enclosed space. It consists of a core conductor, surrounded by pure PTFE dielectric (solid or foamed or air-spaced), covered with a metal shield, and finally encased in an overall jacket.

The shield provides the return path as well as the confinement for RF energy, and is generally grounded. The shield is typically tied to system ground, and the signal is carried as a voltage on the center wire. In the video world, Coax is usually 75 ohms impedance.

Transmission through coaxial cables is called an **unbalanced transmission** because the centre conductor and the shield are not reversible (as opposed to the balanced construction of a twisted pair). Shield can be SPC (silver plated copper) round wire single or double braid, round wire served (helically wrapped), flat foil served (helically wrapped with drain wires), foil and braid combination with drain wires, and other materials.

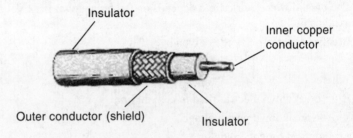

(a) Coaxial cable (Braid Shield)

(b) Round Wire Served Shield

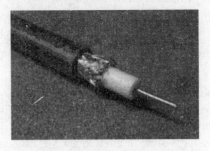

(c) RG-59 coaxial cable

Fig. 3.8. (a, b, and c) Coaxial cables

Jackets provide outer cover to keep the cable clean, smooth and to prevent ingress of moisture into shield, as well as for isolation of shield. Materials used are PTFE (our specialty), VFG (Fibre Glass braid covered with lacquer Varnish, over sintered PTFE tape moisture seal), PVC or PU. High temperature melt extrudible jackets such as FEP (Fluorinated Ethylene Propylene) and PFA (Per Fluoro Alkoxy) are also likely to be available in future.

. The distance between the outer conductor (shield) and inner conductor plus the type of material used for insulating the inner conductor determine the cable properties or impedance. Typical impedances for coaxial cables are 75 ohms for Cable TV, 50 ohms for Ethernet Thinnet and Thicknet. The excellent control of the impedance characteristics of the cable allow higher data rates to be transferred than Twisted Pair cable.

Until the advent of optical fibre, coaxial lines were the standard method of long-distance, high bit-rate communication. They are expensive (but getting cheaper, especially as demand rises), and only used where necessary. Typical attenuation values for coaxial cable are 10dB/Km at 10 KHz, 50dB/Km at 500MHz. For very long-haul routes, repeaters and equalizers are necessary.

There are two types of **coaxial lines**,
1. Rigid (air) coaxial line and
2. Flexible (solid) coaxial line.

The physical construction of both types is basically the same; that is, each contains two concentric conductors. The **rigid coaxial line** consists of a central, insulated wire (inner conductor) mounted inside a tubular outer conductor. This line is shown in figure (3.9). In some applications, the inner conductor is also tubular. Insulating spacers or beads at regular intervals insulates the inner

MICROWAVE TRANSMISSION LINES

conductor form the outer conductor. The spacers are made of Pyrex, polystyrene, or some other material that has good insulating characteristics and low dielectric losses at high frequencies. The electric and magnetic fields in a two-wire parallel line extend into space for relatively great distances and radiation losses occur. However, in a coaxial line no electric or space between the two conductors, resulting in a perfectly shielded coaxial line. Another advantage is that interference from other lines is reduced.

The rigid line has the following disadvantages:

1. It is expensive to construct
2. It must be kept dry to prevent excessive leakage between the two conductors; and
3. Although high frequency losses are somewhat less than in previously mentioned lines, they are still excessive enough to limit the practical length of the line.

Leakage caused by the condensation of moisture is prevented in some rigid line applications by the use of an inert gas, such as nitrogen, helium or argon. It is pumped into the dielectric space of the line at a pressure that can vary from 3 to 35 pounds per square inch. The inert gas is used to dry the line when it is first installed and pressure is maintained to ensure that no moisture enters the line.

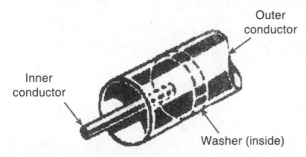

Fig. 3.9. Rigid / Air coaxial line

Flexible coaxial lines (Fig. 3.10) are made with an inner conductor that consists of flexible wire insulated from the outer conductor by a solid, continuous insulating material. The outer conductor is made of metal braid, which gives the line flexibility. Early attempts at gaining flexibility involved using rubber insulators between the two conductors. However, the rubber insulator caused excessive losses at high frequencies.

Because of the high-frequency losses associated with rubber insulators, polyethylene plastic is a solid substance that remains flexible over a wide range of temperatures. It is unaffected by seawater, gasoline, oil, and most other liquids that may be found aboard ship. The use of polyethylene as an insulator results in greater high-frequency losses than the use of air as an insulator. However, these losses are still lower than the losses associated with most other solid dielectric materials.

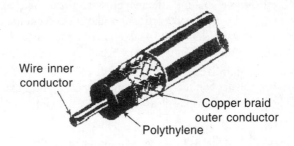

Fig. 3.10. Flexible coaxial lines

Category of Coaxial cable:

Coaxial cables are categorized by their Radio Government (RG) rating. Each RG number denotes a unique set of physical specifications. These different standards are categorized on the basis of physical specification including the dimensions of inner conductor, type of insulator and its thickness, construction and type of shielding.

    a. RG-59: Impedance is 75ohm and used for cable television.

    b. RG-58: Impedance is 50ohm and used for thin Ethernet.

    c. RG-11: Impedance is 50ohm and used for thick Ethernet.

Coaxial connectors

There are numbers of coaxial connectors are available but mostly used is BNC (Bayonet-Neill-concelman).

Four types of BNC connectors exist such as

- BNC-T: Connect the network boards in the PC to the network.
- BNC Terminator: A BNC terminator is a special connector that includes a resistor that is carefully matched to the characteristics of the cable system.
- BNC barrel connector: A BNC barrel connector connects to thin net cables.
- BNC cable connector: It attach cable segment to the T-connectors.

3.3. LOSSES IN TRANSMISSION LINES

Line losses may be any of three types

1. Copper,
2. Dielectric, and
3. Radiation or induction losses

3.3.1 Copper Losses

One type of copper loss is I^2R loss. In RF lines the resistance of the conductors is never equal to zero. Whenever current flowers through one of these conductors, some energy is dissipated in the form of heat. This heat loss is a power loss. With copper braid, which has a resistance higher than solid tubing, this power loss is higher. Another type of copper loss is due to skin effect. When dc flows through a conductor, the movement of electrons through the conductor's cross section is uniform. The situation is somewhat different when ac is applied. The expanding and collapsing fields about each electron encircle other electrons. This phenomenon, called self-induction, retards the movement of the encircled electrons. The flux density at the center is so great that electron movement at this point is reduced. As frequency is increased, the opposition to the flow of current in the center of the wire increases. Current in the center of the wire becomes smaller and most of the electron flow is one the wire surface. When the frequency applied is 100 megahertz or higher, the electron movement in the center is so small that the center of the wire could be removed without any noticeable effect on current. You should be able to see that the effective cross-sectional area decreases as the frequency increases. Since resistance is inversely proportional to the cross-sectional area, the resistance will increase as the frequency is increased. Also, since power loss increases as resistance increases, power losses increase with an increase in frequency because of skin effect.

Copper losses can be minimized and conductivity increased in an RF line by plating the line with silver. Since silver is a better conductor than copper, most of the current will flow through the silver layer. The tubing then serves primarily as a mechanical support.

3.3.2 Dielectric Losses

Dielectric losses result from the heating effect on the dielectric material between the conductors. Power from the source is used in heating the dielectric. The heat produced is dissipated into the

MICROWAVE TRANSMISSION LINES

surrounding medium. When there is no potential difference between two conductors, the atoms in the dielectric material between them are normal and the orbits of the electrons are circular. When there is a potential difference between two conductors, the orbits of the electrons change. The excessive negative charge on one conductor repels electrons on the dielectric toward the positive conductor and thus distorts the orbits of the electrons. A change in the path of electrons requires more energy, introducing a power loss. The atomic structure of rubber is more energy, introducing a power loss. The atomic structure of the atoms of materials, such as polyethylene, distort easily. Therefore, polyethylene is often used as a dielectric because less power is consumed when its electron orbits are distorted.

3.3.3 Radiation and Induction Losses

Radiation and induction losses are similar in that the fields surrounding the conductors cause both. Induction losses occur when the electromagnetic field about a conductor cuts through any nearby metallic object and a current is induced in that object. As a result, power is dissipated in the object and is lost. Radiation losses occur because some magnetic lines of force about a conductor do not return to the conductor when the cycle alternates. These lines of force are projected into space as radiation and this result in power losses. That is, power is supplied by the source, but is not available to the load.

3.4. LENGTH OF A TRANSMISSION LINE

A transmission line is considered to be electrically short when its physical length is short compared to its quarter wavelength $(\lambda/4)$. For example, a line that has a physical length of 3 meters (approximately 10 feet) is considered quite short electrically if it transmits a radio frequency of 30 kilohertz. On the other hand, the same transmission line is considered electrically long if it transmits a frequency of 30,000 megahertz. To show the difference in physical and electrical lengths of the lines mentioned above, compute the wavelength of the two frequencies, taking the 30-kilohertz example first: Now, computing the wavelength for the line carrying 30,000 megahertz: Thus, you can see that a 3-meter line is electrically very long for a frequency of 30,000 megahertz. When power is applied to a very short transmission line, practically all of it reaches the load at the output end of the line. This very short transmission line is usually considered to have practically no electrical properties of its own, except for a small amount of resistance.

Give $\quad \lambda = \dfrac{v}{f}$

when,
λ = Wavelength, v = velocity of rf in free space, f = frequency of transmission,
H_Z = Cycle per second.

$$\lambda = \dfrac{300 \times 10^6 \text{ meters/second}}{30 \times 10^3 \text{ cycle/second}(Hz)}$$

$$\lambda = 300 \times 10^3 \text{ meters/cycle}$$

$\lambda = 10,000$ meters or approximately 6 miles for complete wavelength

$$\lambda = \dfrac{v}{f}$$

$$\lambda = \dfrac{300 \times 10^6 \text{ meters/second}}{30,000 \times 10^6 \text{ cycles/second }(Hz)}$$

$$\lambda = \frac{1}{100} \text{ meter/cycle}$$

λ = .01 meter or approximately .03 foot for a complete wavelength

A transmission line is electrically long when its physical length is long compared to a quarter-wavelength of the energy it is to carry. You must understand that the terms "short" and "long" are relative ones. However, the picture changes considerably when a long line is used. Since most transmission lines are electrically long because of the distance form transmitter to antenna, so the properties of such lines must be considered. Frequently, the voltage necessary to drive a current through a long line is considerably greater than the amount that can be accounted for by the impedance of the load in series with the resistance of the line.

3.5. WAVEGUIDES

A hollow metallic tube of uniform cross section for transmitting electromagnetic waves by successive reflections form the inner of the hollow tube is called a **waveguide**.

Wave-guides are the most efficient way to transfer electromagnetic energy. Wave-guides are essentially coaxial lines without center conductors. In fact, waveguides are better alternate to conventional transmission lines for use at microwave frequencies, in terms of attenuation per unit length experienced by the wave propagating through them.

However, usage has generally limited the term to mean a hollow metal tube or a dielectric transmission line. In this chapter, we use the term waveguide only to mean "hollow metal tube". A waveguide is a special form of transmission line consisting of hollow, metal tube. The tube wall provides distributed capacitance. It is interesting to note that the transmission of electromagnetic energy along a waveguide travels at a velocity somewhat slower than electromagnetic energy traveling through free space.

A waveguide may be classified according to its cross section (rectangular, elliptical, or circular), or according to the material used in its construction (metallic or dielectric).

The waveguide is classified as a transmission line. However, the method by which it transmits energy down its length differs from the conventional methods.

There are various types of Wave guides available which are as follows and shown in figure (3.11):

(*a*) Rectangular waveguide
(*b*) Cylindrical waveguide
(*c*) Elliptical waveguide
(*d*) Ridged waveguide (Single and double ridged waveguides)

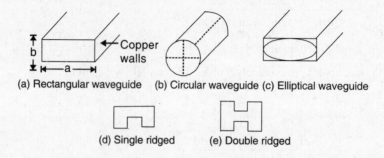

(a) Rectangular waveguide (b) Circular waveguide (c) Elliptical waveguide

(d) Single ridged (e) Double ridged

Fig. 3.11. Waveguides shapes

MICROWAVE TRANSMISSION LINES

The rectangular waveguide is used more frequently than the cylindrical waveguide. The term waveguide can be applied to all types of transmission lines in the sense that they are all used to guide energy from one point to another.

Dielectric waveguides are seldom used because the dielectric losses for all known dielectric materials are too great to transfer the electric and magnetic fields efficiently. The installation of a complete waveguide transmission system is somewhat more difficult than the installation of other types of transmission lines. The radius of bends in the waveguide must measure greater than two wavelengths at the operating frequency of the equipment to avoid excessive attenuation. The cross section must remain uniform around the bend. These requirements hamper installation in confined spaces. If the waveguide is dented, or if solder is permitted to run inside the joints, the attenuation of the line is greatly increased. Dents and obstructions in the waveguide also reduce its breakdown voltage, thus limiting the wave guide's power-handling capability because of possible arc over. Great care must be exercised during installation; one or two careless made joints can seriously inhibit the advantage of using the waveguide.

The two-wire transmission line used in conventional circuits is inefficient for transferring electromagnetic energy at microwave frequencies. At these frequencies, energy escapes by radiation because the fields are not confined in all directions as illustrated in Fig. 3.12(a). Coaxial lines are more efficient than two-wire lines for transferring electromagnetic energy because the fields are completely confined by the conductors, as illustrated in Fig. 3.12(b).

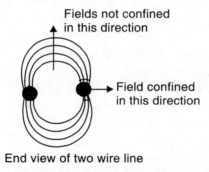

End view of two wire line

Fig. 3.12 (a) Fields confined in two directions only

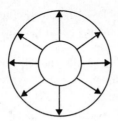

Fig. 3.12 (b) End view of coaxial cable (fields confined in all directions)

3.5.1 Advantages of Waveguide

Waveguides have several advantages over two-wire and coaxial transmission lines. They are as follows:

1. The large surface area of wave guide reduces copper $\left(I^2 R\right)$ losses. Two-wire transmission lines have large copper losses because they have a relatively small surface area. The surface area of the outer conductor of a coaxial cable is large, but the surface area of the inner conductor is relatively small.

2. At microwave frequencies, the current-carrying area of the inner conductor is restricted to a very small layer at the surface of the conductor by an action called skin effect. Skin effect tends to increase the effective resistance of the conductor. Although energy transfer in coaxial cable is caused by electromagnetic field motion, the magnitude of the field is limited by the size of the current-carrying area of the inner conductor. The small size of the center conductor is even further reduced by skin effect and energy transmission by coaxial cable becomes less efficient than by waveguides.

3. **Dielectric losses** are also lower in waveguides than in two-wire and coaxial transmission lines. Dielectric losses two-wire and coaxial lines are caused by the heating of the insulation between the conductors. The insulation behaves as the dielectric of a capacitor formed by the two wires of the transmission line. A voltage potential across the two-wires causes heating of the dielectric and results in a power loss. In practical applications, the actual breakdown of the insulation between the conductors of a transmission line is more frequently a problem than is the dielectric loss. This breakdown is usually caused by stationary voltage spikes or "nodes" which are caused by standing waves. Standing waves are stationary and occur when part of the energy traveling down the line is reflected by an impedance mismatch with the load. The voltage potential of the standing waves at the points of greatest magnitude can become large enough to break down the insulation between transmission line conductors. The dielectric in waveguides is air, which has a much lower dielectric loss than conventional insulating materials. However, waveguides are also subject to dielectric breakdown caused by standing waves. Standing waves in waveguides cause arcing, which decreases the efficiency of energy transfer and can severely damage the waveguide.

4. Also since the electromagnetic fields are completely contained within the waveguide, **radiation losses** are kept very low.

5. **Power-handling capability** is another advantage waveguides. Waveguides can handle more power than coaxial lines of the same size because power-handling capability is directly related to the distance between conductors in a waveguide.

In view of the advantages of waveguides, we may think that waveguides should be the only type of transmission lines used. However, waveguides have certain disadvantages that make them practical for use only as microwave frequencies.

3.5.2 Disadvantages of Waveguide

1. Physical size is the primary lower-frequency limitation of waveguides. The width of a waveguide must be approximately a half wavelength at the frequency of the wave to be transported. For example, a waveguide for use at 1 megahertz would be about 500 feet wide. This makes the use of waveguides at frequencies below 1000 megahertz increasingly impractical. The physical dimensions of the waveguides limit the lower frequency range of any system using waveguides.

2. Waveguides are difficult to install because of their rigid, hollow-pipe shape. Special couplings at the joints are required to assure proper operation.

3. Also, the inside surface of waveguides are often plated with silver or gold to reduce skin effect losses. These requirements increase the costs and decrease the practicality of waveguide systems at any other than microwave frequencies.

3.5.3 Wave equations for electromagnetic fields

In fact, waveguides are better alternate to conventional transmission lines for use at microwave frequencies, in terms of attenuation per unit length experience by the wave propagation through them. Waveguide analysis is based on the solution of wave equation for electromagnetic fields.

Wave equations

Wave guides are analyzed in reference of electric and magnetic fields inside the guide. Basic rules that govern the behavior of these fields are Maxwell's equation, given as following.

$$\nabla \times H = \left[\frac{\partial D}{\partial t}\right] + J \qquad \ldots 3.24$$

$$\nabla \times E = -\frac{\partial B}{\partial t} \qquad \ldots 3.25$$

$$\nabla \cdot E = \rho \qquad \ldots 3.26$$

$$\nabla \times H = 0 \qquad \ldots 3.27$$

Where H and E denote magnetic and electric fields,

$D(=\varepsilon E)$ and $B(=\mu H)$ are electric and magnetic flux densities respectively.

μ is permeability and ε represents permittivity of the medium. **J** and ρ are electric current density and electric charge density, respectively.

For perfect dielectric media containing no charges and no conduction currents, we may substitute $\rho F = 0$ and $J = 0$. Taking curl of equation 3.24 and substituting 3.25, we have $J = 0$,

$$\nabla \times \nabla \times E = -\mu\varepsilon \partial^2 E / \partial t^2 \ldots\ldots\ldots \qquad \ldots 3.28$$

Using vector identity

$$\nabla \times \nabla \times A = (\nabla \cdot A - \nabla^2 A)$$

Or $\qquad \nabla \cdot E - \nabla^2 E = -\mu\varepsilon \partial^2 E / \partial t^2 \qquad \ldots 3.29$

Since $\rho = 0$, we have $\Delta \cdot D$ and $\Delta \cdot E$ equal to zero and equation (3.29) will become

$$\nabla^2 E = -\mu\varepsilon \partial^2 E / \partial t^2 \qquad \ldots 3.30$$

This is the wave equation for the electric field vector **E**.

A similar equation for the magnetic field vector **H** may be derived as

$$\nabla^2 H = -\mu\varepsilon \partial^2 H / \partial t^2 \qquad \ldots 3.31$$

The time variation of **E** is expressed by $e^{j\ddot{u}t}$,

i.e., $\qquad E(r,t) = \text{Re}\{E(r) e^{j\omega t}\} \qquad \ldots 3.32$

$E(r)$ is a complex vector and is not a function of time, where ù is angular frequency in radians. Other variables also follow a similar variation. So equation (30) may be written as

$$\nabla^2 E = -\omega^2 \mu\varepsilon E \qquad \ldots 3.33$$

Solutions of equation (3.33) for two types of waveguides such as rectangular and circular are discussed separately further.

3.5.4 Developing the Waveguide from Parallel Lines

You may better understand the transition from ordinary transmission line concepts to waveguide theories by considering the development of a waveguide from a two-wire transmission line. Fig. (3.13a) shows a section of two-wire transmission line supported on two insulators. At the junction with the line, the insulators must present a very high impedance to ground for proper operation of the line. A low impedance insulator would obviously short-circuit the line to ground, and this is what happens at very high frequencies. Ordinary insulators display the characteristics of the dielectric of a capacitor formed by the wire and ground. As the frequency increases, the overall impedance decreases. A better high-frequency insulator is a quarter-wave section of transmission line shorted at one end. Such an insulator is shown in Fig. (3.13b). The impedance of a shorted quarter-wave section is very high at the open-end junction with the two-wire transmission line. This type of insulator is known as a **metallic insulator** and may be placed anywhere along a two-wire line. Note that quarter-wave sections are insulators at only one frequency. This severely limits the bandwidth, efficiency, and application of this type of two-wire line.

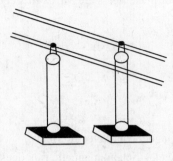

Fig. 3.13. (a)

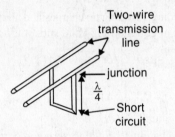

Fig. 3.13(b)

Fig. (3.14) shows several metallic insulators on each side of a two-wire transmission line. As more insulators are added, each section makes contact with the next, and a rectangular waveguide is formed. The lines become part of the walls of the waveguide, as illustrated in fig. (3.15). The energy is then conducted within the hollow waveguide of along the two-wire transmission line.

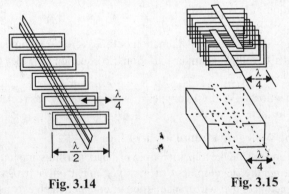

Fig. 3.14   **Fig. 3.15**

The comparison of the way electromagnetic fields work on a transmission line and in a waveguide is not exact. During the change form a two-wire line to a waveguide, the electromagnetic field configurations also undergo many changes. As a result of these changes, the waveguide does not actually operate like a two-wire line that is completely shunted by quarter-wave sections. If it did, the use of a waveguide would be limited to a single-frequency wavelength that was four times the length of the quarter-wave sections. In fact, wave of this length cannot pass efficiently through waveguides. Only a small range of frequencies of somewhat shorter wavelength (higher frequency) can pass efficiently. As shown in Fig. (3.16), the widest dimension of a waveguide is called the "a"

MICROWAVE TRANSMISSION LINES

dimension and determines the range of operating frequencies. The narrowest dimension determines the power-handling capability of the waveguide and is called the "b" dimension.

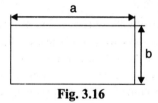

Fig. 3.16

3.5.5 Rectangular Waveguides

Rectangular waveguides are mostly used for power transmission at microwave frequencies. Their physical dimensions are being decided by the frequency of the signal to be transmitted.

Rectangular waveguides are the one of the earliest type of the transmission lines. They are used in many applications. A lot of components such as isolators, detectors, attenuators, couplers and slotted lines are available for various standard waveguide bands between 1 GHz to above 220 GHz.

A rectangular waveguide supports TM and TE modes but not TEM waves because we cannot define a unique voltage since there is only one conductor in a rectangular waveguide. The shape of a rectangular waveguide is as shown in Fig. 3.17. A material with permittivity ε and permeability μ fills the inside of the conductor.

A rectangular waveguide cannot propagate below some certain frequency. This frequency is called the cut-off frequency. (We will discuss this later section).

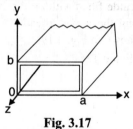

Fig. 3.17

3.5.5.1 Waveguides Modes:

When an electromagnetic wave propagates down a hollow tube, only one of the fields – either electric or magnetic – will actually be transverse to the wave's direction of travel. The other field will "loop" longitudinally to the direction of travel, but still be perpendicular to the other field. Whichever field remains transverse to the direction of travel determines whether the wave propagates in TE mode (Transverse Electric) or TM (Transverse Magnetic) mode Fig. (3.18).

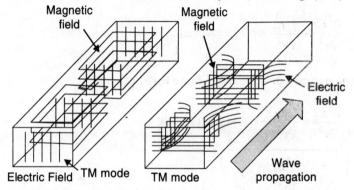

Fig. 3.18. Field pattern in TE and TM modes in waveguide.

Shows the field pattern in TE and TM modes in waveguides, magnetic flux lines are appears as continuous loop and electric flux lines appears as beginning and end lines. Many variations of each mode exist for a given waveguide.

The field pattern is formed from the superposition of two plane wave traveling in different direction. These two waves have the same free space wavelength, and give the standing wave pattern along y required to make the fields vanish at the side guide walls. The whole field pattern is called a "**Mode**". For the TE_{10} mode (transverse electric), the magnetic field is always parallel to the top and bottom guide walls, and turns to become parallel to the side guide walls where it approaches them. The guide wall currents are at right angles to the adjacent magnetic field lines.

3.5.5.2 Propagation of Wave in Rectangular Waveguides

A uniform plane wave reflected in rectangular waveguide is shown in Fig. 3.19.

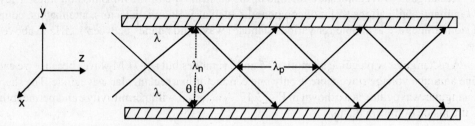

Fig. 3.19. Wave reflection in waveguide.

Consider a rectangular waveguide filled with air dielectric, situated in the rectangular co-ordinate system with width 'b' along the y-axis, breadth 'a' along x-axis and the direction of propagation of wave z-direction as shown in Fig. 3.19.

As per the wave equation of TE and TM,

$$\nabla^2 H_z = -\omega^2 \mu\varepsilon H_z \qquad \ldots 3.34$$

$$\nabla^2 E_z = -\omega^2 \mu\varepsilon E_z \qquad \ldots 3.35$$

Now we know that for TE wave $E_z = 0$ and for TM wave, $H_z = 0$.

Further expanding the wave equation for TM wave, we get

$$\frac{\partial^2 E_z}{\partial x^2} + \frac{\partial^2 E_z}{\partial y^2} + \frac{\partial^2 E_z}{\partial z^2} = -\omega^2 \mu\varepsilon E_z \qquad \ldots 3.36$$

$$\frac{\partial^2 E_z}{\partial x^2} + \frac{\partial^2 E_z}{\partial y^2} + \gamma^2 E_z = -\omega^2 \mu\varepsilon E_z \qquad \ldots 3.37$$

($\because$ wave propagation in 'z' direction so we have operator $\frac{\partial^2}{\partial z^2} = \gamma^2$)

$$\frac{\partial^2 E_z}{\partial x^2} + \frac{\partial^2 E_z}{\partial y^2} + (\gamma^2 + \omega^2 \mu\varepsilon).E_z = 0$$

Now $$\boxed{\frac{\partial^2 E_z}{\partial x^2} + \frac{\partial^2 E_z}{\partial y^2} + \gamma^2 E_z = 0} \qquad \ldots 3.38$$

MICROWAVE TRANSMISSION LINES

Similarly, expanding the wave equation for TE wave,

$$\frac{\partial^2 H_z}{\partial x^2} + \frac{\partial^2 H_z}{\partial y^2} + \frac{\partial^2 H_z}{\partial z^2} = -\omega^2 \mu\varepsilon H_z$$

or

$$\frac{\partial^2 H_z}{\partial x^2} + \frac{\partial^2 H_z}{\partial y^2} + \gamma^2 H_z = -\omega^2 \mu\varepsilon H_z$$

or

$$\boxed{\frac{\partial^2 H_z}{\partial x^2} + \frac{\partial^2 H_z}{\partial y^2} + \gamma^2 H_z = 0} \qquad \text{...3.39}$$

Equation (38) and (39), are the partial differential equation of TM and TE wave, using Maxwell's equation.

The various components such E_x, H_x, E_y and H_y along x and y direction can be find out. These components are as follows:

$$E_x = \frac{-\gamma}{h^2} \cdot \frac{\partial E_z}{\partial x} - \frac{j\omega\mu}{h^2} \cdot \frac{\partial H_z}{\partial y} \qquad (\because \gamma^2 + w^2\mu\varepsilon = h^2) \quad \text{...3.40}$$

$$E_y = \frac{-\gamma}{h^2} \cdot \frac{\partial E_z}{\partial y} + \frac{j\omega\mu}{h^2} \cdot \frac{\partial H_z}{\partial x} \qquad \text{...3.41}$$

$$H_x = \frac{-\gamma}{h^2} \cdot \frac{\partial H_z}{\partial x} + \frac{j\omega\varepsilon}{h^2} \cdot \frac{\partial E_z}{\partial y} \qquad \text{...3.42}$$

$$H_y = \frac{-\gamma}{h^2} \cdot \frac{\partial H_z}{\partial y} - \frac{j\omega\varepsilon}{h^2} \cdot \frac{\partial E_z}{\partial x} \qquad \text{...3.43}$$

Equation (3.40), (3.41), (3.42) and (3.43), are the general equations which will give the relationship of field components within a waveguide.

3.5.5.3 Modes of Waveguides

In the waveguide propagation we have two types of modes, TE (Transverse Electric) and TM (Transverse Magnetic). In wave guide propagation, we have an infinity number of patterns, known as modes. For the existence of mode patterns in the waveguide, it should obey certain physical laws of waveguides.

- To have propagation in the waveguide, the electric field component must always be perpendicular to the surface of the conductor not parallel to the surface of the conductor.
- In other conditions, the magnetic field components must always be parallel to the surface of the conductor not perpendicular to the surface of the conductor.

In general we have many modes of wave such as TE, TM, TEM and hybrid wave. Each mode has some characteristics which are discussed here:

(i) TE (Transverse Electric) wave: In TE wave, only the electric field (E_z) is transverse to the direction of propagation and the magnetic field is not purely transverse. We may write that

$$E_z = 0$$
$$H_z \neq 0$$

The direction of electric and magnetic field components along the three mutually perpendicular directions X, Y and Z are shown in Fig. (3.20).

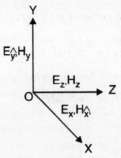

Fig. 3.20 Fields direction

(ii) TM (Transverse Magnetic) waves: In TM mode, only the magnetic field is transverse to the direction of propagation and the electric field is not purely transverse.

$$E_z \neq 0$$
$$H_z = 0$$

(iii) TEM (Transverse Electromagnetic) waves:

The focus so far was on the existence of TM and TE modes in hollow metallic waveguides. Let us investigate the possibility of the existence and propagation of TEM modes in hollow metallic waveguides.

Let us assume that a TEM wave is progressing through a hollow metallic waveguide in a given direction. It is known that In TEM mode, both the fields components, electric and magnetic field component are purely transverse to the direction of propagation and no longitudinal field components associated in the 'z' direction ($E_z = 0$ and $H_z = 0$). For free space propagation, this is not a problem because there are no boundary restrictions in free space transmission. However, this is not the case with waveguide transmissions.

We know that magnetic fields exist in the form of closed loops. When it is assumed that H fields are transverse to the direction of propagation, it also must be remembered that these fields progress along the direction of propagation as closed loops, which alternate in their transverse directions, as shown in Fig. 3.21.

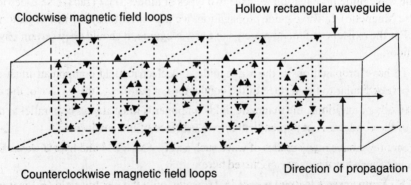

Fig. 3.21. Closed loops of ac magnetic field in a hollow metallic waveguide.

Now, consider the assumption that there exists a TEM wave inside a hollow metallic waveguide. Associated with the TEM wave, magnetic fields exists in the form of transverse closed loops, as shown in Fig. 3.21. It is known that when a magnetic fields in the form of a closed loop inside a metallic conductor, such as waveguide, then as Ampere's law, an electric field must exist or to be induced along the axis of the tube. In other words, if there exists a closed magnetic field around a metallic conducting tube, then will exist an electric field along the axis of the tube. Now to support

MICROWAVE TRANSMISSION LINES

this axial electric field, we must have a conductor existing along the axis of the conducting tube. But such conductors exist only in coaxial cables, and no such conductor exists in hollow metallic waveguides.

In TEM mode, both the fields components, electric and magnetic field component are purely transverse to the direction of propagation and no longitudinal field components associated in the 'z' direction ($E_z = 0$ and $H_z = 0$).

If we substitute the value of the field components in the wave equation (3.40), (3.41), (3.42) and (3.43), we get

$$E_x = 0$$
$$E_y = 0$$
$$H_x = 0$$
$$H_y = 0$$

It means that there is no existence of TEM mode in the wave guide propagation.

Therefore, we conclude that TEM mode of wave propagation is impossible in the hollow metallic wave guides. However, it may be seen that TEM modes exists in the coaxial cable.

3.5.5.4 Propagation of TE Waves in a Rectangular Waveguide

TE mode in a rectangular waveguide is characterized by, $E_z = 0$ because we have assumed that the wave propagation is in the z direction of the rectangular co-ordinate system. Thus for the transmission of energy H_z, it must exist in the waveguide.

The Helmholtz equation (wave equation) for TE wave is

$$\nabla^2 H_z = -\omega^2 \mu \varepsilon H_z$$

Or in partial differentiation form, it may be written as

$$\frac{\partial^2 H_z}{\partial x^2} + \frac{\partial^2 H_z}{\partial y^2} + \frac{\partial^2 H_z}{\partial z^2} = -\omega^2 \mu \varepsilon H_z$$

or

$$\frac{\partial^2 H_z}{\partial x^2} + \frac{\partial^2 H_z}{\partial y^2} + \gamma^2 H_z = -\omega^2 \mu \varepsilon H_z \quad (d^2/d_z^2 = \gamma^2)$$

or

$$\frac{\partial^2 H_z}{\partial x^2} + \frac{\partial^2 H_z}{\partial y^2} + \gamma^2 H_z + \omega^2 \mu \varepsilon H_z = 0$$

or

$$\frac{\partial^2 H_z}{\partial x^2} + \frac{\partial^2 H_z}{\partial y^2} + (\gamma^2 + \omega^2 \mu \varepsilon) H_z = 0$$

or

$$\frac{\partial^2 H_z}{\partial x^2} + \frac{\partial^2 H_z}{\partial y^2} + h^2 H_z = 0 \qquad (\because \gamma^2 + \omega^2 \mu \varepsilon = h^2) \quad \ldots 3.44$$

Equation (3.44) is a partial differential equation can be solved further by assuming a solution $H_z = XY$.

X = Pure function of 'x' only
Y = Pure function of 'y' only

Equation (3.44) can be written as below by substituting the value of H_z, we get

$$Y \frac{d^2 X}{dx^2} + X \frac{d^2 Y}{dy^2} + h^2 \cdot XY = 0$$

Dividing by XY, we get

$$\frac{Y}{XY}\cdot\frac{d^2X}{dx^2}+\frac{X}{XY}\cdot\frac{d^2Y}{dy^2}+h^2\cdot\frac{XY}{XY}=0$$

or
$$\frac{1}{X}\cdot\frac{d^2X}{dx^2}+\frac{1}{Y}\cdot\frac{d^2Y}{dy^2}+h^2=0 \qquad \ldots 3.45$$

$$\frac{1}{X}\cdot\frac{d^2X}{dx^2} \qquad \text{(Pure function of } x \text{ only)}$$

$$\frac{1}{Y}\cdot\frac{d^2Y}{dy^2} \qquad \text{(Pure function of } y \text{ only)}$$

The sum of above components is a constant; hence X and Y are independent variables, since each term must be equal to a constant separately. We use a method of separation of variables to solve the differential equation (45).

Let
$$\frac{1}{X}\cdot\frac{d^2X}{dx^2}=-B^2$$

and
$$\frac{1}{Y}\cdot\frac{d^2Y}{dy^2}=-A^2 \qquad (-B^2 \text{ and } -A^2 \text{ are constants})$$

Substituting in equation (45), we get

$$-B^2 - A^2 + h^2 = 0$$

or $\quad h^2 = A^2 + B^2 \qquad \ldots 3.46$

Now solving for X and Y by separation of variable method,

$$X = C_1 \cos B_x + C_2 \sin B_z \qquad \ldots 3.47$$

$$Y = C_3 \cos A_y + C_4 \sin A_y \qquad \ldots 3.48$$

Hence C_1, C_2, C_3 and C_4 are constants which will be evaluated by applying the boundary conditions.

The complete solution,

$$H_z = XY$$

or
$$\boxed{H_z = \left(C_1 \cos B_x + C_2 \sin B_x\right)\left(C_3 \cos A_y + C_4 \sin A_y\right)} \qquad \ldots 3.49$$

3.5.5.5 Boundary conditions

A rectangular waveguide has four walls as shown in Fig. (3.22) and all the boundary walls of a waveguide will act as a short circuit or ground for electric field components, $E_2 = 0$ all along the four boundary walls of the waveguide. There are four boundary conditions, which are discussed here.

MICROWAVE TRANSMISSION LINES

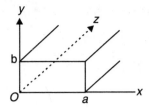

Fig. 3.22

In TE wave,

$E_z = 0$ (all along x and y direction)
$E_x = 0$ (all along bottom and top walls of waveguide)
$E_y = 0$ (all along left and right walls of the waveguide)

1st boundary condition (Bottom wall) :

We know $E_x = 0$ at $Y = 0 \forall\ x \to 0$ to a ($\forall$ stand for all and $x \to 0$ to a mean varying between 0 to a.)

From equation (3.40), we have

$$E_x = \frac{-\gamma}{h^2}\frac{\partial E_z}{\partial x} - \frac{j\omega\mu}{h^2}\frac{\partial H_z}{\partial y}$$

Since $E_z = 0$, than we get

$$E_x = -\frac{j\omega\mu}{h^2}\cdot\frac{\partial H_z}{\partial y}$$

or

$$E_x = -\frac{j\omega\mu}{h^2}\cdot\frac{\partial}{\partial y}\left[(C_1\cos B_x + C_2\sin B_x)(C_3\cos A_y + C_4\sin A_y)\right]$$

or

$$E_x = -\frac{j\omega\mu}{h^2}\cdot\left[(C_1\cos B_x + C_2\sin B_x)(-AC_3\cos A_y + AC_4\sin A_y)\right]$$

By applying 1st boundary condition in the equations,

$E_x = 0$ at $y = 0\ \forall x \to 0$ to a

We get

$$0 = -\frac{j\omega\mu}{h^2}(C_1\cos B_x + C_2\sin B_x)(0 + AC_4)$$

$C_1\cos B_x + C_2\sin B_x \neq 0$ and $A \neq 0$

Then $C_4 = 0$

If we put the value of C_4 in equation (49), we get

$$H_z = (C_1\cos B_x + C_2\sin B_x)(C_3\cos A_y) \qquad \ldots 3.50$$

2nd boundary condition (Top well):

$E_x = 0$ at $y = b\ \forall\ x \to 0$ to a, we know

$$E_x = -\frac{\gamma}{h^2}\cdot\frac{\partial E_z}{\partial x} - \frac{j\omega\mu}{h^2}\frac{\partial H_z}{\partial y}$$

$$0 = 0 - \frac{j\omega\mu}{h^2}\frac{\partial}{\partial y}(C_1 C_3\cos B_x \sin A_y)$$

$$0 = +\frac{j\omega\mu}{h^2} C_1 C_3 A \cos B_x \sin A_y$$

Now applying the 2nd boundary conditions, we get

$$0 = \frac{j\omega\mu}{h^2} \cdot C_1 C_3 \cos B_x \sin Ab$$

or $\quad B_x \neq 0, C_3, C_1 \neq 0$

or $\quad \sin Ab = 0$ or $Ab = n\pi$ (where $n = 0, 1, 2, 3, \ldots, \infty$)

$$\boxed{A = \frac{n\pi}{b}} \quad \ldots 3.51$$

3rd boundary condition (Left-side wall):

According to 3rd boundary conditions,

$$E_y = 0 \text{ at } x = 0 \forall\ y \to 0 \text{ to } b$$

From equation (41), we know that

$$E_y = -\frac{\gamma}{h^2}\frac{\partial E_z}{\partial y} + \frac{j\omega\mu}{h^2}\cdot\frac{\partial H_z}{\partial x}$$

By substituting the value of H_z from equation (40) and $E_z = 0$

$$E_y = 0 + \frac{j\omega\mu}{h^2}\frac{\partial}{\partial x}\left[(C_1 \cos B_x + C_2 \sin B_x)(C_3 \cos A_y)\right]$$

or $\quad E_y = \frac{j\omega\mu}{h^2}\left[(-BC_1 \sin B_x + BC_2 \cos B_x)(C_3 \cos A_y)\right]$

By applying 3rd boundary condition,

$$0 = \frac{j\omega\mu}{h^2}\left[(0 + BC_2)(C_3 \cos A_y)\right]$$

$B \neq 0$, $C_3 \neq 0 \cos A_y \neq 0$, and $C_2 = 0$

By substituting the value of C_2 in equation (50), the equation is reduced to

$$H_z = C_1 C_3 \cos B_x \cos A_y \quad \ldots 3.52$$

4th boundary condition (Left-side wall):

$E_y = 0$ at $x = a \forall\ y \to 0$ to b, we know

$$E_y = -\frac{\gamma}{h^2}\cdot\frac{\partial E_z}{\partial y} + \frac{j\omega\mu}{h^2}\frac{\partial H_z}{\partial x}$$

or $\quad E_y = \frac{j\omega\mu}{h^2}\cdot\frac{\partial}{\partial x}\left[C_1 C_3 \cos B_x \cos A_y\right]$

$(\because$ where $E_z = 0$ and $H_z = C_1 C_3 \cos B_x \cos A_y)$

$$E_y = -\frac{j\omega\mu}{h^2} C_1 C_3 B \sin B_x \cos A_y$$

By applying the 4th boundary condition

$$0 = -\frac{j\omega\mu}{h^2}\cdot C_1 C_3 B \sin Ba \cos A_y\ (y \to 0 \text{ to } b)$$

MICROWAVE TRANSMISSION LINES

$\cos A_y \neq 0,\ C_1,\ C_3 \neq 0$

∴ $\sin Ba = 0$

or $Ba = m\pi$ where $m = 0, 1, 2, 3, \ldots, \infty$

$$\boxed{B = \frac{m\pi}{a}} \qquad \ldots 3.53$$

By substituting to value of A and B from equation 3.51 and 3.53 in equation (52) we get

$$H_z = C_1 C_3 \cos\left[\frac{m\pi}{a}\right] x \cos\left[\frac{n\pi}{b}\right] y \qquad \ldots 3.54$$

or $$H_z = C \cos\left[\frac{m\pi}{a}\right] x \cos\left[\frac{n\pi}{b}\right] y \cdot e^{(j\omega t - \gamma z)}$$

It can be seen that for a TE wave H, has cosine-cosine components as per equation (54).

Field components in TE mode:

$$E_x = -\frac{\gamma}{h^2}\frac{\partial E_z}{\partial x} - \frac{j\omega\mu}{h^2}\cdot\frac{\partial H_z}{\partial y}$$

$E_z = 0$

$$E_x = \frac{j\omega\mu}{h^2}\cdot C\left[\frac{n\pi}{b}\right]\cos\left[\frac{m\pi}{a}\right] x \sin\left[\frac{n\pi}{b}\right] y \cdot e^{(j\omega t - \gamma z)} \qquad \ldots 3.55$$

$$E_y = -\frac{\gamma}{h^2}\frac{\partial E_z}{\partial y} + \frac{j\omega\mu}{h^2}\cdot\frac{\partial H_z}{\partial x} \qquad \ldots 3.56$$

Again since $E_z = 0$, Ist term $= 0$,

$$E_y = -\frac{j\omega\mu}{h^2} C\left[\frac{m\pi}{a}\right]\sin\left[\frac{m\pi}{a}\right] x \cos\left[\frac{n\pi}{b}\right] y \cdot e^{(j\omega t - \gamma z)} \qquad \ldots 3.57$$

Similarly,

$$H_x = -\frac{\gamma}{h^2}\frac{\partial H_z}{\partial x} + \frac{j\omega\varepsilon}{h^2}\cdot\frac{\partial E_z}{\partial y}$$

∴ $$H_x = +\frac{\gamma}{h^2} C\left[\frac{m\pi}{a}\right]\sin\left[\frac{m\pi}{a}\right] x \cos\left[\frac{n\pi}{b}\right] y \cdot e^{(j\omega t - \gamma z)}$$

$$H_y = -\frac{\gamma}{h^2}\frac{\partial H_z}{\partial y} + \frac{j\omega\varepsilon}{h^2}\cdot\frac{\partial E_z}{\partial x}$$

and $$H_y = -\frac{\gamma}{h^2} C\cdot\left[\frac{n\pi}{b}\right]\cos\left[\frac{m\pi}{a}\right] x \sin\left[\frac{n\pi}{b}\right] y \cdot e^{(j\omega t - \gamma z)} \qquad \ldots 3.58$$

3.5.5.6 Propagation of TM Waves in Rectangular Waveguide

The wave equation for TM wave is given by

$$\nabla^2 E_z = -\omega^2 \mu \varepsilon E_z \qquad (H_z = 0)$$

By expanding $\nabla^2 E_z$ in rectangular co-ordinate system, we get

$$\frac{\partial^2 E_z}{\partial x^2} + \frac{\partial^2 E_z}{\partial y^2} + \frac{\partial^2 E_z}{\partial z^2} = -\omega^2 \mu\varepsilon E_z$$

The wave is propagating in the 'z' direction so we have the operator

$$\frac{\partial^2}{\partial z^2} = \gamma^2$$

By putting the value in above equation

$$\frac{\partial^2 E_z}{\partial x^2} + \frac{\partial^2 E_z}{\partial y^2} + \gamma^2 E_z = -\omega^2 \mu\varepsilon E_z$$

or

$$\frac{\partial^2 E_z}{\partial x^2} + \frac{\partial^2 E_z}{\partial y^2} + (\gamma^2 + \omega^2 \mu\varepsilon) E_z = 0$$

$$\gamma^2 + \omega^2 \mu\varepsilon = h^2$$

$$\frac{\partial^2 E_z}{\partial x^2} + \frac{\partial^2 E_z}{\partial y^2} + \gamma^2 E_z = 0 \qquad \ldots 3.59$$

This equation is partial differential equation and further can be solved for various field components.

$$E_z = XY \qquad \text{(Assumed)}$$

X = Pure function of 'x' only
Y = Pure function of 'y' only

$$\frac{\partial^2 E_z}{\partial x^2} = \frac{\partial^2 (XY)}{\partial x^2}$$

or

$$= Y \frac{\partial^2 X}{\partial x^2}$$

$$\frac{\partial^2 E_z}{\partial y^2} = \frac{\partial^2 (XY)}{\partial y^2} = X \frac{\partial^2 Y}{\partial y^2}$$

Put these values in wave equation of TM,

$$Y \frac{\partial^2 X}{dx^2} + X \frac{\partial^2 Y}{\partial y^2} + h^2 XY = 0$$

Dividing that equation by XY, we get

$$\frac{1}{X} \frac{d^2 X}{dx^2} + \frac{1}{Y} \frac{d^2 Y}{dy^2} + h^2 = 0 \qquad \ldots 3.60$$

Let

$$\frac{1}{X} \frac{d^2 X}{dx^2} = -B^2 \qquad (-B^2 = \text{constant})$$

and

$$\frac{1}{Y} \frac{d^2 Y}{dy^2} = -A^2 \qquad (-A^2 = \text{constant})$$

MICROWAVE TRANSMISSION LINES

The solutions for the above 2nd order differential equations are given as

$$X = C_1 \cos Bx + C_2 \sin Bx \qquad \ldots 3.61$$

$$Y = C_3 \cos A\, y + C_4 \sin Ay \qquad \ldots 3.62$$

where C_1, C_2, C_3 and C_4 are constants and can be calculated the values by applying the boundary condition.

By putting the values of $-B^2$ and $-A^2$ in the equation (60), we get

$$-B^2 - A^2 + h^2 = 0$$

$$h^2 = A^2 + B^2$$

The complete solution is given by

$$E_z = X \cdot Y$$

Put the values of X and Y, we get

$$E_z = [C_1 \cos Bx + C_2 \sin Bx][C_3 \cos Ay + C_4 \sin Ay] \qquad \ldots 3.63$$

Boundary conditions: Fig (3.23) show the rectangular waveguide in co-ordinate system, the entire surface of the waveguide acts as a short circuit for electric field. $E_z = 0$, all along the walls of waveguide and since we have four walls as shown in figure (3.23), so we have four boundary conditions.

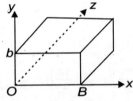

Fig. 3.23

1st boundary condition (Bottom wall): $E_z = 0$ at $y = 0 \,\forall\, x \to 0$ to a

$$E_z = [C_1 \cos Bx + C_2 \sin Bx][C_3 \cos Ay + C_4 \sin Ay] \qquad \ldots 3.64$$

By applying the 1st boundary condition,

$$0 = [C_1 \cos Bx + C_2 \sin Bx] C_3$$

This true for all $x \to 0$ to a

$$C_1 \cos Bx + C_2 \sin Bx \neq 0$$

$$C_3 = 0$$

By putting the value of C_3 in equation (64), we get

$$E_z = [C_1 \cos Bx + C_2 \sin Bx][C_4 \sin Ay]$$

2nd boundary condition (Left side wall): $E_z = 0$ at $x = 0 \,\forall\, y \to 0$ to b

By putting the 2nd boundary conditions in eqn. we get

$$0 = C_1 C_4 \sin Ay \,\forall\, y \to 0 \text{ to } b$$

Since $\sin Ay \neq 0$ and $C_4 \neq 0$

$$C_1 = 0$$

This equation can be reduced to further

$$E_z = C_1 C_4 \sin Bx \sin Ay \qquad \ldots 3.65$$

3rd boundary condition (Top wall) : $E_z = 0$ at $y = b \; \forall \; y \to 0$ to a

By substituting the value of 3rd boundary conditions in equation (65),

$$E_z = 0 = C_2 C_4 \sin Bx \sin Ab$$

Since $\sin Bx \neq 0$, $C_4 \neq 0$, $C_2 \neq 0$

$\sin Ab = 0$ or $Ab =$ multiples of $\pi = n\pi$

$n = 0, 1, 2,...$

$$\boxed{A = \frac{n\pi}{b}}$$ ...3.66

4th boundary conditions (Right side wall) : $E_z = 0$ at $x = a \; \forall \; y \to 0$ to b

By applying the 4th boundary conditions, we get

$$E_z = 0 = C_2 \, C_4 \sin B \, a \sin A_y$$

Since $A_y \neq 0$, $C_2, \, C_4 \neq 0$

So $\sin Ba = 0$

or $Ba = m\pi$ \hspace{2cm} (m is constant and its value 0, 1, 2, 3, ...)

$$B = \frac{m\pi}{a}$$ ...3.67

By putting the values of A and B in

$$E_z = C_2 \, C_4 \sin \left[\frac{m\pi}{a} \right] x \sin \left[\frac{n\pi}{b} \right] \cdot y \cdot e^{-\gamma z} e^{+j\omega t}$$

$e^{j\omega t} =$ Sinusoidal variation w.r.t. to 't'

$e^{-\gamma z} =$ Propagation along 'z' direction

If $C_2 \, C_4 = C$, from

$$\boxed{E_z = C \sin \left[m \frac{\pi}{a} \right] x \sin \left(n \frac{\pi}{b} \right) y \cdot e^{j\omega t - \gamma z}}$$ ...3.68

Field Components in TM mode: By having shown the value of E_z, the value of other field components can be given by following equations

$$H_x = -\frac{\gamma}{h^2} \frac{\partial H_z}{\partial x} + \frac{j\omega \varepsilon}{h^2} \cdot \frac{\partial E_z}{\partial y}$$

or $$H_x = \frac{j\omega \varepsilon}{h^2} \cdot C \left(\frac{n\pi}{b} \right) \sin \left[\frac{m\pi}{a} \right] x \cos \left[\frac{n\pi}{b} \right] y \, e^{j\omega t - \gamma z}$$ ... 3.69

$$H_y = \frac{-\gamma}{h^2} \frac{\partial H_z}{\partial y} + \frac{j\omega \varepsilon}{h^2} \frac{\partial E_z}{\partial x}$$

or $$H_y = \frac{j\omega \varepsilon}{h^2} C \left[\frac{m\pi}{a} \right] \cos \left[\frac{m\pi}{a} \right] x \cdot \sin \left[\frac{n\pi}{b} \right] y \, e^{j\omega t - \gamma z}$$ ... 3.70

$$E_x = -\frac{\gamma}{h^2}\frac{\partial E_z}{\partial x} - \frac{j\omega\mu}{h^2}\frac{\partial H_z}{\partial y}$$

$$E_x = \frac{-\gamma \partial E_z}{h^2 \partial x} \qquad \ldots 3.71$$

(as for a TM wave $H_z = 0$)

$$E_x = -\frac{\gamma}{h^2}\cdot C\left[\frac{m\pi}{a}\right]\cos\left[\frac{m\pi}{a}\right]\cdot x\,\sin\left[\frac{n\pi}{b}\right] y\cdot e^{j\omega t-\gamma z} \qquad .3.72$$

$$E_y = -\frac{\gamma \partial E_z}{h^2 \partial y} + \frac{j\omega\mu}{h^2}\cdot\frac{\partial H_z}{\partial x}$$

$$E_y = -\frac{\gamma}{h^2}\cdot C\left[\frac{n\pi}{b}\right]\sin\left[\frac{m\pi}{a}\right]\cdot x\,\cos\left[\frac{n\pi}{b}\right] y\cdot e^{j\omega t-\gamma z} \qquad \ldots 3.73$$

3.5.5.7 TM Modes in Rectangular Waveguides

We have various modes in TM waves, depending upon the values of m and n. It is represent as TM_{mn} where $m=$ number of half wave length across waveguide width (a).

$n=$ number of half wave length along the waveguide height (b).

TM_{mn} Modes :

TM_{00} mode , $m=0$ and $n=0$

If we put the values of m and n in the field component equation, we find all of them vanish and hence TM_{00} mode cannot exist.

TM_{01} mode – It does not exist

TM_{10} mode – It does not exist

TM_{11} mode – In this mode we have all the four components H_x, H_y, E_x, E_y. All the higher modes available in the rectangular waveguide.

3.5.5.8 Cut off Frequency of Rectangular Waveguide

We know

$$h^2 = A^2 + B^2$$

or $$h^2 = \gamma^2 + \omega^2\mu\varepsilon$$

$$h^2 = \left[\frac{m\pi}{a}\right]^2 + \left[\frac{n\pi}{b}\right]^2$$

or $$\gamma^2 = \left[\frac{m\pi}{a}\right]^2 + \left[\frac{n\pi}{b}\right]^2 - \omega^2\mu\varepsilon$$

or $$\gamma = \sqrt{\left[\frac{m\pi}{a}\right]^2 + \left[\frac{n\pi}{b}\right]^2 - \omega^2\mu\varepsilon}$$

The propagation constant 'γ' is also

$$\gamma = \alpha + j\beta$$

where $\alpha=$ Attenuation constant

$\beta=$ Phase constant

When $\omega^2 \mu\varepsilon < \left[\dfrac{m\pi}{a}\right]^2 + \left[\dfrac{n\pi}{b}\right]^2$ (for lower frequencies)

At lower frequencies the propagations constant 'γ' becomes real and positive.

$\gamma = \alpha$ (There is no phase change, *i.e.*, $\beta = 0$)

In this condition the wave is completely attenuated and hence the wave cannot propagate.

Whereas at higher frequencies 'γ' becomes imaginary and β is changes, means the wave is propagating. Whenever 'γ' becomes zero, the wave propagation starts and the frequency at which 'γ' becomes zero is known as **cut-of frequency**.

$$\gamma = 0 \text{ or } \omega = 2\pi f$$

At $f = f_c$ or $\omega_c = 2\pi f_c$

$$0 = \sqrt{\left[\dfrac{m\pi}{a}\right]^2 + \left[\dfrac{n\pi}{b}\right]^2 - \omega^2{}_c \mu\varepsilon}$$

or $$\omega_c = \dfrac{1}{\sqrt{\mu\varepsilon}}\left[\left(\dfrac{m\pi}{a}\right)^2 + \left(\dfrac{n\pi}{b}\right)^2\right]^{1/2}$$

or $$2\pi f_c = \dfrac{1}{\sqrt{\mu\varepsilon}}\left\{\left[\dfrac{m\pi}{a}\right]^2 + \left[\dfrac{n\pi}{b}\right]^2\right\}^{1/2}$$

$$f_c = \dfrac{1}{2\pi\sqrt{\mu\varepsilon}}\left\{\left[\dfrac{m\pi}{a}\right]^2 + \left[\dfrac{n\pi}{b}\right]^2\right\}^{1/2}$$

$$f_c = \dfrac{c}{2\pi}\left\{\left[\dfrac{m\pi}{a}\right]^2 + \left[\dfrac{n\pi}{b}\right]^2\right\}^{1/2} \qquad \left(c = \dfrac{1}{\sqrt{\mu\varepsilon}}\right)$$

$$f_c = \dfrac{c}{2} \times \dfrac{\pi}{\pi}\left[\left(\dfrac{m}{a}\right)^2 + \left(\dfrac{n}{b}\right)^2\right]^{1/2}$$

or $$\boxed{f_c = \dfrac{c}{2}\left[\left(\dfrac{m}{a}\right)^2 + \left(\dfrac{n}{b}\right)^2\right]^{1/2}} \qquad \ldots 3.74$$

In terms of wavelength it can be written as

$$\lambda_c = \dfrac{c}{f_c}$$

$$\lambda_c = \dfrac{c}{\dfrac{c}{2}\left[\left(\dfrac{m}{a}\right)^2 + \left(\dfrac{n}{b}\right)^2\right]^{1/2}}$$

or $$\boxed{\lambda_{c\,m,n} = \dfrac{2ab}{\sqrt{m^2 b^2 + n^2 a^2}}} \qquad \ldots 3.75$$

MICROWAVE TRANSMISSION LINES

λ_C = Cut of wavelength. ... 3.71

Wavelengths greater than λ_C (**cut of wavelength**) are attenuated and less than λ_C are passed to propagate inside the wave guide.

So, we can say that waveguide acts as **high pass filter**, all the frequency above cut of frequency (f_c) are allowed to propagate and less than f_c are attenuated inside the waveguide.

3.5.5.9 Guide Wavelength (λ_g)

The **guide wavelength** may be defined as the distance travelled by the wave to undergo a phase shift of 2π radians, as shown in Fig. 3.24. It may be written as relative to the phase constant (β).

$$\lambda_g = \frac{2\pi}{\beta} \quad \ldots 3.76$$

where, λ_g = Guide wavelengths
β = Phase constant

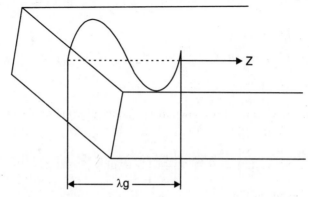

Fig. 3.24. Guide Wavelength (λ_g).

3.5.6 Group and Phase Velocity in the Waveguide

When a wave is reflected from conducting walls it will have two velocities in a direction parallel to the wall. Those two velocities are known as:

(i) Group velocity (V_g) (ii) Phase velocity (V_p)

If the actual velocity of the wave is V_C, than the velocity of the wave in a direction parallel to the conducting surface is V_g and given by

$$V_g = V_C \sin\theta$$

Group velocity can be defined as that if there is modulation in the carrier, the modulated envelope travels at velocity slower than that of carrier and slower than velocity of light also. The velocity of modulation envelope is called the **group velocity.**

Group velocity may be defined as the rate at which the wave propagates through the waveguide. It can be written as

$$\boxed{V_g = \frac{d\omega}{d\beta}} \quad \ldots 3.77$$

Phase velocity can be defined as the rate at which the wave changes its phase in terms of the guide wavelength. If the frequency is f, it follows that the velocity (Phase velocity) with which the wave changes phase in a direction parallel to the conducting surface is given by,

$$V_P = f\lambda_g$$

However, the wavelengths in the wave guide is the length of the cycle and V_p represents the phase velocity

$$V_P = \frac{\lambda_g}{\text{Unit time}} = f \cdot \lambda_g$$

$$= \frac{2\pi f \cdot \lambda_g}{2\pi} = \frac{2\pi f}{2\pi/\lambda_g}$$

So, $\boxed{V_P = \frac{\omega}{\beta}}$ $\qquad (2\pi f = \omega \text{ and } 2\pi/\lambda_g = \beta)$ ...3.78

Derivation for the Phase Velocity and Group Velocity: We know that phase velocity (V_P) is the ratio of ω and β and can be written as

$$V_P = \frac{\omega}{\beta}$$

And also $\quad h^2 = \gamma^2 + \omega^2 \mu\varepsilon$

$$h^2 = A^2 + B^2$$

$$h^2 = \left(\frac{m\pi}{a}\right)^2 + \left(\frac{n\pi}{b}\right)^2$$

In the propagation of wave, it is known that α (attenuation) is zero, so we have

$$\gamma = \alpha + j\beta \qquad (\alpha = 0)$$

$$\gamma^2 = (j\beta^2) = \left[\frac{m\pi}{a}\right]^2 + \left[\frac{n\pi}{b}\right]^2 - \omega^2\mu\varepsilon \qquad \ldots 3.79$$

$\gamma = 0$, $f = f_C$ and $\omega = \omega_C$

$$\left[\frac{m\pi}{a}\right]^2 + \left[\frac{n\pi}{b}\right]^2 = \omega^2\mu\varepsilon \qquad \ldots 3.80$$

Putting the value of $\omega_C^2 \mu\varepsilon$ from equation (79) in equation (80), we get

$$\gamma^2 = \omega_c^2\mu\varepsilon - \omega^2\mu\varepsilon = (j\beta)^2$$

$$\gamma^2 = \beta^2 = \omega^2\mu\varepsilon - \omega_c^2\mu\varepsilon$$

or $\qquad \beta^2 = \mu\varepsilon(\omega^2 - \omega_c^2)$

$$\beta = \sqrt{\mu\varepsilon} \cdot \sqrt{\omega^2 - \omega_c^2}$$

Putting the value of B in $V_P = \dfrac{\omega}{\beta}$

$$V_P = \frac{\omega}{\beta} = \frac{\omega}{\sqrt{\mu\varepsilon}\sqrt{\omega^2 - \omega_C^2}} = \frac{1}{\sqrt{\mu\varepsilon}}\frac{1}{\sqrt{1-(\omega_c/\omega)^2}}$$

$$V_P = c \cdot \frac{1}{\sqrt{1-(\omega_c/\omega)^2}} \qquad \left(\therefore \frac{1}{\sqrt{\mu_c}} = c\right)$$

$$V_P = \frac{c}{\sqrt{1-(f_c/f)^2}} \qquad \ldots 3.81$$

MICROWAVE TRANSMISSION LINES

We know that

$$f = \frac{c}{\lambda_0} \qquad (\lambda_0 = \text{free space wavelength})$$

$$f_c = \frac{c}{\lambda_c} \qquad (\lambda_0 = \text{Cut-off wavelength})$$

$$\frac{f_c}{f} = \frac{c/\lambda_c}{c/\lambda_0} = \frac{\lambda_0}{\lambda_c}$$

By putting the value of $\frac{f_c}{f}$ in eqn. (3.81)

$$\boxed{V_P = \frac{c}{\sqrt{1-(\lambda_0/\lambda_c)^2}}} \qquad \ldots 3.82$$

Group velocity:

$$V_g = \frac{d\omega}{d\beta}$$

$$\beta = \sqrt{\mu\varepsilon(\omega^2 - \omega_c^2)} \qquad \ldots 3.83$$

Now differentiating the equation w.r.t. 'ω',

$$\frac{d\beta}{d\omega} = \frac{1}{2(\omega^2 - \omega_c^2)\mu\varepsilon} \cdot 2\omega\mu\varepsilon$$

$$\frac{d\beta}{d\omega} = \frac{\sqrt{\mu\varepsilon}}{\sqrt{1-(\omega_c/\omega)^2}} = \frac{\sqrt{\mu\varepsilon}}{\sqrt{1-(f_c/f)}}$$

or

$$V_g = \frac{d\omega}{d\beta} = \frac{\sqrt{1-(f_c/f)^2}}{\sqrt{\mu\varepsilon}}$$

or

$$V_g = \frac{1}{\sqrt{\mu\varepsilon}} \cdot \sqrt{1-\left(\frac{\lambda_0}{\lambda_c}\right)^2} \qquad \left(f_c/f = \frac{\lambda_0}{\lambda_c}\right)$$

$$\boxed{V_g = c \cdot \sqrt{1-\left[\frac{\lambda_0}{\lambda_c}\right]^2}} \qquad \ldots 3.84$$

Relationship between V_p and V_g can be given as

$$V_P \times V_g = \frac{c}{\sqrt{1-(\lambda_0/\lambda_c)^2}} \cdot c \cdot \sqrt{1-(\lambda_0/\lambda_c)^2}$$

$$V_P \cdot V_g = c \times c = c^2$$

$$\boxed{V_P \cdot V_g = c^2} \qquad \ldots 3.85$$

Relationship between λ_0, λ_g **and** λ_c **can be given as**

$$V_P = \lambda_g \cdot f = \frac{\lambda_g}{\lambda_0} \cdot c \qquad \left(f = \frac{c}{\lambda_0}\right)$$

$$V_P = \frac{\lambda_g}{\lambda_0} \cdot c = \frac{c}{\sqrt{1-(\lambda_0/\lambda_c)^2}}$$

$$\frac{\lambda_g}{\lambda_0} = \frac{1}{\sqrt{1-(\lambda_0/\lambda_c)^2}}$$

$$\boxed{\lambda_g = \frac{\lambda_0}{\sqrt{1-(\lambda_0/\lambda_c)^2}}} \qquad \ldots 3.86$$

In TE mode the expression for V_P, V_g, λ_g remain same as in TM mode operation,

$$V_P = \frac{c}{\sqrt{1-(\lambda_0/\lambda_c)^2}}$$

$$V_g = c \cdot \sqrt{1-(\lambda_0/\lambda_c)^2}$$

$$\lambda_g = \frac{\lambda_0}{\sqrt{1-(\lambda_0/\lambda_c)^2}}$$

3.5.7 Degenerate Modes

The higher order modes, having the same cut-off frequency are called **degenerate modes**. In a waveguide the possible TE_{mn} and TM_{mn} modes are always degenerate. The dimensions of waveguide are always chosen such that only the desired mode (dominant mode as TE_{10} and TM_{11}) propagate and other higher modes are allowed to propagate.

3.5.8 Dominant Modes in TE Modes

The lowest order TE mode wave in the case of rectangular waveguide is the TE_{10} mode. This mode has the lowest cut-off frequency and is called as **dominant mode** in the rectangular waveguide. The subscripts m and n represent the number of half period variations. The subscripts m and n represent the number of half period variations of the field along the x and y co-ordinates respectively. x co-ordinate is taken to coincides the larger transverse dimension so that the TE_{10} wave has a lower cut-off frequency than TE_{01}. The electric and magnetic field configurations are shown for TE_{10} mode waves in fig. (3.25) for TM waves, the lowest possible value for either 'm' or 'n' is unity. The TM wave obey the waveguide equation and cut-off frequency as given in equation (87),

$$(f_c)_m = \frac{1}{2\pi\sqrt{\mu\varepsilon}} \sqrt{\left[\frac{m\pi}{a}\right]^2 + \left[\frac{n\pi}{b}\right]^2} \qquad \ldots 3.87$$

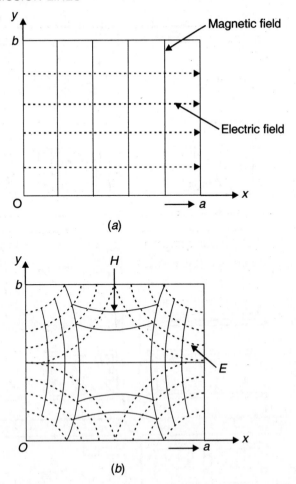

Fig. 3.25. Field configuration in rectangular waveguide in TE_{10} and TM_{11} mode.

From equation (3.87), the lowest cut-off frequency for the TM mode corresponds to $m=1$ and $n=1$. The wave mode which meets these restriction form the dominant mode and is designed as TM_{11}. The lowest order magnetic mode in rectangular waveguide is, therefore significantly different from the dominant electric mode TE_{10} in the rectangular waveguide. Fig. (3.26) shows various modes in rectangular waveguide.

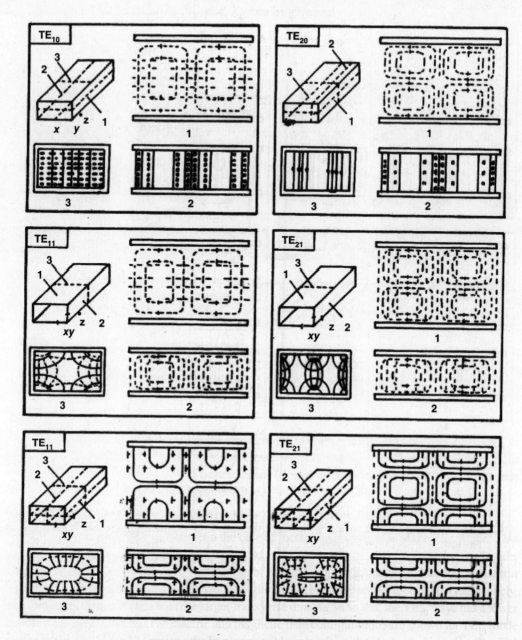

Fig. 3.26. Modes in rectangular waveguide

3.5.9 Mode-Excitations in Rectangular waveguides

It is already seen that modes in rectangular waveguides are designated as TE_{mn} and TM_{mn} modes, respectively. As shown in Fig. 3.27, how a couple of these modes are excited in a rectangular waveguide, the modes are launched by means of coaxial cables. It can be seen that the launching end of the cable will be slightly protruded to act as an antenna inside the waveguide. The protruding edges of the cable are fixed at a distance of $\lambda/2$ from one end of the waveguide so that the desired mode gets launched into it.

MICROWAVE TRANSMISSION LINES

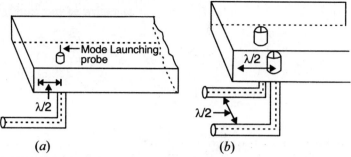

Fig. 3.27. Launching of modes in waveguides (a) TE_{10} mode (b) TE_{20} mode.

3.5.10 Power Transmission in Rectangular Waveguides

The power transmitted through a waveguide and the loss of power in the guide walls may be calculated by using complex poynting theorem. It is considered that the guide is terminated in such a manner that there is no reflection from the receiving end or in other words, the waveguide is infinitely long compared with the wavelength (λ).

The power transmitted through a waveguide is as

$$P = \oint P \cdot ds$$

$$P = \oint \frac{1}{2}(E \times H) \cdot ds \qquad \ldots 3.88$$

The time average power flow through a rectangular waveguide for a lossless dielectric is given by

$$P = \frac{1}{2Z_g} \int_a |E|^2 \, da = \frac{Z_g}{2} \int_a |H|^2 \, da$$

where

$$Z_g = \frac{E_x}{H_y} = -\frac{E_y}{H_x}$$

$$|E|^2 = |E_x|^2 + |E_y|^2$$

$$|H|^2 = |H_x|^2 + |H_y|^2$$

For TE_{mn} modes in waveguide, the average power transmitted is given as

$$P = \frac{\sqrt{1-(f_c/f)^2}}{2\eta} \int_0^b \int_0^a |E_x|^2 + |E_y|^2 \, dx \cdot dy \qquad \ldots 3.89$$

For TM_{mn} modes in waveguide, the average power transmitted through a rectangular waveguide is given by

$$P = \frac{1}{\sqrt{1-(f_c/f)^2}} \cdot \int_0^b \int_0^a (|E_x|^2 + |E_y|^2 \, dx \cdot dy) \qquad \ldots 3.90$$

where η (intrinsic impedance) = $\sqrt{\mu/\varepsilon}$.

3.5.11 Power Losses in Rectangular Waveguide

Losses in a waveguide mainly are two types such as
1. Losses through the guide walls
2. Losses in the dielectric

If the operating frequency is below the cut-off frequency, the propagation constant (γ) will have only be α, attenuation term.

$$\gamma = \alpha + j\beta \qquad (\beta \text{ becomes imaginary})$$

$$\beta = \frac{j2\pi}{\lambda_g}, \quad \lambda_g = \frac{\lambda}{\cos\theta} = \frac{\lambda}{\sqrt{1-(f_c/f)^2}}$$

$$\beta = j\frac{2\pi}{\lambda}\sqrt{(f_c/f)^2 - 1} = j\frac{2\pi f_c}{c}\sqrt{1-(f/f_c)^2}$$

$$\beta = j\alpha$$

The cut-off attenuation constant α is given by

$$\alpha = \frac{54.6}{\lambda_c}\sqrt{1-(f/f_c)^2} \; db/length \qquad \ldots 3.91$$

for $f > f_c$, the waveguide offers very low loss and for $f < f_c$, the attenuation in waveguide is very high and its result in full reflection of the wave.

When the dielectric is an imperfect and non-magnetic in the waveguide the attenuation constant is given by

$$\alpha_{di} = \frac{27.3\sqrt{\varepsilon_R}\tan\delta}{\lambda_0\sqrt{1-(f_c/f)^2}} \; db/length \qquad \ldots 3.92$$

where tan δ = dielectric loss tangent of the dielectric.

In TE_{10} mode the attenuation constant due to the imperfect conducting walls is given by

$$\alpha_c = \frac{R_s}{b\eta_0} = \frac{1+\frac{2}{a}\frac{b}{\left[\frac{f_c}{f}\right]^2}}{\sqrt{1-(f_c/f)^2}} \; N_P/length \qquad \ldots 3.93$$

where, η_0 = intrinsic impedance of free space (377 Ω)

R_S = sheet resistivity in ohm/m²

Then $\qquad R_S = \dfrac{1}{\sigma\delta_S}$

$$\delta_S \text{(Skin depth)} = \frac{1}{\sqrt{\pi f \mu_r \mu_0 \sigma}}.$$

where, f = frequency, μ_r = relative permeability (typically $\mu_r = 1$), μ_0 = permeability of free space ($4\pi \times 10^{-7}$ H/m, σ = conductivity of the metallic walls of waveguide in S/m).

3.5.12 Wave Impedance

Wave impedance in the rectangular waveguide is defined as the ratio of electric field strength in one transverse direction to the strength of the magnetic field along the other transverse direction.

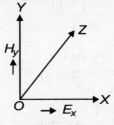

Fig. 3.28

MICROWAVE TRANSMISSION LINES

Wave impedance may be written as

$$Z_z = \frac{E_x}{H_y} = -\frac{E_y}{H_x}$$

as shown in Fig. (3.28).

(a) Wave impedance of TE waves : Wave impedance of TE waves in rectangular waveguide can be given as

$$Z_{TE} = \frac{E_x}{H_y}$$

$$\begin{cases} E_x = -\frac{\gamma}{h^2}\cdot\frac{\partial E_z}{\partial x} - \frac{j\omega\mu}{h^2}\cdot\frac{\partial H_z}{\partial y} \\ H_y = -\frac{\gamma}{h^2}\cdot\frac{\partial H_z}{\partial y} - \frac{j\omega\varepsilon}{h^2}\cdot\frac{\partial E_z}{\partial x} \end{cases}$$ and

By putting the values of E_x and H_y, we get

$$Z_{TE} = \frac{-\frac{\gamma}{h^2}\cdot\frac{\partial E_z}{\partial x} - \frac{j\omega\mu}{h^2}\cdot\frac{\partial H_z}{\partial y}}{-\frac{\gamma}{h^2}\cdot\frac{\partial H_z}{\partial y} - \frac{j\omega\varepsilon}{h^2}\cdot\frac{\partial E_z}{\partial x}}$$

We know that for TE wave propagation $E_z = 0$ and $\gamma = j\beta$.

Further,

$$Z_{TE} = \frac{-\frac{j\omega\mu}{h^2}\cdot\frac{\partial H_z}{\partial y}}{-\frac{\gamma}{h^2}\cdot\frac{\partial H_z}{\partial y}} = \frac{j\omega\mu}{\gamma}$$

$$Z_{TE} = \frac{j\omega\mu}{j\beta} = \frac{\omega\mu}{\beta}$$

We know that $\beta = \sqrt{\omega^2\mu\varepsilon - \omega_c^2\mu\varepsilon}$

or

$$Z_{TE} = \frac{\omega\mu}{\sqrt{\mu\varepsilon}\sqrt{\omega^2-\omega_c^2}} = \frac{\eta}{\sqrt{1-(\omega_c/\omega)^2}}$$

or

$$Z_{TE} = \frac{\eta}{\sqrt{1-(f_c/f)^2}} = \frac{\eta}{\sqrt{1-(\lambda_0/\lambda_c)^2}} \qquad \dots 3.94$$

Wave impedance $Z_{TE} > \eta$, $\lambda_0 < \lambda_c$ for wave to propagate as per the equation (3.61). So the equation (3.94). So the wave impedance for the TE wave is always more than free space impedance.

(b) Wave impedance of TM waves: Wave impedance for a TM waves in rectangular waveguide is given by

$$Z_{TM} = \frac{E_x}{H_y} = \frac{-\frac{\gamma}{h^2}\cdot\frac{\partial E_z}{\partial x} - \frac{j\omega\mu}{h^2}\cdot\frac{\partial H_z}{\partial y}}{-\frac{\gamma}{h^2}\cdot\frac{\partial H_z}{\partial y} - \frac{j\omega\varepsilon}{h^2}\cdot\frac{\partial E_z}{\partial x}}$$

For TM wave $H_z = 0$ and $\gamma = j\beta$

or

$$Z_{TM} = \frac{-\frac{\gamma}{h^2}\cdot\frac{\partial E_z}{\partial x}}{-\frac{j\omega\varepsilon}{h^2}\cdot\frac{\partial E_z}{\partial x}}$$

$$Z_{TM} = \frac{\gamma}{j\omega\varepsilon} = \frac{j\beta}{j\omega\varepsilon}$$

or

$$Z_{TM} = \frac{\beta}{\omega\varepsilon} = \frac{\sqrt{\mu\varepsilon}\ \sqrt{\omega^2 - \omega_c^2}}{\varepsilon \cdot \omega} \quad (\beta = \sqrt{\mu\varepsilon} \cdot \sqrt{\omega^2 - \omega_c^2})$$

$$Z_{TM} = \sqrt{\frac{\mu}{\varepsilon}}\sqrt{1 - \left(\frac{\omega_c}{\omega}\right)^2} = \sqrt{\frac{\mu}{\omega}}\sqrt{1 - \left(\frac{f_c}{f}\right)^2}$$

$$Z_{TM} = \sqrt{\frac{\mu}{\omega}}\sqrt{1 - (\lambda_0/\lambda c)^2} \qquad \ldots 3.95$$

$$Z_{TM} = \eta\sqrt{1 - (\lambda_0/\lambda_c)^2} \qquad \ldots 3.96$$

$$\left(\eta = \sqrt{\frac{\mu}{\varepsilon}}\right)$$

where η = intrinsic impedance.

λ_0 is always $< \lambda_C$ for wave propagations.

$Z_{TM} < n$, This indicates that wave impedance for a TM wave is always less than free space impedance.

3.5.13 Waveguide Input/output Methods

When a small probe is inserted into a waveguide and supplied with microwave energy, it acts as a quarter-wave antenna. Current flows in the probe and sets up an E field such as the one shown in Fig. 3.29. The E lines detach themselves from the probe. When the probe is located at the point of highest efficiency, the E lines set up an E field of considerable intensity. The most efficient place to locate the probe is in the center of the 'a' wall, parallel to the 'b' wall, and one quarter-wavelength from the shorted end of the waveguide, as shown in Fig. 3.30. This is the point at which the E field is maximum in the dominant mode. Therefore, energy transfer (coupling) is maximum at this point. Note that the quarter-wavelength spacing is at the frequency required to propagate the dominant mode.

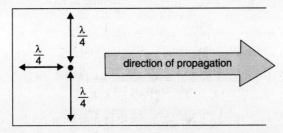

Fig. 3.29. Probe coupling in a rectangular waveguide.

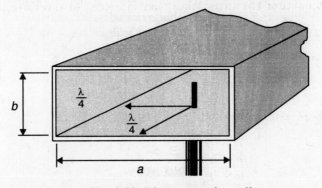

Fig. 3.30. Distances to the walls.

MICROWAVE TRANSMISSION LINES

By probe coupling in a rectangular waveguide first an E field is produced which causes an H field. Removal of energy from a waveguide is simply a reversal of the injection process using the same type of probe.

Fig. 3.31. A probe in the waveguide for 2.7 GHz.

In practice is to handle a small problem, however. The probe also should have a good customization for the cable. This means, it also should be long for $\lambda/4$! Unfortunately, this measure contradicts with the length of the wall 'b', however. This wall is usually longer than $\lambda/4$ only a little.

Because of this the probe gets an additional capacitive reactance. This additional capacity reactance is in the accompanying example created with an opposite screw. This is advisable to get also a tunable resonant frequency also. The form of the probe widened to the top strongly causes a greater bandwidth of the probe.

Loop Coupling in a Waveguide

Another way of injecting energy into a waveguide is by setting up an H field in the waveguide. This can be accomplished by inserting a small loop which carries a high current into the waveguide, as shown in the Fig. 3.32. A magnetic field builds up around the loop and expands to fit the waveguide. If the frequency of the current in the loop is within the bandwidth of the waveguide, energy will be transferred to the waveguide.

Fig. 3.32. BNC-jacks with loops for 1060 MHz.

For the most efficient coupling to the waveguide, the loop is inserted at one of several points where the magnetic field will be of greatest strength. By loop coupling in a rectangular waveguide first a H field is produced which causes an E field.

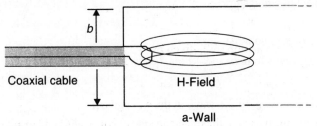

Fig. 3.33. Loop coupling in a rectangular waveguide.

3.6. CIRCULAR WAVEGUIDE

A **circular waveguide** is a cylindrical hollow metallic tabular with uniforms circular cross-section of a finite radius a, as shown in Fig. (3.34).

A circular waveguide of inner radius of 'a' and length 'l' is shown in Fig. (3.34). The value of ϕ varies from 0 to 2π and length varies along z- axis. The wave equation (Helmholtz equation) in cylindrical co-ordinates is given by

$$\frac{1}{r}\frac{\partial}{\partial r}\left(r\frac{\partial \psi}{\partial r}\right) + \frac{1}{r^2}\frac{\partial^2 \psi}{\partial \phi^2} + \frac{\partial^2 \psi}{\partial z^2} = \gamma^2 \psi \qquad \ldots 3.97$$

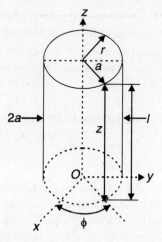

Fig. 3.34. A circular waveguide.

By using separation of variable method, the solution assumed

$$\psi = R(r)\Phi(\phi)Z(z) \qquad \ldots 3.98$$

where, $Z(z)$ = function of z co-ordinate
  $R(r)$ = function of r co-ordinate
  $\Phi(\phi)$ = function of ϕ co-ordinate

By substituting of eqn. (3.98) in eq. (3.97) and resultant is divided by eqn. (3.98) will give the

$$\frac{1}{rR}\frac{d}{dr}\left(r\frac{dR}{dr}\right) + \frac{1}{r^2\Phi}\frac{d^2\Phi}{d\phi^2} + \frac{1}{Z}\frac{d^2Z}{d_z^2} = \gamma^2 \qquad \ldots 3.99$$

The each of the terms in eq. (3.99) is constant because the sum of the three independent terms is a constant.

$$\frac{d^2z}{dz^2} = \gamma_g^2 Z \qquad \ldots 3.100$$

The solution for Z in equation (67), given by

$$Z = Ae^{-\gamma_g z} + Be^{\gamma_g z} \qquad \ldots 3.101$$

γ_g = Propagation constant of the wave in the waveguide.

Putting γ_g^2 for the third term in the left-hand side of eqn. (3.66) and resultant multiplied by r^2, we get

MICROWAVE TRANSMISSION LINES

$$\frac{r}{R} \cdot \frac{d}{dr}\left(r \frac{dR}{dr}\right) + \frac{1}{\Phi} \cdot \frac{d^2\Phi}{d\phi^2} - (\gamma^2 - \gamma_g^2) r^2 = 0 \qquad \ldots 3.102$$

The second term is a pure-function of ϕ only, hence equating by second term to a constant $(-n^2)$, we get

$$\frac{d^2\Phi}{d\phi^2} = -n^2\Phi \qquad \ldots 3.103$$

This equation is also a function of harmonics:

$$\Phi = A_n \sin(n\phi) + B_n \cos(n\phi) \qquad \ldots 3.104$$

Substituting Φ term by $(-n^2)$ in eq. (3.102) and multiplying through by R, we get

$$r \frac{d}{dr}\left(r \frac{dR}{dr}\right) + \left[(K_c r)^2 - n^2\right] R = 0 \qquad \ldots 3.105$$

It is a Bessel's equation of nth order in which.

$$K_c^2 + \gamma^2 = \gamma_g^2 \qquad \ldots 3.106$$

The equation (3.106) is known as **characteristic equation of Bessel's equation** and for a loss-less guide, the characteristic equation reduces to

$$\beta_g = \pm \sqrt{\omega^2 \mu\varepsilon - K_c^2} \qquad \ldots 3.107$$

The solution for the Bessel's equation is

$$R = C_N J_n(K_c r) + D_N N_n(K_c r) \qquad \ldots 3.108$$

where $J_n(K_c r)$ = nth order Bessel's function of the first kind, it is representing a standing wave of $\cos(K_c r)$. For $r < a$ as shown in Fig. (3.34).

$N_n(K_c r)$ = nth order Bessel's function of second kind, it is representing a standing wave of $\sin(K_c r)$. For $r > a$ as shown in Fig. (3.34).

The total solution of the Helmholtz equation in cylindrical co-ordinates can be given as

$$\psi = \left[C_n J_n(K_c r) + D_n N_n(K_c r)\right] \left[A_n \sin(n\phi) + B_n \cos(n\phi) e^{\pm j \beta_h z}\right] \qquad \ldots 3.109$$

At $r = 0$, however $K_c r = 0$, than the function N_n approaches infinity, so $D_n = 0$. It means that at $r = 0$ as the z axis, the field must be finite.

By use of trigonometric, the two sinusoidal terms become

$$A_n \sin(n\phi) + B_n \cos(n\phi) = \sqrt{A_n^2 + B_n^2} \cos\left[n\phi + \tan^{-1}\left(\frac{A_n}{B_n}\right)\right]$$

$$= F_n \cos(n\phi) \qquad \ldots 3.110$$

The final solution of the Helmholtz equation is reduced to

$$\psi = \psi_0 J_n(K_c r) \cos(n\phi) e^{-j\beta_g z} \qquad \ldots 3.111$$

3.6.1 TE Modes in Circular Waveguide

It is assumed that the waves in a circular waveguide are propagating in the positive z direction. In TE_{np} modes in the circular waveguide, $E_2 = 0$.

(where 'n', represents the number of full cycles of fixed variation in revolution through 2π rad of ϕ and ρ, μ presents the number of zero's of $E\phi$ along the radial of a waveguide, but the zero on axis is not included if it exists.)

It means the z components of the H_z must exist in the guide in order to have propagation of EM wave. Helmholtz equation for H_z in a circular wave guide is given by,

$$\nabla^2 H_z = \gamma^2 H_z \qquad \ldots 3.112$$

By using equation (3.78), solutions for this given by

$$H_z = H_{0z} J_n(K_c r)\cos(n\phi)e^{-j\beta_g z} \qquad \ldots 3.113$$

Maxwell's curl equation in frequency domain for a dielectric (lossless) is given by

$$\nabla \times H = j\omega\mu H$$
$$\nabla \times E = j\omega\mu E$$

The complete field equation of the TE_{np} modes in circular waveguide:

$$E_r = E_{0r} J_n\left(\frac{X'_{np} r}{a}\right) \sin(n\phi) e^{-j\beta_g z} \qquad \ldots 3.114$$

$$E_z = 0 \qquad \ldots 3.115$$

$$E_\phi = E_{0\phi} J'_n\left(\frac{X'_{np} r}{a}\right) \cos(n\phi) e^{-j\beta_g z} \qquad \ldots 3.116$$

$$H_r = -\frac{E_{0\phi}}{Z_g} \cdot J_n\left(\frac{X'_{np} r}{a}\right) \cos(n\phi) e^{-j\beta_g z} \qquad \ldots 3.117$$

$$H_\phi = -\frac{E_{0r}}{Z_g} \cdot J_n\left(\frac{X'_{np} r}{a}\right) \sin(n\phi) e^{-j\beta_g z} \qquad \ldots 3.118$$

$$H_z = H_{0z} J_n\left(\frac{X'_{np} r}{a}\right) \cos(n\phi) e^{-j\beta_g z} \qquad \ldots 3.119$$

where, $Z_g = \dfrac{E_r}{H_\phi} = -\dfrac{E_\phi}{H_r}$, have been substitute for the wave impedance in the waveguide and where P = 1, 2, 3, 4, ... and n = 0, 1, 2, 3, 4, ...

Here, $K_c = \dfrac{X'_{np}}{a}$, represents the Eigen values where x'_{np} are the roots of $\dfrac{dJ_n(K_c r)}{d\rho} = 0$ at $r = a$. The value of x'_{np} in TE modes is given in table. 3.1.

Table 3.1 : pth zeroes of $K_c a$ for TE_{np} modes.

P	n = 0	1	2	3
1	3.832	1.841	3.054	4.201
2	7.016	5.331	6.706	8.015
3	10.173	8.536	9.969	11.346
4	13.324	11.706	13.170	

n = It represents the number of full cycles of field variations in revolution through 2π rad of ϕ.

P = It represents the number of zero's of E_ϕ along the radial of a waveguide, but the zero on axis is not included if it exists.

The mode propagation constant in circular waveguide is obtained by

$$\beta_g = \sqrt{\omega^2 - \mu\varepsilon - \left(\frac{X'_{np}}{a}\right)^2} \qquad \ldots 3.120$$

MICROWAVE TRANSMISSION LINES

The cut-off wave number of a mode is that for which the β_g (propagation constant) vanishes.

$$K_c = \frac{X'_{np}}{a} = \omega_c \sqrt{\mu\varepsilon} \qquad \ldots 3.121$$

The cut-off frequency (f_c) for TE modes in a circular waveguide is given by

$$f_c = \frac{X'_{np}}{2\pi a\sqrt{\mu\varepsilon}} \qquad \ldots 3.122$$

The phase velocity (V_P) of TE modes is circular waveguide is given by

$$V_P = \frac{1\sqrt{\mu\varepsilon}}{\sqrt{1-(f_c/f)^2}} \qquad \ldots 3.123$$

The guide wavelength is given by

$$\lambda_g = \frac{\lambda}{\sqrt{1-(f_c/f)^2}} \qquad \ldots 3.124$$

where λ is wavelength in a unbounded dielectric $\left(\frac{V_P}{f}\right)$.

The wave impedance in TE mode is given by

$$Z_g = \frac{\omega\mu}{\beta} = \frac{\sqrt{\mu/\varepsilon}}{\sqrt{1-(f_c/f)^2}} \qquad \ldots 3.125$$

3.6.2 TEM Wave Mode in Circular Waveguide

In the similar manner as in rectangular waveguide, the TEM mode does not exist in circular waveguide also. TEM mode can exist only in the two conductor system where the center conductor is not there. For $E_z = H_z = 0$, it may be proved that all the field components vanish in circular waveguide.

3.6.3 TM Modes in Circular Waveguides

In the TM mode of a circular waveguide, $H_z = 0$, however, the z components of the electric field E_z should exist so that to have energy transmission in the waveguide.

The Helmholtz equation for $\psi = E_z$ in circular waveguide is written as

$$\nabla^2 E_z = \gamma^2 E_z$$

By applying the boundary condition, the field solutions in the circular guide are given as follows:

$$E_z = E_{0z} j_n(K_c r)\cos(n\phi) e^{-j\beta_g z} \qquad \ldots 3.126$$

$$E_r = -\frac{j\beta_g}{K_c^2} \cdot \frac{\partial E_z}{\partial r} \qquad \ldots 3.127$$

$$E_\phi = -\frac{j\beta_g}{K_c^2} \cdot \frac{1}{r}\frac{\partial E_z}{\partial \phi} \qquad \ldots 3.128$$

$$H_r = -\frac{j\omega\varepsilon}{K_c^2} \cdot \frac{1}{r}\frac{\partial E_z}{\partial \phi} \qquad \ldots 3.129$$

$$H_z = 0 \qquad \ldots 3.130$$

$$H_\phi = -\frac{j\omega\varepsilon}{K_c^2} \cdot \frac{\partial E_z}{\partial r} \qquad \ldots 3.131$$

where $K_c^2 = \omega^2 \omega\varepsilon - \beta_g^2$.

By differentiating eqn. (3.126) with respect to z and substitute the result in eqn. (3.127) to (3.131), obtained the field equations of TM_{np} modes in circular guide.

$$E_r = E_{0r} J'_n\left(\frac{X_{np}r}{a}\right) \cos(n\phi) e^{-j\beta_g z^z} \qquad \ldots 3.132$$

$$E_z = E_{0z} J_n\left(\frac{X_{np}r}{a}\right) \cos(n\phi) e^{-j\beta_g z^z} \qquad \ldots 3.133$$

$$E_\phi = E_{0\phi} J_n\left(\frac{X_{np}r}{a}\right) \sin(n\phi) e^{-j\beta_g z^z} \qquad \ldots 3.134$$

$$H_r = \frac{E_{0\phi}}{Z_g} J_n\left(\frac{X_{np}r}{a}\right) \sin(n\phi) e^{-j\beta_g z} \qquad \ldots 3.135$$

$$H_\phi = \frac{E_{0r}}{Z_g} \cdot J'_n\left(\frac{X_{np}r}{a}\right) \cos(n\phi) e^{-j\beta_g z} \qquad \ldots 3.136$$

$$H_z = 0 \qquad \ldots 3.137$$

where $K_c = \dfrac{X_{np}}{a}$ have been replaced.

The value of $n = 0, 1, 2, 3, \ldots$ and $p = 1, 2, 3, 4, \ldots$

$$\beta_g = \sqrt{\omega^2 \mu\varepsilon - \left(\frac{X_{np}}{a}\right)^2} \qquad \ldots 3.138$$

$$K_c = \frac{X_{np}}{a} = \omega_c \sqrt{\mu\varepsilon} \qquad \ldots 3.138$$

$$f_c = \frac{X_{np}}{2\pi a \sqrt{\mu\varepsilon}} \qquad \ldots 3.139$$

$$\lambda_g = \frac{\lambda}{\sqrt{(1 - f_c/f)^2}} \qquad \ldots 3.140$$

$$Z_g = \frac{\beta_g}{\omega_c} = n\sqrt{1 - f_c/f}$$

$$V_g = \frac{\omega}{\beta_g} \cdot \frac{V_p}{\sqrt{1 - (f_c/f)^2}} \qquad \ldots 3.141$$

3.6.4 Dominant Mode in Circular Waveguide

For TE waves, the cut-off waveguide in given by

$$K_c = \frac{X_{np}}{a}.$$

MICROWAVE TRANSMISSION LINES

It should be noted that the dominant mode or the mode of lowest cut-off frequency in a circular waveguide, is the mode of TE_{11}, Which has the smallest value of the product, $K_c a = 1.841$ as shown in table – 3.2.

Table 3.2: pth zeroes of $K_c a$ for TM_{np} modes

P	n = 0	1	2	3
1	2.405	3.832	5.136	6.380
2	5.520	7.106	8.417	9.761
3	8.645	10.173	11.620	13.015
4	11.792	13.324	14.796	

Therefore the dominant mode in a circular waveguide is TE_{11} mode.

Also from these tables, it is noted that $x'_{op} = X_{1p}$ and all the TE_{OP} and TM_{IP} modes are degenerate modes in a uniform circular waveguide. The modes field configurations are shown in the Fig. (3.35).

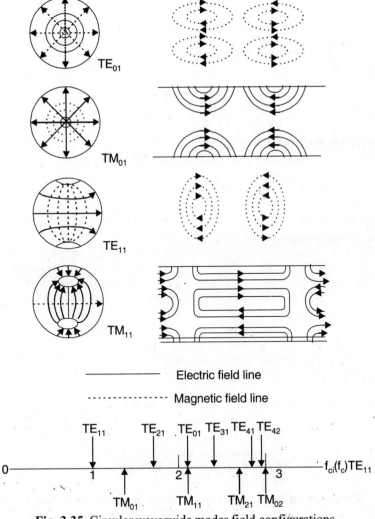

Fig. 3.35. Circular waveguide modes field configurations.

3.6.5 Attenuation in Circular waveguide

It is possible to determine the attenuation in a circular waveguide, in similar manner as we found in rectangular waveguide. For an air field waveguide the attenuation is because of finite conductivity of the waveguide walls and may be given as

$$\alpha = \frac{\text{Power loss per unit length}}{2\,(\text{Average power transmitted})}$$

Average power may be expressed as

$$P_{np} = \frac{1}{2Z_g} \cdot \int_0^{2\pi}\int_0^a \left(|E_\phi|^2 + |E_p|^2\right) r\,dr\,d\phi \qquad \ldots 3.142$$

or

$$P_{np} = \frac{Z_g}{2} \cdot \int_0^{2\pi}\int_0^a \left(|H_r|^2 + |H_\phi|^2\right) r\,dr\,d\phi \qquad \ldots 3.143$$

For TE modes:

$$P_{np} = \frac{\sqrt{1-(f_c/f)^2}}{2\eta} \cdot \int_0^{2\pi}\int_0^a \left(|E_r|^2 + |E_\phi|^2\right) r\,dr\,d\phi \qquad \ldots 3.144$$

Where, $\eta = \sqrt{\mu/\varepsilon}$ (intrinsic impedance).

For TM modes:

$$P_{np} = \frac{1}{\sqrt{1-(f_c/f)^2}} \cdot \int_0^{2\pi}\int_0^a \left(|E_r|^2 + |E_\phi|^2\right) r \cdot dr\,d\phi \qquad \ldots 3.145$$

3.6.6 Power Losses in Circular Waveguide

The power losses in circular waveguide is given by

$$P_{loss} = \frac{R_s}{2} \oint \overline{J_s} \cdot \overline{J_s} \cdot dl \qquad \ldots 3.146$$

The expressions for the attenuation constant in circular waveguide are obtained as following:

Attenuation constant α_{TE} is given by

$$\alpha_{TE} = \frac{R_s}{aZ_0}\left[1-(f_c^2/f^2)\right]^{-1/2}\left[f_c^2/f^2 + \frac{n^2}{{X'}_{np}^2 - n^2}\right] \qquad \ldots 3.147$$

Attenuation constant α_{TM} is given by

$$\alpha_{TM} = \frac{R_s}{aZ_0}\left(\frac{1}{\sqrt{1-(f_c/f)^2}}\right) \qquad \ldots 3.148$$

For TE_{OP} modes attenuation falls off as $f^{-3/2}$ as per

$$\alpha = \frac{R_s}{aZ_0}\left[\frac{f_c^2}{f\left(f^2 - f_c^2\right)^{-1/2}}\right] \qquad \ldots 3.149$$

The rapid decrease of attenuation along with the frequency increment in TE_{01} mode helps is too useful for very long-low loss waveguide communication link. However, there are practical

MICROWAVE TRANSMISSION LINES

problems while operating the waveguide at a frequency above the dominant mode (TE_{11}). All modes above dominant mode results in mode conversions and this conversion will lead to signal distortion.

3.6.7 Power Handling Capacity in Circular Waveguide

The power handling capacity of circular waveguide in TE_{11} mode when the breakdown power for maximum field strength of 30 kV/cm is given by

$$(P_{bd})_{11} \approx 179.a^2 \left[1-\left(f_c^2{}_{11}/f^2\right)\right]^{1/2} kW \qquad ...\ 3.150$$

for TE_{01} mode it is given by

$$(P_{bd})_{01} \approx 1805.a^2 \left[1-\left(f_c^2{}_{01}/f^2\right)\right]^{1/2} kW \qquad ...\ 3.151$$

where a is the radius of circular waveguide in C_m and f_{11} and $f_{c\,01}$ are the cut-off frequency of TE_{11} and TE_{01} modes respectively.

3.6.8 Standard Dimensions of Circular Waveguides

The standard dimensions for circular waveguide are recommended by the EIA (Electronic Industries Association) as given by table 3.3.

Table 3.3: Characteristics by Standard Circular Waveguide

EIA[a] designation WC[b]	Inside diameter in cm (in.)	Cutoff frequency for air-filled waveguide in GHz	Recommended frequency rated for TE_{11} mode
992	25.184 (9.915)	0.698	0.80-1.10
847	21.514 (8.470)	0.817	0.94-1.29
724	18.377 (7.235)	0.957	1.10-1.52
618	15.700 (6.181)	1.120	1.29-1.76
528	13.411 (5.280)	1.311	1.51-2.07
451	11.458 (4.511)	1.534	1.76-2.42
385	9.787 (3.853)	1.796	2.07-2.83
329	8.362 (3.292)	2.102	2.42-3.31
281	7.142 (2.812)	2.461	2.83-3.88
240	6.104 (2.403)	2.880	3.31-4.54
205	5.199 (2.047)	3.381	3.89-5.33
175	4.445 (1.750)	3.955	4.54-6.23
150	3.810 (1.500)	4.614	5.30-7.27
128	3.254 (1.281)	5.402	6.21-8.51
109	2.779 (1.094)	6.326	7.27-9.97
94	2.382 (0.938)	7.377	8.49-11.60
80	2.024 (0.797)	8.685	9.97-13.70
69	1.748 (0.688)	10.057	11.60-15.90
59	1.509 (0.594)	11.649	13.40-18.40
50	1.270 (0.500)	13.842	15.90-21.80
44	1.113 (0.438)	15.794	18.20-24.90
38	0.953 (0.375)	18.446	21.20-29.10

33	0.833 (0.328)	21.103	24.30-33.20
28	0.714 (0.281)	24.620	28.30-38.80
25	0.635 (0.250)	27.683	31.80-43.60
22	0.556 (0.219)	31.617	36.40-49.80
19	0.478 (0.188)	36.776	42.40-58.10
17	0.437 (0.172)	40.227	46.30-63.50
14	0.358 (0.141)	49.103	56.60-77.50
13	0.318 (0.125)	55.280	63.50-87.20
11	0.277 (0.109)	63.462	72.70-99.70
9	0.239 (0.094)	73.552	84.80-116.0

[a]Electronic Industry Association
[b]Circular Waveguide

3.6.9 Applications of Circular Waveguide

However the circular waveguides are easy to manufacture than the rectangular waveguide but still they are not preferred due to following reasons:

(*i*) They occupy more space in comparison with the rectangular waveguide.

(*ii*) Change of polarization will take place due to roughness or discontinuities in the circular waveguide

(*iii*) Due to infinite number of modes, it is very difficult to separate these modes, so it is difficult to avoid interference to the dominant mode of circular waveguide.

In some of the applications we find that circular waveguides and more useful than the rectangular waveguide. The few applications of circular waveguide are given as:

(*i*) TE_{01} mode is used for long distance wave propagation because of low attenuation for the frequencies above 10 GHz.

(*ii*) In the radars, circular waveguide are used to couple the energy to the horn antenna feeder from the rotary joint.

(*iii*) For the short and medium distance broadband communication, circular waveguides are used.

3.6.10 Excitation of Modes in Circular Waveguide

We have already discussed that, TE modes have no z components of an electric field TM modes have no z components of magnetic field. For TM wave propagation, a device is inserted in a circular guide in such a way that it excites only z component of electric field. For TE wave propagation, a device is placed in a circular waveguide in such a manner that only the z components of magnetic intensity exist.

The various methods of excitation for different modes in circular waveguide are shown in the Fig. (3.36)

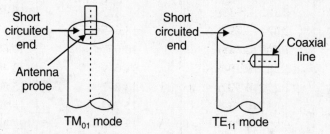

Fig. 3.36. Methods of exciting in TE_{11} and TM_{01} modes in circular waveguide.

MICROWAVE TRANSMISSION LINES

3.7. PLANAR TRANSMISSION LINES

With the induction of microwave integrated circuits technology in microwave engineering, a modern transmission lines, referred as planar transmission line become more suitable and are used extensively.

A planar geometry implies that the characteristics impedance of the element can be determined from the dimensions in a single plane.

Planar transmission line is the first step to fabrication of microwave integrated circuit where the entire transmission line components can be fabricated in one step by thin film or photolithographic techniques.

These processes are similar to those used for making printed circuits cards for low frequency electronic circuits. Several configurations for planar transmission lines have been realized. Some of these lines support TEM modes and other hybrid or higher modes. Commonly used planar transmission lines are as follows:

(*i*) Stripline (*ii*) Microstripline
(*iii*) Slot line (*iv*) Coplanar lines etc.

Some principle features of these lines are described below:

(*i*) Stripline:

A conventional **striplines** is a balanced line in which a flat conducting strip is placed symmetrically between two large ground planes, with the space between the ground planes filled with a homogeneous dielectric as shown in Fig. 3.37

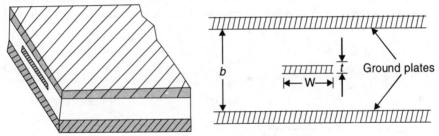

Fig. 3.37. Stripline.

TEM is the dominant mode of propagations in a stripline. The stripline can also support higher order TM and TE modes but are not so popular. Ground planes will be at zero potential with field lines concentrated near the central conductor as shown in Fig. 3.38

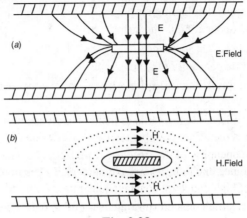

Fig. 3.38

As one moves in the mid plane away from the strip, the field decays rapidly. This suggests that the geometry could be terminated in the transverse direction without affecting the transmission characteristics.

The characteristics equation of the stripline is given as

$$Z_0 = 1/V_p \cdot C$$

Where,

Z_0 = characteristics impedance of the stripline

C = shunt capacitance of a unit length of line

V_p = phase velocity along the stripline.

   = $V_p = c/\varepsilon_r$ [where c is the velocity of EM waves in free space and ε_r is the relative permittivity of the dielectric medium.]

The calculations of the above parameters are based on solution of two dimensional Laplace's equation subjected to the boundary conditions determined by the geometry of the line.

Further higher order TEM modes are not allowed to propagate in coaxial system and which is ensured by keeping the distance (d) between the two ground planes less than $\lambda_d/2$, where λ_d is the wavelength in the dielectric medium.

Microstrip lines

A microstrip line is a type of open transmission line that consists of a single dielectric substrate with ground plane on one side and a strip on the other face. The top ground plane of stripline configuration is not present in this case. A cross-sectional view of a microstrip is shown in Fig. 3.39

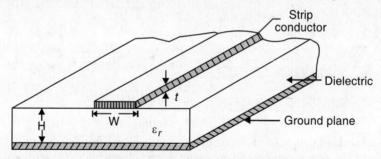

Fig. 3.39. Geometry of strip line.

Where,

W = width of the upper conducting strip

t = thickness of the upper conducting strip

H = height of the dielectric substrate

ε_r = relative dielectric constant of the substrate

The following features of microstriplines makes more popular over any other transmission lines:

1. There is an easy access to top surfaces makes it convenient to mount discrete devices and also minor adjustments are possible after the fabrication of the circuit.
2. Because of the openness of the structure, care has to be taken to minimize the radiation loss or interference due to necessary conductors. To ensure that fields are confined near the strip, use of high dielectric constant substrate is required.
3. Use of high dielectric constant substrate is useful to reduce the circuit dimensions.
4. Both AC and DC signals can be transmitted.

MICROWAVE TRANSMISSION LINES

5. Microwave circuit can be designed with microstrip within a frequency range of few GHz to tens of GHz.
6. Line wavelength reduces to one third of its free space value, because of the substrate field. Hence distributed component diameters are relatively small.
7. Microstrip line structure is rugged and can withstand large power and voltage.

The mode of propagation through a microstrip is assumed to be quasi-TEM, i.e., both electric and magnetic fields in the direction of propagation are negligible compared to the TEM fields.

The distributions of the electric and magnetic fields in a microstripline are shown in Fig.3.40

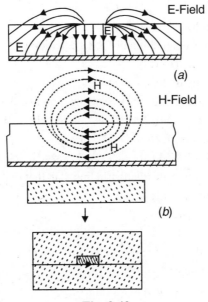

Fig. 3.40

Because of the concentration of electric field in the dielectric region below the strip most of the energy of the wave is concentrated there. The distribution of the electric field lines indicates the 'E' lines approach the air-dielectric interface obliquely, and thus there are at least two components of electric field.

From Fig.3.40, we note that some of the electric field lines are within the dielectric and some fringe are in air. The filling factor 'q' defines the electric field line distribution and it becomes unity when all the electric field lines are in the substrate.

Effective dielectric constant of microstripline is given as

$$E_{eff} = 1 + q(\varepsilon_r - 1) \qquad \ldots 3.152$$

Let us consider free space dielectric constant $\varepsilon_r = 1$,

Therefore $\quad q = E_{eff} - 1/\varepsilon_r - 1 \leq 1$

Where, q is the filling factor.

The effective dielectric constant reduced the phase velocity and characteristics impedance according to following relations:

$$V_p = 1/\sqrt{(Lc)} = c/\sqrt{(E_{eff}/\varepsilon_0)} \qquad \ldots 3.153$$

And $\quad Z_0 = \sqrt{(L/c)} = Z_{oa}/\sqrt{(E_{eff}/\varepsilon_0)} \qquad \ldots 3.154$

Where L is the physical length of the microstripline and c is the velocity of light ($c = 2.99793 \times 10^8$ m/s) in the space. The characteristics impedance of this air filled 'microstrip' line is Z_{oa}.

The guide wavelength becomes (λ_g);
$$\lambda_g = V_p/f = c/f \sqrt{(E_{eff}/\varepsilon_0)} \qquad \ldots 3155$$
the propagation function,
$$\gamma = \alpha + j\beta ;$$
$$\gamma = \sqrt{[(r_g - \omega^2 Lc) + j\omega(r_g + r)]}$$
$$\gamma = j\omega \sqrt{Lc} \; [1 + j\omega(Lg + r/-\omega^2 Lc)]^{1/2}$$

Expanding binomially and neglecting higher order terms
$$\alpha = j\beta = j\omega\sqrt{(Lc)} \; [1 + \tfrac{1}{2} j\omega(Lg + r)/-\omega^2 Lc]$$
$$= Lg + r/2(Lc)^{1/2} + j\omega(Lc)^{1/2} \qquad \ldots 3.156$$

By comparing real part of LHS with that of RHS equation (3.156), we get,
$$\alpha = Lg + r/2 \sqrt{(Lc)} \qquad \ldots 3.157$$
where r = surface resistivity of the conductor.

Now α = attenuation loss due to dielectric + attenuation loss due to conductor
$$\alpha = \alpha_d + \alpha_c = Lg + r/2\sqrt{(LC)} \; db/\text{unit length}$$
therefore, $\alpha_d + \alpha_c = g/r \sqrt{L/c} + r/z \sqrt{c/L}$
$$= gZ_0/2 + r/2Z_0 \qquad \ldots 3.158$$

Attenuation loss is more in narrow microstrip line than that of dielectric.

The loss tangent of the material is given by
$$\tan \delta_d = \sigma_d/\omega\varepsilon_r \qquad \ldots 3.159$$
where,

σ_d = conductivity of the dielectric substance.

δ_d = skin depth

The Q of a microstrip line is calculated as,
$$Q = \beta/2.\alpha = 2\pi/\lambda_g . 1/2d = \pi/\alpha.\lambda_g = 27.3/\alpha.\lambda_g \; db$$
$$Q = 27.3 \; [\sqrt{(E_{eff}/\varepsilon_r)}/\alpha.c] \; f \; db \qquad \ldots 3.160$$

The dielectric substrate must have the following properties:

1. The dielectric substrate should have high thermal conductivity.
2. The dielectric substrate should have low dissipation factor.
3. Minimum variation of dielectric constant.
4. Dimensional stability.
5. Uniformity of thickness.
6. The surface should be polished.

Limitations of Microstrip

1. The circuit dimensions at higher frequency are very small and that result in fabrication problems.
2. A slight change in ε_r due to temperature variation or batch to batch variation changes the impedance and guide wavelength considerably.
3. Although thinner substrates permit high frequency operation but its 'Q' factor is low.
4. In the microstrip, the conductor loss increases with an increase in frequency. Maximum frequency range may go up to 50 GHz safely.

MICROWAVE TRANSMISSION LINES

Fig. 3.41 shows the different types of microstrip line. These are different types of microstrip lines such as embedded microstrip, suspended microstrip, parallel strip line, inverted microstrip and coplanar strip line.

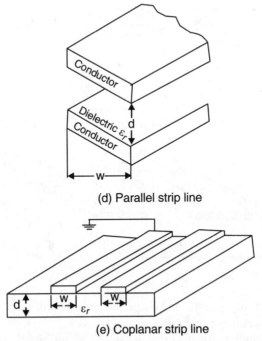

(d) Parallel strip line

(e) Coplanar strip line

Fig. 3.41. The different types of microstrip line.

Slot line

Although microstrip is the most commonly used in transmission structure for microwave integrated circuits, a slot line is an alternative transmission structure. A slot line structure consists of a slot etched from the conducting layer on one surface of a dielectric substrate with the opposite side surface being bare. Its structure is thus complementary to that of microstrip.

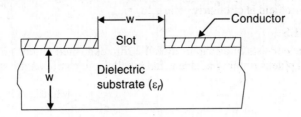

Fig. 3.42. The structure of slot line.

Fig. 3.43, illustrated approximate electric and magnetic field distribution in slot line structure. It is to be noted that the magnetic field has a component in the direction of propagation also. Thus the mode of propagation is a transverse electric mode, TE mode and not a TEM mode.

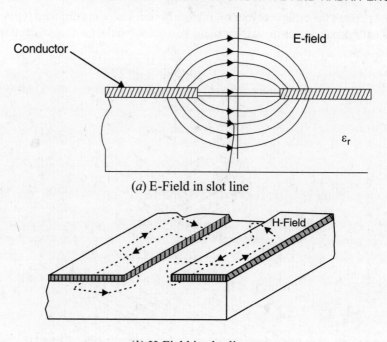

(a) E-Field in slot line

(b) H-Field in slot line

Fig. 3.43. Approximate electric and magnetic field distribution in slot line structure.

Main feature of a slot line are as follows:
1. In a slot line, both the conductors are in one plane, and therefore shunt mounting of components (active or passive) across the line is very convenient.
2. Higher impedance can be easily realized by increasing the slot width than a microstrip.
3. Slot line is more dispersive than microstrip, i.e., variation of Z_0 and λ_0/λ_g with frequency is larger than microstrip.
4. Co-axial line to microstrip transition is relatively easier to fabricate than coaxial line to slot line microstrip.
5. Resonance isolator may be designed by properly selecting a location in the slot line where the magnetic field is circularly polarized.

COPLANAR –LINES

A coplanar line consists of a strip of metallic film deposited on the surface of the dielectric slab with two ground planes running adjacent and parallel to the strip on the same side, as shown in Fig. 3.44.

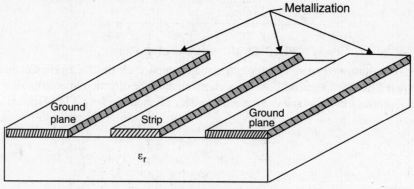

Fig. 3.44. Coplanar lines.

MICROWAVE TRANSMISSION LINES

Main features of coplanar lines are as follows:
1. Both series and shunt mounted components can be easily incorporated than microstrip and slot lines.
2. Microstrip to coplanar and coaxial to coplanar transitions can be done easily.
3. Both balanced and unbalanced modes are possible in coplanar line. Thus this transmission line finds extensive application, such as in balanced mixer, balanced modulator, etc.
4. Coplanar has elliptically polarized magnetic field regions, such characteristics is used in design of several ferrite components.

It has a limitation at higher frequency because of the open nature of the structure results in a considerable radiation.

Microwave integrated circuits (MIC's)

Microwave integrated circuits (MIC's) are the integrated circuits (IC's) used at microwave frequency regions. Thin or thick film passive components are connected to active silicon planar chip by means of fine wires of conducting material. Because of its high reliability, low weight, small size and reproducibility, they are mostly used in cellular, trans-receiver system, pagers etc.

Selection of dielectric Substrate material

Preparation of an MIC starts with the selection of dielectric substrate. Several factors that need to be considered in selecting a particular substrate are its thermal, mechanical, electronic and economical aspects, which influence the decision process leading to the correct choice for a specific type of MIC and their application.

The most important criterion which should be considered is as follow:
1. The mechanical strength and thermal conductivity should be sufficient for the desired application.
2. The surface finish of the substrate should be sufficiently good so that conductor loss may be minimized.
3. The cost of the substrate should be justified.
4. The frequency ranges involved must be taken into consideration because this will influence the thickness and permittivity of the dielectric substance selected.
5. The technology chosen should be thin or thick film.
6. The substrate should be readily available with sufficient surface area.

Substrate materials:

There is some of the substrate materials often used in MIC application.
(*i*) Plastic substrate.
(*ii*) Alumina substrate.

***Plastic substrate*:** Plastic substrate materials are used in circuits at frequency below a few GHz and array of antenna for frequencies up to and beyond 20GHz. Plastic substrate are such as Irradiated polyolefin, rexolite, RT-duroid materials are normally used.

Alumina substrate: alumina substrate is very popular for manufacturing circuits at any frequency up to 10GHz.

Conductor material: The ideal conductor materials used for MIC's should have high conductivity, low temperature coefficient of resistance, good adhesion to the substrate, good etch ability and can be deposited or electroplated easily. Metals like gold, silver, copper, aluminium, chromium are best suited.

Resistive material: The desired properties of resistive material are low temperature coefficient of resistance, good stability and should have adequate heat dissipation capability. The sheet resistance should be in the range of 10 to 1000Ω unit length. Suitable materials are NiCr, CrSio, Cr etc.

Dielectric material: A dielectric material should have low dielectric loss at higher frequencies and should withstand a very high value of voltage. Also it should have good reproducibility capacity and have the ability to undergo processes without developing pin holes. Commonly used dielectric materials are SiO_2, Si_2N_4, Al_2O_3 etc.

Advantages of MIC's

MIC's are preferred more over the conventional microwave circuits due to following reasons:

(*i*) Small size and weight.
(*ii*) Reproducibility and reliability.
(*iii*) Reduction is cost.
(*iv*) Improvements in performance.

All the above mentioned reasons are discussed further.

(*i*) Reduction in size and weight :

There is large reduction in cross section for MIC transmission lines. A cross-band wave guide has a size of 1 × 0.5 inch whereas cross-section required for a 50Ω microstrip on alumina substrate is of the order of 0.25 × 0.25 inch. Using high dielectric constant substrates reduce the guide wavelength by a factor equal to the square root of 2.5 to 3.00 in obtained, hence the size of the distributed elements of the circuits reduces in the same ratio.

By using lumped elements, a further reduction in size is achieved. The lumped elements are an order of magnitude smaller than distributed elements. In MIC's large complicated circuits can be produced by photo etching on a single metallization, hence it elements inter-connections which saves considerable volume occupied by connectors. Also this fabrication process allows the designer to position the input output ports at the designed location. Thus the amount of hardware, which conventional microwave circuits use for bends etc., is saved.

In MIC's it is possible to use encapsulated device chips. These device chips occupy much smaller space than packaged devices.

(*ii*) Reproducibility and reliability:

Thin film deposition and photo etching technologies used in MIC's fabrication leads to a very uniform product. Increased reliability of MIC's may be achieved

(*a*) Microwave semiconductor devices used in MIC's have longer life as compared to that of tubes.

(*b*) More durable connections between device and circuits are obtained because of improved bonding techniques. Possibility of relative motion among various parts of the circuits is less; therefore the circuits' construction is more rigid.

(*iii*) Reduction in cost:

In conventional microwave circuits, time consuming and expensive operation of precision machining has been replaced in the case of MIC's with techniques like photo-etching which are much quicker and cheaper. The cost of fabrication of MIC's is to a considerable extent independent of the circuit complexity. For example, etching out a microstrip section or a filter circuit requires about the same effort. Once the mask is prepared large scale production does not require any additional tools.

MIC's can use semiconductor devices in chip form. Chips are much cheaper that the packaged devices.

MICROWAVE TRANSMISSION LINES

(iv) Improvement in performance:

In several conditions, the circuit performance may be improved by adopting MIC technique. In MIC's transverse dimension of the microwave structure are compatible with the size of the device chips while in the conventional microwave circuits, device mounting structures like posts etc., introduce additional reactance which limit the circuit performance. Elimination of this reactance in MIC's increases the circuit bandwidth and device importance level. Combination of different microstrip lines in the same circuit introduces another degree of flexibility in circuits design. These factors improve the performance of MIC's.

Limitations of MIC's:

(i) Microstrip resonators have low Q in comparison to waveguide resonator.

(ii) As there is no tuning screw or variable short, so after fabrication readjustment is not possible. Hence the circuits have to be designed fairly accurately taking all the parameters into account.

(iii) MIC's have low power handling capability i.e., approximately 10 watts.

(iv) Accurate design of circuits pose problems for precise characterization of semiconductor devices requiring computer aided design so that the performance of MIC's can be known before it is actually made.

SUMMARY

A **Transmission line** is a device designed to guide electrical energy from one point to another. It is used to transfer the output RF energy of a transmitter to an antenna.

- **Types of Transmission Line**

    Generally there are six types of transmission mediums, which are as:

    1. Parallel-line
    2. Twisted pair
    3. Shielded pair
    4. Coaxial line
    5. Strip type substrate transmission lines and
    6. Wave guides.

- **Losses in Transmission Lines**

    Line losses may be any of three types

    1. Copper,
    2. Dielectric, and
    3. Radiation or induction losses

- **Microwave Transmission Lines**

The conventional open wire transmission lines are not suitable for microwave frequency transmissions due to high radiation loss. At microwave frequencies following transmission lines are generally employed.

1. Multi conductor lines:
    (a) Co-axial lines
    (b) Coplanar lines
    (c) Strip lines
    (d) Micro strip lines

2. Waveguides:
   (a) Rectangular wave guides
   (b) Circular wave guides
   (c) Elliptical waveguides
   (d) Ridge wave guides
3. Open-boundary structures:
   (a) Open wave-guides
   (b) Dielectric rods

- **Waveguides**

A hollow metallic tube of uniform cross section for transmitting electromagnetic wave by successive reflections from the inner walls of the hollow tube is called a **waveguide**.

The **waveguide** is classified as a transmission line. However, the method by which it transmits energy down its length differs from the conventional methods. There are various types of Wave guides available which are as follows:

   (a) Rectangular waveguide
   (b) Cylindrical waveguide
   (c) Elliptical waveguide
   (d) Ridged waveguide (Single and double ridged waveguides)

A wave may be classified according to its cross section (rectangular, elliptical, or circular), or according to the material used in its construction (metallic or dielectric).

- **Advantages of Waveguide**
  1. The large surface area of wave guides greatly reduces **copper (I^2R) losses.**
  2. At microwave frequencies, the current-carrying area of the inner conductor is restricted to very small layer at the surface of the conductor by an action called **skin effect**.
  3. **Dielectric losses** are also lower in waveguide than in two-wire and coaxial transmission lines.
  4. **Radiation losses** are kept very low.
  5. **Power-handing capability** is another advantage of waveguides.

- **Disadvantages of Waveguide**
  1. **Physical size** is the primary lower-frequency limitation of waveguides
  2. Waveguides are **difficult to install**.
  3. **Inside surface of waveguide** are often plated with silver of gold to reduce skin effect losses.
  - Rectangular waveguides are mostly used for power transmission at microwave frequencies. Their physical dimensions are being decided by the frequency of the signal to be transmitted.
  - A rectangular waveguide supports TM and TE modes but not TEM waves.
  - A rectangular waveguide cannot propagate below some certain frequency.
  - The field pattern is formed from the superposition of two plane waves traveling in different directions.
  - The whole field pattern is called a "**Mode**".
  - **Modes of Waveguides**
    (i) TE (Transverse Electric) wave
    $$E_z = 0$$
    $$H_z \neq 0$$

MICROWAVE TRANSMISSION LINES

(ii) TM (Transverse Magnetic) waves

$$E_z \neq 0$$
$$H_z = 0$$

(iii) TEM (Transverse Electromagnetic) waves

$$E_z = 0 \quad \text{and} \quad H_z = 0$$

It means that there is no existence of TEM mode in the waveguide propagation.

- In waveguide propagation, we have an infinity number of patterns, known as **modes**.
- To have propagation in the waveguide, the electric field component must always be perpendicular to the surface of the conductor not parallel to the surface of the conductor.
- In other conditions, the magnetic field components must always be parallel to the surface of the conductor not perpendicular to the surface of the conductor.

- **Propagation of TE Waves in a Rectangular Waveguide**

    TE mode in a rectangular waveguide is characterized by $E_z = 0$, because we have assumed that the wave is propagating in the z direction of the rectangular co-ordinate system. This for the transmission of energy H_z must exist in the waveguide.

- **Boundary Conditions**

    A rectangular waveguide has four walls and all the boundary walls of a waveguide will act as a short circuit or ground for electric field components, $E_z = 0$ all along the four boundary walls of the waveguide. There are four boundary conditions, which are discussed here.

    1st boundary condition (Bottom wall): $E_x = 0$ at $Y = 0$ $\forall x \to 0$ to a $\forall$ stands for all and $x \to 0$ to a mean varying between 0 to a.

    2nd boundary conditions: $E_x = 0$ at $y = b$ $\forall x \to 0$ to a.

    $$A = \frac{n\pi}{b}$$

    3rd boundary conditions: $E_y = 0$ at $x = 0$ $\forall y \to 0$ to b.

    4th boundary conditions: $E_y = 0$ at $x = a$ $\forall y \to 0$ to b, we know

    $$B = \frac{m\pi}{a}$$

- *Field components in TE mode*

    $$E_x = \frac{j\omega\mu}{h^2} \cdot C\left[\frac{n\pi}{b}\right] \cos\left[\frac{n\pi}{a}\right] x \cdot \sin\left[\frac{n\pi}{b}\right] y \cdot e^{(j\omega t - \gamma z)}$$

    $$E_y = \frac{j\omega\mu}{h^2} C\left[\frac{m\pi}{a}\right] \sin\left[\frac{m\pi}{a}\right] x \cdot \cos\left[\frac{n\pi}{b}\right] y e^{(j\omega t - \gamma z)}$$

    $$H_x = \frac{\gamma}{h^2} C\left[\frac{m\pi}{a}\right] \sin\left[\frac{m\pi}{a}\right] x_1 \cdot \cos\left[\frac{n\pi}{b}\right] y e^{(j\omega t - \gamma z)}$$

    $$H_y = \frac{\gamma}{h^2} C \cdot \left[\frac{n\pi}{b}\right]^2 \cos\left[\frac{m\pi}{a}\right] x \cdot \sin\left[\frac{n\pi}{b}\right] y \cdot e^{(j\omega t - \gamma z)}$$

- **Propagation of TM Waves in Rectangular Waveguide**
    Boundary conditions:

    1st boundary condition (Bottom wall): $\qquad E_z = 0$ at $y = 0$ $\forall x \to 0$ to a.

2nd boundary conditions (Left side wall): $E_z = 0$ at $x = 0$ $\forall\ y \to 0$ to b.

3rd boundary conditions (Top wall)l : $E_z = 0$ at $y = b$ $\forall\ x \to 0$ to a.

$$A = \frac{n\pi}{b}$$

4th boundary conditions: $E_z = 0$ at $x = a$ $\forall\ y \to 0$ to b,

- **Field Components in TM mode:**

$$H_x = \frac{j\omega\mu}{h^2} \cdot C\left(\frac{n\pi}{b}\right) \sin\left[\frac{m\pi}{a}\right] x \cos\left[\frac{n\pi}{b}\right] y\, e^{j\omega t - \gamma z}$$

$$H_y = \frac{j\omega\mu}{h^2} C\left[\frac{m\pi}{a}\right] \cos\left[\frac{m\pi}{a}\right] x \cdot \sin\left[\frac{n\pi}{b}\right] y\, e^{j\omega t - \gamma z}$$

$$E_x = -\frac{\gamma}{h^2}\frac{\partial E_z}{\partial x}$$

$$E_y = \frac{\gamma}{h^2} \cdot C\left[\frac{n\pi}{b}\right] \sin\left[\frac{m\pi}{a}\right] \cdot x \cos\left[\frac{n\pi}{b}\right] y \cdot e^{j\omega t - \gamma z}$$

- We have various modes in TM waves, depending upon the values of m and n. It is represent as TM_{mn} where m = number of half wave length across waveguide width (a).

n = number of half wave length along the waveguide height (b).

Cut off Frequency of Rectangular Waveguide:

Whenever 'γ' becomes zero, the wave propagation starts and the frequency at which 'γ' becomes zero is known as **cut-off frequency**.

$$f_c = \frac{c}{2}\left[\left[\frac{m}{a}\right]^2 \left[\frac{n}{b}\right]^2\right]^{1/2}$$

$$\lambda_{c\ mn} = \frac{2ab}{\sqrt{m^2 b^2 + n^2 a^2}}$$

Wavelengths greater than λ_C (**cut of wavelength**) are attenuated and less λ_C are passed to propagate inside the wave guide.

Waveguide acts as **high pass filter**, all the frequency above cut of frequency (f_c) are allowed to propagate and less than f_c are attenuated inside the waveguide.

- **Guide Wavelength (λ_g)**

The **guide wavelength** can be defined as the distance traveled by the wave to undergo a phase shift of 2π radians. It may be written as relative to the phase constant (β),

$$\lambda_g = \frac{2\pi}{\beta}$$

- When a wave is reflected from conducting walls it will have two velocities in a direction parallel to the wall. Those two velocities are known as:

  (i) Group velocity (V_g)   (ii) Phase velocity (V_p)

- **Group velocity** can be defined as that if there is modulation in the carrier, the modulated envelope travels at velocity slower than that of carrier and slower than velocity of light also. The velocity of modulation envelope is called the **group velocity**.

MICROWAVE TRANSMISSION LINES

- Group velocity may be defined as the rate at which the wave propagates through the waveguide. It can be written as

$$V_g = \frac{d\omega}{d\beta}$$

- **Phase velocity** can be defined as the rate at which the wave changes its phase in turns of the guide wavelength.

$$V_P = f\lambda_g$$

$$V_P = \frac{\omega}{\beta}$$

$$V_P = \frac{c}{\sqrt{1-(f_c/f)^2}}$$

$$V_P = \frac{c}{\sqrt{1-(\lambda_0/\lambda_c)^2}}$$

$$V_g = c \cdot \sqrt{1-\left[\frac{\lambda_0}{\lambda_c}\right]^2}$$

$$V_P \cdot V_g = c^2$$

$$\lambda_g = \frac{\lambda_0}{\sqrt{1-(\lambda_0/\lambda_c)^2}}$$

- In TE mode the expression for V_P, V_g, λ_g remain same as in TM mode operation,

$$V_P = \frac{c}{\sqrt{1-(\lambda_0/\lambda_c)^2}}$$

$$V_g = c \cdot \sqrt{1-(\lambda_0/\lambda_c)^2}$$

$$\lambda_g = \frac{\lambda_0}{\sqrt{1-(\lambda_0/\lambda_c)^2}}$$

- The higher order modes, having the same cut-off frequency are called **degenerate modes**.
- The lowest order TE mode wave in the case of rectangular waveguide is the TE_{10} mode. This mode has the lowest cut-off frequency and is called as **dominant mode** in the rectangular waveguide.

$$\bullet \quad P = \frac{\sqrt{1-(f_c/f)^2}}{2\eta} \int_0^b \int_0^a |E_x|^2 + |E_y|^2 \, dx \cdot dy$$

$$\bullet \quad P = \frac{1}{\sqrt{1-(f_c/f)^2}} \cdot \int_0^b \int_0^a |E_x|^2 + |E_y|^2 \, dx \, dy$$

- Losses in a waveguide mainly are two types such as
  1. Losses through the guide walls
  2. Losses in the dielectric

- The cut-off attenuation constant α is given by

$$\alpha = \frac{54.6}{\lambda_c} \alpha = \frac{54.6}{\lambda_c}\sqrt{1-(f/f_c)^2} \text{ db/length}$$

$$\alpha_{di} = \frac{54.6\sqrt{\varepsilon_R}\ \tan\delta}{\lambda_0\sqrt{1-(f/f_c)^2}} \text{ db/length}$$

- **Wave impedance** in the rectanguide waveguide is defined as the ratio of the electric field strength in one transverse direction to the strength of the magnetic field along the other transverse direction.

$$Z_z = \frac{E_x}{H_y} = -\frac{E_y}{H_x}$$

$$Z_{TE} = \frac{\eta}{\sqrt{1-(f_c/f)^2}} = \frac{\eta}{\sqrt{1-(\lambda_0/\lambda_c)^2}}$$

- **Wave impedance of TM waves:**

$$Z_{TM} = \frac{E_x}{H_y}$$

$$Z_{TM} = \eta\ \sqrt{1-(\lambda_0/\lambda_c)^2}$$

- **A circular waveguide** is a cylindrical hollow metallic tabular with uniforms circular.
- In TE_{np} modes in the circular waveguide, $E_z = 0$. It means the z components of the H_z must exist in the guide in order to have propagation of EM wave.

$$\beta_g = \sqrt{\omega^2\mu\varepsilon - \left(\frac{X'_{np}}{a}\right)^2}$$

$$K_c = \frac{X'_{np}}{a} = \omega_c\sqrt{\mu\varepsilon}$$

$$f_c = \frac{X'_{np}}{2\pi a\sqrt{\mu\varepsilon}}$$

$$V_P = \frac{1}{\sqrt{1-(f_c/f)^2}}\sqrt{\mu\varepsilon}$$

$$\lambda_g = \frac{\lambda}{\sqrt{1-(f_c/f)^2}}$$

$$Z_g = \frac{\omega\mu}{\beta} = \frac{\sqrt{\mu/\varepsilon}}{\sqrt{1-(f_c/f)^2}}$$

- In TM of a circular waveguide $H_z = 0$. However, the z components of the electric field E_z should exist so that to have energy transmission in the waveguide.
- Field equations of TM_{np} modes in circular guide.

$$E_r = E_{0r}\ J'_n\left(\frac{X_{np}r}{a}\right)\cos(n\phi)e^{-j\beta_g z}$$

$$E_z = E_{0z} J_n\left(\frac{X_{np}r}{a}\right) \cos(n\phi) e^{-j\beta_g z}$$

$$E_\phi = E_{0\phi} J_n\left(\frac{X_{np} r}{a}\right) \sin(n\phi) e^{-j\beta_g z}$$

$$H_r = \frac{E_{0\phi}}{Z_g} J_n\left(\frac{X_{np}r}{a}\right) \sin(n\phi) e^{-j\beta_g z}$$

$$H_\phi = \frac{E_{0\phi}}{Z_g} \cdot J'_n\left(\frac{X_{np}r}{a}\right) \cos(n\phi) e^{-j\beta_g z}$$

$$H_z = 0$$

$$\beta_g = \sqrt{\omega^2 \mu\varepsilon - \left(\frac{X_{np}}{a}\right)^2}$$

$$K_c = \frac{X_{np}}{a} = \omega_c \sqrt{\mu\varepsilon}$$

$$f_c = \frac{X_{np}}{2\pi a \sqrt{\mu\varepsilon}}$$

$$\lambda_g = \frac{\lambda}{\sqrt{(1-f_c/f)^2}}$$

$$Z_g = \frac{\beta_g}{\omega_c} = n\sqrt{1 - f_c/f}$$

$$V_g = \frac{\omega}{\beta_g} \cdot \frac{V_p}{\sqrt{1-(f_c/f)^2}}$$

- The dominant mode or the mode of lowest cut-off frequency in a circular waveguide is the mode of TE_{11}.

- $$\alpha = \frac{\text{Power loss per unit length}}{2(\text{Average power transmitted})}$$

- $$P_{np} = \frac{\sqrt{1-(f_c/f)^2}}{2\eta} \cdot \int_0^{2\pi}\int_0^a [|E_r|^2 + |E_\phi|^2] r\, dr\, d\phi$$

- $$P_{np} = \frac{1}{2n\sqrt{1-(f_c/f)^2}} \cdot \int_0^{2\pi}\int_0^a [|E_r|^2 + |E_\phi|^2] r \cdot dr\, d\phi$$

- The power losses in circular waveguide is given by

$$P_{loss} = \frac{R_s}{2} \oint \bar{J}_s \cdot \bar{J}_s \cdot dl$$

- **Attenuation constant** α_{TE} is given by

$$\alpha_{TE} = \frac{R_s}{a Z_0} \left[1-(f_c^2/f^2)\right]^{-1/2} \left[f_c^2/f^2 + \frac{n^2}{X'^2_{np} - n^2}\right]$$

- **Attenuation constant** α_{TM} is given by

$$\alpha_{TM} = \frac{Rs}{aZ_0}\left(\frac{1}{\sqrt{1-(f_c/f)^2}}\right)$$

$$(P_{bd})_{11} \approx 179.\ a^2\left[1-(f_{c11}^2/f^2)\right]^{1/2} kW$$

$$(P_{bd})_{01} \approx 1805\ a^2\left[1-(f_{c01}^2/f^2)\right]^{1/2} kW$$

- **Applications of Circular Waveguide**
    (i) For long distance wave propagation of the frequency a have 10 GHz. TE_{01} mode is used because of low attenuation.
    (ii) In the radars, circular waveguide are used to couple the energy to the horn antenna feeder from the rotary joint.
    (iii) For the short and medium distance broadband communication, circular waveguides are used.

SOLVED PROBLEMS

Problem 1. *An air-filled rectangular waveguide of inside dimensions 7 × 3.5 cm operates in TE_{10} mode.*
  (a) *Find the cut-off frequency.*
  (b) *Determine the guided wavelength at 3.5 GHz.*
  (c) *Determine the phase velocity of the wave in the guide at the same frequency.*

Solution: (a) Cut-off frequency $f_c = \dfrac{c}{2a}$

$$f_c = \frac{3 \times 10^8}{2 \times 7 \times 10^2} = 2.14\ GHz$$

$$f_c = 2.14\ GHz$$

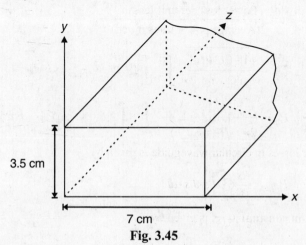

Fig. 3.45

MICROWAVE TRANSMISSION LINES

(b) Guided wavelength (λ_g)

$$\lambda_g = \frac{\lambda_0}{\sqrt{1-(f_c/f)^2}}$$

$$\lambda_g = \frac{3\times10^8}{\sqrt{1-(2.14/3.5)^2}} = 3.78\times10^8 \text{ m/s}$$

$$\boxed{\lambda_g = 3.78\times10^8 \text{ m/s}}$$

(c) Phase velocity of rectangular waveguide

$$V_P = \frac{c}{\sqrt{1-(f_c/f)^2}} = \frac{3\times10^8}{\sqrt{1-(2.14/3.5)^2}} = 3.78\times10^2 \text{ m/s}$$

$$\boxed{V_P = 3.78\times10^2 \text{ m/s}}$$

Problem 2. *A rectangular waveguide with the dimension f_0 5 cm × 2 cm is used to propagate TM_{11} mode at 9 GHz. Calculate the wave impedance and cut-off wavelength.*

Solution: Given, $a = 5$ cm and $b = 2$ cm
Operating frequency = 9 GHz

$$\lambda = \frac{c}{f} = \frac{3\times10^8 \text{ m/s}}{9\times10^9 \text{ Hz}} = 3.33 \text{ cm}$$

TM_{11} mode, $m = 1$ and $n = 1$

Cut-off wavelength (λ_c) = $\dfrac{2}{\sqrt{(m/a)^2+(n/b)^2}}$

or

$$\lambda_c = \frac{2ab}{\sqrt{a^2+b^2}}$$

By putting the values, we get

$$\lambda_c = \frac{2\times5\times2}{\sqrt{5^2+2^2}} = 3.714 \text{ cm}$$

$$\lambda_c = 3.714 \text{ cm}$$

Wave impedance

$$(Z_0) = 120\,\pi\sqrt{1-(\lambda/\lambda_0)^2}$$

$$= 120\,\pi\sqrt{1-\left(\frac{3.333}{3.714}\right)^2}$$

$$\boxed{Z_0 = 325 \text{ }\Omega}$$

Problem 3. *A rectangular waveguide in TE10 mode having dimensions $a = 2.26$ cm and $b = 1$ cm. If $\lambda_g = 4$ cm, calculate:*

(a) *Cut-off frequency*
(b) *Frequency operation.*

Solution: (a) Cut-off frequency $(f_c) = \dfrac{1}{2a\sqrt{\omega\varepsilon}} = \dfrac{c}{2a}$

$$f_c = \dfrac{3\times 10^{10} \text{ cms}}{2\times 2.26} = 6.64 \text{ GHz}$$

$$\boxed{f_c = 6.64\, GHz}$$

We know, $\quad \beta = \dfrac{2\pi}{\lambda_g}$

or $\quad \beta = \dfrac{2\pi}{4}$

$$\beta_{10} = \sqrt{\omega^2 \mu\varepsilon - \left(\dfrac{\pi}{a}\right)^2}$$

$$\beta_{10}^2 = \left(\dfrac{\pi}{a}\right)^2 = \left(\dfrac{2\pi f}{c}\right)^2$$

$$f = \dfrac{c^2}{4\pi^2}\left[\beta^2 + \left(\dfrac{\pi}{a}\right)\right]$$

$$\boxed{f = 10\, GHz}$$

(c) For higher order mode is TE_{20}, where $m=2$ and $n=0$

$$f_{20} = \dfrac{c}{2\pi}\sqrt{\left(\dfrac{2\pi}{a}\right)^2} = \dfrac{c}{a}$$

$$\boxed{f_{20} = \dfrac{c}{a} = 13.27\, GHz}$$

Problem 4. *A rectangular waveguide with dimensions of $a = 3$ cm, $b = 1.5$ cm, $\mu = 1$ and $\varepsilon = 2.25$. Determine the followings*

(a) *Cut-off wavelength and cut-off frequency for TE_{10} and TM_{11}*

(b) *Guide wavelength and wave impedance at 4.0 GHz.*

Solution: (a) Cut-off wavelength $(\lambda_c) = \dfrac{1}{\sqrt{\left[\dfrac{m}{2a}\right]^2 + \left[\dfrac{n}{2b}\right]^2}}$

For TE_{10} mode : $m=1, n=0$

$$\lambda_c = 2a$$

or $\quad \lambda_c = 6\, cm$

For TM_{11} mode : $m=1$ and $n=1$

$$\lambda_c = \dfrac{2ab}{\sqrt{a^2+b^2}} = \dfrac{2\times 3\times 1.5}{\sqrt{(3)^2 + (1.5)^2}} = 2.68 \text{ cm}$$

Cut-off frequencies

$$f_c = \dfrac{c}{\lambda_c \sqrt{\varepsilon\mu}}$$

MICROWAVE TRANSMISSION LINES

Substitute $\varepsilon = 2.25$; $\mu = 1$

$$f_{c\,10} = \frac{c}{\lambda_c \sqrt{\varepsilon \mu}}$$

$$f_{c\,10} = \frac{3 \times 10^{10}}{6 \times \sqrt{2.25 \times 1}}$$

$$\boxed{f_{c\,10} = 3.33 \ GHz}$$

$$f_{c\,11} = \frac{3 \times 10^{10}}{2.68\sqrt{2.25 \times 1}}$$

$$\boxed{f_{c\,11} = 7.46 \ GHz}$$

(b) Guided wavelength $(\lambda_g) = \dfrac{\lambda_0}{\sqrt{\mu\varepsilon - (\lambda_0/\lambda_c)^2}}$

at 4.0 GHz, $\lambda = 7.5$ cm, $\lambda_c = 6$ cm (TE$_{10}$ mode)

$$\lambda_g = \frac{7.5}{\sqrt{2.25 - (7.5/6)}}$$

$$\boxed{\lambda_g = 9.05 \ cm}$$

Wave impedance (Z_0)

$$= \frac{377 b/a \ \sqrt{\dfrac{\mu}{\varepsilon}}}{\sqrt{1 - \left[\dfrac{f_c}{f_0}\right]^2}} = 377 \frac{b}{a} \sqrt{\frac{\mu}{\varepsilon} \frac{\lambda_g}{\lambda}}$$

$$= 377 \times \frac{1.5}{3.0} \cdot \sqrt{\frac{1}{2.25}} \times \frac{9.05}{7.5/\sqrt{2.25}} = 2.27 \ \Omega$$

$$\boxed{Z_0 = 227 \ \Omega}$$

Problem 5. *An air filled hollow rectangular conducting waveguide has cross-section of 8 × 10cm. How many TE mode will this waveguide propagate at frequencies below 4 GHz. How these modes are designated and determine their cut-off frequencies.*

Solution: Maximum operating frequency = 4 GHz

$$\lambda = \frac{c}{f} = \frac{3 \times 10^8}{4 \times 10^9} = 0.075 \text{ meters}$$

The cut-off frequencies (f_c) in terms of waveguide dimensions are given by

$$\lambda_c = \frac{2}{\sqrt{\left(\dfrac{m}{a}\right)^2 + \left(\dfrac{n}{b}\right)^2}}$$

$$\lambda_c = \frac{2}{\sqrt{\left[\dfrac{m}{0.10}\right]^2 + \left[\dfrac{n}{0.08}\right]^2}} = \frac{2}{\sqrt{\dfrac{n^2}{0.0064} + \dfrac{m^2}{0.01}}}$$

$$= \frac{2}{\frac{10}{8}\sqrt{n^2 \times 100 + 64m^2}} = \frac{1.6}{\sqrt{100\ n^2 + 64m^2}}$$

$$\lambda_c = \frac{0.16}{\sqrt{n^2 + 0.64\,m^2}}$$

Now for,

(i) TE_{10} mode, $m = 1, n = 0$

$$\lambda_c = \frac{0.16}{\sqrt{0 + 0.64 \times 1}} = \frac{16}{18} = 0.2 \text{ mtrs}$$

(ii) TE_{01} mode, $m = 0, n = 1$

$$\lambda_c = \frac{0.16}{\sqrt{0 + 1^2}} = 0.16 \text{ mtrs}$$

(iii) TE_{11} mode, $m = 1, n = 1$

$$\lambda_c = 0.125 \text{ meters}$$

(iv) TE_{20} mode, $m = 2$ and $n = 0$

$$\lambda_c = \frac{0.16}{\sqrt{0.64 \times 2^2 + 0^2}} = \frac{0.16}{160} = 0.1 \text{ mtrs}$$

$$\lambda_c = 0.1 \text{ meters}$$

(v) TE_{02} mode, $m = 0$ and $n = 2$

$$\lambda_c = \frac{0.16}{\sqrt{0.64 \times 0 + 2^2}} = \frac{0.16}{2}$$

$$\lambda_c = 0.08 \text{ meters}$$

(vi) TE_{21} mode, $m = 2, n = 1$

$$\lambda_c = 0.0874 \text{ mtrs}$$

(vii) TE_{12} mode, $m = 1, n = 1$

$$\lambda_c = 0.07428 \text{ mtrs}$$

It is seen that only first six modes will be propagated and the last mode being less than 0.075 meter (λ_c) will not be transmitted.

Cut-off frequencies of all the propagating modes:

$$\lambda_c = \frac{c}{f_c}$$

$$f = \frac{c}{\lambda_c} \text{ Hz}$$

(i) TE_{10} mode, $f_c = \frac{3 \times 10^8}{0.2} = 1500 \text{ MHz}$

(ii) TE_{01} mode, $f_c = \frac{3 \times 10^8}{0.16} = 1875 \text{ MHz}$

(iii) TE_{11} mode, $f_c = \frac{3 \times 10^8}{0.125} = 2400 \text{ MHz}$

MICROWAVE TRANSMISSION LINES

(iv) TE_{20} mode, $f_c = \dfrac{3\times 10^8}{0.1} = 3000 \ MHz$

(v) TE_{02} mode, $f_c = \dfrac{3\times 10^8}{0.08} = 3750 \ Hz$

(vi) TE_{21} mode, $f_c = \dfrac{3\times 10^8}{0.0847} = 3541.91 \ MHz$

Problem 6. *An air filled rectangular waveguide has dimensions, a = 8 cm and b = 4 cm. Find the cut-off frequencies for the modes of TE_{10}, TE_{20} and TE_{11} and the ratio of the guide velocity to the velocity in free space for each these modes, of $f = 3/2 f_c$.*

Solution: Given $a = 0.08$ m; $b = 0.04$ m

$$f = 1.5 \ f_c$$

$$f_c = \dfrac{c}{2ab}\sqrt{(mb)^2 + (na)^2}$$

Now putting the values, we have

$$f_c = \dfrac{3\times 10^8}{2\times 0.04 \times 0.08} \cdot \sqrt{(m\times 0.04)^2 + (n\times 0.08)^2}$$

$$= \dfrac{1.5\times 10^8}{0.08} \cdot \sqrt{m^2 + 4n^2}$$

$$f_c = 18.75\times 10^8 \sqrt{m^2 + 4n^2}$$

(i) For TE_{10} mode, $m = 1$ and $n = 0$

$$f_c = 18.75\times 10^8 \sqrt{1^2 + 4\times 0}$$

$$f_c = 1875 \ MHz$$

(ii) For TE_{20}, $m = 2$ and $n = 0$

$$f_c = 18.75\times 10^8 \sqrt{(2)^2 + (0)^2} = 3750 \ MHz$$

(iii) For TE_{11} mode, $m = 1$ and $n = 1$

$$f_c = 18.75\times 10^8 \sqrt{1^2 + 4\times 1^2} = 18.75\times 10^8 \sqrt{5}$$

$$f_c = 3521.80 \ MHz$$

Now since $f = 1.5 f_c$ then

$$\dfrac{\text{Guide velocity}(V_g)}{\text{Velocity in free space }(V_P)} = \dfrac{1}{\sqrt{1-\left(\dfrac{f_c}{f}\right)^2}}$$

$$= \dfrac{1}{\sqrt{1-\left(\dfrac{f_c}{1.5 \ f_c}\right)^2}}$$

$$= \dfrac{1}{\sqrt{1-(0.66)^2}} = \dfrac{1}{\sqrt{1-0.4399}} = \dfrac{1}{\sqrt{0.5601}} = 1.3362$$

$$\dfrac{V_g}{V_p} = 1.3362$$

Problem 7. *A hollow rectangular waveguide has dimensions a = 4 cm, b = 2 cm. Calculate amount of attenuation, if the frequency is* 3 GHz.

Solution: The critical frequency (f_c) of the waveguide is given by:

$$f_c = \frac{c}{2ab}\sqrt{(mb)^2 + (na)^2}$$

For TE_{10} mode,

$$f_c = \frac{c}{2ab}\sqrt{(mb)^2 + 0} = \frac{c}{2a} \cdot m$$

$$f_c = \frac{3 \times 10^8 \times 100 \times 1}{2 \times 4} = 0.375 \times 10^{10} \text{ Hz}$$

$$\boxed{f_c = 3.75 \text{ GHz}}$$

Since f_c is 3.75 GHz which is more than the frequency 3 GHz. There will not be any propagation and waves get attenuated.

The amount of attenuation is given by

$$P = \sqrt{\left[\frac{m\pi}{a}\right]^2 + \left[\frac{n\pi}{b}\right]^2 - \omega^2\mu\varepsilon} = \alpha + j\beta$$

$$\alpha = \sqrt{\left[\frac{m\pi}{a}\right]^2 + \left[\frac{n\pi}{b}\right]^2 - \omega^2\mu_0\varepsilon_0} \qquad (\because \beta = 0 \text{ since wave is attenuated})$$

For TE_{10} mode

$$\alpha = \sqrt{\left[\frac{m\pi}{a}\right]^2 - \omega^2\mu_0\varepsilon_0}$$

$$\mu_0 = 4\pi \times 10^7 \text{ H/m}$$

$$\omega = 2\pi f$$

$$\varepsilon_0 = \frac{10^{-9}}{36\pi} \text{ F/m}$$

$$f = 3 \text{ GHz}$$

$$\alpha = \sqrt{\left[\frac{1 \times 3.14 \times 100}{4}\right]^2 - \frac{\left(2\pi \times 3 \times 10^9\right)^2 \times 4\pi \times 10^{-7} \times 10^{-9}}{36\pi}}$$

$$= \sqrt{\left(\frac{314}{4}\right)^2 - \frac{4\pi^2 \times 10^{18} \times 10^{-16}}{9}}$$

$$= \sqrt{6162.25 - 3943 - 84} = \sqrt{2218.41}$$

$\alpha = 47.1$ neper / metre (1 neper = 8.686 db, 1 db = 0.115 neper)

Hence, α 47.1 × 8.686 db = 409.1106 dB.

$$\boxed{\alpha = 409.1106 \text{ dB}}$$

Problem 8. *Calculate the values of cut-off frequency and guide wavelength in an air filled rectangular waveguide; with internal dimensions 7.62 cm 2.54 cm for the normal H_{10} mode at a frequency of* 30 GHz.

MICROWAVE TRANSMISSION LINES

Solution: Since $\lambda_c = \dfrac{2}{\sqrt{\left(\dfrac{m}{a}\right)^2 + \left(\dfrac{n}{b}\right)^2}}$

For H_{10} mode, $m = 1$ and $n = 0$

$$\lambda_c = \dfrac{2}{\sqrt{\left[\dfrac{m}{a}\right]^2}} = \dfrac{2a}{m}$$

$$\lambda_c = \dfrac{2 \times 7.62}{100 \times 1} = 0.1524 \text{ meter}$$

$$\lambda_0 = \dfrac{c}{f} = \dfrac{3 \times 10^8}{30 \times 10^9} = \dfrac{1}{100} = 0.01 \text{ meter}$$

$$\lambda_g = \dfrac{\lambda_0}{\sqrt{1 - \left[\dfrac{\lambda_0}{\lambda_c}\right]^2}} = \dfrac{0.01}{\sqrt{1 - \left[\dfrac{0.01}{0.1524}\right]^2}} = \dfrac{0.01}{\sqrt{1 - (0.06)^2}}$$

$$\lambda_g = \dfrac{0.01}{\sqrt{1}} = \dfrac{0.01}{1} = 0.01 \text{ meter}$$

$$\boxed{\lambda_g = 0.01 \text{ meter}}$$

Problem 9. *A circular waveguide of internal diameter 10 cm is excited by the dominant mode, i.e., TE_{11} of 9 GHz. Calculate the breakdown power of that waveguide.*

Solution: Power breakdown in TE_{11} mode

$$= 1790\, a^2 \left[1 - \left[\dfrac{f_{c_{11}}}{f}\right]^2\right]^{1/2} \text{ kW}$$

$$(P_{bd})_{TE_{11}} = 1790\, a^2 \left[1 - \left[\dfrac{f_{c_{11}}}{f}\right]^2\right]^{1/2} \text{ kW}$$

$$\lambda_0 = \dfrac{3 \times 10^{10}}{9 \times 10^9} = 3.333 \text{ cm}$$

$$\lambda_{c\,11} = \dfrac{\pi d}{(h_{mn})_{mn}} = \dfrac{\pi \times 10}{1.841} = 17.05 \text{ cm}$$

$$\lambda_{c\,11} = \dfrac{3 \times 10^{10}}{17.05} = 1.75 \text{ GHz}$$

By putting the values of a, f_{c11} and f, we have

$$(P_{bd})_{TE_{11}} = 1790 \times (5)^2 \left[1 - \left(\dfrac{1.75 \times 10^9}{9 \times 10^9}\right)^2\right]^{1/2} \text{ kW}$$

$$= 1790 \times 25 [0.97]^{1/2} \text{ kW}$$

$$\boxed{(P_{bd})_{TE_{11}} = 44069.8 \text{ kW}}$$

Problem 10. *Given a rectangular waveguide 3 × 1 cm operating at a frequency of 9 GHz in dominant mode. Calculate the maximum power handling capacity of the waveguide if the maximum potential gradient of the signal is 3 kV/ cm.*

Solution: $\lambda = \dfrac{c}{f} = \dfrac{3\times 10^{10}}{9\times .10^{9}} = 3.333$ cm

$\lambda_{c\,10} = 2a = 2\times 3 = 6$ cm

$\lambda_g = \dfrac{\lambda_0}{\sqrt{1-(\lambda_0/\lambda_c)^2}} = \dfrac{3.33}{\sqrt{1-(3.33/6)^2}} \approx 5\,\text{cm}$

Power handling capacity is given by

$$P = (6.63\times 10^{-4})E_{max}^2\, ab\left[\dfrac{\lambda_0}{\lambda_g}\right]$$

$$P = (6.63\times 10^{-4})(3\times 10^3)^2 (3\times 1)\left[\dfrac{3.33}{5}\right]$$

$$\boxed{P = 11.922\ kW}$$

Problem 11. *A circular waveguide operating in the TE_{11} mode at a frequency of 9 GHz with a maximum field strength of 500 V/ cm. The internal diameter is 5 cm. Determine the maximum power.*

Solution: Operating frequency $f = 9\times 10^9$ Hz

$$\lambda_0 = \dfrac{c}{f} = \dfrac{3\times 10^{10}}{9\times .10^9} = 3.333\ \text{cm}$$

TE_{11} mode (dominant mode),

$$\lambda_c = \dfrac{2\pi}{1.841} = \dfrac{\pi d}{1.841} = \dfrac{\pi \times 5}{1.841}$$

$$\boxed{\lambda_c = 8.533\ \text{cm}}$$

$$\lambda_g = \dfrac{\lambda_0}{\sqrt{1-(\lambda_0/\lambda_c)^2}} = \dfrac{3.333}{\sqrt{1-(3.333/8.533)^2}}$$

$$\boxed{\lambda_g = 3.62\ \text{cm}}$$

P_{max} of circular waveguide is given by,

$$P_{max} = 0.498\times E_{max}^2\cdot d^2\left[\dfrac{\lambda_0}{\lambda_g}\right]$$

$$= 0.498\times (500)^2 \times 5^2\left[\dfrac{3.333}{3.62}\right]$$

$$= 124500\times 23 = 2863500\ \text{watts}$$

$$\boxed{P_{max} = 2.863500\times 10^6\ \text{watts}}$$

REVIEW QUESTIONS

1. How are waveguide different from two wire transmission lines? Discuss the similarities and dissimilarities.
2. Verify that a TEM wave cannot propagate in a waveguide by making use of Maxwells equation.
3. Explain the various modes in rectangular wave guide.
4. Derive the wave equation for a TM wave and obtain all the field components in a rectangular waveguide. **(MDU-2004)**
5. Derive the wave equation for a TE wave and obtain all the field components in a rectangular waveguide.
6. What do you understand by the terms cut-off wavelength, guide wavelength, phase velocity and group velocity. Give their mathematical derivations and their inter relationships.
7. Derive the wave equation for a TM wave and obtain all the field components in a circular waveguide. **(MDU-2004)**
8. Derive the wave equation for a TE wave and obtain all the field components in a circular waveguide.
9. Show the field configurations of TE_{10} mode in a rectangular waveguide. **(MDU-2006)**
10. What is characteristics impedance? **(MDU-2004)**
11. Explain propagation in TEM mode. **(MDU-2004)**
12. Discuss the power losses in circular waveguide. **(MDU-2004)**
13. Explain the advantages of waveguide over coaxial cable.
14. Explain the waveguide.
15. Explain the mode of propagation of waveguide.
16. Explain the circular waveguide.
17. Explain the strip line and microstrip line.
18. What do subscripts 'm' and 'n' denote in rectangular and circular waveguides? Why is TM_{01} mode not possible in rectangular waveguide whereas it does exist in circular waveguide? **(UPTU-2004)**
19. When the dominant mode is propagated in an air filled standard rectangular waveguide, the guide wavelengths at a frequency of 9 GHz, 4 cm. Calculate width of the guide. **(UPTU-2003, 2004)**
20. An air filled rectangular waveguide of cross-section 1 cm × 2 cm is operating in TE_{10} mode at a frequency of 12 GHz. What is the maximum power handling capacity of the guide, if the dielectric strength of the medium is 3×10^6 V/m? **(UPTU-2003, 2004)**
21. Discuss the attenuation characteristics of circular waveguide. Why does TE_{10} mode show a very low loss at higher frequencies? **(UPTU-2003, 2004)**
22. An air filled circular waveguide having inner radius of 1 cm, is excited in dominant mode at 10 GHz. Find the :
    (*i*) Cut-off frequency of dominant mode
    (*ii*) Guide waveguide
    (*iii*) Wave impedance and
    (*iv*) Bandwidth for operations in dominant mode only. **(UPTU-2003, 2004)**
23. Given the physical structure and field distribution of a microstrip line. Why can a pure TEM mode not be propagated in a microstrip line? **(UPTU-2003, 2004)**

24. What do subscripts of the modes TE_{nml} and TM_{nml} of a rectangular cavity designate? Write an expression for the resonant frequency of TM_{nml} mode in a rectangular cavity.
(UPTU-2002, 2003)

25. Give any one method of exciting TE_{01} mode in an absorption type rectangular cavity.
(UPTU-2002, 2003)

26. What problem are associated with the simple form of an adjustable short circuit? How these problems overcome in variable short circuit? **(UPTU-2002, 2003)**

27. Calculate the cut-off frequency of the dominant mode in a 2.5 cm diameter, teflon filled ($E_r = 2.3$) circular waveguide. What is its maximum operating frequency if the possibility of higher mode propagation is to be avoided? Include a 5% safety factor. **(UPTU-2002, 2003)**

28. Show that a metal rectangular waveguide is a high pass filter. Derive the formula used.
(UPTU-2002, 2003)

29. What are degenerate modes? Explain why TEM mode cannot exist in metallic waveguides.
(UPTU-2002, 2003)

30. Discuss in brief, the methods of excitation of TE_{10}, TE_{20} and TM_{11} modes in a rectangular metal waveguide. **(UPTU-2002, 2003)**

31. Which mode in circular metal waveguide has got highest cut-off wavelength? What do the subscripts 'm' and 'n' indicate in TE_{mn} mode of a circular waveguide. Give two important applications of this waveguide. **(UPTU-2002, 2003)**

32. What is a microstrip line? How does its characteristics impedance change with charge in width to height ratio? Give a reason for using lower dielectric constant substrate in place of alumina at higher microwave frequencies. **(UPTU-2002, 2003)**

33. A rectangular cavity resonators in the TM_{111} mode at 5.0 GHz. Given $a = 8$ cm and $b = 6$ cm, calculate the resonant frequencies for the TE_{101}, TE_{102} and TE_{111} modes. Assume that the cavity is air filled. **(UPTU-2002, 2003)**

34. A rectangular waveguide of cross-section 5 cm × 2 cm is used to propagate TM_{11} mode at 10 GHz. Determine the cut-off wavelength and the characteristic impedance. **(MDU-2006)**

35. Find the resonant frequencies of first two lowest modes of an air filled rectangular cavity of dimensions 5 cm x 2.5 cm. **(MDU-2006)**

36. A rectangular waveguide has dimensions 2.5 x 5 cms. Determine the guide wavelength, phase constant β and phase velocity V_P at a wavelength of 4-5 cm for the dominant mode.
(MDU-2004)

37. Show that a TEM wave cannot propagate in a waveguide by making use of Maxwell's equations. **(MDU-2003)**

38. Deduce the expressions for cut-off wavelength of a waveguide system. Explain the physical significance of cutt-off wavelength. **(MDU-2003)**

39. What are the difference between the waveguides and the transmission lines? Explain how a waveguide acts as a high-pass filter. **(MDU-2002)**

40. A rectangular waveguide measures 3 x 4.5 cm internally ($a = 4.5$, $b = 3.0$) and has a 9 GHz signal propagated in it. Calculate the cut-off wavelength, the guide wavelengths, the group and phase velocities and the characteristic wave impedance for TE_{10} mode and TM_{11} mode.
(MDU-2001)

41. Highlights the differences in propagation and general behavior between TE and TM modes in a rectangular waveguide. **(MDU-2001)**

42. A rectangular waveguide has $a = 4$ cm and $b = 3$ cm. Find all the TE modes, which will propagate at 5000 MHz. **(MDU-2001)**

43. A circular waveguide whose internal radius is 2.5 cm and operates on 8 GHz signal propagation in it is the TE_{11} mode. Determine the cut-off wavelength, the guide length and the wave impedance of the circular waveguide.

CHAPTER 4

MICROWAVE CAVITY RESONATORS

OBJECTIVES
- Introduction
- Advantages of Cavity Resonators
- Types of Cavity Resonators
- Operation of Microwave Cavity Resonator
- Field Expressions
- Coupling to Cavity
- Tuning of Cavities
- Applications of Cavity Resonantors.
- Field Expression for Rectangular Cavity Resonator
- Quality factor(Q) of Cavity Resonators
- Transmission line Resonators

4.1. INTRODUCTION

In ordinary electronic equipment, a resonant circuit consists of a coil and a capacitor that are connected either in series or in parallel. The resonant frequency of the circuit is increased by reducing the capacitance, the inductance, or both. As you carry on reducing the values of capacitance and inductance, a point will reached where the value of inductance and the capacitance can not be reduced further. This is the highest frequency at which a conventional circuit can oscillate. The upper limit for a conventional resonant circuit is between 2000 and 3000 MHz. At these frequencies, the inductance may consist of a coil of one-half turn, and the capacitance may simply be the stray capacitance of the coil. Tuning a one-half turn coil is very difficult and tuning stray capacitance in even more difficult. In addition such a circuit will handle only very small amounts of current.

By definition, **a resonant cavity** is any space completely enclosed by conducting walls that can contain oscillating electromagnetic fields and possess resonant properties.

4.2. ADVANTAGES OF CAVITY RESONATORS

The cavity has many advantages and uses at microwave frequencies.
1. Resonant cavities have a very high Q
2. It can be built to handle relatively large amounts of power.
3. The high Q gives these devices a narrow band pass and allows very accurate tuning.
4. Simple, rugged construction is an additional advantage.

4.3. TYPES OF CAVITY RESONATORS

Although cavity resonators, built for different frequency ranges and applications, have a variety of shapes, the basic principles of operation are the same for all. The specific shape and size are decided by the following factors:

1. Resonant frequency required to be produced by the cavity.
2. Desired value of the quality (Q) factor.
3. Mechanical considerations.
4. Required value of the shunt conductance.

There are two types of cavity resonators.

1. Rectangular cavity resonators
2. Circular cavity resonators

One example of a cavity resonator is the rectangular box shown in Fig. 4.1.

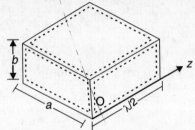

(a) Rectangular waveguide cavity resonator

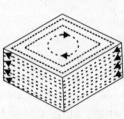 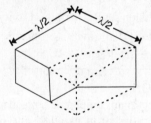

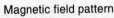

Magnetic field pattern   Electric field pattern

(b) Rectangular waveguide cavity resonator showing Field patterns of a simple mode

Fig. 4.1. (a) Rectangular waveguide cavity resonator; (b) Rectangular waveguide cavity resonator showing Field patterns of simple mode.

4.4. OPERATION OF MICROWAVE CAVITY RESONATOR

The microwave cavity resonator is similar to a tuned circuit at low frequencies having a resonant frequency

$$fo = 1/2\pi\sqrt{LC}$$

The cavity resonator can resonant at only one particular frequency as a parallel resonate circuit shown in Fig. (4.2)

It may be thought of as a section of rectangular waveguide closed at both ends by conducting plates. The frequency at which the resonant mode occurs has the distance between the end plates is $\lambda_g/2$. The magnetic and electric field patterns in the rectangular cavity are shown in Fig. 4.1 (b)

The rectangular cavity is only one of many cavity devices that are useful as high-frequency resonators.

MICROWAVE CAVITY RESONATORS

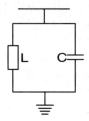

Fig. 4.2

Fig. (4.3) shows the development of a cylindrical resonant cavity from an infinite number of quarter-wave sections of transmission line. In Fig. 4.3(a) the $\lambda_g/4$ Section is shown to be equivalent to a resonant circuit with a very small amount of inductance and capacitance. Three $\lambda_g/4$ sections are joined in parallel in Fig. 4.3(b).

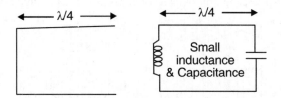

(*a*) Quarter-wave section equivalent to LC circuit

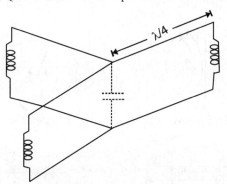

(*b*) Quarter-wave lines joined

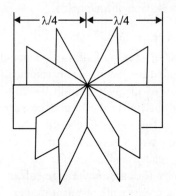

(*c*) Cylindrical resonant cavity being formed from quarter-wave sections

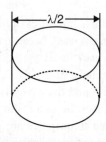

(*d*) Cylindrical resonant cavity

Fig. 4.3 (*a, b, c, d*) Development of a cylindrical resonant cavity.

Note that although the current-carrying ability of several $\lambda_g/4$ sections is greater than that of any one section, the resonant frequency is unchanged. This occurs because capacitance in parallel increases the total capacitance by the same proportion. Thus, the resonant frequency remains the same as it was for one section. The increase in the number of current paths also decreases the total resistance and increases the Q of the resonant circuit. Fig. 4.3 (c) shows an intermediate step in the development of the cavity. Fig. 4.3 (d) shows a completed cylindrical resonant cavity with a diameter of $\lambda_g/2$ at the resonant frequency.

There are two variables that determine the primary frequency of any resonant cavity. The first variable is physical size. In general, if the size the size of cavity is smaller, the higher it's resonant frequency.

The second controlling factor is the shape of the cavity. Fig. 4.4 illustrates several cavity shapes that are commonly used. According to the definition of a resonant cavity, any completely enclosed conductive surface, regardless of its shape, can act as a cavity resonator.

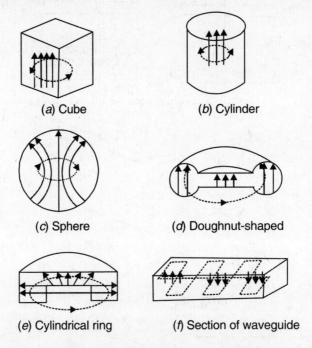

Fig. 4.4. Several types of cavities.

Although the simplest cavity resonators may be spheres, cylinders or rectangular but such cavities are often not used, because their various resonant frequencies are harmonically related. It is a serious drawback in all conditions in which pulses of energy are fed to a cavity. It is desired that cavity to maintain sinusoidal oscillations through the flywheel effect but because such pulses contain harmonics and the cavity is able to oscillate at harmonics frequencies, the output is still in the form of pulses. So as a result the most practical cavities have odd shapes to ensure that the various oscillating frequencies are not harmonically related and harmonics are attenuated. Fig. 4.4(d, e) shows some typical irregularly shaped resonators. They are known as **re-entrant resonators**, that is, resonators are so shaped that one of the walls re-enter the resonant shape.

MICROWAVE CAVITY RESONATORS

4.5. FIELD EXPRESSIONS

Cavity resonators are energized in the same manner as waveguides and have a similar field distribution. If the cavity shown in Fig. (4.5) were energized in the TE mode, the electromagnetic wave would reflect back and forth along the Z axis and form standing waves. These standing waves would form a field configuration within the cavity that would have to satisfy the same boundary conditions as those in a waveguide. Modes of operation in the cavity are described in terms of the fields that exist in the X, Y, and Z dimensions. Three subscripts are used; the first subscript indicates the number of $\lambda_g/2$ along the X axis; the second subscript indicates the number of $\lambda_g/2$ along the Y axis; and the third subscript indicates the number of $\lambda_g/2$ along the Z axis.

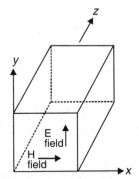

Fig. 4.5. Rectangular cavity resonator.

4.6. COUPLING TO CAVITY

Energy can be inserted or removed from a cavity by the same methods that are used to couple energy into and out of waveguides. The operating principles of probes, loops, and slots are the same whether used in a cavity or a waveguide. Therefore, any of the three methods can be used with cavities to inject or remove energy.

4.7. TUNING OF CAVITY

The resonant frequency of a cavity can be varied by changing any of three parameters: cavity volume, cavity capacitance, or cavity inductance. Changing the frequencies of a cavity is known as **Tuning**. It is important to examine the effects of such tuning and also loading on the bandwidth and Q of the cavity resonator.

"Q" may be defined as the ratio of the resonant frequency to the bandwidth. However, it is more useful to base the definitions of Q on more fundamental relations,

$Q = 2\pi$ {Energy stored / Energy lost each cycle}

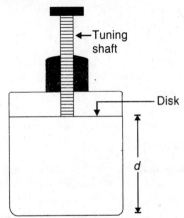

Fig. 4.6. Cavity tuning by volume.

The mechanical methods of tuning a cavity may vary with the application, but all methods use the same electrical principles.

A mechanical method of tuning a cavity by changing the volume (*volume tuning*) is illustrated in Fig. (4.6). Varying the distance d will result in a new resonant frequency because the inductance and the capacitance of the cavity are changed by different amounts. If the volume is decreased, the resonant frequency will be higher. The resonant frequency will be lower if the volume of the cavity is made larger.

Capacitive tuning of a cavity is shown in Fig. 4.7(a). An adjustable slug or screw is placed in the area of maximum E lines. The distance d represents the distance between two capacitor plates. As the slug is moved in the distance between the two plates becomes smaller and the capacitance increases. The increase in capacitance causes a decrease in the resonant frequency. As the slug is moved out, the resonant frequency of the cavity increases.

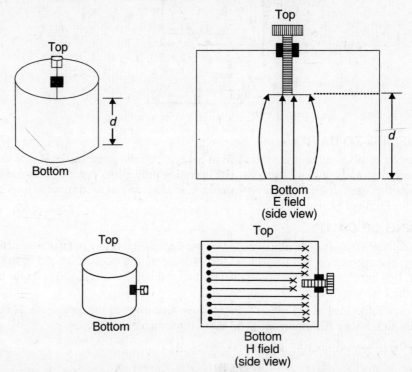

Fig. 4.7. (*a*) Method of changing the resonant frequency of a cavity by varying the capacitance
(*b*) Method of changing the resonant frequency of a cavity by varying the inductance

Inductive tuning is accomplished by placing a nonmagnetic slug in the area of maximum H lines, as shown in Fig. 4.7(b). The changing H lines induce a current in the slug that sets up an opposing H field. The opposing field reduces the total H field in the cavity, and therefore reduces the total inductance. As you reduce the inductance by moving the slug in, raises the resonant frequency and increasing the inductance by moving the slug out, lower down the resonant frequency.

4.8. APPLICATION OF CAVITY RESONANTORS

Resonant cavities are widely used in the microwave range. For example, most microwave tubes and transmitting devices use cavities in some form to generate microwave energy. Cavities are also used to determine the frequency (as a wave meter) of the energy traveling in a waveguide, since conventional measurement devices do not work well at microwave frequencies.

MICROWAVE CAVITY RESONATORS

4.9. FIELD EXPRESSION FOR RECTANGULAR CAVITY RESONATOR

4.9.1 Field expressions for TE_{mnp} and TM_{mnp} Modes in a Rectangular Cavity Resonation

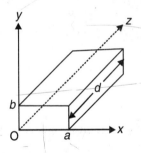

Fig. 4.8. Rectangular cavity resonator.

We know that for TM waves
$$H_z = 0 \text{ and } E_z \neq 0$$
And the wave equation is given as

$$\frac{\partial^2 E_z}{\partial x^2} + \frac{\partial^2 E_z}{\partial y^2} + \gamma^2 E_z = -\omega^2 \mu E_z \qquad \ldots 4.1$$

Let $\gamma^2 + \omega^2 \mu \varepsilon = h^2$

The above equation (1) may rewritten as

$$\frac{\partial^2 E_z}{\partial x^2} + \frac{\partial^2 E_z}{\partial y^2} \gamma^2 E_z = 0 \qquad \ldots 4.2$$

Solve this equation (2) by using separation of variables method.
We get

$$E_z = (C_1 \cos B_x + C_2 \sin B_x)(C_3 \cos A_y + C_4 \sin A_y) \qquad \ldots 4.3$$

Apply the boundary conditions to get the values of constants, C_1, C_2, C_3 and C_4.
We get the following equation for TM_{mnp} waves

$$E_z = C \sin\left(\frac{m\pi}{a}\right) x \sin\left(\frac{n\pi}{b}\right) y \cos\left(\frac{p\pi}{d}\right) . ze^{j\omega t - \gamma z}$$

or $\boxed{E_z(TM_{mnp}) = C \sin\left[\frac{m\pi}{a}\right] x \sin\left[\frac{n\pi}{b}\right] y \cos\left[\frac{p\pi}{d}\right] \cdot ze^{j\omega t - \gamma z}}$ ...4.4

where m = 0, 1, 2, 3, ...
It represents the number of half wave variation in the x-direction.
  n = 0, 1, 2, 3, ...
It represents the number of half wave variation in the y-direction.
  p = 0, 1, 2, 3, ...
It represents the number of half wave variation in the z-direction.

(ii) For TE waves: It is known from Maxwell's equation that

$$\nabla^2 H_z = -\omega^2 \mu \varepsilon H_z$$

For TE wave to propagate, $E_z = 0$ and $H_z \neq 0$

$$\frac{\partial^2 H_z}{\partial x^2} + \frac{\partial^2 H_z}{\partial y^2} + \frac{\partial^2 H_z}{\partial z^2} = -\omega^2 \mu\varepsilon H_z \qquad \ldots 4.5$$

$$\frac{\partial^2}{\partial z^2} = \gamma^2 \qquad \text{(It is an operator)}$$

But

$$\frac{\partial^2 H_z}{\partial x^2} + \frac{\partial^2 H_z}{\partial y^2} + \gamma^2 H_z + \omega^2 \mu\varepsilon H_z = 0$$

or

$$\frac{\partial^2 H_z}{\partial x^2} + \frac{\partial^2 H_z}{\partial y^2} + H_z(\gamma^2 + \omega^2 \mu\varepsilon) = 0 \qquad \ldots 4.6$$

If we put, $\gamma^2 + \omega^2 \mu\varepsilon = h^2$

$$\frac{\partial^2 H_z}{\partial x^2} + \frac{\partial^2 H_z}{\partial y^2} + h^2 H_z = 0 \qquad \ldots 4.7$$

Equation (7) is a partial differentiation of 2nd order can be solve by using separation method and apply, the boundary conditions,

$$H_z = C\cos\left[\frac{m\pi}{a}\right] \cdot x \cos\left[\frac{n\pi}{b}\right] y \cdot \sin\left[\frac{p\pi}{d}\right] z e^{(j\omega t - \gamma z)}$$

For TE_{mnp} mode,

$$\boxed{H_z = C\cos\left[\frac{m\pi}{a}\right] \cdot x \cos\left[\frac{n\pi}{b}\right] y \cdot \sin\left[\frac{p\pi}{d}\right] z e^{(j\omega t - \gamma z)}} \qquad \ldots 4.8$$

4.9.2 RESONANT FREQUENCY (f_0) IN A RECTANGULAR CAVITY RESONATOR

For rectangular waveguide

$$h^2 = \gamma^2 + \omega^2 \mu\varepsilon$$

$$h^2 = A^2 + B^2 = \left[\frac{m\pi}{a}\right]^2 + \left[\frac{n\pi}{b}\right]^2$$

and

$$\gamma^2 + \omega^2 \mu\varepsilon = \left[\frac{m\pi}{a}\right]^2 + \left[\frac{n\pi}{b}\right]^2$$

or

$$\gamma^2 + \omega^2 \mu\varepsilon = \left[\frac{m\pi}{a}\right]^2 + \left[\frac{n\pi}{b}\right]^2$$

$$\omega^2 \mu\varepsilon = \left[\frac{m\pi}{a}\right]^2 + \left[\frac{n\pi}{b}\right]^2 - \gamma^2$$

$$\gamma = j\beta \qquad \text{(If wave propagation exists)}$$

or

$$\gamma^2 = j^2 \beta^2$$

$$\gamma^2 = -\beta^2$$

MICROWAVE CAVITY RESONATORS

$$\therefore \quad \omega^2 \mu\varepsilon = \left[\frac{m\pi}{a}\right]^2 + \left[\frac{n\pi}{b}\right]^2 + \beta^2 \qquad \ldots 4.9$$

In cavity resonator, if wave propagation is existing than there must be a phase change corresponding to a given guide wave length that is $\beta = 2\pi/\lambda_g$.

$$\beta = p\pi/d \quad \text{(Condition for the resonator to resonate)} \qquad \ldots 4.10$$

where d = length of the cavity resonator

p = indicates the number of half wave variation of electric or magnetic field along be z-direction. (P = 1, 2, 3,..., ∞).

Now $\quad \beta = \dfrac{P\pi}{d}$

than $\quad f = f_0$ and $\omega = 2\pi f_0 = \omega_0$

By substituting the value of β in equation (9), we get

$$\omega^2 \mu\varepsilon = \left[\frac{m\pi}{a}\right]^2 + \left[\frac{n\pi}{b}\right]^2 + \left[\frac{p\pi}{d}\right]^2$$

or $\quad (2\pi f_0)^2 \mu\varepsilon = \left[\dfrac{m\pi}{a}\right]^2 + \left[\dfrac{n\pi}{b}\right]^2 + \left[\dfrac{p\pi}{d}\right]^2$

or $\quad f_0 = \dfrac{1}{2\pi\sqrt{\mu\varepsilon}} \left\{ \left[\dfrac{m\pi}{a}\right]^2 + \left[\dfrac{n\pi}{b}\right]^2 + \left[\dfrac{p\pi}{d}\right]^2 \right\}^{1/2}$

$$\boxed{f_0 = \dfrac{C}{2}\left[\left(\dfrac{m}{a}\right)^2 + \left(\dfrac{n}{b}\right)^2 + \left(\dfrac{p}{d}\right)^2\right]^{1/2}} \qquad \ldots 4.11$$

The resonant frequency (f_0) is the same for the TE_{mnp} and TM_{mnp} mode in a rectangular cavity resonator.

4.9.3 FIELD EXPRESSION FOR TE_{nmp} AND TM_{nmp} MODES IN CIRCULAR RESONATOR

(1) TE wave in circular resonator : Fig. (4.9) shows the cylindrical cavity is a section of circular wave guide of length 'd' and radius 'a'.

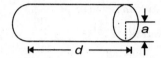

Fig. 4.9. Circular waveguide.

We know that

$$H_z = A' J_n(\rho h) \cos(n\phi') \, e^{j\omega t - \gamma z}$$

where $\quad A' = \sqrt{(A'_n)^2 + (B'_n)^2}$

and $\quad n\phi' = n\phi - \tan^{-1} \dfrac{B'_n}{A'_n}$

The wave traveling in positive 'z' direction given by

$$H_z = A' J_n(\rho h) \cos(n\phi') e^{j(\omega t - \beta z)} \qquad \ldots 4.12$$

And the wave traveling in the negative 'z' direction

$$H_z = A' J_n(\rho h) \cos(n\phi') e^{j(\omega t + \beta z)} \qquad \ldots 4.13$$

Assume A^+ and A^- are the amplitude components of the wave in positive 'a' and negative 'z' direction

$$\left[A^+ e^{-\beta z} + A^- e^{\beta z} \right] \cos(n\phi') = H_z \qquad \ldots 4.14$$

When we take $A^- = A^+$, than $E\phi$ and $E\rho$ vanish at $z = 0$

$$A^+ e^{-\beta d} + A^- e^{\beta d} = -2jA^+ \sin \beta d$$

To make $\sin \beta z$ vanish,

$$\beta d = p\pi. (p = 1, 2, 3, \ldots)$$

$$\beta = \dfrac{p\pi}{d}$$

Equation (13) may written as

$$H_z = -2jA J_n(\rho h) \cos(n\phi') \sin\left[\dfrac{p\pi}{d}\right] z e^{j(\omega t - \gamma z)}$$

or $\quad H_z = C J_n(\rho h) \cos(n\phi') \left[\sin \dfrac{p\pi}{d}\right] z e^{j\omega t - j\beta z} \quad (C = -2jA)$

or $\quad \boxed{H_z = C J_n(\rho h) \cos(n\phi') [A^+ e^{-j\beta z} + A^- e^{+j\beta z}]} \qquad \ldots 4.15$

(2) **TM wave**: Similarily for TM wave, we can solve the field equation as given below

$$E_z = C J'_n(\rho h)[A'_n \cos n\phi + B'_n \sin n\phi] \qquad \ldots 4.16$$

where $\quad A' = \sqrt{(A'_n)^2 + (B'_n)^2}$

$$n\phi' = n\phi - \tan^{-1} \dfrac{B'_n}{A'_n}$$

Therefore,

$$E_z = A' J'_n(\rho h) \cos(n\phi') \qquad \ldots 4.17$$

The wave is propagating in +ve direction 'z',

$$E_z = A' J'_n(\rho h) \cos(n\phi') e^{j(\omega t - \beta z)} \qquad \ldots 4.18$$

If wave is propagating in –ve direction 'z' and, $\gamma = j\beta$.

than $\quad E_z = A J'_n(\rho h) \cos(n\phi') e^{j(\omega t + \beta z)}$

or $\quad E_z = J'_n(\rho h) [A^+ e^{-\beta z} + A^- e^{\beta z}] \cos(n\phi') \qquad \ldots 4.19$

To have $E_z = 0$ at $z = 0$ and $z = d$,

Choose $\quad A^+ = -A^-$ or $A^- = -A^+$

Put the values in eq. (19).

$$0 = J'_n(\rho h) A^+ [e^{-\beta z} - e^{\beta z}] \cos(n\phi')$$

But $e^{-\beta z} - e^{\beta z} = -2j \sin \beta_z$

At $z = 0$, $E_z = 0$ and $z = d$

$$0 = J'_n(\rho h) A^+ (-2j \sin \beta d) \cos(n\phi')$$

$A^+ \neq 0$ $\cos(n\phi') = 0$

$\sin \beta_d = 0$ or $\beta_d = p\pi$

$$\beta = \frac{p\pi}{d} \qquad \text{(where } p = 1, 2, 3, \ldots)$$

$$E_z = J'_n(\rho h) \cos(n\phi') \sin\left[\frac{p\pi}{d}\right] ze^{j(\omega t - \beta_z)} \quad \text{(Along +ve 'z' direction)} \ldots 4.20$$

$$E_z = J'_n(\rho h) \cos(n\phi') \sin\left[\frac{p\pi}{d}\right] ze^{j(\omega t + \beta_z)} \quad \text{(Along the –ve 'z' direction)} \ldots 4.21$$

4.9.4 Expression for Resonant Frequency (f_0) in Circular Cavity Resonator

As shown in fig. (3.10), a = radius of the waveguide, d = length of cylindrical waveguide.

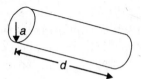

Fig. 4.10. Circular cavity resonator.

Now for resonance $\beta = \dfrac{p\pi}{d}$

We know that

$$h^2 = \gamma^2 + \omega^2 \mu\varepsilon = \lambda_{nm}^2$$

$$= \left[\frac{P_{nm}}{a}\right]^2$$

or $\quad \omega^2 \mu\varepsilon = \left[\dfrac{P_{nm}}{a}\right]^2 - \gamma^2 \qquad \ldots 4.22$

$$\omega_0^2 \mu\varepsilon = \left[\frac{P_{nm}}{a}\right]^2 + \left[\frac{p\pi}{d}\right]^2 \qquad (\gamma = j\beta \text{ and } \omega = \omega_0)$$

$$f_0 = \frac{1}{2\pi\sqrt{\mu\varepsilon}} \left[\left(\frac{P_{nm}}{a}\right)^2 + \left(\frac{p\pi}{d}\right)^2\right]^{1/2}$$

or

$$\boxed{f_0 = \frac{C}{2\pi} \left[\left(\frac{P_{nm}}{a}\right)^2 + \left(\frac{p\pi}{d}\right)^2\right]^{1/2} \quad [\text{For TM}_{nmp} \text{ mode}]} \qquad \ldots 4.23$$

$$\boxed{f_0 = \frac{C}{2\pi} \left[\left(\frac{P'_{nm}}{a}\right)^2 + \left(\frac{p\pi}{d}\right)^2\right]^{1/2} \quad [\text{For TE}_{nmp} \text{ mode}]} \qquad \ldots 4.24$$

4.10. QUALITY FACTOR (Q) OF CAVITY RESONATORS

The quality factor 'Q' is a measure of selectivity of frequency of a resonant or antiresonant circuit and it is defined by the following equation

$$Q = \frac{\omega_0 W}{p} = 2\pi \frac{\text{Maximum energy stored in Tank circuit}}{\text{Energy dissipated per cycle}} \qquad ...4.25$$

where ω_0 = Resonant frequency
W = Maximum energy stored
p = Average power loss
Q = Quality factor

From the abovementioned relation, it is observed that higher the Q factor of a resonant circuit, the higher the amount of energy stored in its tank circuit storage elements (such as inductors and capacitors) and the smaller the amount of energy dissipating in it, the dissipating elements are resistors. It means that high value of Q must have the product LC to be much higher than R. The term 2π has been just added to get a convenient expression at lower frequencies. As it is known that Q of an LCR circuit is given by

$$Q = 2\pi \times (1/2) LI^2_{max} / I^2 RT \qquad ...4.26$$

Where, $(1/2) LI^2_{max}$ the maximum amount of energy stored in the inductor and $I^2 RT$ is the energy dissipated in the resistors of the tank circuit. Since $I = I_{max}/\sqrt{2}$, we get

$$Q = 2\pi \times (1/2) LI^2_{max} / (1/2) I^2_{max} RT$$
$$= 2\pi \times L/RT$$
$$= 2\pi \, f \times L/R$$
$$\boxed{Q = \omega L/R,} \qquad ...4.27$$

Where $f = 1/T$ is the resonant frequency and $\omega = 2\pi f$ is angular resonant frequency of the tank circuit.

At resonant frequency, the magnetic and electric energies are equal in time quadrative; when magnetic energy is maximum, the electric energy is zero and vice-versa. The total energy is stored in the cavity resonator is determined by integrating the energy density over the volume of the resonator,

$$W_e = \int_v \frac{\varepsilon}{2} / E /^2 \, dv = W_m = \int_v \frac{\mu}{2} / H /^2 \, dv = W \qquad ...4.28$$

Here, W = Maximum stored energy; $|E|$ and $|H|$ are peak values of the field intensities.

The average power loss can be evaluated by integrating the power density over the inner surface of the resonator. Thus

$$P = \frac{R_s}{2} \int_s / H_t /^2 \, da \qquad ...4.29$$

where R_s = surface resistance of the resonator and H_t = peak value of the tangential magnetic intensity.

By substituting the value of W and P from the above equation (28) and (29) respectively in eq. (25),

$$Q = \frac{\omega \mu \int_v / H_t /^2 \, dv}{R_s \int_s / H_t /^2 \, da} \qquad ...4.29$$

As the peak value of the magnetic intensity is to its tangential and normal components by

$$/H^2/ = /H_t/^2 + /H_n/^2 \qquad \ldots 4.30$$

H_n = Peak value of the normal magnetic intensity, the value of $/H_l/^2$ at the resonator width is approximately twice the value of $/H/^2$ averaged over the volume, so the Q of a cavity resonator can be expressed

$$Q = \frac{\omega_\mu (\text{volume})}{2R_s (\text{surface areas})} \qquad \ldots 4.31$$

An unloaded resonator can be represented by either a series or a parallel resonant circuit. The resonant frequency (f_0) and the unloaded (Q_0) of a cavity resonator are given as

$$f_0 = \frac{1}{2\pi\sqrt{LC}} \qquad \ldots 4.32$$

$$Q_0 = \frac{\omega_0 L}{R} \qquad \ldots 4.33$$

If cavity is coupled by ideal $N : 1$ transformer and a series inductance L_s to a generator having internal impedance Z_g, than the coupling circuit and its equivalent are shown in fig. (4.11).

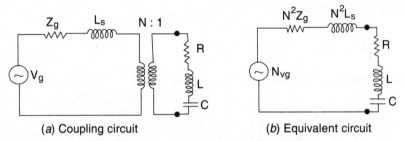

(a) Coupling circuit (b) Equivalent circuit

Fig. 4.11. A cavity coupled to a generator.

The loaded Q_l of the system may be expressed as

$$Q_l = \frac{\omega_0 L}{R + N^2 Z_g} \qquad (\text{for} /N^2 L_s/ << /R + N^2 zg/) \qquad \ldots 4.34$$

The coupling coefficient of the system may be defined as

$$K = \frac{N^2 Z_g}{R} \qquad \ldots 4.35$$

The loaded Q_l would become,

$$Q_l = \frac{\omega_0 L}{R(1+k)} = \frac{Q_0}{1+k} \qquad \ldots 4.36$$

or

$$\frac{1}{Q_l} = \frac{1}{Q_0} + \frac{1}{Q_{ext.}} \qquad \ldots 4.37$$

$Q_{ext.} = Q_0 / k = \omega_0 L / (kR)$ is the external Q.

Types of coupling coefficients:
(i) Critical coupling ($k = 1$) (ii) Over coupling ($k > 1$)
(iii) Under coupling ($k < 1$)

(i) Critical coupling: If the resonator is matched to the generator, i.e., $k = 1$.

The loaded $Q_l = \dfrac{1}{2} Q_{ext.} = \dfrac{1}{2} Q_0$ ...4.38

(ii) Over coupling: If $k > 1$, the cavity terminals are at a voltage maximum in the input line at resonance. The normalized impedance at the voltage maximum is the standing wave ratio ρ.

$$k = \rho$$

The loaded Q_l is given by

$$Q_l = \dfrac{Q_0}{1+\rho}$$ ...4.39

(iii) Under coupling: If $k < 1$, the cavity terminals are at a voltage minimum and the input terminal impedance is equal to the reciprocal of the standing wave ratio.

It is given by

$$k = \dfrac{1}{\rho}$$

The Q_l is expressed by

$$Q_l = \dfrac{\rho}{\rho+1} Q_0$$ ...4.40

Fig. (4.12) shows the relationship of the coupling coefficient k and the standing wave ratio.

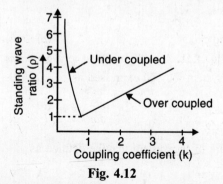

Fig. 4.12

4.11. TRANSMISSION LINE RESONATOR

It was discussed in earlier section that short sections of transmission lines shorted at one end may be employed to provide capacitive and inductive reactance's by suitable selection of their lengths. The capacitive and inductive reactance's may be combined together to obtain a distributed circuit equivalent of a parallel resonant LC circuit.

MICROWAVE CAVITY RESONATORS

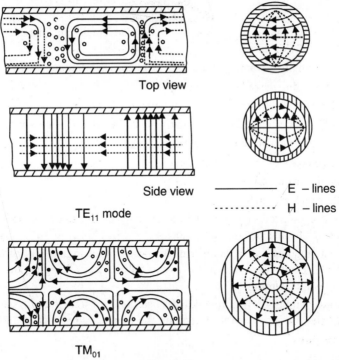

Fig. 4.13

Let us consider a shorted transmission line of length L_1, as shown in Fig. 4.13. When L_1 is less than $\lambda/4$, the input susceptance is inductive and is given by

$$B_L = Y_0 \cot \beta L_1 \qquad ...4.41$$

Where, B_L = input inductive susceptance.

For resonance, a capacitive susceptance, $B_c (= -B_L)$, is required. This may be obtained by another shorted line length L_2, such that

$$B_c = -Y_0 \cot \beta L_1$$
$$B_c = Y_0 \cot (\pi - \beta L_1)$$
$$B_c = Y_0 \cot \beta L_2) \qquad ...4.42$$

which gives,

$$\beta L_2 = \pi - \beta L_1;$$
or $\quad \beta (L_1 + L_2) = \pi$
or
$$(L_1 + L_2) = \lambda / 2 \qquad ...4.43$$

The equation (43) shows that if the length of a transmission line in $\lambda/2$ and short circuits are placed at both the ends, the resulting structure will behave like a resonant circuit at frequency where its length is $\lambda/2$. Similar to any LC resonators, the transmission line resonator is capable of storing electromagnetic energy. In a lumped element LC resonator, the energy is stored in the magnetic field of the inductor and the electric field of the capacitor alternately. In a transmission line resonator, the electric and magnetic fields are distributed all over the resonator and the energy storage is not localized.

Resonators of the other types than the transmission line resonators are also designed.

SUMMARY

- **A resonant cavity** is any completely enclosed by conducting walls that can contain oscillating electromagnetic fields and possess resonant properties.
- The cavity has many advantages and uses at microwave frequencies.
  1. Resonant cavities have a very high Q
  2. It can be built to handle relatively large amounts of power.
  3. The high Q gives these devices a narrow band pass and allows vety accurate tuning.
  4. Simple, rugged construction is an additional advantage.
- There are two types of cavity resonators.
  1. Rectangular cavity resonators
  2. Circular cavity resonators
- Energy can be inserted or removed from a cavity by the same methods that are used to couple energy into and out of waveguides.
- The resonant frequency of a cavity can be varied by changing any of three parameters: cavity volume, cavity capacitance, or cavity inductance. Changing the frequencies of a cavity is known as **Tuning**.
- Most microwave tubes transmitting devices use cavity in some form to generate microwave energy. Cavities are also used to determine the frequency (as a wave meter).
- **Field Expressions**
  (*i*) For TM waves

$$E_z(TM_{mnp}) = C \sin\left[\frac{m\pi}{a}\right] x \sin\left[\frac{n\pi}{b}\right] y \cos\left[\frac{p\pi}{d}\right] z e^{j\omega t - \gamma z}$$

  (*ii*) For TE waves

$$H_z = C \cos\left[\frac{m\pi}{a}\right] x \cdot \cos\left[\frac{n\pi}{b}\right] \cdot y \sin\left[\frac{p\pi}{d}\right] z e^{j(\omega t - \gamma z)}$$

$$f_0 = \frac{C}{2}\left[\left(\frac{m}{a}\right)^2 + \left(\frac{n}{b}\right)^2 + \left(\frac{p}{d}\right)^2\right]^{1/2}$$

- The resonant frequency (f_0) is the same for the TE_{mnp} and TM_{mnp} mode in a rectangular cavity resonator.
- **Field Expression**
  (1) TE wave in circular resonator

$$H_z = CJ_n(\rho h) \cos(n\phi') [A^+ e^{-j\beta_z} + A^- e^{+j\beta_z}]$$

  (2) TM wave

$$E_z = J'_n(\rho h) \cos(n\phi') \sin\left[\frac{p\pi}{d}\right] z\, e^{j(\omega t - \beta_z)} \quad \text{(Along +ve 'z' direction)}$$

$$E_z = J'_n(\rho h) \cos(n\phi') \sin\left[\frac{p\pi}{d}\right] z\, e^{j(\omega t + \beta_z)} \quad \text{(Along the -ve 'z' direction)}$$

- **Expression for Resonant Frequency (f_0) Circular Cavity Resonator**

$$f_0 = \frac{C}{2\pi}\left[\left(\frac{P_{mn}}{a}\right)^2 + \left(\frac{p\pi}{d}\right)^2\right]^{1/2} \quad \text{[For } TM_{nmp} \text{ mode]}$$

$$f_0 = \frac{C}{2\pi}\left[\left(\frac{P'_{mn}}{a}\right)^2 + \left(\frac{p\pi}{d}\right)^2\right]^{1/2} \qquad \text{[For TE}_{nmp}\text{ mode]}$$

- The quality factor 'Q' is a measure of selectivity of a frequency of a resonant or antiresonant circuit and it is defined by the following equation

$$Q = \frac{\omega_0 W}{P} = 2\pi \frac{\text{maximum energy stored}}{\text{energy dissipated per cycle}}$$

- Types of coupling coefficients:
  (i) Critical coupling ($k=1$)   (ii) Over coupling ($k>1$)
  (iii) Under coupling ($k<1$)

REVIEW QUESTIONS

1. What do you understand by cavity resonator?
2. Derive the equations for resonant frequencies for a rectangular cavity resonator.
3. Derive the field expressions for a rectangular cavity resonator.
4. Derive the field expressions for a circular cavity resonator.
5. Discuss the quality factor of a cavity resonator and explain the term unloaded Q, loaded Q, critical coupled Q, under Q and over coupled Q related to a cavity resonator.
6. Discuss the various ways of coupling energy to a resonator.
7. Define the excitation of cavities.
8. A circular waveguide with the radius of 3 cms and is used as a cavity resonator for TM_{011} mode at 10 GHz by inserting two perfectly conducting plates at its two end-plates.
9. A circular cavity resonator of dimensions $a=3$ cms, $b=2$ cms and length is 4 cms, calculate the resonant frequency if it is operating with the mode of TE_{101}.
10. Obtain the resonant frequency of a circulator resonator with the dimensions of diameter $=12.5$ cms and length is 5 cms for TM_{012} mode.
11. Show that a rectangular cavity may be viewed as a rectangular waveguide shorted at both ends. Also find the resonance condition.
12. What is a cavity resonator?
13. What is a re-entrant cavity?
14. What do you understand by 'Q' of a cavity resonator?
15. Find expressions for 'Q' of a rectangular cavity resonator.
16. Describe various ways of coupling energy to a resonator.

CHAPTER 5

MICROWAVE COMPONENTS AND DEVICES

OBJECTIVES
- Introduction
- Scattering Parameters
- Waveguide Plumbing
- Microwave T-junctions (E-plane, H-plane, Hybrid Junction, Magic –T)
- Hybrid Ring
- Waveguide Impedance Matching
- Waveguide Terminations
- Waveguide Bends
- Waveguide Twist
- Flexible Waveguide
- Waveguide Joints (Rotating and Permanent joints)
- Directional Coupler
- Hybrid Coupler
- Waveguide and Microwave Attenuators
- Ferrites devices (Ferrite Attenuator, Circulators, Isolator, Gyrator)
- Waveguide Transition
- Microwave Filters

5.1. INTRODUCTION

Low frequency transmission of electrical signals send through wires is essentially a one-dimensional operation in time. Wires may be connected either by soldering or some other similar techniques. But when we have to deal with microwave frequencies, the picture is something different. In microwave region, we know that electrical signals get propagated in the form of EM waves, which spread both in space and time domain. Hence, they are essentially four dimensional (x,y,z and t) entities. Ordinary (one-dimensional) wires are not suitable for the propagation of the microwave frequencies. Instead, we have to use three-dimensional transmission lines such as waveguides for this purpose.

MICROWAVE COMPONENTS AND DEVICES

Transmission of microwave energy or RF energy may be simulated to the flow of water through pipes. Waveguides are essentially pipes. As that pipes are available as short length pieces of approximately 5 meter each. To construct a 50-m long pipeline, 10 such individual pipes are to be interconnected. The individual pipes pieces are joined by means of coupling, flanges, etc. We use tees, to tap the water and to change the direction of water, we use bends, corners etc. Similarly, we use various microwave components to control the flow of RF energy and power, such as waveguides Terminations, Waveguide Bends, Waveguide Twist, Flexible Waveguide, Waveguide Joints, waveguide junctions, directional couplers, ferrite devices etc.

Microwave systems generally consist of various microwave components connected to each other by waveguide or some transmission lines. All the microwave components must have lower insertion loss, low standing wave ratio and low attenuation to achieve the required transmissions of microwave signal. In this chapter we will be discussing various microwave components and devices used in RF plumbing such as Waveguide Terminations, Waveguide Bends, Waveguide Twist, Flexible Waveguide, Waveguide Joints, waveguide junctions, directional couplers, ferrite devices etc.

5.2. SCATTERING PARAMETERS

In a microwave circuit, the incoming wave is "scattered" by the circuit and its energy is partitioned between all the possible outgoing waves on all the other transmission lines connected to the circuit. The scattering parameters are fixed properties of the (linear) circuit which describe how the energy couples between each pair of ports or transmission lines connected to the circuit.

S-parameters can be defined for any collection of linear electronic components, whether or not the wave view of the power flow in the circuit is necessary. They are algebraically related to the impedance parameters (z-parameters), also to the admittance parameters (y-parameters) and to a notional characteristic impedance of the transmission lines.

An n-port microwave network has n arms into which power can be fed and from which power can be taken. In general, power can get from any arm (as input) to any other arm (as output). There are thus n incoming waves and n outgoing waves. We also observe that power can be reflected by a port, so the input power to a single port can partition between all the ports of the network to form outgoing waves.

Associated with each port is the notion of a "reference plane" at which the wave amplitude and phase is defined. Usually the reference plane associated with a certain port is at the same place with respect to incoming and outgoing waves.

The n incoming wave complex amplitudes are usually designated by the n complex quantities a_n, and the n outgoing wave complex quantities are designated by the n complex quantities b_n. The incoming wave quantities are assembled into an n-vector A and the outgoing wave quantities into an n-vector B. The outgoing waves are expressed in terms of the incoming waves by the matrix equation B = SA, where S is an n by n square matrix of complex numbers called the **"scattering matrix"**. It completely determines the behavior of the network. In general, the elements of this matrix, which are termed **"scattering-parameters"** or "scattering coefficient", are all frequency-dependent.

For example, the matrix equations for a 2-port are

$$b_1 = s_{11}a_1 + s_{12}a_2$$
$$b_2 = s_{21}a_1 + s_{22}a_2$$

And the matrix equations for a 3-port are

$$b_1 = s_{11}a_1 + s_{12}a_2 + s_{13}a_3$$
$$b_2 = s_{21}a_1 + s_{22}a_2 + a_{23}a_3$$
$$b_3 = s_{31}a_1 + s_{32}a_2 + a_{33}a_3$$

The wave amplitudes a_n and b_n are obtained from the port current and voltages by the relations

$$a = (V + ZoI)/(2\,(2Zo)^{1/2}) \text{ and}$$
$$b = (V - ZoI)/(2\,(2Zo)^{1/2}).$$

Here, "a" refers to "a_n" if "V" is Vn and "I" "I_n" for the nth port.

5.2.1 TWO PORT SCATTERING MATRIX

In the case of a microwave network having two ports only, an input and an output, the s-matrix has four s-parameters as shown in Fig. (5.1) & (5.2), designated

$$\begin{array}{cc} s_{11} & s_{12} \\ s_{12} & s_{22} \end{array}$$

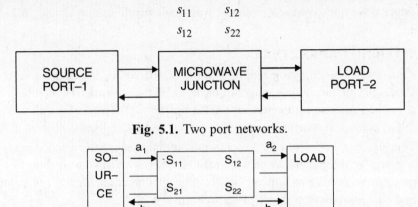

Fig. 5.1. Two port networks.

Fig. 5.2.

These four complex quantities actually contain eight separate numbers; the real and imaginary parts, or the modulus and the phase angle, of each of the four complex scattering parameters.

Let us consider the physical meaning of these s-parameters. If the output port 2 is terminated, that is, the transmission line is connected to a matched load impedance giving rise to no reflections, and then there is no input wave on port 2. The input wave on port 1 (a_1) gives rise to a reflected wave at port 1 ($s_{11}\,a_1$) and a transmitted wave at port 2 which is absorbed in the termination on 2. The transmitted wave size is ($s_{21}\,a_1$).

If the network has no loss and no gain, the output power must equal the input power and so in this case $|s_{11}|^2 + |s_{21}|^2$ must equal unity. We see therefore that the sizes of S_{11} and S_{21} determine how the input power splits between the possible output paths.

In general, the s-parameters tell us how much power "comes back" or "comes out" when we "throw power at" a network. They also contain phase shift information.

5.2.2 PROPERTIES OF S-MATRIX

1. The **symmetry of s-matrix:** A reciprocal s-matrix has symmetry about the leading diagonal. Many networks are reciprocal. In the case of a 2-port network, that means that $s_{21} = s_{12}$ and interchanging the input and output ports does not change the transmission properties. A transmission line section is an example of a reciprocal 2-port. A dual directional coupler is an example of a reciprocal 4-port. In general for a reciprocal n-port

MICROWAVE COMPONENTS AND DEVICES

2. S-Matrix is always a **square matrix** $(n \times n)$
3. S-Matrix is always a **unitary matrix** [S]. [S*] = [I]
   [I] = Identity matrix or unit matrix
   [S*] = complex conjugates of [S]
4. The sum of the products of any column or row multiplied by the complex conjugates of any other column or row is always zero.

5.3. WAVEGUIDE PLUMBING

Since waveguides are really only hollow metal pipes, the installation and the physical handling of waveguides have many similarities to ordinary plumbing. In light of this fact, the bending, twisting, joining, and installation of waveguides is commonly called **waveguide plumbing**. Naturally, waveguides are different in design from pipes that are designed to carry liquids or other substances. The design of a waveguide is determined by the frequency and power level of the electromagnetic energy it will carry.

5.4. MICROWAVE JUNCTIONS OR WAVEGUIDE JUNCTIONS

Microwave Tees are similar to the tees used in hydraulic systems. They are formed by joining two rectangular waveguide sections to form into the shape of the English letter *T*. This is used to tap the microwave power flowing through a main line into auxiliary line.

You may have assumed that when energy traveling down a waveguide reaches a junction, it simply divides and follows the junction. This is not strictly true. Different types of junctions affect the energy in different ways. Since waveguide junctions are used extensively in most systems, you need to understand the basic operating principles of those most commonly used.

1. The E-Type
2. The H-Type.
3. Hybrid Junctions (Magic - T – Junction and Hybrid Ring)

5.4.1 E-Type T Junction.

An E-type T junction is illustrated in Fig. (5.3 & 5.4). It is made up of a longer piece of rectangular waveguide section to which a shorter piece of rectangular waveguide section is jointed, as shown. Flanges are made an integral part of the arms for making interconnections an easy task.

It is called an **E-type T junction** because the junction arm extends from the main waveguide in the same direction as the E field in the waveguide. Ports 1 and port-2 are collinear arms and port-3 is the E- arm as shown in Fig. (5.3).

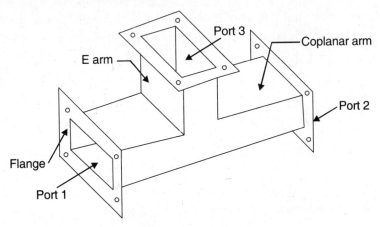

Fig. 5.3. E-type T Junction.

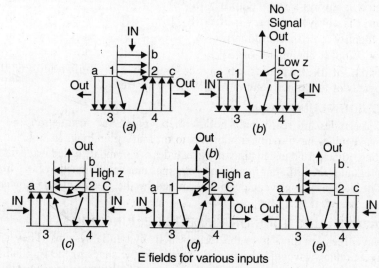

E fields for various inputs

Fig. 5.4. E fields in an E-type T junction.

Fig. (5.4), illustrates cross-sectional views of the E-type T junction with inputs fed into the various arms. For simplicity, the magnetic lines that are always present with an electric field have been omitted.

In Fig. 5.4 (a), the input is fed into arm b and the outputs are taken from the arms a and c arms. When the E field arrives between points 1 and 2, point 1 becomes positive and point 2 becomes negative. The positive charge at point 1 then induces a negative charge on the wall at point 3. The negative charge at point 2 induces a positive charge at point 4. These charges cause the fields to form 180 degrees out of phase in the main waveguide; therefore, the outputs will be 180 degrees out of phase with each other.

In Fig. 5.4 (b), two in-phase inputs of equal amplitude are fed into the arms a and c. The signals at points 1 and 2 have the same phase and amplitude. No difference of potential exists across the entrance to the b arm, and no energy will be coupled out.

However, when the two signals fed into the a and c arms are 180 degrees out of phase, as shown in Fig. 5.4 (c), points 1 and 2 have a difference of potential. This difference of potential induces an E field from point 1 to point 2 in the b arm, and energy is coupled out of this arm.

Views 5.4 (d) and (e) illustrate two methods of obtaining two outputs with only one input.

Scattering Matrix of *E*-Type *T*-Junction:

E-type *T*-junction is the three port device, the scattering matrix will be the order of 3 × 3. (Refer Fig. 5.3)

$$[S] = \begin{bmatrix} S_{11} & S_{12} & S_{13} \\ S_{21} & S_{22} & S_{23} \\ S_{31} & S_{32} & S_{33} \end{bmatrix}$$

If an input energy is fed at port (3), than outputs at ports (1) and (2) are out of phase by 180°. The scattering coefficient

$$S_{23} = -S_{13}$$

From symmetric property $\quad S_{ij} = S_{ji}$

MICROWAVE COMPONENTS AND DEVICES

$$S_{12} = S_{21}$$
$$S_{13} = S_{31}$$
$$S_{23} = S_{32}$$

By substituting the above properties in [S] matrix, we get

$$[S] = \begin{bmatrix} S_{11} & S_{12} & S_{13} \\ S_{12} & S_{22} & S_{13} \\ S_{13} & S_{13} & 0 \end{bmatrix} \qquad \ldots 5.1$$

$S_{33} = 0$, if the port (3) is perfectly matched.
Now by using unitary property, i.e., $[S].[S]^* = [I]$

$$\begin{bmatrix} S_{11} & S_{12} & S_{13} \\ S_{12} & S_{22} & S_{13} \\ S_{13} & S_{13} & 0 \end{bmatrix} \cdot \begin{bmatrix} S^*_{11} & S^*_{12} & S^*_{13} \\ S^*_{12} & S^*_{22} & S^*_{13} \\ S^*_{13} & S^*_{13} & 0 \end{bmatrix} = \begin{bmatrix} 1 & 0 & 0 \\ 0 & 1 & 0 \\ 0 & 0 & 1 \end{bmatrix}$$

Multiplying we get,

R_1C_1 (Row 1 : Column 1): $S_{11} \cdot S^*_{11} + S_{12} \cdot S^*_{12} + S_{13} \cdot S^*_{13} = 1$

$$|S_{11}|^2 + |S_{12}|^2 + |S_{13}|^2 = 1 \qquad \ldots 5.2$$

Similarity, R_2C_2: $|S_{12}|^2 + |S_{22}|^2 + |S_{13}|^2 = 1$ ... 5.3

R_3C_3: $|S_{13}|^2 + |S_{13}|^2 = 1$ ... 5.4

R_3C_1: $S_{13}S^*_{11} - S_{13}S^*_{12} = 0$ ... 5.5

By equations (2) and (3), we get

$$S_{11} = S_{22} \qquad \ldots 5.6$$

By equation (4)

$$2|S_{13}|^2 = 1$$
$$|S_{13}|^2 = 1/2 \qquad \ldots 5.7$$
$$S_{13} = \frac{1}{\sqrt{2}}$$

From equation (5)

$$S_{13}\left(S^*_{11} - S^*_{12}\right) = 0$$

Since $S_{13} \neq 0$, $S^*_{11} - S^*_{12} = 0$ ... 5.8

$$S_{11} = S_{12} = S_{22}$$

By putting the values from equation (6, 7 and 8) in the equation (2)

$$|S_{11}|^2 + |S_{11}|^2 + \frac{1}{2} = 1$$

$$2|S_{11}|^2 = \frac{1}{2} \quad \text{or} \quad S_{11} = \frac{1}{2} \qquad \ldots 5.9$$

By substituting the values from (7, 8 and 9) in equation (1)

$$[S] = \begin{bmatrix} 1/2 & 1/2 & 1/\sqrt{2} \\ 1/2 & 1/2 & -1/\sqrt{2} \\ 1/\sqrt{2} & -1/\sqrt{2} & 0 \end{bmatrix} \quad \ldots 5.10$$

Since $|b| = [S][a]$

$$\begin{bmatrix} b_1 \\ b_2 \\ b_3 \end{bmatrix} = \begin{bmatrix} 1/2 & 1/2 & 1/\sqrt{2} \\ 1/2 & 1/2 & -1/\sqrt{2} \\ 1/\sqrt{2} & -1/\sqrt{2} & 0 \end{bmatrix} \begin{bmatrix} a_1 \\ a_2 \\ a_3 \end{bmatrix} \quad \ldots 5.11$$

$$b_1 = \frac{1}{2} \times a_1 + \frac{1}{2} \times a_2 + \frac{1}{\sqrt{2}} \times a_3 \quad \ldots 5.12$$

$$b_2 = \frac{1}{2} \times a_1 + \frac{1}{2} \times a_2 + \frac{1}{\sqrt{2}} \times a_3 \quad \ldots 5.13$$

$$b_3 = \frac{1}{\sqrt{2}} \times a_1 + \frac{1}{\sqrt{2}} \times a_2 \quad \ldots 5.14$$

Let us consider the various conditions in *E*-plane *T*-junction.

Condition 1: If we feed equal inputs at port (1) and port (2), there is no output at port (3).

$a_1 = a_2 = a$ (equal inputs feed at port (1) and port (2))

$a_3 = 0$ (No input at port (3))

Substitute these values in eq. (12, 13 and 14), we get

$$b_1 = \frac{a}{2} + \frac{a}{2}; \quad b_2 = \frac{a}{2} + \frac{a}{2}$$

$$b_3 = \frac{1}{\sqrt{2}} a - \frac{1}{\sqrt{2}} a = 0$$

Condition 2 : If we feed the input at port (3) than output equally divideds between port (1) and port (2) with a phase difference of 180° between the outputs.

$a_1 = a_2 = 0$ (No inputs at port (1) and port (2))

$a_3 \neq 0$ (Input feed to port (3))

Now substitute the above values in eq. (12, 13 and 14), we get

$$b_1 = \frac{1}{\sqrt{2}} a_3, \quad b_2 = -\frac{1}{\sqrt{2}} a_3; \quad b_3 = 0$$

Hence E-plane Tee also acts as 3 db splitter.

MICROWAVE COMPONENTS AND DEVICES 133

Condition 3 : If $a_1 \neq 0$ (Input feed to port (1))

$a_2 = a_3 = 0$ (No input at port (2) and (3))

Hence substitute the above values in eq. (12, 13 and 14), we get

$$b_1 = \frac{a_1}{2}, \quad b_2 = \frac{a_1}{2}; \quad b_3 = -\frac{a_1}{\sqrt{2}}$$

5.4.2 H-TYPE T JUNCTION

An H-type T junction is illustrated in Fig. 5.5. It is called an **H-type T junction** because the long axis of the "H" arm is parallel to the plane of the magnetic lines of force in the waveguide. An H.Plane T-junction it formed by cutting a rectangular slot along the width of the main guide and attaching another waveguide, the side arm called H-arm as shown in Fig. 5.5. The port-1 and port-2 of the main waveguide are called coplaner ports.

As all the arms of H-plane T lie in plane of magnetic field, the magnetic field divides itself in to the arms and therefore this is also called a current junction.

If port-1 & 2 are terminated with idential loads then power coming through in put port-3 will be divided equally towards each load in phase. But if two signals of equal amplitude and in same phase are fed in to two side ports (1 & 2), they will be added together and combined power comes out through third arm, port-3.

Again, for simplicity, only the E lines are shown in the figure 5.6. Each X indicates an E line moving away from the observer. Each dot indicates an E line is moving toward the observer.

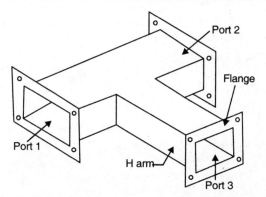

Fig. 5.5. Construction of H-type T- Junction

In view (1) of Fig. 5.6, the signal is fed into arm "b" and in-phase outputs are obtained from the "a" and "c" arms.

In view (2), in-phase signals are fed into the arms "a" and "c", the output signal is obtained from the "b" arm because the fields add at the junction and induce E lines into the "b" arm.

If 180-degree-out-of-phase signals are fed into arms "a" and "c", as shown in view (3), no output is obtained from the arm because the opposing fields cancel at the junction.

If a signal is fed into the "a" arm, as shown in view (4), output will be obtained from the "b" and "c" arms. The reverse is also true. If a signal is fed into the "c" arm, output will be obtained from the "a" and "b" arms.

Since the power divided and hence current gets divided between the main arm and the central arm, which is similar to the division of current in a parallel circuit, this tee is also called the parallel or shunt tee.

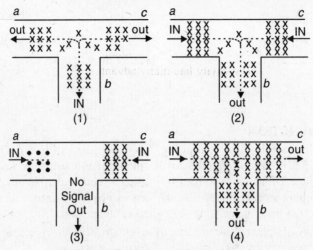

Fig. 5.6. E fields in an H-type junction.

H-type T-junction: With the help of scattering matrix, the properties of the *H-plane T-junction* can be defined. (Refer Fig. 5.5)

The order of scattering matrix is 3 × 3 because 3 possible inputs and output ports.

$$[S] = \begin{bmatrix} S_{11} & S_{12} & S_{13} \\ S_{21} & S_{22} & S_{23} \\ S_{31} & S_{32} & S_{33} \end{bmatrix} \qquad \ldots 5.15$$

From the symmetric property of 'S' matrix,

$$S_{ij} = S_{ji}$$

∴ $S_{12} = S_{21}, \; S_{13} = S_{31}, \; S_{23} = S_{32}$

$S_{13} = S_{23}$ (Because of plane of symmetry of the junction)

$S_{33} = 0$ (Port (3) is perfectly matched to the junction)

By substituting the S-parameters value in eq. (15), we get

$$[S] = \begin{bmatrix} S_{11} & S_{12} & S_{13} \\ S_{12} & S_{22} & S_{13} \\ S_{13} & S_{13} & 0 \end{bmatrix}$$

By using the unitary property

$$[S][S]^* = [I]$$

$$\begin{bmatrix} S_{11} & S_{12} & S_{13} \\ S_{12} & S_{22} & S_{13} \\ S_{13} & S_{13} & 0 \end{bmatrix} \cdot \begin{bmatrix} S^*_{11} & S^*_{12} & S^*_{13} \\ S^*_{12} & S^*_{22} & S^*_{13} \\ S^*_{13} & S^*_{13} & 0 \end{bmatrix} = \begin{bmatrix} 1 & 0 & 0 \\ 0 & 1 & 0 \\ 0 & 0 & 1 \end{bmatrix}$$

$R_1 C_1 : |S_{11}|^2 + |S_{12}|^2 + |S_{13}|^2 = 1$ ... 5.16

$R_2 C_2 : |S_{12}|^2 + |S_{22}|^2 + |S_{13}|^2 = 1$ ... 5.17

$R_3 C_1 : S_{13} S^*_{11} + S_{13} S^*_{12} = 0$ ... 5.18

$R_3 C_3 : |S_{13}|^2 + |S_{13}|^2 = 1$ ... 5.19

or $\quad 2|S_{13}|^2 = \Rightarrow |S_{13}| = \dfrac{1}{\sqrt{2}}$ ... 5.20

By using eq. (16) and (17),

or $\quad |S_{11}|^2 = |S_{22}|^2$

$\quad\quad S_{11} = S_{22}$ ... 5.21

By solving eq. (18)

$$S_{13}\left(S^*_{11} + S^*_{12}\right) = 0$$

Since $\quad S_{13} \neq 0, S^*_{11} + S^*_{12} = 0$

$\quad\quad S^*_{11} = -S^*_{12}$

or $\quad S_{11} = -S_{12}$ or $S_{11} = -S_{12}$ ... 5.22

By using these values in eq. (16), we get

$$|S_{11}|^2 + |S_{11}|^2 + \left[\dfrac{1}{\sqrt{2}}\right]^2 = 1$$

or $\quad 2|S_{11}|^2 = \dfrac{1}{2}$

or $\quad S_{11} = \dfrac{1}{2}$ ... 5.23

$S_{12} = -S_{11}$

$S_{12} = -1/2$ and $S_{22} = \dfrac{1}{2}$

By putting the values of S_{11}, S_{12}, S_{13} and S_{22}, in following

$$[S] = \begin{bmatrix} S_{11} & S_{12} & S_{13} \\ S_{12} & S_{22} & S_{13} \\ S_{13} & S_{13} & 0 \end{bmatrix}$$

$$[S] = \begin{bmatrix} 1/2 & -1/2 & 1/\sqrt{2} \\ -1/2 & 1/2 & 1/\sqrt{2} \\ 1/\sqrt{2} & 1/\sqrt{2} & 0 \end{bmatrix}$$ ... 5.24

It is known that $[b] = [S][a]$

$$\begin{bmatrix} b_1 \\ b_2 \\ b_3 \end{bmatrix} = \begin{bmatrix} 1/2 & -1/2 & 1/\sqrt{2} \\ -1/2 & 1/2 & 1/\sqrt{2} \\ 1/\sqrt{2} & 1/\sqrt{2} & 0 \end{bmatrix} \begin{bmatrix} a_1 \\ a_2 \\ a_3 \end{bmatrix}$$

$$b_1 = \dfrac{1}{2}a_1 - \dfrac{1}{2}a_2 + \dfrac{1}{\sqrt{2}}a_3$$ ... 5.25

$$b_2 = -1/2 a_1 - \dfrac{1}{2}a_2 + \dfrac{1}{\sqrt{2}}a_3$$ ... 5.26

$$b_3 = \frac{1}{\sqrt{2}} a_1 + \frac{1}{\sqrt{2}} a_2 \qquad \ldots 5.27$$

Let us consider the various conditions in H-plane T-junction

Condition 1 : If $a_3 \neq 0$ \hfill (Input given to port (3) only)

$$a_1 = a_2 = 0 \qquad \text{(No input)}$$

Substituting the values of a_1, a_2 and a_3 in eqn. (25), (26) and (27)

$$b_1 = \frac{a_3}{\sqrt{2}}; \; b_2 = \frac{a_3}{\sqrt{2}} \text{ and } b_3 = 0$$

The amount of power coming out from port-1 and port-3 are equal and half of the input at port-3.

If P_3 be the power input at port-3 then this power divides equally between port-1 and port-2 in phase, i.e, $P_1 = P_2$
But. $P_3 = P_1 + P_2$
or $\quad P_3 = 2P_1 = 2P_2$

The amount of power coming out of port-1 or port-2 in 3db down with respect to input power port-3 hence H-plane T-junction is also called as db splitter. Incase of TE_{10} mode in allowed to propagate through the Port-3, there is no change in the phase of outputs at port-1 and port-2.

The amount of power coming out of port-1 and port-2 due to output at port-3 is given as

$$P_1 = P_2 = 10 \log_{10} P_1/P_3$$
$$= 10 \log_{10} P_1/2P_1$$
$$= 10 \log_{10} (1/2)$$
$$= -10 \log_{10}^2$$
$$P_1 = P_2 = -10(0.3010), \text{ i.e. approximately } -3db.$$

Condition 2: If $a_1 = a_2 = a$ \hfill (Equal inputs fed to port (1 & 2))

$$a_3 = 0 \qquad \text{(No input at port (3))}$$

Substitute the values of a_1, a_2, and a_3, in eq. (25), (26) and (27)

$$b_1 = 0, \; b_2 = 0, \; b_3 = \frac{a}{\sqrt{2}} + \frac{a}{\sqrt{2}}$$

The output at port (3) in addition of the inputs fed from port (1) and port (2) are combined in phase.

5.4.3 HYBRID JUNCTIONS

Hybrid junctions are four port networks in which the signal incident at any one of the ports gets divided between two other ports, with the fourth ports isolated from the rest of the three ports. It is supposed that all the output ports are connected to perfect matching terminations. There are two types of hybrid junctions, Magic-T and Hybrid ring which are discussed here.

5.4.3.1 MAGIC-T HYBRID JUNCTION

A simplified version of the magic-T hybrid junction is shown in Fig. (5.7). The **magic-T** is a combination of the H-type and E-type T junctions. This is four port hybride T-junction combine the power dividing property of both E-plane and H-plane T, and has the benifit of being completely matched at all its ports. Rectangular slots are cut both side along the width and breadth of a long wage guide and side arms are attached. In this fig., ports/ arms 1, 2, and 3 forms the H-plane T and ports/arms 1, 2, and 4 forms E-plane T. Ports 1 & 2 are collinear arms;

MICROWAVE COMPONENTS AND DEVICES

port-3 is the H-arm and port-4 is the E-arm. It is necessary to have such the device because of the difficulty of obtaining a parfectly matched three-port T-junction. The magic-T hybrid junction is designed for TE_{10} mode of propagation. The most common application of this type of junction is as the mixer section for microwave radar receivers.

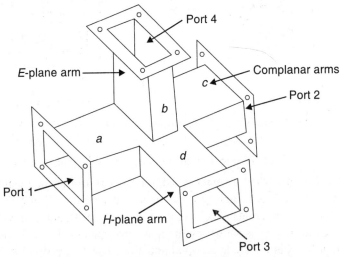

Fig. 5.7. Magic-T hybrid junction.

As with the E-type T-junction, if a signal is fed into the port (4) of the magic- T, it will divide into two out-of-phase components. As shown in Fig. (5.8), view (A), these two components will move into the (1) and (2) arms. The signal entering the (4) arm will not enter the (3) arm because of the zero potential existing at the entrance of the (3) arm. The potential must be zero at this point to satisfy the boundary conditions of the (4) arm. This absence of potential is illustrated in views (B) and (C) where the magnitude of the E field in the (4) arm is indicated by the length of the arrows. Since the E lines are at maximum in the center of the (4) arm and minimum at the edge where the (3) arm entrance is located, no potential difference exists across the mouth of the (3) arm.

In summary, when an input is applied to arm (4) of the magic-T hybrid junction, the output signals from arms (1) and (2) are 180 degrees out of phase with each other, and no output occurs at the (3) arm.

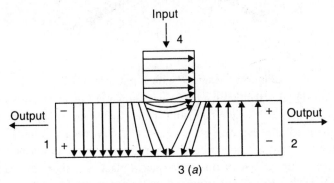

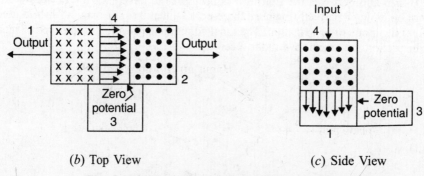

(b) Top View (c) Side View

Fig. 5.8(a) Magic-T with input to arm (4) (Front view).
  (b) Magic-T with input to arm (4) (Top view).
  (c) Magic-T with input to arm (4). (Side view).

The action that occurs when a signal is fed into the (d) arm of the magic-T is illustrated in Fig. (5.9). As with the H-type T junction, the signal entering the (d) arm divides and moves down the (a) and (c) arms as outputs which are in phase with each other and with the input. The shape of the E fields in motion is shown by the numbered curved slices. As the E field moves down the (d) arm, *points 2* and *3* are at an equal potential. The energy divides equally into arms (a) and (c), and the E fields in both arms become identical in shape. Since the potentials on both sides of the (b) arm are equal, no potential difference exists at the entrance to the (b) arm, resulting in no output.

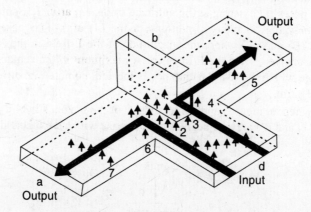

Fig. 5.9. Magic-T with input to arm (d).

When an input signal is fed into the (a) arm as shown in Fig. (5.10), a portion of the energy is coupled into the (b) arm as it would be in an E-type T junction. An equal portion of the signal is coupled through the (d) arm because of the action of the H-type junction. The (c) arm has two fields across it they are out of phase with each other. Therefore the fields cancel, resulting in no output at the (c) arm. The reverse of this action takes place if a signal is fed into the (a) arm, resulting in outputs at the (b) and (c) arms and no output at the (d) arm.

MICROWAVE COMPONENTS AND DEVICES

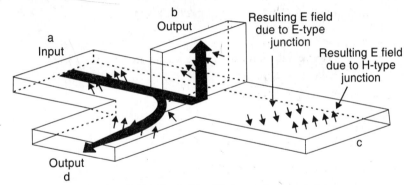

Fig. 5.10 Magic-T with input to arm (a)

E-H Type T-Junction: By using the properties of E and H type T-junction; its scattering matrix may be obtained. (Refer fig. 5.11)

- Scattering matrix of magic Tee (E-H plan) is the order of because of 4-ports.

$$[S] = \begin{bmatrix} S_{11} & S_{12} & S_{13} & S_{14} \\ S_{21} & S_{22} & S_{23} & S_{24} \\ S_{31} & S_{32} & S_{33} & S_{34} \\ S_{41} & S_{42} & S_{43} & S_{44} \end{bmatrix} \qquad \ldots 5.28$$

- $S_{24} = -S_{14}$ (E-plane Tee section)
- $S_{23} = S_{13}$ (H-plane Tee section)
- $S_{34} = S_{43} = 0$ (Because of geometry of the junction an input at port (3) cannot come out of port (4))

- $S_{ij} = S_{ji}$ (Symmetric property)

- $S_{12} = S_{21}$, $S_{13} = S_{31}$, $S_{23} = S_{32}$
- $S_{34} = S_{43}$, $S_{24} = S_{42}$, $S_{41} = S_{14}$

- $S_{33} = S_{44} = 0$ (Port (3) and port (4) are perfectly matched, to the junction)
- Substituting the values in equation (28)

$$[S] = \begin{bmatrix} S_{11} & S_{12} & S_{13} & S_{14} \\ S_{12} & S_{22} & S_{13} & -S_{14} \\ S_{13} & S_{13} & 0 & 0 \\ S_{14} & -S_{14} & 0 & 0 \end{bmatrix} \qquad \ldots 5.28.1$$

- $[S][S]^* = [I]$ (By unitary property)

$$\begin{bmatrix} S_{11} & S_{12} & S_{13} & S_{14} \\ S_{12} & S_{22} & S_{13} & -S_{14} \\ S_{13} & S_{13} & 0 & 0 \\ S_{14} & -S_{14} & 0 & 0 \end{bmatrix} \begin{bmatrix} S^*_{11} & S^*_{12} & S^*_{13} & S^*_{14} \\ S^*_{12} & S^*_{22} & S^*_{13} & -S^*_{14} \\ S^*_{13} & S^*_{13} & 0 & 0 \\ S^*_{14} & -S^*_{14} & 0 & 0 \end{bmatrix} = \begin{bmatrix} 1 & 0 & 0 & 0 \\ 0 & 1 & 0 & 0 \\ 0 & 0 & 1 & 0 \\ 0 & 0 & 0 & 1 \end{bmatrix}$$

Solving above matrix equation, we get

$$R_1 C_1 = |S_{11}|^2 + |S_{12}|^2 + |S_{13}|^2 + |S_{14}|^2 = 1 \qquad \ldots 5.29$$

$$R_2 C_2 = |S_{12}|^2 + |S_{22}|^2 + |S_{13}|^2 + |S_{14}|^2 = 1 \qquad \ldots 5.30$$

$$R_3 C_3 = |S_{13}|^2 + |S_{13}|^2 = 1 \qquad \ldots 5.31$$

$$R_4 C_4 = |S_{14}|^2 + |S_{14}|^2 = 1 \qquad \ldots 5.32$$

From the above equation (31) and (32),

$$S_{13} = \frac{1}{\sqrt{2}} \qquad \ldots 5.33$$

$$S_{14} = \frac{1}{\sqrt{2}} \qquad \ldots 5.34$$

By equation (29) and (30),

$$S_{11} = S_{22} \qquad \ldots 5.35$$

By putting all there values in eqn. (29)

$$|S_{11}|^2 + |S_{12}|^2 + \frac{1}{2} + \frac{1}{2} = 1$$

$$|S_{11}|^2 + |S_{12}|^2 = 0$$

$$S_{11} = S_{12} = 0$$

$$S_{22} = 0$$

From equation (30),

- In Magic-T, if port (3) and port (4) are perfectly matched to the junction than the port (1) and port (2) are also perfectly matched to the junction. In Magic-T all the four ports are perfectly matched to the junction.

By putting the 'S'-parameter in the eq. (28.1), we get

$$[S] = \begin{bmatrix} 0 & 0 & 1/\sqrt{2} & 1/\sqrt{2} \\ 0 & 0 & 1/\sqrt{2} & -1/\sqrt{2} \\ 1\sqrt{2} & 1/\sqrt{2} & 0 & 0 \\ 1/\sqrt{2} & -1/\sqrt{2} & 0 & 0 \end{bmatrix} \qquad \ldots 5.36$$

We know that

$$[b] = [S][a]$$

$$\begin{bmatrix} b_1 \\ b_2 \\ b_3 \\ b_4 \end{bmatrix} = \begin{bmatrix} 0 & 0 & 1/\sqrt{2} & 1/\sqrt{2} \\ 0 & 0 & 1/\sqrt{2} & -1/\sqrt{2} \\ 1\sqrt{2} & 1/\sqrt{2} & 0 & 0 \\ 1/\sqrt{2} & -1/\sqrt{2} & 0 & 0 \end{bmatrix} \begin{bmatrix} a_1 \\ a_2 \\ a_3 \\ a_4 \end{bmatrix}$$

$$b_1 = \frac{1}{\sqrt{2}}(a_3 + a_4) \quad b_3 = \frac{1}{\sqrt{2}}(a_1 + a_2)$$

$$b_2 = \frac{1}{\sqrt{2}}(a_3 - a_4) \quad b_4 = \frac{1}{\sqrt{2}}(a_1 - a_2)$$

Condition 1 : If $a_3 \neq 0$ \hfill (Input at port (3))

$$a_1 = a_2 = a_4 = 0 \qquad \text{(No inputs at port (1), (2) and (4))}$$

MICROWAVE COMPONENTS AND DEVICES

$$b_1 = \frac{a_3}{\sqrt{2}}, \quad b_2 = \frac{a_3}{\sqrt{2}}, \quad b_3 = b_4 = 0$$

This is the property of H-type Tee junction.

Condition 2 : If $\quad a_1 = a_2 = a_3 = 0$ \hfill (No inputs at port (1), (2) and (3))

$\quad a_4 \neq 0$ \hfill (Input at port (4))

$$b_1 = \frac{a_4}{\sqrt{2}}, \quad b_2 = -\frac{a_4}{\sqrt{2}}, \quad b_3 = b_4 = 0 \quad \text{(Property of } E\text{-type Tee junction)}$$

Condition 3 : If $\quad a_3 = a_4$

$\quad a_1 = a_2 = 0$

Than $\quad b_1 = \frac{1}{\sqrt{2}}(2a_3), \; b_2 = 0, \; b_3 = b_4 = 0$

This is the additive property. At ports (3) and (4) inputs are equals and results in an output at port (1).

Condition 4 : If $\quad a_2 = a_3 = a_4 = 0$ \hfill (No inputs at port (2), (3) and (4))

$\quad a_1 \neq 0$ \hfill (Input at port (1))

$$b_1 = b_2 = 0, \quad b_3 = b_4 = \frac{a_1}{\sqrt{2}}$$

When input power is fed into port (1) but there is no output at port (2), even though they are collinear ports. This is known *magic of magic-tee*. Sameway, an input at port (2) cannot come out at port (1).

Similarly E and H arm ports are isolated ports.

Condition 5 : If $\quad a_1 = a_2, \; a_3 = a_4 = 0$

$$b_1 = b_2 = b_4 = 0; \quad b_3 = \frac{1}{\sqrt{2}}(2a_1)$$

Equal inputs fed to port (1) and port (2) results an output at port (3) (addition property) and there is no outputs at ports (1), (2) and (4) same as condition 3.

Advantages of Magic-T

1. Power delivered to one of the output ports becomes independent of the termination at the other output port because of the decoupling property of the output ports.
2. In three port junction's devices such as the *E-* or *H-plane tees,* power division between ports depends on terminations existing at the respective output ports, whereas in case of magic tees all the output ports are perfectly matched. Hence power division between the ports is independent of the terminations.

Disadvantages of Magic-T

When a signal is applied to any arm of a magic-T, the flow of energy in the output arms is affected by reflections. Reflections are caused by impedance mismatching at the junctions. These reflections are the cause of the two major disadvantages of the magic-T.

1. The reflections represent a power loss since all the energy fed into the junction does not reach the load which the arms feed.
2. The reflections produce standing waves that can result in internal arcing.

Thus the maximum power a magic-T can handle is greatly reduced.

Reflections can be reduced by using some means of impedance matching that does not destroy the shape of the junctions. One method is shown in Fig. (5.11).

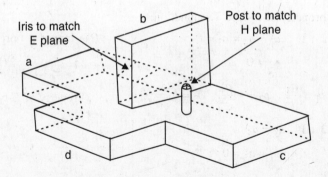

Fig. 5.11. Magic-T impedance matching.

A post is used to match the H plane, and an iris is used to match the E plane. Even though this method reduces reflections, it lowers the power-handling capability even further.

5.4.3.2 APPLICATIONS OF MAGIC-T

1. Impedance Measurement using Magic-T
2. Microwave discriminator and microwave bridges etc.
3. **As the mixer section for microwave radar receivers**: The most common application of this type of junction is as the mixer section for microwave radar receivers. A mixer in which the RF signal and local oscillator output are fed into E and H arms as shown in Fig. 5.12. Half of the power of RF signal received through the antenna and the local oscillator power goes to the mixer and generate the IF frequency.

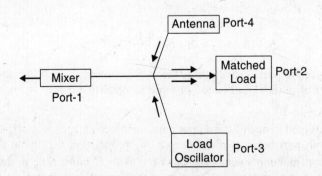

Fig. 5.12. As the mixer section for microwave radar receivers.

4. Magic -T as Decoupling Device or Duplexer

Fig. 5.13 shows the application of magic tee as a decoupling device or may be referred as duplexer. Let us assume that system is transmitting, connect the transmitter to port-1. The signal generated by the transmitter is fed to the antenna through ports 1 and 3. The magic tee device is such that part of this signal power is sent to the matching termination, where it gets dissipated, but no part will be sent to the local receiver. By using the magic tee hybrid junction, one can isolate a transmitter from exciting its own receiver.

MICROWAVE COMPONENTS AND DEVICES

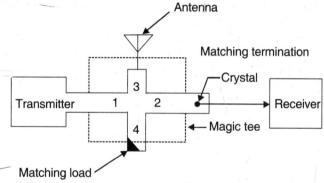

Fig. 5.13. Magic tee as a coupling device or Duplexer.

Assume now that system is receiving, the input is fed into the device through the antenna connected to port-3. This signal is detected and being rectified by the crystal diode connected to port-2, and applied to the receiver connected, as shown. The matching termination is required to prevent any type of feedback through which oscillations may build up in the system.

5. Magic –T in Automatic Frequency Control:

Magic –T in automatic frequency control of reflex klystron or magnetron shown in Fig. 5.14.

There are two Magic-T's (magic-tees A and B) used for automatic frequency control of magnetron or reflex klystron, as shown in Fig. 5.14.

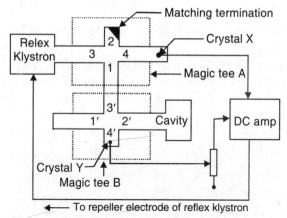

Fig. 5.14. Magic –T in automatic frequency control of reflex klystron.

Fig. 5.15. Magic-T.

The reflex klystron, whose frequency is to be controlled, is connected to arm 3 of magic tee A. Arm 1 of magic tee A is connected to arm 3' of B. A matching termination is connected to arm 2 of A. Crystal diodes (point-contact diodes) X and Y are connected to arm 4 and 4', respectively. The outputs of the crystal diodes are fed into differential amplifier (DC), whose output is connected to drive the repeller electrode of the reflex klystron.

Let us assume that, initially the system operates in equilibrium state. In this condition, the output voltages of crystal diodes, fed to the differential amplifier input terminals, which produces an output voltage proportional to the difference between the two. This output voltage of the differential amplifier is use as the repeller voltages of the reflex klystron, which in turn controls its operating frequency. A potentiometer is used at the output section of crystal Y, to obtain the initial equilibrium state.

Suppose that the due to some reason the operating frequency of the reflex klystron changes slightly. Then the output of crystal diode X changes in relation to that of crystal Y. Now the differential amplifier amplifies this difference in the crystal output voltages, and uses it as the repeller control voltage to regulate the frequency variation.

The same procedure may be use to regulate the operating frequency of magnetrons. We can replace the magnetron instead of reflex klystron, and the output of the differential amplifier is used as corrective voltage for electronic tuning of the magnetron.

5.5. HYBRID RING

A type of hybrid junction that overcomes the power limitation of the magic-T is the hybrid ring, also called a RAT RACE. The hybrid ring, illustrated in Fig. (5.16), view (A), is actually a modification of the magic-T.

It is constructed of rectangular waveguide molded into a circular pattern. The arms are joined to the circular waveguide to form E-type T junctions. View (B) shows, in wavelengths, the dimensions required for a hybrid ring to operate properly.

The hybrid ring also has the characteristics of hybrid T. When the wave is incident into port-1, it will not appear at port-3 because the difference of phase shifts for the wave traveling in clockwise and anticlockwise direction is 180 degree. Hence the waves are canceled at port-3. Similarly, the waves fed into ports-2 will not appear at port-4 and so on.

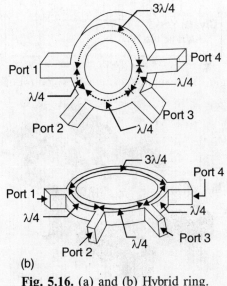

Fig. 5.16. (a) and (b) Hybrid ring.

MICROWAVE COMPONENTS AND DEVICES

The scattering matrix for an ideal hybrid ring may be expressed as

$$[S] = \begin{bmatrix} 0 & s_{12} & 0 & s_{14} \\ s_{21} & s_{21} & s_{23} & 0 \\ 0 & s_{32} & 0 & s_{34} \\ s_{41} & 0 & s_{43} & 0 \end{bmatrix}$$

Applications of Hybrid Ring

The hybrid ring is used primarily in high-powered radar and communications systems to perform two functions. During the transmit period, the hybrid ring couples microwave energy from the transmitter to the antenna and allows no energy to reach the receiver. During the receive cycle, the hybrid ring couples energy from the antenna to the receiver and allows no energy to reach the transmitter. Any device that performs both of these functions is called a **Duplexer**. A duplexer permits a system to use the same antenna for both transmitting and receiving. Since the only common application of the hybrid ring is as a duplexer.

5.6. WAVEGUIDE IMPEDANCE MATCHING

Reflected waves are generally to be avoided in waveguide for exactly the same reasons that they are avoided in transmission lines. Waveguide transmission systems are not always perfectly impedance matched to their load devices. The standing waves that result from a mismatch cause a power loss, a reduction in power-handling capability, and an increase in frequency sensitivity.

One method of achieving this result in a waveguide is to arrange matters so that the load impedance that is used will completely absorbs the incident fields exactly as they arrive, so that there is nothing left over to be reflected; this corresponds to characteristics impedance termination in a transmission line.

A second approach to the problem is to create a reflected wave near the load that is equal in magnitude but opposite in phase from the wave reflected by the load; in this way the two reflected wave cancel each other.

Most commonly both methods of impedance matching are used simultaneously. That is, the system is initially so arranged that the load provides as good an impedance match as is possible to obtain with reasonable efforts, and then what reflected wave still remains is eliminated by the use of an impedance matching system that introducing a neutralizing reflection.

Several waveguide arrangements have been designed for a controllable reflection. Some of these are analogous to the impedance matching arrangements used in transmission lines, while others are unique to waveguides.

5.6.1 STUB GUIDE OR T- SECTION

The waveguide analogue of the stub line is the stub guide or T-section as shown in Fig. 5.17. The arrangements at fig. (a) and (b) are often referred to as E and H stubs, respectively.

In Fig. 5.17 (a), the reactance at the input of the stub guide is effectively in series with the equivalent transmission line of the guide, while with the stub shown in Fig. 5.17 (b), the reactance introduced by the stub is in shunt in the equivalent transmission line circuit of the guide, this is shown in Fig. 5.17 (c & d), respectively. The magnitude of the reflection introduced by such a stub guide is controlled by the position of the short circuiting plunger in the stub guide. The phase of the reflected wave produced by the stub is determined by the position of the stub in relation to a minimum of the standing-wave pattern existing in the absence of the stub. Thus, to eliminate a reflected wave using a single stub, it is necessary to be able to vary not only the effective length of the stub, but also its distance to the load.

The latter requirement makes a single stub arrangement unsatisfactory in system that must be adjusted by trial and error, since there is no simple way that the position of the stub can be continuously varied. When trial and error adjustments is required, one can, however, employ two waveguide stubs spaced approximately $n\lambda/8$, where n is odd, to give the waveguide equivalent of the two stub tuner.

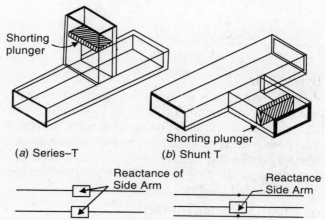

(a) Series–T (b) Shunt T

(c) Transmission line equivalent of (a) (d) Transmission line equivalent of (b)

Fig. 5.17. Waveguides provides with tuning stubs together with equivalent transmission-line circuits.

5.6.2 PROBE OR SCREWS

An alternative to the waveguide stub is an adjustable screw or probe that projects into the waveguide in a direction parallel to the electric field, as shown in Fig. 5.18.

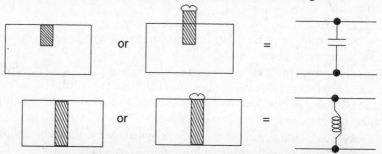

Fig. 5.18. Conducting screws. (Penetrating), Conducting screws. (Extending through).

Screws made from conductive material can be used for impedance-changing devices in waveguides. Figure **5.18**; illustrate two basic methods of using screws. Such an arrangement has the same effect as shunting a capacitive load across the equivalent transmission line of the waveguide, with the susceptance of this capacitive load increasing with penetration in to the guide to the point where the equivalent penetration is a quarter of a wavelength. When the equivalent penetration is exactly a quarter wavelengths, the probe becomes resonant. The system then acts as though a series resonant circuit of low resistance was connected in shunt with the waveguide; thus at exact resonance the probe acts as a shunt of very low resistance.

A screw which only partially penetrates into the waveguide acts as a shunt capacitive reactance. When the post or screw extends completely through the waveguide, making contact with the top and bottom walls, it acts as an inductive reactance.

Thus to extent to which such a probe projects into the waveguide determines the magnitude of the compensating reflection, while the position of the probe with respect to the standing wave pattern that is to be eliminated determine the phasing of the reflected wave. When it is necessary

MICROWAVE COMPONENTS AND DEVICES

that the axial position of the probe or screw be adjustable experimentally, this can be achieved by providing the guide with a longitudinal slot located in the middle of the broadside, as shown dotted in Fig. 5.19.

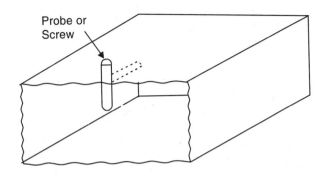

Fig. 5.19. Adjustable screw or probe for producing an adjustable reflection for impedance matching purposes.

5.6.3 WAVEGUIDE IRISES AND WINDOW

Another impedance matching system consists of a thin metallic barrier, or 'Window', placed at right angles to the axis of the waveguide is shown in **fig. 5.20**.

Impedance-changing devices are therefore placed in the waveguide to match the waveguide to the load. These devices are placed near the source of the standing waves.

Fig. (5.20) illustrates three devices, called **irises** that are used to introduce inductance or capacitance into a waveguide. An iris is nothing more than a metal plate that contains an opening through which the waves may pass. The iris is located in the transverse plane.

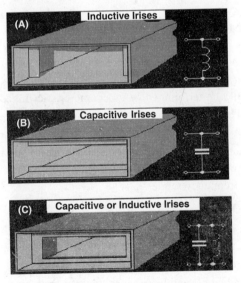

Fig. 5.20. Waveguide irises.

An inductive iris and its equivalent circuit are illustrated in Fig. 5.20 (A). The iris places a shunt inductive reactance across the waveguide that is directly proportional to the size of the opening. Notice that the edges of the inductive iris are perpendicular to the magnetic plane.

The shunt capacitive reactance, illustrated in Fig. 5.20 (B), basically acts the same way. Again, the reactance is directly proportional to the size of the opening, but the edges of the iris are perpendicular to the electric plane.

The iris, illustrated in Fig. 5.20 (C), has portions across both the magnetic and electric planes and forms an equivalent parallel-LC circuit across the waveguide. At the resonant frequency, the iris acts as a high shunt resistance. Above or below resonance, the iris acts as a capacitive or inductive reactance.

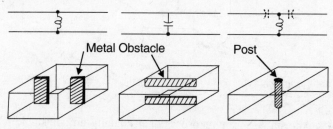

(a) Inductive window (b) Capacitive window (c) Post (Inductive)

Fig. 5.21. Examples of obstacles used in waveguides to introduce reflection, with equivalent transmission-line systems.

5.7. WAVEGUIDE TERMINATIONS

Electromagnetic energy is often passed through a waveguide to transfer the energy from a source into space. As previously mentioned, the impedance of a waveguide does not match the impedance of space, and without proper impedance matching, standing waves cause a large decrease in the efficiency of the waveguide. Any abrupt change in impedance causes standing waves, but when the change in impedance at the end of a waveguide is gradual, almost no standing waves are formed.

Terminations, or loads, are matched to characteristic impedance. Matched loads provide a termination designed to absorb all the incident power with very little reflection, effectively terminating the line or port in its characteristic impedance. Terminations are used in a wide variety of measurement systems; any port of a multi-port microwave device that is not involved in the measurement should be terminated in its characteristic impedance in order to ensure an accurate measurement. Terminations are also used in devices such as directional couplers and isolators. High power versions are used in transmitter applications as dummy loads.

When selecting a termination for a given connector style, one must determine the amount of power that needs to be absorbed and the acceptable level of reflection (VSWR) that can be tolerated over the given frequency range.

5.7.1 WAVEGUIDE HORNS

Gradual changes in impedance can be obtained by terminating the waveguide with a funnel-shaped Horn, such as the three types illustrated in Fig. 5.22. The type of horn used depends upon the frequency and the desired radiation pattern.

They have several advantages over other impedance-matching devices, such as their large bandwidth and simple construction.

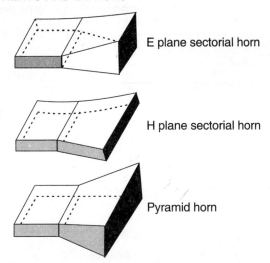

Fig. 5.22. Waveguide horns. E plane sectorial horn, H plane sectorial horn, pyramid horn.

5.7.2 DUMMY LOAD

A waveguide may also be terminated in a resistive load that is matched to the characteristic impedance of the waveguide. The resistive load is most often called a **Dummy load**, because its only purpose is to absorb all the energy in a waveguide without causing standing waves. There is no place on a waveguide to connect a fixed termination resistor; therefore, several special arrangements are used to terminate waveguides. One method is to fill the end of the waveguide with a graphite and sand mixture, as illustrated in Fig. 5.23(a).

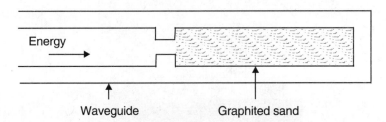

Fig. 5.23. (*a*) Terminating waveguides.

When the fields enter the mixture, they induce a current flow in the mixture which dissipates the energy as heat. Another method Fig. 5.23 (b) is to use a high-resistance rod placed at the center of the E field. The E field causes current to flow in the rod, and the high resistance of the rod dissipates the energy as a power loss, again in the form of heat.

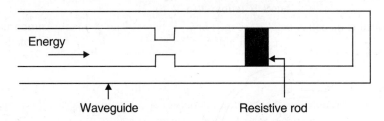

Fig. 5.23. (*b*) Terminating waveguides.

Still another method for terminating a waveguide is the use of a block of highly resistive material, as shown in of Fig. 5.23 (c). The plane of the block is placed perpendicular to the magnetic lines of force. When the H lines cut through the block or wedge, current flows in the wedge and causes a power loss. As with the other methods, this loss is in the form of heat. Since very little energy reaches the end of the waveguide, reflections are minimum.

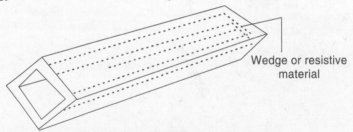

Fig. 5.23. (*c*) Terminating waveguides

All of the terminations discussed so far are designed to radiate or absorb the energy without reflections. In many instances, however, all of the energy must be reflected from the end of the waveguide. The best way to accomplish this is to permanently weld a metal plate at the end of the waveguide, as shown in Fig. 5.23 (d).

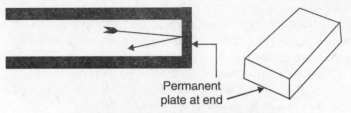

Fig. 5.23. (*d*) Terminating waveguides

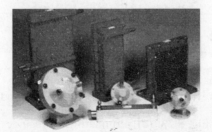

Fig. 5.23. (*e*) Terminating waveguides

One can change the guide impedance as required by varying the height, b, of the guide, using a gradual taper as shown in fig. 5.24 (a), to avoid introducing a reflection. A variation consists in tapering only the center portion of the waveguide to form a ridge, as in fig. 5.24(b).

Non-reflecting Loads in systems involving waveguides it is often necessary, a particularly in measurement work, to provide a termination that will completely absorb any wave going down the guide, irrespective of the exact frequency of this wave, and without any adjustment being required.

This result is most conveniently achieved by absorbing the wave in a lossy section tapered so gradually as to introduce no reflection. Examples of such sections are shown in Fig. 5.24; these involves lossy vanes, or wedges of lossy dielectric or iron dust core material, tapered on the centering edge, and having a sufficient length to absorbs an entering wave almost completely.

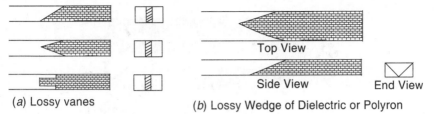

(a) Lossy vanes (b) Lossy Wedge of Dielectric or Polyron

Fig. 5.24. Examples of Non-reflecting Loads

5.8. WAVEGUIDE BENDS

The size, shape, and dielectric material of a waveguide must be constant throughout its length for energy to move from one end to the other without reflections. Any abrupt change in its size or shape can cause reflections and a loss in overall efficiency. When such a change is necessary, the bends, twists, and joints of the waveguides must meet certain conditions to prevent reflections.

Waveguides may be bent in several ways that do not cause reflections.

Gradual bends

One way is the gradual bend shown in fig. (5.25). This **gradual bend** is known as an E bend because it distorts the E fields. The E bend must have a radius greater than two wavelengths to prevent reflections.

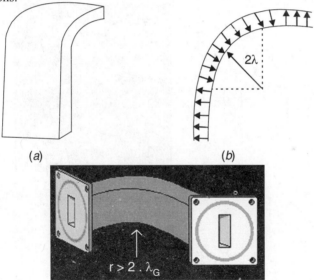

Fig. 5.25. (*a* & *b*) Gradual E Bend

Another common bend is the gradual H bend (fig. 5.26). It is called an H bend because the H fields are distorted when a waveguide is bent in this manner.

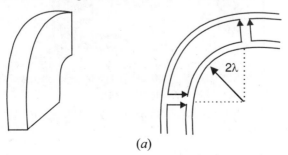

(*a*)

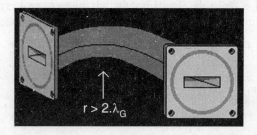

(b)

Fig. 5.26. (a & b) Gradual H bend.

Again, the radius of the bend must be greater than two wavelengths to prevent reflections. Neither the E bend in the "a" dimension nor the H bend in the "b" dimension changes the normal mode of operation.

Sharp bends

A sharp bend in either dimension may be used if it meets certain requirements. Consider the two 45-degree bends in fig. (5.27); the bends are $\lambda/4$ apart. The reflections that occur at the 45-degree bends cancel each other, leaving the fields as though no reflections have occurred.

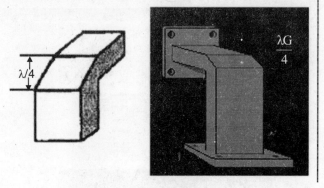

(a) H-Plane Bend

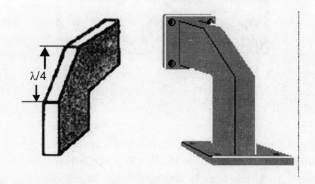

(b) E-Plane Bend

Fig. 5.27. (a & b) Sharp bends—45 degree.

MICROWAVE COMPONENTS AND DEVICES

5.9. WAVEGUIDE TWIST

Sometimes the electromagnetic fields must be rotated so that they are in the proper phase to match the phase of the load. This may be accomplished by twisting the waveguide as shown in Fig. (5.28). The twist must be gradual and greater than 2λ.

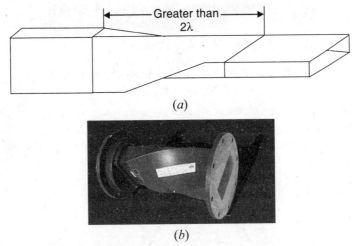

Fig. 5.28. Waveguide twist

5.10. FLEXIBLE WAVEGUIDE

The flexible waveguide shown in Fig. (5.29) allows special bends which some equipment applications might require. It consists of a specially wound ribbon of conductive material, most commonly brass, with the inner surface plated with chromium. Power losses are greater in the flexible waveguide because the inner surfaces are not perfectly smooth. Therefore, it is only used in short sections where no other reasonable solution is available.

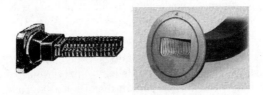

Fig. 5.29. Flexible waveguides.

5.11. WAVEGUIDE JOINTS

Since an entire waveguide system cannot possibly be fabricated into one piece, the waveguide must be constructed in sections and the sections connected with joints. The three basic types of waveguide joints are

1. The Permanent,
2. The Semi permanent, and
3. The Rotating joints.

Permanent Joint

Since the **permanent** joint is a factory-welded joint that requires no maintenance, only the semi permanent and rotating joints will be discussed.

Sections of waveguide must be taken apart for maintenance and repair.

Semi Permanent Joint

A semi permanent joint, called a **Choke joint,** is most commonly used for this purpose. The choke joint provides good electromagnetic continuity between sections of waveguide with very little power loss. A cross-sectional view of a choke joint is shown in Fig. 5.30(a) and 5.30 (b).

Fig. 5.30. (a) Choke joint.

The pressure gasket shown between the two metal surfaces forms an airtight seal. Notice in Fig. (5.30b) that the slot is exactly $\lambda/4$ from the "a" wall of the waveguide. The slot is also $\lambda/4$ deep, as shown in Fig. 5.30 (a), and because it is shorted at point (1), high impedance results at point (2). Point (3) is $\lambda/4$ from point (2). The high impedance at point (2) results in a low impedance, or short, at point (3). This effect creates a good electrical connection between the two sections that permits energy to pass with very little reflection or loss.

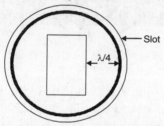

Fig. 5.31(b) Choke joint

There is therefore a (HF) electrically conducting connection between the two waveguide sections. A galvanic connection of the two hollow leaders or the flanges isn't necessary at this. Therefore it is possible to mount a seal between the two flanges to lock the inside of the waveguide air tightly.

Rotating Joint

Whenever a stationary rectangular waveguide is to be connected to a rotating antenna, a **rotating joint** must be used. A circular waveguide is normally used in a rotating joint. Rotating a rectangular waveguide would cause field pattern distortion. The rotating section of the joint, illustrated in Fig. (5.32), uses a choke joint to complete the electrical connection with the stationary section. The circular waveguide is designed so that it will operate in the TM_{01} mode. The rectangular sections are attached as shown in the illustration to prevent the circular waveguide from operating in the wrong mode.

Distance "o" is $3/4\ \lambda$ so that high impedance will be presented to any unwanted modes. This is the most common design used for rotating joints, but other types may be used in specific applications.

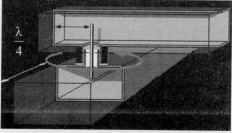

Fig. 5.32. (a) Rotating joint with a coaxial section.

MICROWAVE COMPONENTS AND DEVICES

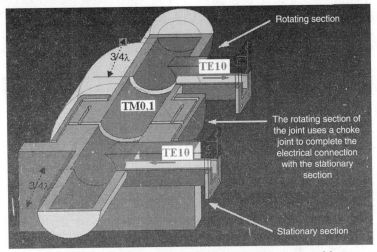

Fig. 5.32. (*b*) Rotating joint with a circular waveguide.

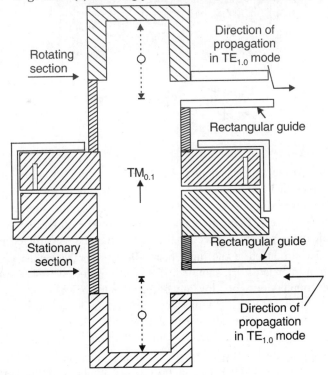

Fig. 5.32. (*c*) Rotating joint.

5.12. DIRECTIONAL COUPLER

The basic directional coupler is a four port junction that is used in a wide variety of microwave systems to satisfy almost any requirement for sampling incident and reflected microwave power conveniently and accurately with minimal disturbance to the transmission line.

The **directional coupler** is a four port device that provides a method of sampling energy from within a waveguide for measurement or use in another circuit. It consists of a primary waveguide and secondary waveguide as shown in Fig. (5.33).

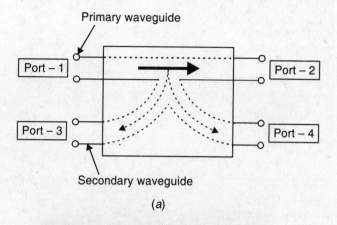

Fig. 5.33. (*a* & *b*) Directional coupler (Four-port network) (b) Directional couplers.

Most of the directional couplers sample energy traveling in one direction (incident power) only. However, directional couplers can be constructed that sample energy in both directions (incident and reflected power). These are called **Bidirectional couplers** and are widely used in radar and communications systems. Directional couplers may be constructed in many ways.

When power is introduced at the input port, all of the power appears at the output port except for the portion intended to be sampled. If power is reflected back from the output port, the ideal directional coupler does not allow any of the reflected power to appear on the secondary line. Regrettably, the ideal directional coupler does not exist in our world. Consequently, a small amount of backward power will be coupled to the secondary line 180° out of phase from the incident wave canceling power on the secondary line and adding uncertainty to the measurement.

It may be noted that, even though directional couplers are four-port devices, this principle may be extended to couple power from a given port to any number of other ports. Such couplers may be assuming several shapes and sizes. One such coupler is the star coupler, used in optical fiber technology to couple signal from a given port to various other ports.

When all the 4-ports are terminated with their characteristics impedances, than there is:
1. Free transmission of power between port-1 and port-2, without any reflection.
2. There is no transmission of power port-1 & 3 and port-2 & 4 because there no coupling exists between these two pairs of ports. The structure of the directional coupler decides the degree of coupling.

It is assumed that power is coupled from port-1 to port-2 in the primary waveguide. A small portion of the power is coupled to the port-4 for the measurement.

MICROWAVE COMPONENTS AND DEVICES

There should not be any power output at port-3.

P_1 = Incident power at port-1
P_r = Received power at port-2
P_f = Forward power at port-4
P_b = Backward power at port-3

The performance of a directional coupler can be expressed with the following parameters.

(i) **Coupling Factor (C):** The coupling factor is a measure of the ratio of the incident power (P_i) in the primary waveguide to the forward power (P_f) in the secondary waveguide measured in dB. It is a measure of how much of the incident power is being sampled.

$$\boxed{\text{Coupling Factor (C)} = 10 \log_{10} P_i/P_f \text{ (In dB)}} \qquad \text{...5.37}$$

(ii) **Directivity (D):** The directivity of a directional coupler is the measure of forward power (P_f) to the backward power (P_b) measured in dB.

$$\boxed{\text{Directivity (D)} = 10 \log_{10} P_f/P_b \text{ (In dB)}} \qquad \text{... 5.38}$$

It is the measure of how well the forward traveling wave in the primary waveguide couples only to a particular port of the secondary waveguide. An ideal directional coupler has infinite directivity. The typical value of well designed directional coupler is 30 to 35 dB only.

(iii) **Isolation (I):** The Isolation in direction coupler is defined as the ratio of incident power (P_i) to the backward power (P_b) measured in dB.

$$\boxed{\begin{array}{l}\text{Isolation (I)} = 10 \log_{10} P_i/P_b \text{ (In dB)} \\ \text{Isolation (I)} = \text{Coupling Factor (C)} + \text{Directivity (D)}\end{array}} \qquad \begin{array}{l}\text{... 5.39} \\ \text{...5.40}\end{array}$$

Isolation between two ports of a passive device is the amount of attenuation that a signal from a source of characteristic impedance Z_o applied to one port undergoes when measured at the other port terminated in Z_o.

(iv) **Insertion Loss:** Another important consideration when specifying a directional coupler is to ensure the device has minimal mainline insertion loss. Through virtue of their design, coaxial air-line couplers offer the lowest possible loss when inserted in a transmission path. Generally, the **insertion loss** of a coupler (or any microwave device for that matter) becomes more significant at higher frequency, namely because loss increases with frequency and higher frequency power sources are considerably more expensive. Accordingly, the criteria of low insertion loss will prevent precious power from being wasted on measurement components. It relates the total output power from all the ports relative to the input power. The output power is less than input power due to two main reasons:

(a) Absorption of some part of input power and
(b) Reflected due to mismatching

$$\boxed{\text{Insertion Loss} = 10 \log_{10} (P_f + P_r + P_b)/P_i \text{ (In dB)}} \qquad \text{...5.41}$$

(v) **Coupling Flatness:** This is the variation in coupling over the frequency range specified. The **frequency sensitivity** or "flatness" of a coupler is a measure of how coupling varies over a given frequency range. Optimum coupling frequency response is achieved by "centering" the design within the specified band of interest. Typical coupling flatness for a quarter-wavelength coupler operating over an octave band is within ± 0.75 dB of nominal. When operating over frequency bands greater than an octave, the flatness tolerance may need to be relaxed due to the inherent characteristics of coupling roll-off.

(vi) Impedance: This is the nominal characteristic impedance (Z_o) for the device.

(vii) VSWR: Voltage Standing Wave Ratio (VSWR) is a measure of the impedance of a device relative to Z_o.

It can be expressed as VSWR = 1 + |P| / 1 - |P|, where |P| is the magnitude of the reflection coefficient at the frequency of interest.

(viii) Amplitude Balance: The difference in attenuation between two or more output signals fed from a common input generally expressed as a maximum variation.

(ix) Phase Balance: The difference in phase between two or more output signals fed from a common input generally expressed as a maximum variation relative to the nominal phase difference between the paths. This nominal phase difference may be 0, 90, or 180°.

When specifying a directional coupler having a coupling factor greater than 20 dB (3, 6 or 10 dB), consideration should also be given to the theoretical insertion loss caused by power coupling from the mainline. **Table 1** illustrates the amount of additional loss the device exhibits as a function of the proximity of the two transmission lines. It should also be noted that dual directional couplers exhibit twice the loss of single directional models because there are two secondary lines drawing power from the mainline.

Table 1

Theoretical Mainline Insertion Loss Due to Coupling Factor (dB)							
Coupling Factor	3 dB	6 dB	10 dB	20 dB	30 dB	40 dB	50 dB
Single Directional Coupler	3.01	1.2560	0.4560	0.0436	0.0043	0.0004	0.00004
Dual Directional Coupler	6.02	2.5120	0.9120	0.0872	0.0086	0.0008	0.00008

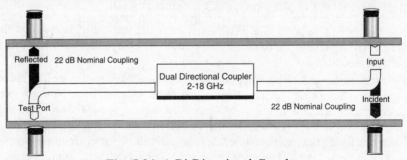

Fig. 5.34. A Bi-Directional Coupler.

5.12.1 TYPES OF DIRECTIONAL COUPLER

There are several types of directional couplers available such as shown in Fig. (5.35).

1. Two-Hole Directional Coupler
2. Bethe or Single-Hole Directional Coupler
3. Four-Hole Directional Coupler
4. Reverse coupling Directional Coupler (Schwinger coupler)

The degree of coupling is depending upon the size and position of the hole in the wave guide walls. Only the very commonly used two hole directional coupler which is described.

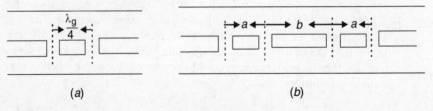

(a)   (b)

MICROWAVE COMPONENTS AND DEVICES

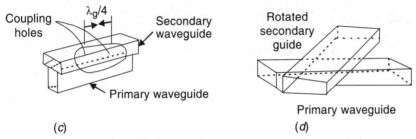

(c)   (d)

Fig. 5.35. Types of directional couplers.

(*a*) **Two-hole directional coupler** (*b*) **Four-hole directional coupler** (*c*) **Schwinger directional coupler** (*d*) **Bethe-hole directional coupler.**

5.12.2 TWO-HOLE DIRECTIONAL COUPLER

Fig. (5.36) illustrates Two - hole directional coupler, having two portions of the incident wave front in a waveguide. The waves travel down the waveguide in the direction indicated and enter the coupler section through both holes. Since both portions of the wave travel the same distance, they are in phase when they arrive at the pickup probe. Because the waves are in phase, they add together and provide a sample of the energy traveling down the waveguide. The sample taken is only a small portion of the energy that is traveling down the waveguide. The magnitude of the sample, however, is proportional to the magnitude of the energy in the waveguide. The absorbent material is designed to ensure that the ratio between the sample energy and the energy in the waveguide is constant. Otherwise the sample would contain no useful information. The ratio is usually stamped on the coupler in the form of an attenuation factor.

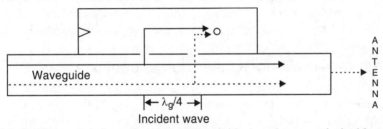

Fig. 5.36. Incident wave in a directional coupler designed to sample incident waves.

The effect of a directional coupler on any reflected energy is illustrated in Fig. (5.37). Note that these two waves do not travel the same distance to the pickup probe. The wave represented by the dotted line travels $\lambda/2$ further and arrives at the probe 180 degrees out of phase with the wave represented by the solid line.

The spacing between the centers of the hole should be
$$L = (2n+1)\lambda_g / 4$$
where n is any positive integer.

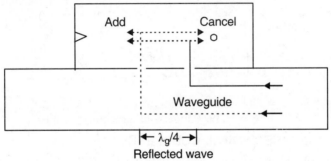

Fig. 5.37. Reflected wave in a directional coupler.

Because the waves are 180 degrees out of phase at the probe, they cancel each other and no energy is induced in the pickup probe. When the reflected energy arrives at the absorbent material, it adds and is absorbed by the material. A directional coupler designed to sample reflected energy is shown in Fig. (5.38).

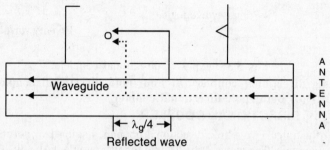

Fig. 5.38. Directional coupler designed to sample reflected energy.

The absorbent material and the probe are in opposite positions from the directional coupler designed to sample the incident energy. This positioning causes the two portions of the reflected energy to arrive at the probe in phase, providing a sample of the reflected energy. The sampled transmitted energy, however, is absorbed by the absorbent material.

This is most popular waveguide directional coupler. It is also be used for direct Standing-Wave Ratio (SWR) measurements if the absorbing attenuator is replaced by a detecting device, for measuring the components in the auxiliary guide that are proportional to the reflected wave in the main waveguide. Such type of directional coupler is called as **reflectometer.** But it is very difficult to match two detectors so it is preferable to use, two separate directional couplers to form the reflectometer.

5.12.3 BETHE OR SINGLE-HOLE DIRECTIONAL COUPLER

Fig. (5.39) illustrates a single hole directional coupler. In the single hole direction coupler directivity is improved because of single hole for coupling rather than the distance between the two holes. The amount of coupling is determined by the angle θ between the axis of the two waveguides and also the operating frequency and the size of the hole.

The power is coupled through the port-1 to the output probe (Coaxial) and the power entering through the port-2 is absorbed by the matched load.

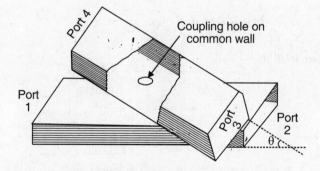

Fig. 5.39. Single hole directional coupler.

5.12.4 MULTIHOLE DIRECTIONAL COUPLERS

The principle of operation is same as that of two hole directional coupler and wide frequency of operation may be achieved by varying the size of the hole. Fig. (5.40) illustrated the construction of multihole directional coupler. The holes are drilled in the narrow side of the waveguide to have high power operation and the distances between two successive holes are $\lambda_g/4$.

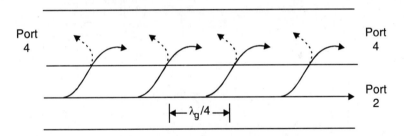

Fig. 5.40. Multihole coupler.

A three port directional coupler can be achieved by terminating a matched load in the isolated port that is port-4, as shown in fig. 5.41(a) and 5.41 (b).

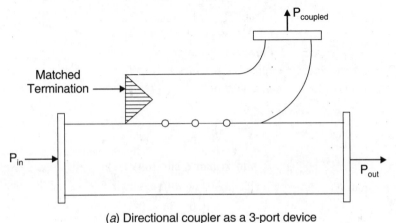

(a) Directional coupler as a 3-port device

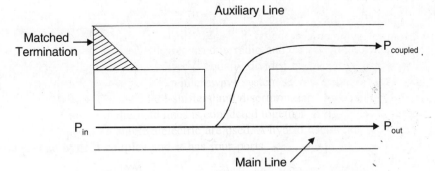

(b) Power coupling in a directional coupler used as 3 port

Fig. 5.41. Set-up of three-port directional coupler.

5.12.5 BIDIRECTIONAL COUPLER

A simple bidirectional coupler for sampling both transmitted and reflected energy can be constructed by mounting two directional couplers on opposite sides of a waveguide, as shown in Fig. (5.42).

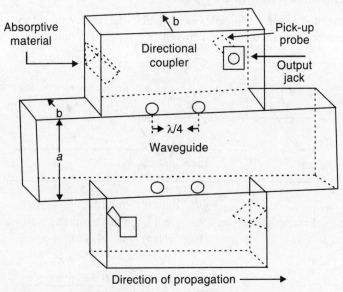

Fig. 5.42 Bi-directional coupler

S-Matrix of a Directional Coupler

Directional coupler has a four port network, so the scattering matrix is also 4 × 4 matrix.

$$[S] = \begin{bmatrix} S_{11} & S_{12} & S_{13} & S_{14} \\ S_{21} & S_{22} & S_{23} & S_{24} \\ S_{31} & S_{32} & S_{33} & S_{34} \\ S_{41} & S_{42} & S_{43} & S_{44} \end{bmatrix} \qquad \ldots 5.42$$

All the four ports are completely matched to the junction. Thus the diagonal elements are zero's and

$$S_{11} = S_{22} = S_{33} = S_{44} = 0 \qquad \ldots 5.43$$

There is no coupling between port (1) and port (3) and between port (2) and port (4).

$$S_{13} = S_{31} = S_{24} = S_{42} = 0 \qquad \ldots 5.44$$

By using the symmetric property, $S_{ij} = S_{ji}$

$$S_{23} = S_{32}; \quad S_{13} = S_{31}, \quad S_{24} = S_{42}$$

$$S_{34} = S_{43}; \quad S_{41} = S_{14} \qquad \ldots 5.45$$

Substituting in eq. (42), the values of scattering parameters as per eq. (43, 44 & 45), we get

$$[S] = \begin{bmatrix} 0 & S_{12} & 0 & S_{14} \\ S_{12} & 0 & S_{23} & 0 \\ 0 & S_{23} & 0 & S_{34} \\ S_{14} & 0 & S_{34} & 0 \end{bmatrix} \qquad \ldots 5.46$$

We know $\qquad [S][S^*] = I$

$$\begin{bmatrix} 0 & S_{12} & 0 & S_{14} \\ S_{12} & 0 & S_{23} & 0 \\ 0 & S_{23} & 0 & S_{34} \\ S_{14} & 0 & S_{34} & 0 \end{bmatrix} \begin{bmatrix} 0 & S_{12}^* & 0 & S_{14}^* \\ S_{12}^* & 0 & S_{23}^* & 0 \\ 0 & S_{23}^* & 0 & S_{34}^* \\ S_{14}^* & 0 & S_{34}^* & 0 \end{bmatrix} = \begin{bmatrix} 1 & 0 & 0 & 0 \\ 0 & 1 & 0 & 0 \\ 0 & 0 & 1 & 0 \\ 0 & 0 & 0 & 1 \end{bmatrix} \qquad \ldots 5.47$$

$R_1C_1: |S_{12}|^2 + |S_{14}|^2 = 1$ ... 5.48

$R_2C_2: |S_{12}|^2 + |S_{23}|^2 = 1$ ... 5.49

$R_3C_3: |S_{23}|^2 + |S_{34}|^2 = 1$ ... 5.50

$R_1C_3: S_{12}{}^*S_{23}{}^* + S_{14}S_{34}{}^* = 0$

By using equation (48 and 49), we get

$S_{14} = S_{23}$ ... 5.51

$S_{12} = S_{34}$ ... 5.52

It is assumed that S_{12} is real and positive ('P')

$S_{12} = S_{34} = P = S_{34}{}^*$ ... 5.53

$P S_{23}{}^* + S_{23}P = 0$

$P[S_{23} + P_{23}{}^*] = 0$

Since $P \neq 0$, $S_{23} + S_{23}{}^* = 0$

$S_{23} = jq = S_{14}$

$S_{12} = S_{34} = P$

$S_{23} = S_{14} = jq \qquad (P^2 + q^2 = 1)$

where q is positive real number.

By putting these values in eq. (46), the S matrix of a directional coupler is reduced to

$$[S] = \begin{bmatrix} 0 & P & 0 & jq \\ P & 0 & jq & 0 \\ 0 & jq & 0 & P \\ jq & 0 & P & 0 \end{bmatrix} \qquad ... 5.54$$

5.13. HYBRID COUPLER

A hybrid coupler is a passive device used in radio and telecommunications. It is a type of directional coupler where the input power is equally divided between two output ports. **Hybrid coupler** are used as microwave device in microwave system such as balanced amplifier, balanced mixers, attenuators, modulators, phase-shifters and discriminators. Hybrid coupler consisting of four parallel strip lines with alternate lines is connected together. A single layer of metallization, a single ground plane and a single dielectric are used. A hybrid coupler as shown in Fig. (5.43) is called a **Lange hybrid coupler** and it has four ports.

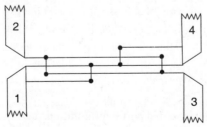

Fig. 5.43. A Lange hybrid couplers.

When a wave incident in port (1), the power is divided equally in to port (2) and port (4) but nothing into port (3). The hybrid coupler has many parameters which are as following:

(i) Directivity is over 27 dB,
(ii) An insertion loss of less than 0-13 dB,
(iii) Return loss of over 25 dB.

There are two types of Lange hybrid coupler:
(i) 90° (quadrature) hybrids also known as 3 dB directional couplers and
(ii) 180° hybrids

5.13.1 HYBRID COUPLER (3 dB, 90°)

A 3 dB, 90° hybrid coupler is a four-port device that is used either to equally split an input signal with a resultant 90° phase shift between output ports or to combine two signals while maintaining high isolation between the ports.

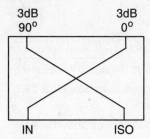

Fig. 5.44. The basic configuration of a hybrid coupler.

The basic configuration of a hybrid coupler is shown in Fig. 5.44 which illustrates two cross-over transmission lines over a length of one-quarter wavelength, corresponding with the center frequency of operation. When power is introduced at the **IN** port, half the power (3dB) flows to the **0°** port and the other half is coupled (in the opposite direction) to the 90° port. Reflections from mismatches sent back to the output ports will flow directly to the **ISO** port or cancel at the input. This is why hybrids are so widely used to split high power signals in applications where unwanted reflections could easily damage the driver device.

3 dB, 90° degree hybrids are also known as *quadrature* hybrids because a signal applied to any input, will result in two equal amplitude signals that are quadrant (90° apart). It also makes no difference which port is the input because the relationship at the outputs remains the same as these devices are electrically and mechanically symmetrical. This configuration ensures a high degree of isolation between the two output ports and the two input ports without unwanted interaction between them.

5.13.2 A 180° HYBRID RING COUPLERS

180° hybrid ring couplers (also called "rat race" couplers) are four-port devices used to either equally split an input signal or to sum two combined signals. An additional benefit of the hybrid ring is to alternately provide equally-split but 180 degree phase-shifted output signals.

The center conductor ring is 1½ wavelengths in circumference (or six ¼ wavelengths) and each port is separated by 90°. This configuration creates a lossless device with low VSWR, excellent phase & amplitude balance, high output isolation and match output impedances. The low loss, airline construction also makes the device a perfect choice for combining high power mixed signals.

Fig. 5.45 shows all four possible port configurations and the resultant phase relationships at the outputs of the device. Again, it makes no difference which port is the input because the device is electrically and mechanically symmetrical.

MICROWAVE COMPONENTS AND DEVICES

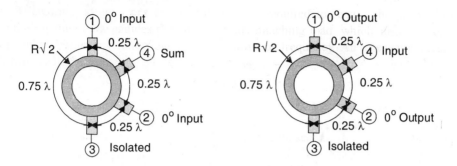

Fig. 5.45. a 0° (in-phase) Power Combiner 0° (in-phase) Power Divider.

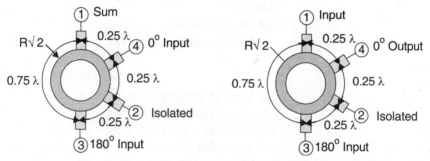

Fig. 5.45. b 180° Power Combiner 180° Power Divider.

Lange hybrid couplers are used in modern microwave circuits design such as balanced amplifier for high power and large-bandwidth applications.

A Lange hybrid coupler used in balanced amplifier circuitry is shown in fig. (5.46).

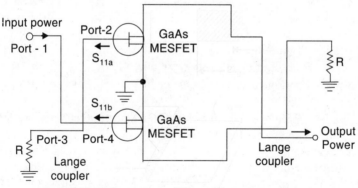

Fig. 5.46. Balanced amplifier with Lange couplers.

Two 3 dB and 90° Lange coupler are connected with cascaded double stage GaAs MESFET chips. Following three equations shows the relationship.

$$S_{11} = \frac{1}{2}(S_{11a} - S_{11b}) \qquad \ldots 5.55$$

$$S_{12} = \frac{1}{2}(S_{22a} - S_{22b}) \qquad \ldots 5.56$$

$$\text{Gain} = |S_{21}|^2 = \frac{1}{4}|S_{21a} + S_{21b}|^2 \qquad \ldots 5.57$$

We can express the VSWR's of the balanced amplifier as following,

$$VSWR = \frac{1+|S_{11}|}{1-|S_{11}|} \text{ for the input port} \quad \ldots 5.58$$

For the output port VSWR

$$VSWR = \frac{1+|S_{22}|}{1-|S_{22}|} \quad \ldots 5.59$$

Theoretically, the amplifier will be balanced and its VSWR should be unity if the two GaAs MESFET chips are identical. However, the characteristics of the two GaAs MESFET chips may not be identical, in that case the amplifier will not be balanced and manual tuning will be required to balance it.

Fig. 5.47. 3 dB Hybrid Dividers, 3 dB Hybrid Couplers & 180° hybrid ring couplers.

5.14 WAVEGUIDE AND MICROWAVE ATTENUATORS.

When a waveguide is excited at a wavelength greater than cutoff, the behavior is entirely different from a behavior at wavelengths less than cutoff. In particular, the electric and magnetic fields now decay exponentially with distance at a very much more rapid rate than is for energy losses in the walls. The rate of attenuation, moreover, depends only on the ratio λ/λ_c of the free space wavelength to the cut-off wavelength; unlike waves shorter than the cutoff wavelength the attenuation is independent of the material of the guide walls. The exact law of attenuation can be derived by application of the fundamental field equation, and is

Attenuation in dB per unit length = $\alpha = (54.6/\lambda_c)\sqrt{1 - (\lambda_c/\lambda)^2}$

When the actual wavelength is much greater than cutoff ($\lambda \gg \lambda_c$), than

$$\alpha = (54.6/\lambda_c)$$

Here λ is the free-space wavelength and λ_c is the cutoff wavelength, measured in the same units of length used in expressing the attenuation.

Waveguides operated at wavelengths greater than cutoff, termed **waveguide attenuators**, are often used as attenuators in signal generators. The usual arrangement for this purpose is illustrated in Fig. 5.48, involves exciting the guide, which may be either circular or rectangular, with a coil, the axis of which is at right angles to the axis of the guide. The pickup system then consists of a similar coil with its axis parallel to the axis of the exciting coil. Such an arrangement uses the TE_{10} mode in the rectangular case and the TE_{11} mode when the guide is circular. The output of such an attenuator is varied by adjusting between the pickup coil and the exciting coil.

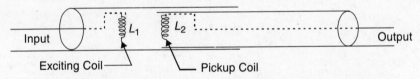

Fig. 5.48. Schematics diagram of typical waveguide attenuator.

MICROWAVE COMPONENTS AND DEVICES

Microwave attenuators are passive devices required to reduce the power level of the signal. Sometimes, waveguides requires perfectly matching load, which absorb incoming wave completely without any reflections and insensitive to frequency. An attenuator must have the following characteristics:

(i) Less VSWR at the required frequency.

(ii) Frequency response should be constant.

(iii) Broadband accuracy.

Attenuator is a two port reciprocal network and its scattering matrix is given as

$$S = \begin{bmatrix} 0 & S_{12} \\ S_{21} & 0 \end{bmatrix} \qquad \ldots 5.60$$

There is no reflection coefficient across both input and output and hence $S_{11} = S_{22} = 0$. From the measurement of the transmission coefficient S_{21}, or S_{12}, get the amount of attenuation α.

There are mainly two types of attenuators such as:

(i) Fixed attenuators

(ii) Variable attenuators

(i) Fixed attenuators: Fixed attenuators are used where a fixed amount of attenuation is desired. The most common resistive terminations is a length of lossy dielectric fitted in at the end of the waveguide and it is tapered gradually with the sharp end pointed at the incoming wave so as not to cause reflections. The lossy vane may sometimes occupy the whole width of the waveguide or just at the center of the waveguide end. The overall lengths of vane is about two wavelength (2λ) and the length of tapered section that may be single or double and has the length about $\lambda/2$. Generally it is made up of dielectrics slab such as glass with the carbon filters or aquadag coating outside. Fig. (5.49) shows the waveguide fixed attenuators.

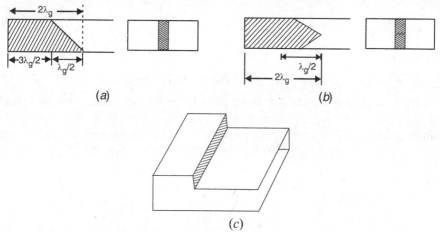

Fig. 5.49. Fixed attenuator (a) Single taper (b) Double taper (c) Stepped load.

(ii) Variable attenuators: Variable attenuators in waveguide use movable vane or movable flap type as shown in fig. (5.50). Rotary type is used for the circular waveguide. The vane may be made movable and can be used as variable attenuator. In this type of attenuator it is tapered at both ends and positioned at the middle instead at the end of waveguide. It may be moved from the center of the waveguide, where it will provide maximum attenuation to the edges, where attenuation is relatively reduced considerably. The mounting rods are made perpendicular to the electric field and they are placed $\lambda_g/2$ apart to minimize the reflections.

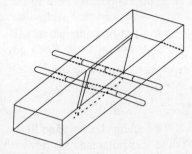

Fig. 5.50. Movable vane attenuator.

The **flap attenuator** consists of a resistive element or disc instead of the moving vane attenuator. A resistive element is mounted on a hinged arm, allowing it to descend in to the center of the waveguide through a suitable slot. The depth of insertion of the flap determining the amount of attenuation and the dielectric may be shaped to make the attenuation very linearly with depth of insertion.

The amount of attenuation is frequency resistive and it is to be calibrated against a standard attenuator. A maximum of 90 dB attenuation is possible with VSWR of 105.

A precision type variable attenuator is the **rotary attenuator.** It consists of two rectangular to circular waveguide tapered transitions, together with an intermediate section of circular waveguide which is free to rotate. The construction of rotary type variable attenuator is shown in the Fig. (5.51).

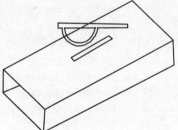

Fig. 5.51. Flap attenuator.

A resistive rotary vane attenuator provides precision attenuation with an accuracy of ±2–1% of the indicated attenuation over the entire range of frequency. It has total three vanes, the central vane rotating type placed in the circular waveguide in the central section, tapered at both ends. Two vanes are placed in the rectangular section as shown in the Fig. (5.52).

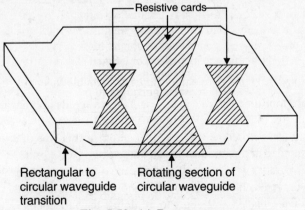

Fig. 5.52. (*a*) Rotary attenuator.

MICROWAVE COMPONENTS AND DEVICES

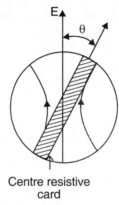

Fig. 5.52. (*b*) Position of centre card w.r.t. E-field.

When central rotary section of a circular waveguide consisting one or more resistive card and when all the three cards or vanes are aligned their plane at 90° to the direction of electric field than the attenuation is zero. When the center card is made parallel to the E-field lines, there is a complete absorption or maximum attenuation will take place. But when the center resistive card is oriented at an angle 'θ' the direction of the electric field than the absorption is proportional to $\sin^2\theta$ and the attenuation produced is given by

$$\alpha = -20\log\left(\sin^2\theta\right)$$

$$\alpha = -40\log\left(\sin\theta\right) \qquad \ldots 5.61$$

It is clear with the eqn. (61), that the attenuation is depends only on the angle θ. A piece of glass coated with metal film at one side is replaced as the resistive card for more accurate attenuation. The only disadvantage of this type of variable attenuator is that it is frequency sensitive and it produces a considerable amount of phase shift.

(*iii*) **High power attenuators:** The attenuators which we have study so far cannot be used at very high powers, because of heat production. When the power is dissipated at the attenuator vanes or pads, the heat is produced which limit the uses of these attenuators at high powers.

A waveguide hybrid junction is used in the high power microwave attenuator as shown in Fig. (5.53). The one point of the hybrid is terminated in a variable short circuit in series with one of the two connecting lines of a cascade arrangement of two hybrids. The high power is normally dissipated in a high power load such as water and circulating air.

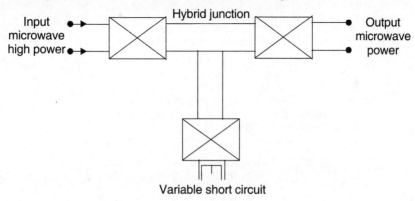

Fig. 5.53. High power microwave attenuator.

(*iv*) **Coaxial attenuators:** Various typical coaxial attenuators are shown in fig. (5.54). A type of coaxial attenuators is shown as **T-section type,** it consists of the central conductor made of glass tube having two resistive metalized film section on the either side of circular disc. The another type of coaxial attenuator consists of higher resistive main attenuating section in the centre and lowers resistive matching sections on either side of the central conductor, is known as **Distributed type** coaxial attenuator.

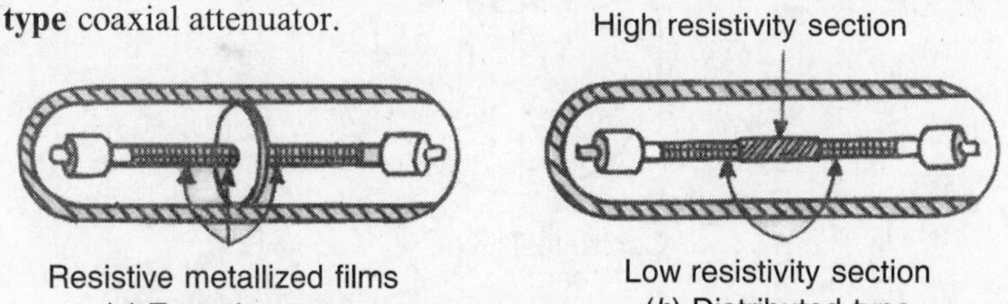

Fig. 5.54. Coaxial attenuators.

5.15. FERRITES DEVICES

A **Ferrite** is a device that is composed of material that causes it to have useful magnetic properties and, at the same time, high resistance to current flow. The primary material used in the construction of ferrites is normally a compound of iron oxide with impurities of other oxides added. The compound of iron oxide retains the properties of the ferromagnetic atoms, and the impurities of the other oxides increase the resistance to current flow. This combination of properties is not found in conventional magnetic materials. Iron, for example, has good magnetic properties but a relatively low resistance to current flow. The low resistance causes eddy currents and significant power losses at high frequencies Ferrites, on the other hand, have sufficient resistance to be classified as semiconductors.

The compounds used in the composition of ferrites can be compared to the more familiar compounds used in transistors. As in the construction of transistors, a wide range of magnetic and electrical properties can be produced by the proper choice of atoms in the right proportions.

The magnetic property of any material is a result of electron movement within the atoms of the material. Electrons have two basic types of motion. The most familiar is the Orbital movement of the electron about the nucleus of the atom. Less familiar, but even more important, is the movement of the electron about its own axis, called electron spin.

You will recall that magnetic fields are generated by current flow. Since current is the movement of electrons, the movement of the electrons within an atom create magnetic fields. The magnetic fields caused by the movement of the electrons about the nucleus have little effect on the magnetic properties of a material. The magnetic fields caused by electron spin combine to give material magnetic properties. The different types of electron movement are illustrated in Fig. (5.55).

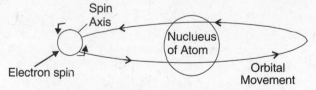

Fig. 5.55. Two types of electron movement.

In most materials the spin axes of the electrons are so randomly arranged that the magnetic fields largely cancel out and the material displays no significant magnetic properties. The electron spin axes within some materials, such as iron and nickel, can be caused to align by applying an external magnetic field. The alignment of the electrons within a material causes the magnetic fields to add, and the material then has magnetic properties.

MICROWAVE COMPONENTS AND DEVICES

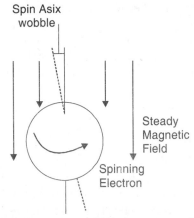

Fig. 5.56. Electron wobbles in a magnetic field.

In the absence of an external force, the axis of any spinning object tends to remain pointed in one direction. Spinning electrons behave the same way. Therefore, once the electrons are aligned, they tend to remain aligned even when the external field is removed. Electron alignment in a ferrite is caused by the orbital motion of the electrons about the nucleus and the force that holds the atom together. When a static magnetic field is applied, the electrons try to align their spin axes with the new force. The attempt of the electrons to balance between the interaction of the new force and the binding force causes the electrons to wobble on their axes, as shown in Fig. (5.56). The wobble of the electrons has a natural resonant wobble frequency that varies with the strength of the applied field. Ferrite action is based on this behavior of the electrons under the influence of an external field and the resulting wobble frequency.

5.15.1 FARADAY ROTATION IN FERRITE

When microwave energy is passed through a piece of ferrite in a magnetic field, another effect occurs. If the frequency of the microwave energy is much greater than the electron wobble frequency, the plane of polarization of the wavefront is rotated. This is known as the **Faraday rotation effect** and is illustrated in Fig. (5.57).

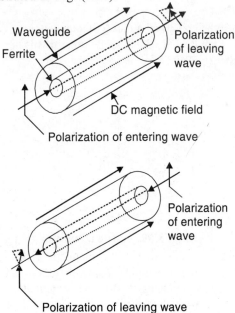

Fig. 5.57. Faraday rotation.

A ferrite rod is placed along the axis of the waveguide, and a magnetic field is set up along the axis by a coil. As a wavefront enters the section containing the ferrite, it sets up a limited motion in the electrons. The magnetic fields of the wavefront and the wobbling electrons interact, and the polarization of the wavefront is rotated. The amount of rotation depends upon the length of the ferrite rod. The direction of rotation depends upon the direction of the external magnetic field and can be reversed by reversing the field. The direction of rotation will remain constant, no matter what direction the energy in the waveguide travels, as long as the external field is not changed.

5.15.2 Applications of Ferrite in Microwave System

Ferrites have long been used at conventional frequencies in computers, television, and magnetic recording systems. The use of ferrites at microwave frequencies is a relatively new development and has had considerable influence on the design of microwave systems. In the past, the microwave equipment was made to conform to the frequency of the system and the design possibilities were limited. The unique properties of ferrites provide a variable reactance by which microwave energy can be manipulated to conform to the microwave system.

At present, ferrites are used as Gyrators, Load Isolators, Circulator, Phase Shifters, Variable Attenuators, Modulators, and switches in microwave systems

Fig. 5.58. Waveguide isolators and circulators.

5.15.3 Ferrite Attenuators

A ferrite attenuator can be constructed that will attenuate a particular microwave frequency and allow all others to pass unaffected. This can be done by placing a ferrite in the center of a waveguide, as shown in Fig. (5.59).

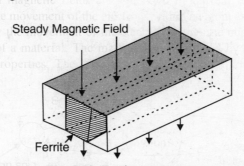

Fig. 5.59. Ferrite attenuator.

The ferrite must be positioned so that the magnetic fields caused by its electrons are perpendicular to the energy in the waveguide. A steady external field causes the electrons to wobble at the same frequency as the energy that is to be attenuated. Since the wobble frequency is the same as the energy frequency, the energy in the waveguide always adds to the wobble of the electrons. The spin axis of the electron changes direction during the wobble motion and energy is used. The force causing the increase in wobble is the energy in the waveguide. Thus, the energy in the waveguide is attenuated by the ferrite and is given off as heat. Energy in the waveguide that is a different frequency from the wobble frequency of the ferrite is largely unaffected because it does not increase the amount of electron wobble. The resonant frequency of electron wobble can be varied over a limited range by changing the strength of the applied magnetic field.

5.15.4 FERRITE ISOLATORS

An **RF isolator** is a two-port ferromagnetic passive device which is used to protect other RF components from excessive signal reflection. Isolators are common place in laboratory applications to separate a device under test (DUT) from sensitive signal sources.

An Isolator is a ferrite device that has two ports, and it can be constructed so that it allows microwave energy to pass in one direction but blocks energy in the other direction in a waveguide. In fig. (5.60) it is shown that the microwave energy is fed through the port-1 of an isolator and through port-2 of isolator a load is connected. It is the property of isolator that it allows the energy to travel through isolator and reach to the load with minimum attenuations and provides maximum attenuations to the energy traveling from load to the source. Hence generator appears to be matched for any loads in presence of isolator so that there should not be any change in frequency and power output due to variations in load.

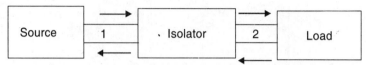

Fig. 5.60. Construction and principle of ferrite isolator.

This isolator is constructed by placing a piece of ferrite off-center in a waveguide, as shown in fig. (5.61). A magnetic field is applied by the magnet and adjusted to make the electron wobble

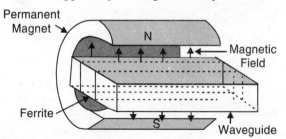

Fig. 5.61. Ferrites isolator.

frequency of the ferrite equal to the frequency of the energy traveling down the waveguide. Energy traveling down the waveguide from left to right will set up a rotating magnetic field that rotates through the ferrite material in the same direction as the natural wobble of the electrons. The aiding magnetic field increases the wobble of the ferrite electrons so much that almost all of the energy in the waveguide is absorbed and dissipated as heat. The magnetic fields caused by energy traveling from right to left rotate in the opposite direction through the ferrite and have very little effect on the amount of electron wobble. In this case the fields attempt to push the electrons in the direction opposite the natural wobble and no large movements occur. Since no overall energy exchange takes place, energy traveling from right to left is affected very little.

Operation

When the TE_{10} wave is entering to the port-1 (shown in fig. 5.62) which is passing through the resistive card is not attenuated. The wave is passing through the 45-degree twist, which will shift the wave by 45-degree in anticlockwise direction. Further a 45-degree shift in clockwise direction is provided by the ferrite rod which is placed in the isolator. The wave comes out from the port-2 with the same polarization as the wave entered through port-1 without any attenuation.

The resistive cord which is placed along the larger dimension of the waveguide such that it absorbs any wave whose plane of polarization is parallel to the plane of resistive card and it does not absorbs any energy whose plane of polarization is perpendicular to its own plane.

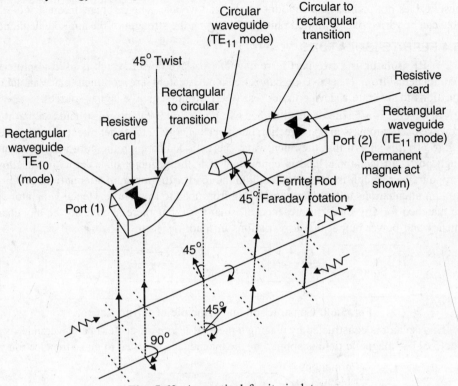

Fig. 5.62. A practical ferrite isolator.

Now if the TE_{10} wave is entered through the port-2, it passes through the resistive cards without any absorption because the plane of polarization of the wave is perpendicular to the plane of the resistive card. The wave get shifted by 45-degree in clockwise direction due to the ferrite rod and again the 45-degree twist provides the shift by 45-degree in clockwise direction.

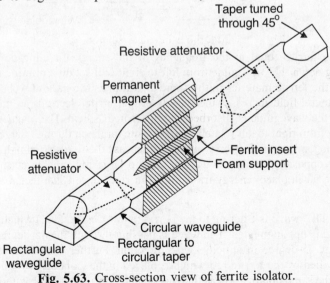

Fig. 5.63. Cross-section view of ferrite isolator.

Now when the wave reaches to the resistive card the plane of polarization of the wave is parallel to the resistive card and hence the wave is completely absorbed by the resistive card which result zero output at port-1. Fig. (5.63) shows the cross-section of the ferrite isolator and Fig. (5.64) shows the Faraday rotation in the isolator.

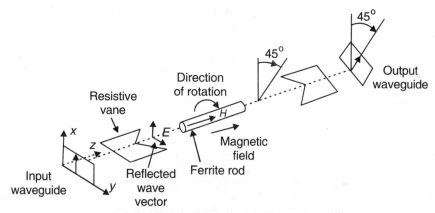

Fig. 5.64. Faraday rotation of isolator.

Fig. 5.65. Isolator and Circulator.

5.15.5 FERRITE CIRCULATORS

A **circulator** is a three-port ferromagnetic passive device used to control the direction of signal flow in a circuit and is a very effective, low-cost alternative to expensive cavity duplexers in base station and in-building mesh networks.

A **Ferrite circulator** is a ferrite device which often has four ports as shown in fig. (5.66).

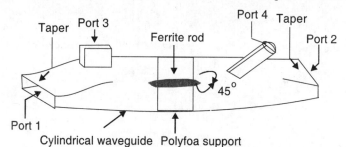

Fig. 5.66. Ferrite circulator.

Although a circulator has no restriction on the number of ports but a four port circulator is commonly used. It is a microwave device in which the wave is connected only to the next clockwise port. For example, port-1 energy is connected to the port-2 only and port-2 energy is connected to the port-3 only and port-3 energy is connected to the port-4 and so on.

The main applications of circulators are duplexers in radar and parametric amplifier.

Construction and Operations

A four port Faraday rotation circulator is shown in Fig. (5.67) when power is fed through port-1 is TE_{10} made, is converted to the TE_{11} made in circulator waveguide, passes port-3 unaffected because the electric field is not significantly cut, is rotated by 45° by the ferrite which is inserted, continues pass port-4 for the same reason that is passed port-3 and finally emerges from port-2.

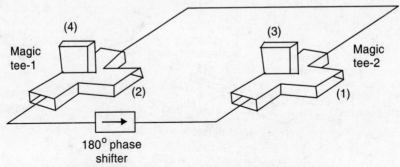

Fig. 5.67. A four-port circulator using Magic-T.

Power entering through port-2 will have plane of polarisation already shifted by 45° with respect to port-1. This power passes port-4 unaffected because again the electric field is not significantly cut. This wave is again rotated by 45° due to ferrite rod, in the clockwise direction. Now further this power is aligned to port-3 suitably to emerge out. Same way port-3 is connected only to port-4 and port-4 is to port-1. This type of circulator is power limited to the same extent as the Faraday rotation isolator, but it is suitable as a low power device. As a result, the sequence of power flow is designated as $1 \to 2 \to 3 \to 4 \to 1$.

Various types of microwave circulator are in use; however, their principles of operation remain the same. Fig. (5.67) shows a four port circulator constructed of two magic T and a phase shifter. The purpose of phase shifter is to provide 180° phase shift.

A lossless, fully matched and reciprocal four port circulator has as 'S' matrix of the form as:

$$[S] = \begin{bmatrix} 0 & S_{12} & S_{13} & S_{14} \\ S_{21} & 0 & S_{23} & S_{24} \\ S_{31} & S_{32} & 0 & S_{34} \\ S_{41} & S_{42} & S_{43} & 0 \end{bmatrix} \quad \ldots (62)$$

By using the properties of 'S' parameters, the final S matrix can be simplified to

$$S = \begin{bmatrix} 0 & 0 & 0 & 1 \\ 1 & 0 & 0 & 0 \\ 0 & 1 & 0 & 0 \\ 0 & 0 & 1 & 0 \end{bmatrix} \quad \ldots (63)$$

5.15.6 Circulators Used as Duplexer

A circulator can be used as a duplexer for a radar antenna system as shown in fig. (5.68).

The energy is feed through the port-1, *i.e.,* from the radar T_X is connected to the port-2 and to the port 4 and 2. Similarly, when the energy is fed through the port-2 (energy received by the antenna) is coupled to the port (3), *i.e.,* radar receiver. Duplexer provides the isolation between the receiver and the transmitter and enables to use a single antenna for transmission and reception.

MICROWAVE COMPONENTS AND DEVICES

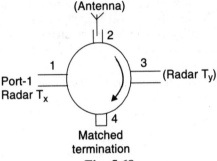

Fig. 5.68

5.15.7 THREE PORT CIRCULATOR

The most frequently used ferrite component is a 3-port junction circulator in waveguide or stripline configurations as shown in fig 5.69 and fig 5.70, respectively. In the waveguide configuration, we have a cylindrical disc of ferrite placed symmetrically with respect to the three ports as shown in fig 5.69. In stripline configurations there are two ferrite cylinders, one on either side of the strips. A magnetizing field is applied parallel to the axis of the ferrite cylinders.

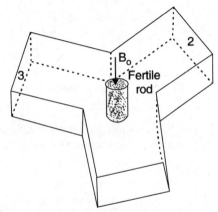

Fig. 5.69. 3-port junction circulator in waveguide.

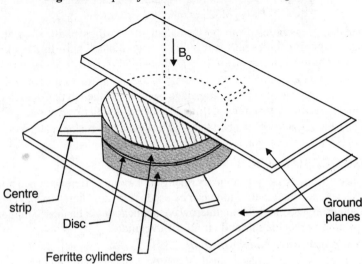

Fig. 5.70

A three port circulator is a symmetrical 'Y' type junctions of three identical waveguides with an axially magnetized ferrite position placed at the center. Fig-5.71 shows a typical three ports circulator. The ferrite position is magnetized by stator B_0 field along the axis. It provides the necessary non-reciprocal property. The junctions can be matched by placing suitable tuning element in each arm.

It is an essential component used to isolate the input and output in negative resistance amplifier. Three port circulators are also used to couple a transmitting to various receivers.

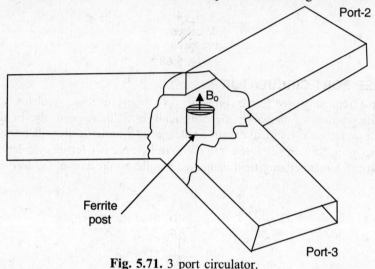

Fig. 5.71. 3 port circulator.

5.15.8 GYRATOR

A gyrator is a non-reciprocal ferrite device. It has two ports, a relative phase shift of 180 degree in forward direction and 0 degree phase shift in reverse direction. A schematic symbol is shown in Fig. 5.72.

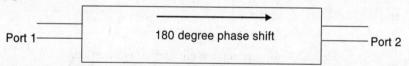

Fig. 5.72. Symbol of gyrator.

When signal is transmitted from port-1 to port-2, it offers phase shift of 180 degree (π radians) and when signal is fed to port-2 it offers 0 degree phase shift to the signal. Hence it is also as differential phase shifter device.

Gyrators consist of a circular to rectangular waveguide transition both at dominant mode. A twin circular ferrite rod tapered at both ends is located inside the circular waveguide surrounded by permanent magnets which generates dc magnetic field for the operation of ferrite.

A rectangular waveguide twisted by 90 degree is connected by input and reduces attenuations and also smooth attenuation of polarized wave.

5.16. WAVEGUIDE TRANSITION

When it becomes necessary to join wavelengths having different dimensions or different cross-sectional shapes, waveguide transition can be used. Transition between waveguides of various size, shape and mode are used in different microwave applications. The most common are dominant mode transitions among rectangular and circular waveguides.

Rectangular to Circular waveguide transition

Some reflections will take place but they can be reduced if the taper or transition section is made gradual, as shown for the rectangular to circular transition.

MICROWAVE COMPONENTS AND DEVICES

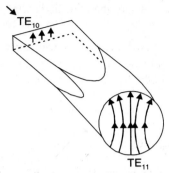

Fig. 5.73. Waveguide transition to convert the TE_{10} mode in rectangular guide to TE_{11} mode in the circular guide.

The tapered transition in part a converts the TE_{10} mode in rectangular guide to TE_{11} mode in the circular guide as shown in Fig. 5.73. Reflection can be reduced by making the taper length long composed to the operating wavelength.

Transition between a dominant mode and a higher mode is shown in following fig.-5.74. This figure shows an efficient method of conversion between the TE_{10} rectangular guide mode and TE_{01} mode in circular guide.

The diameter 'D' of the circular waveguide must be sufficiently large to ensure that the TM_{01} mode is above cut off and not the next higher mode.

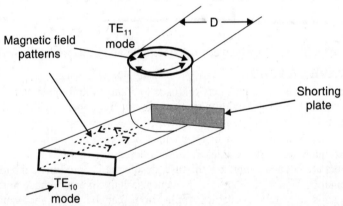

Fig. 5.74. Waveguide transition to convert the TE_{10} mode in rectangular guide to TE_{01} mode in the circular guide.

Circular to rectangular waveguide transition

The TE_{10} mode enters the circular guide creating a TE_{11} mode wave travels directly towards output while the remainder propagates to shorting plate.

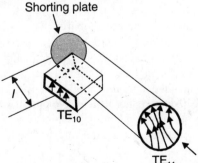

Fig. 5.75. Circular TE_{11} mode to TE_{10} mode rectangular transitions.

By adjusting the length between shorting plate and rectangular guide, the two waves arrive at the output in phase resulting in efficient conversion between the TE_{10} and TE_{11} modes.

Transitions between coaxial cable and rectangular waveguide

A transition between waveguide and coaxial cable is needed in some of the systems. The basic form of transition is shown in the fig.-5.76, where the center conductor of the coaxial cable extends into the waveguide. The current 'J_y' on the center conductor will cause and H-field that circulates around the current flow. The reactance of the transition of the transition is governed by the length and shape of the filament.

The reflection from the transition in to waveguide can then be designed to be very low at a given frequency. No power is dissipated in the transition, if the conductors are assumed to be perfect. The transition is a lossless reciprocal two-port and so the reflection seen by the coaxial cable is equal in magnitude to that in wave guide. The bandwidth over which the reflection can be maintained below a specified level is limited by the frequency dependent nature of 'J_x'.

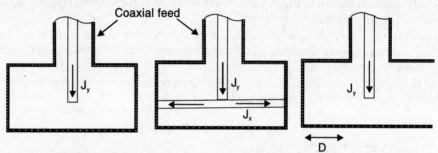

Fig. 5.76. Transitions between coaxial cable and rectangular waveguide.

5.17. MICROWAVE FILTERS

A microwave filter is a two port network used to control the frequency response at a certain point in a microwave system by providing transmission at frequencies within the passband of the filter and attenuation in the stopband of the filter.

The ideal filter is a network which provides perfect transmission in certain band pass regions for all frequencies and provides infinite attenuation in certain stop band regions. In practice ideal filters are not achievable but the filter design is to approximate the ideal requirements to within an acceptable tolerance. Filters are frequency selective network made of reactive elements. There are four types of filters available. Applications can be found in virtually any type of microwave communication, radar, or test and measurement system.

Type of filters:

(i) Low pass filter
(ii) High pass filter
(iii) Band pass filter
(iv) Band stop filter

The response curves of these filters are shown in Fig. (5.77).

At low frequencies the main components used for filter design are ideal capacitor and inductors. These reactive components have very simple frequency characteristics and a complete synthesis technique has been developed to design the filters using these components. At microwave frequencies, the filter designing is much more complicated, where distributed parameter elements are used. These filters are realized by using microwave circuit elements such as suitable length of waveguide, which replace all inductors and capacitors of low frequency filter.

There are two methods of designing a filter at low frequency such as

(i) Image parameter method
(ii) Insertion loss method.

MICROWAVE COMPONENTS AND DEVICES

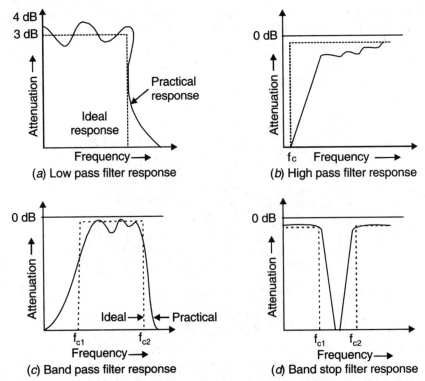

Fig. 5.77. Response curves of various filters.

The insertion loss method is preferable because it allows a high degree of control over the amplitude and phase characteristics in the pass band and stop band. The insertion loss method mainly using following steps:

(i) Determine the low pass "prototype" network for the desired. The insertion loss characteristics, normalized to 1 ohm termination and a cutoff frequency of 1 rad/sec.

(ii) Now transform the prototype structure into high pass, band stop or band pass structures using suitable frequency and impedance from formations.

Generally a microwave filter is built using a high 'Q' cavity and to get desired response more than one cavity may be cascaded, such filter is known as **multi cavity filter.** Irises or slots are used to couple the cavities and tuning is done by capacitive screws. Tuning of cavity to a particular frequency depends upon the penetration depth of capacitive screw. But to have high 'Q' of cavity the depth of penetration is kept minimum. The fig. (5.78) shows the construction of multi cavity filters.

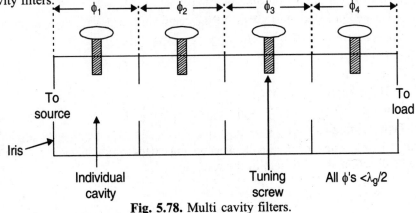

Fig. 5.78. Multi cavity filters.

Equivalent lumped circuit of a waveguide filter comprises of cavities and irises shown in figure (5.78). The factors which affects the characteristics of the filter are as following:

(*i*)   Resonant frequency and loaded Q of each cavity.
(*ii*) Suspectance of each slot irises.
(*iii*) Input and output impedances of filter.
(*iv*) The degree of coupling between neighboring cavities.

Fig. 5.79. Waveguide filters.

SUMMARY

- In a microwave circuit, the incoming wave is "scattered" by the circuit and its energy is partitioned between all the possible outgoing waves on all the other transmission lines connected to the circuit. The scattering parameters are fixed properties of the (linear) circuit which describe how the energy couples between each pair of ports or transmission lines connected to the circuit.

- **Properties of S-Matrix**
  1. Symmetry of s-matrix
  $$s_{ij} = s_{ji}$$
  2. S-Matrix is always a **square matrix** *(n × n)*
  3. S-Matrix is always a **unitary matrix** $[S] \cdot [S^*] = [I]$
  4. The sum of the products of any column or row multiplied by the complex conjugates of any other column or row is always zero.

- You may have assumed that when energy traveling down a waveguide reaches a junction, it simply divides and follows the junction.
- The T Junction is the most simple of the commonly used waveguide junctions.
- T junctions are divided into following basic types,
  1. The E-Type
  2. The H-Type.
  3. E-H Type (Hybrid) or Magic - T – Junction
  4. Rat - Race Junction or Hybrid Ring
- **E-Type T junction**

  An E-Type T junction because the junction arm extends from the main waveguide in the same direction as the E field in the waveguide.

MICROWAVE COMPONENTS AND DEVICES

- **S-Matrix for E-Plane**

$$\begin{bmatrix} b_1 \\ b_2 \\ b_3 \end{bmatrix} = \begin{bmatrix} 1/2 & 1/2 & 1/\sqrt{2} \\ 1/2 & 1/2 & -1/\sqrt{2} \\ 1/\sqrt{2} & -1/\sqrt{2} & 0 \end{bmatrix} \begin{bmatrix} a_1 \\ a_2 \\ a_3 \end{bmatrix}$$

- The various conditions in E-plane T-junction.

  Condition 1: If we feed equal inputs at port (1) and port (2), there is no output at port (1).

  Condition 2: If we feed the input at port (3) than output equally divideds between port (1) and port (2) with a phase difference of 180° between the outputs.

  $$a_1 = a_2 = 0$$
  $$a_3 \neq 0$$

  Condition 3: If $a_1 \neq 0$

  $$a_2 = a_3 = 0$$

  $$b_1 = \frac{a_1}{2}; \quad b_2 = \frac{a_1}{2}; \quad b_3 = -\frac{a_1}{2}$$

- It is called an **H-type T junction** because the long axis of the "b" arm is parallel to the plane of the magnetic lines of force in the waveguide.

- **S-Matrix for H-Plane**

$$\begin{bmatrix} b_1 \\ b_2 \\ b_3 \end{bmatrix} = \begin{bmatrix} 1/2 & -1/2 & 1/\sqrt{2} \\ -1/2 & 1/2 & 1/\sqrt{2} \\ 1/\sqrt{2} & 1/\sqrt{2} & 0 \end{bmatrix} \begin{bmatrix} a_1 \\ a_2 \\ a_3 \end{bmatrix}$$

Condition 1: If $a_3 \neq 0$

$$a_1 = a_2 = 0$$

The amount of power coming out from port (1) and port (2) are equal and half of the input at port (3). The *H*-plane *T*-junction is called as 3 dB splitter.

Condition 2: If $a_1 = a_2 = a$

$$a_3 = 0$$

- The output at port (3) is addition of the inputs fed from port (1) and port (2) are combined in phase.

  The **magic-T** is a combination of the H-type and E-type T junctions.

- **S-Matrix for Magic-T**

$$\begin{bmatrix} b_1 \\ b_2 \\ b_3 \\ b_4 \end{bmatrix} = \begin{bmatrix} 0 & 0 & 1/\sqrt{2} & 1/\sqrt{2} \\ 0 & 0 & 1/\sqrt{2} & -1/\sqrt{2} \\ 1/\sqrt{2} & 1/\sqrt{2} & 0 & 0 \\ 1/\sqrt{2} & -1/\sqrt{2} & 0 & 0 \end{bmatrix} \begin{bmatrix} a_1 \\ a_2 \\ a_3 \\ a_4 \end{bmatrix}$$

Condition 1: If $a_3 \neq 0$

$$a_1 = a_2 = a_4 = 0$$

This is the property of *H*-type Tee junction.

Condition 2: If $a_1 = a_2 = a_3 = 0$

$$a_4 \neq 0$$

(Property of *E*-type Tee junction)

Condition 3 : If $a_3 = a_4$

$$a_1 = a_2 = 0$$

This is the additive property. At ports (3) and (4) inputs are equals and results in an output at port (1).

Condition 4 : If $a_2 = a_3 = a_4 = 0$

$$a_1 \neq 0$$

When input power is feed to into port (1) but there is no output at port (2), even through they are collinear ports. This is known magic of magic. Same way, an input at port (2) cannot come out at port (1).

Condition 5 : It $a_1 = a_2$, $a_3 = a_4 = 0$

Equal inputs fed to port (1) and port (2) results an output at port (3) (addition property) and there is no outputs at ports (1), (2) and (4) same as condition 3.

- **Applications of Magic-T**
    1. The most common application of this type of junction is as the mixer section for microwave radar receivers.
    2. Magic-T as Duplexer.
    3. Impedance Measurement using Magic-T.
    4. Microwave discriminator and microwave bridges etc.
- A type of hybrid junction that overcomes the power limitation of the magic-T is the hybrid ring, also called a RAT RACE.
- The hybrid ring also has the characteristics of hybrid T.
- The hybrid ring is used primarily in high-powered radar and communications systems to perform two functions. Device that performs both of these functions is called a **Duplexer.**
- A waveguide may also be terminated in a resistive load that is matched to the characteristic impedance of the waveguide. The resistive load is most often called a **Dummy load.**
- When a change is necessary, the bends, twists, and joints of the waveguides must meet certain conditions to prevent reflections.
- Waveguides may be bent in several ways that don't cause reflections. This **gradual bend** is known as an E bend because it distorts the E fields.
- A **sharp bend** in either dimension may be used if it meets certain requirements. Consider the two 45-degree bends in fig. the bends are $\lambda/4$ apart.
- The three basic types of waveguide joints are:
- 1. The Permanent, 2. The Semi permanent, and 3. The Rotating joints.
- The **directional coupler** is a four port device that provides a method of sampling energy from within a waveguide for measurement or use in another circuit.
- The performance of a directional coupler can be expressed with the following parameters.

Coupling Factor (C) = $10 \log_{10} P_i/P_f$

Directivity (D) = $10 \log 10 \ P_f/P_b$

Isolation (I) = $10 \log_{10} P_i/P_b$

Isolation (I) = Coupling Factor (C) + Directivity (D)

Insertion Loss = $10 \log_{10} (P_f + P_r + P_b)/P_i$

MICROWAVE COMPONENTS AND DEVICES

There are several types of directional couplers available such as shown.
1. Two-Hole Directional Coupler
2. Bethe or Single-Hole Directional Coupler
3. Four Hole Directional Coupler
4. Reverse coupling Directional Coupler (Schwinger coupler)

- **S-Matrix of a Directional Coupler**

S matrix of a directional coupler is reduced to

$$[S] = \begin{bmatrix} 0 & P & 0 & jq \\ P & 0 & jq & 0 \\ 0 & jq & 0 & P \\ jq & 0 & P & 0 \end{bmatrix}$$

- **Hybrid coupler** are used as microwave device in microwave system such as balanced amplifier, balanced mixers, attenuators, modulators, phase-shifters and discriminators.
- **Microwave attenuators** are passive devices required to reduce the power level of the signal.

  Two types of attenuators such as:

  (*i*) Fixed attenuators (*ii*) Variable attenuators

  Filters are frequency selective network made of reactive elements. There are four types of filters available.
- **Types of filters:** (*i*) Los pass filter (*ii*) High pass filter (*iii*) Band pass filter and (*iv*) Band stop filter.
- A **Ferrite** is a device that is composed of material that causes it to have useful magnetic properties and, at the same time, high resistance to current flow.
- If the frequency of the microwave energy is much greater than the electron wobble frequency, the plane of polarization of the wave front is rotated. This is known as the **Faraday rotation effect.**
- At present, ferrites are used as Gyrators, Load Isolators, Circulator, Phase Shifters, Variable Attenuators, Modulators, and switches in microwave systems
- A **ferrite attenuator** can be constructed that will attenuate a particular microwave frequency and allow all others to pass unaffected.
- An Isolator is a ferrite device that has two ports, and it can be constructed so that it allows microwave energy to pass in one direction but blocks energy in the other direction in a waveguide.
- A **circulator** is a ferrite device often has four port.
- The main applications of circulators are duplexers in radar and parametric amplifier.
- A lossless, fully matched and reciprocal four port circulator has as 'S' matrix of the form as:

$$[S] = \begin{bmatrix} 0 & S_{12} & S_{13} & S_{14} \\ S_{21} & 0 & S_{23} & S_{24} \\ S_{31} & S_{32} & 0 & S_{34} \\ S_{41} & S_{42} & S_{43} & 0 \end{bmatrix}$$

- A circulator can be used as a duplexer for a radar antenna system.

REVIEW QUESTIONS

1. What do you understand by microwave components and list various components used in RF plumbing.
2. Define a microwave junction. Derive the scattering matrix relation between the input and output of $n \times n$ junction.
3. Describe the operation of E-plane, H-plane and hybrid plane. Why E-H plane Tee is referred as magic-Tee. Derive the scattering matrix for magic-Tee.
4. Explain the principle of operation of directional coupler with the neat diagram. Discuss various types of directional coupler.
5. Describe the operations of 2-hole directional coupler. Give the various parameters of the directional coupler.
6. Explain the working of vane and flap attenuator.
7. Describe the functioning of Rat-Race junction.
8. What do you understand by ferrite devices and discuss the properties of ferrites.
9. Discuss Faraday rotation.
10. Explain the working of ferrite isolator with the diagram.
11. Explain the working of ferrite circulator and write down the applications.
12. Write short notes on :
    - (i) Waveguide joints, bends and twist
    - (ii) Scattering matrix
    - (iii) Phase shifters
    - (iv) Microwave attenuator
    - (v) Irises and tuning screws
    - (vi) Detectors
    - (vii) Microwave filters
    - (viii) Directions coupler
    - (ix) Coupling factor
    - (x) Directivity in directional coupler
    - (xi) Hybrid coupler
    - (xii) Waveguide transition

CHAPTER 6

MICROWAVE TUBES

OBJECTIVE
- Introduction to Microwave Tubes
- Frequency Limitations of Conventional Tubes at Microwave Frequency
- Microwave Tubes
- Klystrons
- Two Cavity Klystron Amplifier
- Microwave Tubes
- Klystrons
- Two Cavity Klystron Amplifiers
- Multi-cavity Klystron
- Reflex Klystron
- Travelling-Wave Tube (TWT)
- Backward Wave Oscillator
- Microwave Cross Field Tubes
- Magnetron Oscillator

6.1 INTRODUCTION TO MICROWAVE TUBES

Microwave frequency is used in both radar and communications applications. The fact that the frequencies are very high and the wavelengths very short present special problems in circuit design. Components that were previously satisfactory for signal generation and amplification use are no longer useful in the microwave region. The theory of operation for these components is discussed in this chapter.

The efficiency of conventional tubes is largely independent of frequency up to a certain limit. When frequency increases beyond that limit, several factors combine to rapidly decrease tube efficiency. Tubes that are efficient in the microwave range usually operate on the theory of VELOCITY MODULATION, a concept that avoids the problems encountered in conventional tubes. Velocity modulation is more easily understood if the factors that limit the frequency range of a conventional tube are thoroughly understood. Therefore, the frequency limitations of conventional tubes will be discussed before the concepts and applications of velocity modulation are explained.

6.2 FREQUENCY LIMITATIONS OF CONVENTIONAL TUBES

Conventional vacuum tubes such as triode, tetrodes and pentode are less beneficial signal sources at frequency above 1 GHz because of following characteristics of ordinary vacuum tubes.

These characteristics are:

1. Interelectrode capacitance,
2. Lead inductance,
3. Electron transit time and
4. Gain-Bandwidth product limitations.

6.2.1 Interelectrode Capacitance

The interelectrode capacitances in a vacuum tube, at low or medium radio frequencies, produce capacitive reactance that is so large that no serious effects upon tube operation are noticeable. However, as the frequency increases, the capacitive reactance (X_c) small enough to materially affect the performance of a circuit.

Capacitive reactance $(X_C) = 1/2\pi f C$

Consider a triode valve, whose equivalent diagram is shown in Fig. (6.1). In triode tube three interelectrode capacitances (IEC's) are there which are Cgp, Cgk and Cpk.

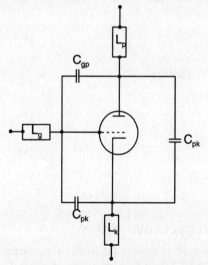

Fig. 6.1. Equivalent circuits at UHF.

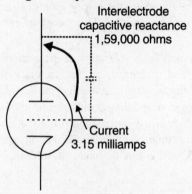

Fig. 6.2. (*a*) Interelectrode capacitance in a triode vacuum tube at 1 MHz.

MICROWAVE TUBES

For example, in Fig. 6.2(a), a 1-picofarad capacitor has a reactance of 159,000 ohms at 1 MHz. If this capacitor was the interelectrode capacitance between the grid and plate of a tube, and the rf voltage between these electrodes was 500 v, then 3.15 mA of current would flow through the interelectrode capacitance. Current flow in this small amount would not seriously affect circuit performance. On the other hand, at a frequency of 100 MHz the reactance would decrease to approximately 1,590 ohms and, with the same voltage applied, current would increase to 315 mA (view (b). Current in this amount would definitely affect circuit performance.

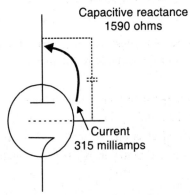

Fig. 6.2(b). Interelectrode capacitance in a triode vacuum tube at 100 MHz.

A good point to remember is that the higher the frequency, or the larger the interelectrode capacitance, the higher will be the current through this capacitance. The circuit in Fig. 6.2(c), shows the interelectrode capacitance between the grid and the cathode (Cgk) in parallel with the signal source. As the frequency of the input signal increases, the effective grid-to-cathode impedance of the tube decreases because of a decrease in the reactance of the interelectrode capacitance.

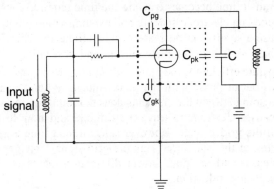

Fig. 6.2(c). Interelectrode capacitance in a vacuum tube.

If the input signal frequency is 100 MHz or greater, than the reactance of the grid-to-cathode capacitance (Cgk) is so small that most of the signal is short-circuited within the tube. Since the interelectrode capacitances are effectively in parallel with the tuned circuits, as shown in views (a), (b), and (c), they will also affect the frequency at which the tuned circuits resonate.

We can minimize the effect of IEC by reducing the IEC's (Cgp, Cgk, and Cpk). These can be reduced by two ways either by decreasing the area of the electrodes by using smaller electrodes or by increasing the distance between electrodes.

6.2.2 Lead Inductance

Another frequency-limiting factor is the **Lead Inductance** of the tube elements. As frequency increases, the inductive reactance $(X_L) = 2\pi\, fL$, increases and hence the voltage appearing at the active electrodes are lesser than the voltages at the base pins. Since the lead inductances within a tube are effectively in parallel with the interelectrode capacitance, the net effect is to raise the frequency limit. However, the inductance of the cathode lead is common to both the grid and plate circuits. This provides a path for degenerative feedback which reduces overall circuit efficiency.

By decreasing the L the effect of Lead Inductance can be minimized. The L can be decreased by using larger sized short leads without base pins. However by reducing the length and increasing the Area of base pins leads to less power handling capability.

6.2.3 Transit Time

A third limitation caused by tube construction is Transit Time. **Transit Time** is the time required for electrons to travel from the cathode to the plate.

Transit Time $(t) = d/v$

where, d = distance between anode and cathode electrodes

t = transit time and

v = velocity of electrons

Under equilibrium condition, the Static energy (eV) = Kinetic energy $(1/2\ mv^2)$.

$$t = \frac{d}{\sqrt{2eV/m}}$$

At low frequencies, the transit time is negligible compared to the signal's period. While some small amount of transit time is required for electrons to travel from the cathode to the plate, the time is insignificant at low frequencies. In fact, the transit time is so insignificant at low frequencies that it is generally not considered to be a hindering factor. However, at high frequencies, transit time becomes an appreciable portion of a signal cycle and begins to hinder efficiency. For example, a transit time of 1 nanosecond, which is not unusual, is only 0.001 cycle at a frequency of 1 megahertz. The same transit time becomes equal to the time required for an entire cycle at 1,000 MHz. Transit time depends on electrode spacing and existing voltage potentials. Transit times in excess of 0.1 cycles cause a significant decrease in tube efficiency. This decrease in efficiency is caused, in part, by a phase shift between plate current and grid voltage.

If the tube is to operate efficiently, the plate current must be in phase with the ground-signal voltage and 180 degrees out of phase with the plate voltage. When transit time approaches 1/4 cycle, this phase relationship between the elements does not hold true. A positive swing of a high-frequency grid signal causes electrons to leave the cathode and flow to the plate. Initially this current is in phase with the grid voltage. However, since transit time is an appreciable part of a cycle, the current arriving at the plate now lags the grid-signal voltage. As a result, the power output of the tube decreases and the plate power dissipation increases. Another loss of power occurs because of Electrostatic induction.

The electrons forming the plate current also electrostatically induce potentials in the grid as they move past it. This electrostatic induction in the grid causes currents of positive charges to move back and forth in the grid structure. This back and forth action is similar to the action of hole current in semiconductor devices. When transit-time effect is not a factor (as in low frequencies), the current induced in one side of the grid by the approaching electrons is equal to the current induced on the other side by the receding electrons. The net effect is zero since the currents are in opposite directions and cancel each other. However, when transit time is an appreciable part of a cycle, the number of electrons approaching the grid is not always equal to the number going away. As a result, the induced currents do not cancel. This uncancelled current produces a power loss in the grid that is considered resistive in nature. In other words, the tube acts as if a resistor were connected between the grid and the cathode. The resistance of this imaginary resistor decreases rapidly as the frequency increases. The resistance may become so low that the grid is essentially short-circuited to the cathode, preventing proper operation of the tube.

MICROWAVE TUBES

Several methods are available to reduce the limitations of conventional tubes, but none work well when frequency increases beyond 1,000 MHz. Interelectrode capacitance can be reduced by moving the electrodes further apart or by reducing the size of the tube and its electrodes. Moving the electrodes apart increases the problems associated with transit time, and reducing the size of the tube lowers the power-handling capability. You can see that efforts to reduce certain limitations in conventional tubes are compromises that are often in direct opposition to each other. The net effect is an upper limit of approximately 1,000 MHz, beyond which conventional tubes are not practical.

6.2.4 Gain-Bandwidth Product Limitations

In ordinary vacuum valves the optimum gain is normally achieved by resonating the output circuit as shown in Fig. (6.3).

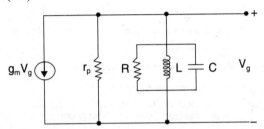

Fig. 6.3. Equivalent output-tuned circuit of a pentode.

It is assumed that $r_p \gg \omega L_k$ the load voltage is given by

$$V_L = g_m \cdot V_g / \{G + j\ [\omega\ C - 1/(\omega L)]\} \qquad \ldots 6.1$$

where, V_L = Load voltage, $G = 1/r_p + 1/R$

r_p = Plate resistance
R = Load resistance

L (Inductance) and C (capacitance) are tuning elements.
The resonant frequency is given by

$$f_r = 1/2\pi/\sqrt{LC} \qquad \ldots 6.2$$

The maximum voltage gain A_{max} at resonance frequency given by

$$A_{max} = g_m / G$$

Since the bandwidth is measured at the half power point mean G is given by

$$G = \omega C - 1/(\omega L)$$

The roots of the quadratics equation give the extreme frequencies ω_1 and ω_2

$$\omega_1 = G/2C - \sqrt{(G/2c)^2 + 1/LC} \qquad \ldots 6.3$$

$$\omega_2 = G/2C + \sqrt{(G/2c)^2 + 1/LC} \qquad \ldots 6.4$$

Bandwidth $= \omega_2 - \omega_1 = G/C$ (for $(G/2c)^2 \gg 1/LC$) $\qquad \ldots 6.5$

Hence the gain bandwidth product of the circuit is given by

$$A_{max}\ (BW) = g_m / C \qquad \ldots 6.6$$

The gain bandwidth product is thus independent of frequency. For a given tube, a higher gain can be achieved at the expense of a narrower bandwidth. This restriction is applicable to a resonant frequency only. In microwave devices this limitations can be overcome by use of reentrant cavities and slow wave tubes for a larger gain over larger bandwidth.

6.3 MICROWAVE TUBES

The conventional vacuum tubes, such as triodes, tetrodes and pentodes are still used as signal sources of low power output at low microwave frequencies.

Microwave tubes are generally categorized in two different groups as shown in Fig. (6.4), one is linear beam tubes and other is cross field tubes. Klystrons, TWT and

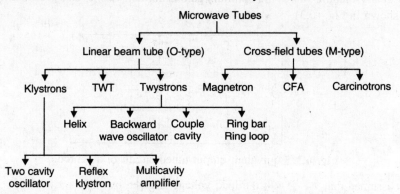

Fig. 6.4. Microwave vacuum-tube family tree.

Twystrons are linear beam or 'o' type tubes in which the accelerating electric field is in the same direction as the static magnetic field used to focus the beam of electrons. Magnetron, cross field Amplifier and carinotrons are crossed field type tubes or M-Type, where the static magnetic field is perpendicular to the electric field. The basic principle of operation of the microwave tubes is the transfer of power from a dc voltage source to a source of ac voltage by means of a current density modulated electron beam. This can be achieved by accelerating electron in a static electric field and retarding them in an ac field. The density modulation of the electrons beam facilities more electrons to be retarded by ac field and than accelerated by ac field which therefore make's possible a net energy to passed to the ac electric field.

In this chapter the operating principles of some mostly used microwave tubes such as Klystrons, Reflex Klystrons, Travelling wave tubes (TWT), Magnetrons and Backward oscillator etc., are discussed.

6.4 KLYSTRONS

A Klystron is a vacuum tube that can be operated either as an oscillator or as an amplifier of power at microwave frequencies. Two basic configurations, of Klystron tubes such as one is called **reflex klystron,** used as a low power microwave oscillator and another is called **multicavity klystron,** used as a low power microwave amplifier.

Klystrons are velocity modulated tubes that are used in Radar and communication systems as oscillators and amplifiers.

Klystrons make use of the transit time effect by varying the velocity of an electron beam. A Klystron uses one or more special cavities, which modulate the electric field around the axis of the tube.

6.4.1 Two Cavity Klystron Amplifiers

The two cavity klystron amplifiers are widely used for microwave amplification.

The idea of klystron amplifier was introduced by Varian brothers in 1939.

MICROWAVE TUBES

Construction

The schematic diagram of a two cavity amplifier is shown in Fig. (6.5). It consists of an electrons gun, buncher cavity, catcher cavity and collector as shown in Fig. (6.5).

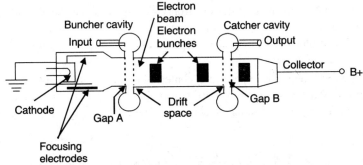

Fig. 6.5. The schematic diagram of a two cavity amplifier.

Electrons emitted by the cathode are focused by one or more focusing electrodes placed in front of the cathode. The electrons beam then passes through the buncher and catcher cavity and finally collected by the collector. By using coupling loop the RF power can be coupled through the buncher and catcher cavities.

Operation

Klystron amplification, power output, and efficiency can be greatly improved by the addition of intermediate cavities between the input and output cavities of the basic klystron. Additional cavities serve to velocity-modulate the electron beam and produce an increase in the energy available at the output.

It is seen that a high-velocity electron beam is formed, focused and sent down a long glass tube to a collector electrode. The beam passes through gap-A in the buncher cavity, to which the input *RF* signal to be amplified is applied, and it is then allowed to drift freely, without any influence from *RF* fields until it reaches gap B at the catcher cavity. The oscillation will be excited in the second cavity which is of a power much higher than those in the buncher cavity, so that a large output can be achieved. Collector electrode will collect the beam.

The effect of the gap voltage upon individual electrons may be investigated. Assume the condition when there is no voltage across the gap, the electrons passes it are unaffected and continue to the collector with the same constant velocities, which they had before approaching the gap as shown in Fig. (6.5). After an input has been fed to the buncher cavity, an electron will pass gap A at the time when the voltage across this gap is zero and going positive, assume this be the **reference electron 'y'** This electron 'y' is unaffected by the gap, and thus it is shown with the same slope on the Apple gate diagram in Fig. (6.8). Another electron, let it be 'Z' passes a gap slightly later than 'y' Electron 'Z' is slightly accelerated by the now positive voltage across gap 'A' and given enough time, it will catch up with the reference electron 'y'.

As shown in Fig. (6.8), it has enough time to catch electron 'y' easily before gap B is approached. Another electron 'x' passes gap 'A' slightly before the reference electron 'y'. Although it passed the gap 'A' before electron 'y', it was retarded by the negative voltage than present across the gap. It has an excellent chance of catching electron x before gap 'B'.

Fig. (6.8) shows the Apple Gate diagram for Klystron Amplifier. As electron pass the buncher gap, they are velocity modulated by the *RF* voltage existing across this gap. This velocity modulation is not sufficient to allow amplification by the Klystron. Electrons have the opportunity of catching up with other electrons in the drift space. When an electron catches up with another one, it may simply pass it and forge ahead. It may exchange energy with the slower electron, giving it some of its excess velocity, and the two bunches together and move on with the average velocity of the beam. Now, as the beam progresses farther down the drift tube, so the bunching becomes more complete, as more and more of the faster electron catch up with bunches ahead. Eventually, the current passes the catcher gap in quite pronounced bunches and therefore varies cyclically with time. This variation in current density is known as **current modulation** and which enables the Klystron to have sufficient gain.

The bunching will occur once per cycle, centering on the reference electron as shown in the apple gate diagram. In the Klystron, a little *RF* power applied to the buncher cavity results in large beam current pulse, being applied to the catcher cavity with a significant power gain as the result.

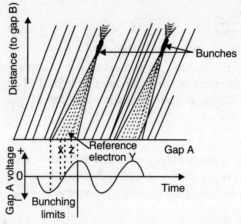

Fig. 6.8. Applegate diagram for Klystron amplifier.

Mathematical Analysis of a Klystron Amplifier: Due to the potential difference V_0 between the anode and cathode, the electrons form a high current density beam with the axial velocity 'v_0' which is obtained by equating Kinetic Energy (K.E.) and Potential energy (P.E.):

$$v_0 = \sqrt{\frac{2eV_0}{m}} = 0.593 \times 10^6 \sqrt{V_0} \, m/\sec \qquad \ldots 6.7$$

Refer Fig. (6.9) Let us assumed the drift space length is 'L' and the *RF* input signal to be amplified by the Klystron be V_S,

$$V_S = V_1 \sin \omega t \qquad \ldots 6.7$$

where, V_1 = amplitude of the signal ($V_1 \gg V_0$)

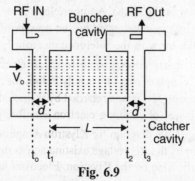

Fig. 6.9

MICROWAVE TUBES

The energy of the electron at the time of leaving buncher cavity can be given by

$$\frac{1}{2}mv_1^2 = e(V_0 + V_1 \sin \omega t_1)$$

or

$$v_1 = \sqrt{\frac{2e(V_0 + V_1 \sin \omega t_0)}{m}}$$

$$= \sqrt{\frac{2eV_0}{m}} \cdot \sqrt{1 + \frac{V_1}{V_0} \sin \omega t_1}$$

$$v_1 = v_0 \cdot \left[1 + \frac{V_1}{V_0} \sin \omega t_1\right]^{1/2} \qquad \left(\because v_0 = \sqrt{\frac{2eV_0}{m}}\right)$$

Now by expanding binomially and neglecting higher powers of $\sin \omega t$, we may get

$$v_1 = v_0 \cdot \left[1 + \frac{V_1}{V_0} \cdot \sin \omega t_1\right] \qquad \ldots 6.9$$

Eq. (9) is known as the **equations of velocity modulation**

$$\omega t_1 = \omega t_0 + \theta_g / 2$$

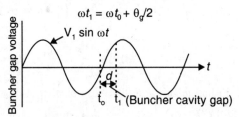

Fig. 6.10

θ_g = phase angle of the *RF* input voltage during which the electron is accelerated.

$$\theta_g = \omega t = \omega(t_1 - t_0) = \frac{\omega d}{v_0} \qquad \ldots 6.10$$

Electron Beam Bunching Process
At $+ \pi/2$, the maximum velocity occurs, so that

$$v_{1(max)} = v_0 \left[1 + \frac{V_1}{2V_0}\right] \qquad \ldots 6.11$$

At $- \pi/2$ the maximum velocity occurs, so that

$$v_{1(min)} = v_0 \left[1 + \frac{V_1}{2V_0}\right] \qquad \ldots 6.12$$

Let the distance in the drift space at which the bunching occurs from the buncher grid time t_1 is L_1 as shown in Fig. (6.11)

$$L_1 = v_0(t_1 - t_0) \qquad \ldots 6.13$$

The distance L_1 at $t - \pi/2$ at is given as

$$L_1 \text{ at } t - \pi/2 = v_{min}(t_1 - t - \pi/2) \qquad \ldots 6.14$$

The distance L_1 at $t + \pi/2$ at is given as

$$L_1 \text{ at } t + \pi/2 = v_{max}(t_1 - t + \pi/2) \qquad \ldots 6.15$$

$$t_1 - \pi/2 = t_0 - \pi/2\omega$$
$$t_1 + \pi/2 = t_0 + \pi/2\omega$$

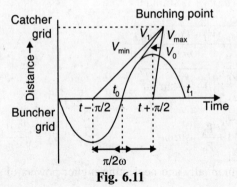

Fig. 6.11

From the above equations, we get,

$$L_1 \text{ at } t - \pi/2 = v_0 \left[1 - \frac{V_1}{2V_0}\right] \left[t_1 - t_0 + \pi/2\omega\right] \qquad \ldots 6.16$$

$$L_1 \text{ at } t + \pi/2 = v_0 \left[1 + \frac{V_1}{2V_0}\right] \left[t_1 - t_0 - \pi/2\omega\right] \qquad \ldots 6.17$$

$$\therefore \quad L_1 = v_0(t_1 - t_0) + v_0 \left[\frac{\pi}{2\omega} - \frac{V_1}{2V_0}(t_1 - t_0) - \frac{V_1}{2V_0} \cdot \frac{\pi}{2\omega}\right]$$

If the distance has to be same for those electrons at $-\pi/2, 0, +\pi/2$ bunchers, L_1 for all should be equal to $v_0(t_1 - t_0)$,

$$\pi/2\omega - \frac{V_1}{2V_0}(t_1 - t_0) - \frac{V_1}{2V_0} \cdot \frac{\pi}{2\omega} = 0$$

or $\quad -(t_1 - t_0) = \frac{\pi}{2\omega}\left[\frac{V_1}{2V_0} - 1\right] \cdot \frac{2V_0}{V_1}$

or $\quad -(t_1 - t_0) = \frac{\pi}{2\omega} - \frac{\pi V_0}{\omega V_1}$

$\therefore \quad t_1 - t_0 \approx \frac{\pi V_0}{\omega V_1}$ \qquad since $\frac{V_0}{V_1}$ very high, so $\pi/2\omega$ may be neglected)

By substituting this in equ. (13), we get

$$L_1 = v_0 \left(\frac{\pi}{\omega} \cdot \frac{V_0}{V_1}\right) \qquad \ldots 6.18$$

It should be noted that the mutual repulsion of the space charge is neglected but the qualitative results are similar to the preceding representation when the effects of repulsion are included.

The maximum bunching occurs for a value of π, since RF signal changes from $-\pi/2$ to $-\pi/2$ i.e., π. The value of $\pi = 3.682$ may be replaced in eq. (18) and we get,

$$L_{max} = 3.682 \frac{v_0 V_0}{\omega V_1} \qquad \ldots 6.19$$

MICROWAVE TUBES

If β, is the beam coupling coefficient of input cavity, which is given as

$$\beta = \frac{\sin(\theta_g/2)}{(\theta_g/2)} \qquad \ldots 6.20$$

where θ_g is the average gap transit angle $= \dfrac{\omega d}{v_0}$

L_{max} may be rewritten as

$$L_{max} = 3.682 \frac{v_0 V_0}{\omega \beta V_1} \qquad \ldots 6.21$$

Power Output (P_{out}) at the Catcher Cavity

RF voltage $= V_2 \sin \omega t_2$

Energy delivered by the electron to the bunch is given by

$$= -e \cdot V_2 \sin \omega t_2$$

In a cycle, the average energy delivered to the RF field,

$$P_{av} = \frac{1}{2\pi} \int_{\omega t_1}^{\omega t_2} (-eV_2 \sin \omega t_2) \, d\omega t_1 \qquad \ldots 6.22$$

The transit time for velocity modulated electron in the field free space between cavities, is given as

$$T = t_2 - t_1 = \frac{L}{v_1} = \frac{L}{\left[v_0\left(1+\dfrac{V_1}{V_0}\right)\sin \omega t_1\right]^{1/2}}$$

or

$$T = \frac{L}{v_0} = \left[1 - \frac{V_1}{2V_0}\sin \omega t_1\right] \qquad \ldots 6.23$$

Multiplying the above equation by ω, we get,

$$\omega T = \omega(t_2 - t_1) = \frac{\omega L}{v_0} = \left[1 + \frac{V_1}{2V_0} \cdot \sin \omega t_1\right] \qquad \ldots 6.24$$

In the above equation,

$$\frac{L}{v_0} = T_0 \text{ (transit time without RF voltage)}$$

V_1 in buncher cavity and

$$\frac{\omega L}{v_0} = \omega T_0 = \theta_0 = 2\pi N$$

Q_0 = The transit angle without RF voltage
N = The number of electron transit cycles in drift-space.

In the Klystron the bunching parameter, X is defined as

$$X = \frac{V_1}{2V_0}\theta_0$$

X is a dimensionless quantity and it is proportional to the input power.

The average power can be rewritten using equation (22),

$$P_{av} = -\frac{eV_2}{2\pi} \int_0^{2\pi} \sin(\omega t_1 + T) \, d\omega t_1$$

or
$$P_{av} = -\frac{eV_2}{2\pi} \int_0^{2\pi} \sin\left[\omega t_1 + \theta_0 \mp \left[1 - \frac{V_1}{2V_0}\sin \omega t_1 \mp\right]\right] d\omega t_1$$

The solution of this equation which is a Bessel function, can be given as
$$P_{av} = -eV_1 J_1(X) \sin \theta_0 \qquad \ldots 6.25$$
where $J_1(X)$ is the Bessel function of the first order for the argument X
Now energy transferred the N electrons transit cycles can be given by
Energy transferred $= N.P_{av} = -N.eV_2 J_1(X) \sin \theta_0$
$$N_e = I_0 = \text{Output current}$$
Energy transferred $= -I_0 V_1 J_1(X) \sin \theta_0 \qquad \ldots 6.26$
$X = 1.84$; where fundamental components (I_f) has its maximum amplitude.
The maximum value of $J_1(X) = 0.58$ for $X = 1.84$
for the Bessel function table
The maximum energy transfer may be written as
$$P_{av} = -I_0 V_2 (0.58) \sin \theta_0$$
where, $\sin \theta_0 = -1$ where $\theta_0 = 2n\pi - \pi/2$
∴ The power output at catcher cavity is
The power output, $P_{out} = P_{max} = 0.58 I_0 V_2 \qquad \ldots 6.27$

Efficiency of Klystron

Efficiency $(\eta) = \dfrac{P_{out}}{P_{in}}$

$P_{out} = 0.58\ I_0 V_2$

$P_{in} = I_0 V_0$ (It is the dc input)

$$\eta = \frac{0.58\ I_0 V_2}{I_0 V_0} = 0.58 \frac{V_2}{V_0} \qquad \ldots 6.28$$

The electronic efficiency of the Klystron amplifier is already defined by the above equation, in which the power losses to the beam loading and cavity wall are included. As V_2 is always less than V_0, it means the maximum efficiency can be obtained is 58%. In practice, the efficiency of a Klystron amplifier is in the range of 15 to 30%. Since the efficiency is a function of the transit angle and in maximum for a catcher gap transit angle of $\theta_g = \theta_0 = -\pi/2$ and is zero for $\theta_g = 2\pi$. Fig. (6.12) shows the maximum efficiency of Klystron as a function of catcher transit angle.

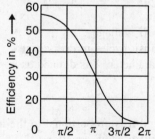

Fig. 6.12. Maximum efficiency versus transit angle.

MICROWAVE TUBES

Typical Characteristics

1. Frequency : 250 MHz to 100 GHz
2. Power : 10 kW to 500 kW for CW, 300 MW for pulsed
3. Power Gain : 15 db to 70 db and nominal is 60 db.
4. Noise figure : 15-20 db
5. Efficiency : 50% maximum (30-40% nominal)
6. Bandwidth : It is very limited (10-60 MHz)

Applications

1. It is mainly used as power output tubes in the
    (i) UHF TV Transmitters
    (ii) Satellite Ground Station
    (iii) RADAR Transmitters
    (iv) Troposphere scatters transmitters
2. As power oscillator between frequency ranges of 5 to 50 GHz.

6.4-2 Multi-cavity Klystron

Very frequently one or more additional cavities are inserted between the catcher and buncher cavities as illustrated in Fig. (6.13) shows a three-cavity system.

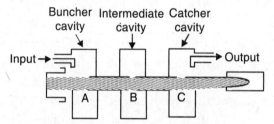

Fig. 6.13. Three-cavity Klystron.

Here oscillations are excited in the middle cavity by the partially bunched electron stream passing gap B; in this way a voltage is produced across B that also acts on the electron stream. By detuning the additional cavity so that its gap offers an impedance having an inductive component (i.e., resonant frequency slightly above the signal frequency) the phase of the voltage across B is related to the electron stream at B in such a manner as to cause further velocity modulation. This very considerably **increases the voltage amplification** of the tube, and likewise raises the efficiency that can be obtained. It is also possible to **increase the bandwidth** of a klystron amplifier by employing one or more intermediate cavities that are appropriately detuned; this expedient is used in all power klystrons designed for television service.

Klystron Bandwidth

The klystron is a narrow bandwidth device because of the use of cavity resonators. If all the cavities are tuned to the same frequency then it is referred as synchronous tuning. The bandwidth will be the order of 0.25% to 0.50%.

Stagger tuned klystrons have each cavity in a multicavity device turned to a slightly different frequency. The overall achievable gain is slightly reduced and the bandwidth is increased to the 2.5% to 3% range. In this case, gain is a trade off for increased bandwidth. In order to increase bandwidth, the catcher cavity is mixture to a frequency slightly higher than the buncher cavities. The bandwidth is increased to the 15% to 25% range but the gain reduces about 10db. The loss of gain may be compromised in many situations because of the increased frequency flexibility.

It is also possible to build a klystron with variable frequency tuning. All tuning methods involve varying the cavity dimensions.

Performance and Application of Multicavity Klystron Amplifier: The multicavity klystron finds its principal use as a power amplifier at frequencies of the order of 500 MHz and higher. By properly coordinating d-c anode voltage, the drift-space distance, and the signal amplitude in such a way as to achieve maximum bunching at the catcher position, efficiencies of the order of 40 percent can be achieved. Continuous wave powers of the order of 15 kw have been developed in commercial tubes designed for television service at frequencies of the order of 900 MHz. Pulsed powers of 30,000 kw have been obtained from klystron tubes operating at 3000 MHz; this represents the highest power that has ever been obtained from a vacuum tube.

The power gain of klystron power amplifiers is considerable, values of 30 db being easily possible in tubes of the three-cavity type. The bandwidth obtainable in klystron tubes is adequate for television applications. The output voltage is also very nearly proportional to the input voltage up to about 80 per cent of full output power, so that the klystron is a fairly linear power amplifier.

Klystron amplifiers do not, however, find important use as amplifiers of relatively weak microwave signals. This is because the noise figure of the klystron tubes that are available exceeds 25 db. As a result, travelling-wave tubes are much superior for the amplification of small microwave signals.

Comparison of Klystron and TWT

Klystron	TWT
1. Klystron circuit is resonant type.	1. The microwave circuit is non-resonant.
2. Field is stationary and only beams travels.	2. Field travels along with beam.
3. The interaction of electron in the klystron occurs only at the gaps of a few resonant cavities	3. The interaction of electron beam and RF field in the TWT is continuous over the entire length of the circuit.
4. In klystron wave is not propagating	4. The wave in TWT is a propagation wave.
5. Klystron uses cavities for input and Output circuits.	5. TWT uses non-resonant wave circuits For input and output.
6. Low power output	6. High power output
7. In klystron each cavity operates independently.	7. In coupled cavity TWT there is coupling effect between the cavities.
8. Short life	8. Long life
9. Narrow band device due to use of resonant wave circuit	9. Wide band device due to use of resonant cavity.
10. Noise figure is high due to use of resonant cavities.	10. Noise figure is low as it is not resonant.

6.4.3 Reflex Klystron

Another tube based on velocity modulation, and used to generate microwave energy, is the **reflex klystron** (repeller klystron). The reflex klystron contains a reflector plate, referred to as the repeller, instead of the output cavity used in other types of klystrons. The electron beam is modulated as it was in the other types of klystrons by passing it through an oscillating resonant cavity, but here the similarity ends. The feedback required to maintain oscillations within the cavity is obtained by reversing the beam and sending it back through the cavity. The electrons in the beam are velocity-modulated before the beam passes through the cavity the second time and will give up the

MICROWAVE TUBES

energy required to maintain oscillations. The electron beam is turned around by a negatively charged electrode that repels the beam. This type of klystron oscillator is called a **reflex klystron** because of the reflex action of the electron beam.

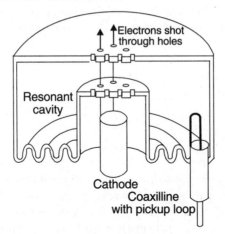

Fig. 6.14

Three power sources are required for reflex klystron operation:

1. Filament power,
2. Positive resonator voltage (often referred to as beam voltage) used to accelerate the electrons through the grid gap of the resonant cavity
3. Negative repeller voltage used to turn the electron beam around.

The electrons are focused into a beam by the electrostatic fields set up by the resonator potential (U_2) in the body of the tube.

The Fig. (6.15) shows a circuit diagram with a reflex klystron using a "doghnut" shaped cavity resonator.

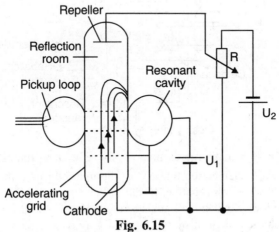

Fig. 6.15

Principle of Working: The *reflex klystron,* or *reflex oscillator* as it is sometimes called, is a form of klystron oscillator that requires only a single resonant cavity. Since it has an efficiency of only a few per cent the reflex klystron is essentially a low-power device, typically being used to generate 10 to 500 mw. The reflex klystron is particularly satisfactory for use in the frequency range 1000 to 25,000 MHz.

The structure of the reflex klystron is illustrated in Fig. (6.16). The tube consists of a cathode, a focusing electrode at cathode potential, a coaxial line or reentrant cavity resonator that also serves as an anode, and a repeller electrode that is at a moderate negative voltage with respect to the cathode. The cathode is so shaped in relation to the focusing electrode and anode that an electrode beam is formed that passes through a gap in the resonator as shown, and travels toward the repeller. Because the repeller has a negative potential with respect to the cathode, it turns these electrons back toward the anode when they have reached some point such as *a* in the repeller space; these returning electrons then pass through the gap a second time.

Mechanism of Operation : A qualitative understanding of the operation of the reflex klystron oscillator can be understood by assuming that oscillations already exist in the resonant cavity, and the examining the mechanism whereby the action of the electron beam sustains these oscillations.

The radio-frequency voltage produced across the gap by the cavity oscillation acts on the electrons travelling toward the repeller, causing the velocity of the electrons that emerge from the gap into the repeller space to vary with time in accordance with the radio-frequency voltage. That is, the electron stream entering the repeller space is velocity modulated. This causes the electrons passing through the gap at different parts of the radio-frequency cycle to take different lengths of time to return to the gap. The result is that when the electrons return through the gap they tend to do so in bunches.

The variation of position with time for some typical electrons in the anode-repeller space is shown in Fig. (6.16).

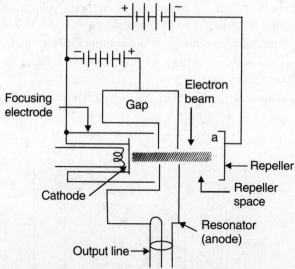

Fig. 6.16. Schematic representation of a reflex-Klystron oscillator.

Here *a* corresponds to an electron that passes through the gap at the instant when the gap voltage is zero and just becoming negative. In the distance-time coordinate system as shown in figure, this electron follows a parabolic path, corresponding to the height-time curve of a ball thrown upward and returned to earth by the force of gravity. This is to be expected, since the electric field in the anode-repeller space acts like a gravitational field. A second electron *b* that passes through the gap just before electron *a* is accelerated by the voltage across the gap, and enters the repeller space with greater velocity than did the first, or reference electron *a*. Electron *b* accordingly penetrates farther toward the repeller against the retarding field, and as a result takes a longer time to return to the anode, just as a ball thrown upward with greater velocity takes longer to return to earth. As a result, this second electron follows path *b* shown in Fig. (6.17) and tends to arrive at the anode on its return path at the same time as the reference electron because its earlier start is more or less compensated for by increased transit time.

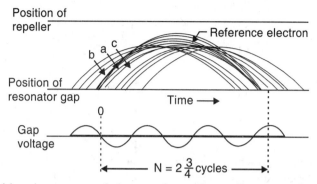

Fig. 6.17. Position-time curves of electrons in anode-repeller space of reflex Klystron.

In a similar manner, an electron passing through the anode gap slightly later than the reference electron will encounter a negative or retarding field across the gap, and so will emerge from the anode with less velocity than reference electron a. This third electron will then follow trajectory c, and will return to the anode more quickly than electron a, just as a ball thrown upward with less velocity returns to earth more quickly. Electron c hence tends to return to the anode at about the same time as electron a, since the later start of electron c is more or less compensated for by the fact that its transit time is less.

The end result of this action is that the returning electrons pass through the gap in bunches that occur once per cycle centered on a reference electron such as a in Fig. (6.17). If these bunches return at such a time during the radio-frequency cycle that the returning electrons are slowed down by the alternating gap voltage, then energy is delivered to the gap voltage and hence to the oscillations in the cavity. Examination of Fig. (6.17), plus a little reflection, will show that this condition occurs when the transit time N in the repeller space in cycles is

$$N = n + \frac{3}{4} \qquad \ldots 6.29$$

Here n can be any integer, including zero; in Fig. (6.17) one has $N = 2\frac{3}{4}$ cycle.

Thus to generate oscillations, the frequency of the resonant system is tuned to the desired value, and then the negative voltage on the repeller electrode is adjusted to give a transit time N that approximates the value called for by Eq. (29). It will be noted that the more negative the repeller voltage the more quickly will the electrons passing into the repeller space be returned to the gap, and hence the less will be the value N of the transit time.

Electronic Admittance of the Gap: The interaction that takes place between the returning electrons and the alternating voltage across the gap is equivalent, as far as the resonant cavity is concerned, to shunting admittance across the gap. This is illustrated schematically in Fig. (6.18), where this admittance Y_e, commonly called the *electronic admittance,* is represented as a conductance G shunted by a Suspectance B_e. The magnitude of the electronic admittance Y_e with increasing repeller-space transit time N can be represented by a spiral starting from the origin for $N = 0$, and expanding outward with increasing N (that is, with less negative repeller voltage), as illustrated in Fig. (6.19). The size of the spiral, i.e., the magnitude of the admittance Y_e for any given tube and beam current, depends upon the amplitude of the voltage across the gap, being maximum when the gap voltage is zero, and shrinking with increasing voltage. The phase angle of the admittance Y_e is, however, determined only by the transit time N, and is independent of the gap voltage; since $N = 0$ is vertically upward, the phase angle of Y_e is $360° \times (1/4 - N)$.

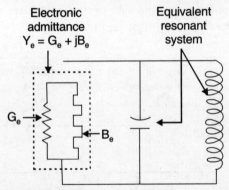

Fig. 6.18. Schematic representation of the resonant system of a reflex-Klystron oscillator, showing the relationship of the electronic admittance to the resonant system.

Thus a position of the electronic admittance spiral such as p or p_1 has a radial direction from the origin of Fig. (6.19) (including number of turns from the origin) that is determined by N. The position on the spiral is hence controlled by varying the repeller voltage.

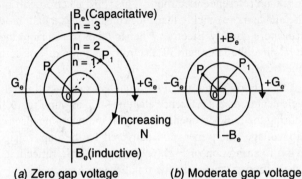

(a) Zero gap voltage   (b) Moderate gap voltage

Fig. 6.19. Polar diagram showing variation of the electronic admittance of the gap of a reflex Klystron as the repeller-space transit time N is increased.

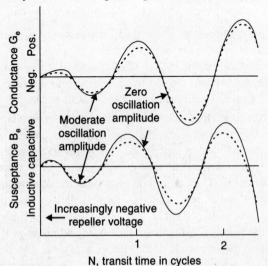

Fig. 6.20. Conductance and Suspectance components of the electronic admittance of a reflex Klystron for two amplitude of gap voltage.

MICROWAVE TUBES

On, the other hand, the absolute magnitude of the admittance corresponding to a point on the spiral such as p or p_1 that is, the distance or from the origin to the point in question (see Fig. 6.19), is determined both by the amplitude of the oscillations in the resonator, and by the number of turns N of p (or) from the beginning of the spiral. When plotted in rectangular coordinates, the conductance and Suspectance components of the electronic admittance have the character illustrated in Fig. (6.18), here the dotted curves correspond to a larger gap voltage than do the solid curves.

Oscillations will be generated in a reflex klystron whenever the electronic conductance G_e is negative and has an absolute magnitude for zero gap voltage that is more than the shunt conductance developed by the cavity across the gap. When this is the case oscillations will build up in amplitude until the magnitude of the negative electronic conductance is reduced to the point where it just equals the conductance of the cavity.

Operating Characteristics of Reflex Klystron Oscillators: Oscillations are obtained from a reflex klystron only for combinations of anode voltage and repeller voltage that give a favorable transit time. The situation existing in a typical klystron at a given frequency is illustrated in Fig. (6.21). Each shaded area corresponds to oscillations at a particular transit-time mode n. If the frequency of the oscillations is changed appreciably, the pattern still has the same general character shown in Fig. (6.21), but the locations of the regions of oscillation are shifted. This is because with a new frequency, a different transit time in seconds is required to give the same transit time in cycles.

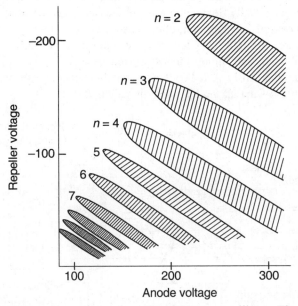

Fig. 6.21. Repeller mode pattern of a reflex Klystron oscillator. The shaded areas correspond to those combinations of voltages for which oscillations occur.

When the resonant frequency of the cavity, and the anode voltage at which the tube is operated, are both kept constant, then the amplitude of the oscillations obtained from a reflex klystron varies with the repeller voltage as shown in Fig. (6.22). The different oscillating regions correspond to different values of n in Fig. (6.21). i.e., to different transit-time modes. Oscillations have the maximum amplitude when the transit time is exactly $n+ 3/4$ cycles, corresponding to maximum possible negative conductance at a given oscillation amplitude. As the transit time departs from this optimum condition, the negative conductance tends to be less and the oscillations have progressively smaller amplitude. When the negative conductance for zero gap voltage is less than the shunt conductance of the cavity, oscillations cease.

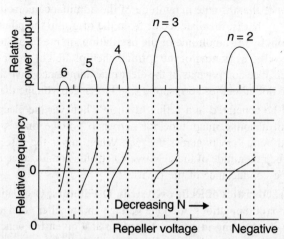

Fig. 6.22. Variation of output power and frequency of reflex oscillator as a function of repeller voltage for the tube of Fig. (6.21).

Frequency—Electronic Tuning: The frequency of the oscillations obtained from a reflex klystron is determined primarily by the resonant frequency of the cavity. When the resonant cavity is included in the evacuated portion of the tube, tuning can be accomplished over a moderate frequency range by mechanically flexing the walls of the cavity and/or simultaneously varying the gap spacing. Reflex klystrons are, however, often constructed so that the resonant system is external to the tube, as illustrated in Fig. (6.23). Under these conditions the frequency can be adjusted by means of a tuning plunger, and tuning ranges as great as 2 to 1 may be obtained provided the repeller voltage is simultaneously varied so that the repeller space transit time measured in cycles is kept approximately constant as the frequency is varied.

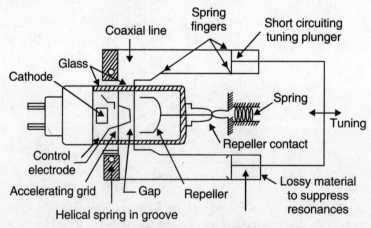

Fig. 6.23. Schematic representation of a reflex-Klystron oscillator with external resonant system supplied by a coaxial line.

The electronic Suspectance B_e also has an effect on the frequency as a result of the fact that it is a part of the equivalent resonant system (see Fig. 6.18). This electronic Suspectance depends upon the repeller-space transit time, as shown in Fig. (6.20). Since the transit time depends on both the anode and repeller voltages, these voltages affect the generated frequency. This is known as *electronic tuning* and is illustrated in Fig. (6.22); it provides a means for making slight corrections in the frequency and for achieving frequency modulation.

MICROWAVE TUBES

The mechanism whereby electronic tuning operates can be correlated with Fig. (6.19) by drawing on the admittance spiral for zero gap voltage the line corresponding to the *negative* of the resonator gap admittance Y_r, as shown in Fig. (6.24).

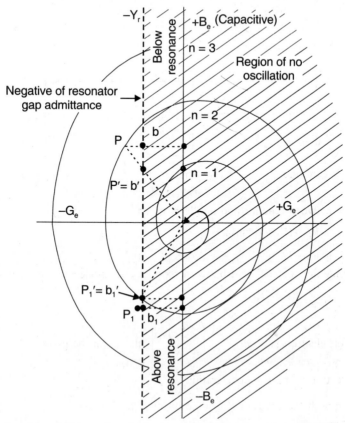

Fig. 6.24. Relationship of electronic admittance spiral to the admittance of the resonator gap.

Any repeller voltage that places the operating point on a part of the spiral to the left of this line is a possible oscillatory condition. For example, consider point p corresponding to $N = 1.9$. Here oscillations start up at a frequency corresponding to the point b on the resonant-circuit line. As the oscillations increase in amplitude, the point p moves toward the origin along a radial line since the electronic admittance spiral diminishes in amplitude with increasing amplitude of oscillations. As this happens, point b moves downward and the frequency increases, until finally equilibrium is established at an amplitude and frequency such that p and b have both moved to $p' = b'$ as shown.

Electronic tuning obtained in this way by varying the repeller voltage is a very important feature of the reflex klystron. It provides a means of obtaining fine tuning, and also a means of introducing frequency modulation. For example, by varying the repeller voltage between values corresponding to transit times represented by p and p_1 in Fig. (6.24) will cause the frequency to vary between values indicated by b' and b'_1. A total frequency variation of the order of 1 per cent can be obtained by electronic tuning in a typical case.

Mathematical Analysis of Reflex Klystron

Velocity Modulation: The analysis of a reflex klystron is similar to that of a two cavity klystron refer Fig. (6.25) electron in the velocity

$$v_0 = \sqrt{\frac{2eV_0}{m}}$$

$$v_1 = v_0\sqrt{\frac{1+V_1}{V_0}}\ \sin \omega t$$

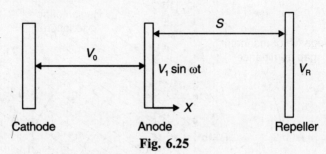

Cathode   Anode   Repeller

Fig. 6.25

The voltage between repeller and anode is given by

$$= V_R - (V_0 + V_1 \sin \omega t)$$

$$= V_R - V_0 \qquad\qquad (V_1 \ll V_0) \qquad\qquad \ldots 6.30$$

The retarding electrostatic field between repeller and anode may be expressed as

$$E = -\left[\frac{V_R - V_0}{S}\right] \qquad\qquad \ldots 6.31$$

The force on electron is $-eE = +e\left[\dfrac{V_R - V_0}{S}\right]$ and it can be given by

Force of electron = mass × acceleration = $m\dfrac{d_x^2}{dt^2}$

By equating both the equations of force on electron, we get

$$m\frac{d_x^2}{dt^2} = +\frac{e}{S}[V_R - V_0]$$

$$\frac{d_x^2}{dt^2} = +\frac{e}{mS}(V_R - V_0)$$

By integrating the above equation, we get

$$\frac{dx}{dt} = \frac{e}{mS}(V_R - V_0)\ t + C \qquad\qquad \ldots 6.32$$

$$t_1 = t, \frac{dx}{dt} = v_1$$

$$v_1 = \frac{e}{mS}(V_R - V_0)t_1 + C$$

or $\qquad C = \dfrac{e}{mS}\cdot(V_R - V_0)t_1$

Substituting of the 'C' in equation (32), we get,

$$\frac{dx}{dt} = \frac{e}{mS}(V_R - V_0)(t - t_1) + v_1$$

MICROWAVE TUBES

By integrating this equation, we get

$$x = \frac{e}{2mS}(V_R - V_0)(t - t_1)^2 + v_1 t + C_1 \quad \ldots 6.33$$

where $x = 0$ at the point of return from repeller space;
$t = t_2$

$$0 = \frac{e}{2mS}(V_R - V_0)(t_2 - t_1)^2 + v_1 t_2 + C_1$$

or $\quad C_1 = -\frac{e}{2mS}(V_R - V_0)(t_2 - t_1)^2 + v_1 t_2$

By substituting value of C_1 in equation (33), we get

$$x = \frac{e}{2mS}(V_R - V_0)[(t - t_1)^2 - (t_2 - t_1)^2] + v_1(t - t_2)$$

Again, $t = t_1$ and $x = 0$

$$-\frac{e}{2mS}(V_R - V_0)(t - t_1)^2 - v_1(t_2 - t_1) = 0$$

The round trip transit time $(t_2 - t_1)$ is given by

$$(t_2 - t_1) = \frac{-2mSv_1}{e(V_R - V_0)} \quad \ldots 6.34$$

The transit angle 'ωt' is given as transit angle at time 't'.

$$\omega(t_2 - t_1) = \frac{-2mSv_1}{e(V_R - V_0)}$$

or $\quad \omega t_2 = \omega t_1 - \frac{2mSv_1 \omega}{e(V_R - V_0)} \quad \ldots 6.35$

We know $\quad v_1 = v_0\left(1 + \frac{V_1}{V_0}\sin \omega t\right)^{1/2}$

$$v_1 \approx v_0\left(1 + \frac{V_1}{V_0}\sin \omega t\right) \quad (\because V_1 \ll V_0)$$

By substituting the value of v_1 from the above equation in the eq.(35), we get

$$\omega t_2 = \omega t_2 - \frac{2mS\omega}{e(V_R - V_0)} \cdot v_0\left[1 + \frac{V_1}{2V_0}\sin \omega t\right] \quad \ldots 6.36$$

$$-\frac{2mS\omega v_0}{e \ (V_R - V_0)} = \omega T'_2 = X = \theta'_0 \quad \ldots 6.37$$

$\theta'_0 =$ round trip dc transit angle of centre of bunch electron

Let, $\frac{V_1}{2V_0} \theta'_0 = X' =$ Bunching Parameter of Reflex Klystron Oscillator

Putting the of X' in eq. (36), we get

$$\omega t_2 = \omega t_1 + \theta'_0\left[1 + \frac{V_1}{2V_0}\sin \omega t\right] \quad \ldots 6.38$$

where, V_0 = electron gun anode voltage
V_R = Repeller voltage
S = Distance between cavity gap and repeller electrode
v_0 = Velocity of electron in gun
V_1 = Velocity due to RF voltage in addition to the electron accelerating voltage V_0
t_0 = Time for electron entering cavity gap at $x = 0$
t_1 = Time for same electron leaving cavity gap at $x = d$
t_2 = Time for same electron centered by retarding field at $x = d$.

Relation between Repeller voltage (V_R) and Accelerating voltage (V_0) : When,
$$V_1 << V_0, \text{ than}$$
$$\omega t_2 = \omega t_1 + \theta'_0$$

And for the maximum transfer of energy, the modes are $1\dfrac{3}{4}$ cycles apart.
The optimum value of θ'_0 is
$$\theta'_0 = (2\pi n - \pi/2)$$
By using eq. (37), we get
$$\theta'_0 = -\frac{2mS\omega}{e(V_R - V_0)} \cdot v_0 \quad \text{or} \quad v_0 = -\frac{e(V_R - V_0)}{2mS\omega} \cdot \theta'_0$$

$$v_0^2 = \frac{e^2(V_R - V_0)^2 (\theta'_0)^2}{4\omega^2 m^2 S^2} \qquad \dots 6.39$$

It is known that
$$\frac{1}{2} m v_0^2 = eV_0$$

or $$V_0 = \frac{m}{2e} v_0^2 \qquad \dots 6.40$$

By putting the value of v_0^2 from eqn. (39) in the above equation, we get
$$V_0 = \frac{m}{2e} \cdot \frac{e^2 (V_R - V_0)^2 (\theta'_0)^2}{4\omega^2 m^2 S^2}$$

or $$\frac{V_0}{(V_R - V_0)^2} = \frac{m}{2e} \frac{e^2}{4\omega^2 m^2 S^2} [2\pi n - \pi/2]^2$$

or it may be rewritten
$$\frac{V_0}{(V_R - V_0)^2} = \frac{1}{8} \cdot \frac{1 \cdot e}{\omega^2 S^2 m} [2\pi n - \pi/2]^2 \qquad \dots 6.41$$

Electronic Tuning of Reflex Klystron
The relationship between v_0 and V_R is given by the equation
$$\frac{V_0}{(V_R - V_0)^2} = \frac{1}{8} \cdot \frac{1 \cdot e}{\omega^2 S^2 m} [2\pi n - \pi/2]^2$$

MICROWAVE TUBES

The equation may be rewritten as

$$(V_R - V_0)^2 = \frac{8mS^2 V_0}{[2\pi n - \pi/2]^2 \cdot e} \cdot \omega^2$$

Now differentiating V_R with respect to ω, we get

$$2(V_R - V_0)^2 \frac{dV_R}{d\omega} = \frac{16mS^2 V_0}{[2\pi n - \pi/2]^2 \cdot e} \cdot \omega^2$$

$$\frac{dV_R}{d\omega} = \frac{8mS^2 V_0 \omega}{e[2\pi n - \pi/2]^2} \cdot \frac{1}{(V_R - V_0)}$$

By resubstituting the value of $(V_R - V_0)$ as

$$V_R - V_0 = \sqrt{\frac{8mS^2 V_0 \omega^2}{e(2\pi n - \pi/2)^2}}$$

$$\frac{dV_R}{d\omega} = \frac{8mS^2 V_0 \omega}{e(2\pi n/\pi/2)^2} \times \sqrt{\frac{e(2\pi n/\pi/2)^2}{8mS^2 V_0 \omega^2}}$$

$$= \sqrt{\frac{8mS^2 V_0}{e}} \cdot \frac{1}{2\pi n - \pi/2}$$

$$\therefore \frac{dV_R}{d\omega} = \frac{2\pi S}{(2\pi n - \pi/2)} \sqrt{\frac{8m V_0}{e}} \qquad \ldots 6.42$$

By analyzing the above relationship for electronic tuning of reflex klystron, it is revealed that the repeller voltage is crucial for frequency of oscillation. The variation in frequency is quite sensitive to repeller voltage adjustments and it may draw heavy current and get overheated. So as a precaution, application of repeller voltage prior to applicating anode voltage and connection of a protective diode across the klystron, so that repeller voltage should never become positive.

Maximum Theoretical Efficiency of Reflex Klystron

The maximum theoretical efficiency is given by

$$\text{Efficiency } (\eta) = \frac{2X' J_1(X)}{(2n\pi - \pi/2)} \qquad \ldots 6.43$$

The factor $X' J_1(X')$ reaches maximum value of 1.252 at $X' = 2.408$ and $J_1(X') = 0.52$. The maximum power is obtained when $n = 2$ of $1\frac{3}{4}$ mode.

By putting the values in above equation, we may get the maximum theoretical efficiency,

$$\eta_{max} = \frac{2(2.408) \ (0.52)}{2\pi(2) - \pi/2} = 22.78\% \qquad \ldots 6.44$$

But the practical value it's approximately 20%.

Typical Operating Characteristics of a Reflex Klystron

Frequency: 4 GHz to 200 GHz
Power Output: 10 mW to 2000 mW
Efficiency: Theoretically—22.70%
          Practically—10-20%
Repeller voltage: 15 to 300 volts

Cavity voltage: 275 to 300 volts

Tuning range: Electronically—1%

Mechanically—30%

Applications

The common applications of reflex klystron are:

(*i*) As local oscillator in microwave receiver

(*ii*) As a microwave signal source

(*iii*) As a pump source for parametric amplifiers

(*iv*) As an oscillator in frequency modulation of low power microwave link.

6.5 THE TRAVELLING-WAVE TUBE (TWT)

The traveling-wave tube is an amplifier that makes use of distributed interaction between an electron beam and a traveling wave. It is particularly suitable for amplification of very high frequency, such as 3000 MHz and higher.

The physical construction of a typical traveling-wave tube is shown in Fig. (6.26). Here an electron gun, normally of the Pierce type, produces a pencil like beam of electrons having a velocity that typically corresponds to an accelerating voltage of the order of 1500 volts. This beam is shot through a long, loosely wound helix, and is collected by an electrode at anode potential, as shown. An axial magnetic focusing field is provided to prevent the beam from spreading, and to guide it through the center of the helix.

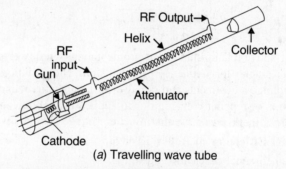

(*a*) Travelling wave tube

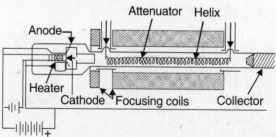

(*b*) Tube mounted in magnetic focusing field

Fig. 6.26. Schematic diagram of travelling-wave tube.

The signal to be amplified is applied to the end of the helix adjacent to the electron gun. Under appropriate operating conditions an amplified signal then appears at the other end of the helix. Simple coaxial input and output couplings are illustrated in Fig. (6.26); other arrangements, for example, waveguide coupling, may be employed if desired.

Mechanism of Operation: The applied signal propagates around the turns of the helix and produces an electric field at the center of the helix that is directed along the helix axis. Since the

MICROWAVE TUBES

velocity with which the signal propagates along the helix wire approximates the velocity of light if the frequency is not too low, the axial electric field due to the signal advances with a velocity that is very closely the velocity of light multiplied by the ratio of helix pitch to helix circumference. When the velocity of the electrons traveling through the helix approximates the rate of advance of the axial field, an interaction takes place between this moving axial electric field and the moving electrons, which is of such a character that on the average the electrons deliver energy to the wave on the helix. This causes the signal wave on the helix to become larger as the output end of the helix is approached; i.e., amplification is then obtained.

The mechanism of energy conversion by which the electrons deliver energy to the signal can be understood by reference to Fig. (6.27).

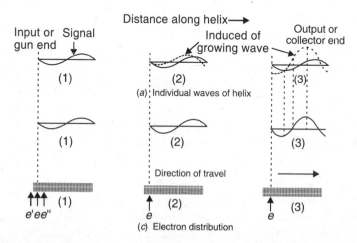

Fig. 6.27. Schematic diagram showing how the electrons bunch in a travelling-wave tube in a manner that causes the electron bunches to deliver energy to the wave.

Here the solid line in Fig. (6.27a) shows the distribution of the electric field strength along the helix axis produced by the signal in the absence of an electron beam. This distribution assumes that the polarities are so chosen that a positive field accelerates the electrons, and the entire distribution can be regarded as traveling toward the right. Consider now the group of electrons near the input end of the helix in the vicinity of e (see 1 of Fig. 6.27c) and assume that the axial electric field is zero at this point and is negative in the direction of the output end of the tube. An electron located exactly at e is then unaffected by the signal on the helix, as this electron encounters zero axial electric field. However, electron e' just to the left of e encounters a positive axial field and so is accelerated slightly thus tending to catch up with electron e. Similarly, electron e'' that was originally just to the right of e encounters a negative or decelerating field, and so slows down and tends to be overtaken by electron e. The electrons centered about e are thus velocity modulated. This action is similar to that occurring about the reference electron in a klystron tube, except that the velocity modulation continues to take place as the electrons travel toward the collector, instead of being produced at only one localized position.

After the electrons centered on e have traveled some distance down the tube, they begin to be bunched about electron e as shown at 2 of Fig. (6.27). If the velocity of the electrons at the input to the helix is the same as the velocity corresponding to the solid curve in Fig. 6.27(a), then e in 2 is still located at the zero of the solid wave, as shown. However, the presence of the bunch of electrons centered on e induces a second wave on the helix which produces an axial electric field that lags the solid curve of Fig. 6.27(a) by a quarter wavelengths as shown. The resultant electric field that is produced along the helix axis by the combined action of the two waves on the helix, shown in Fig. 6.27(b), is located a small fraction of a wavelength, closer to the gun end of the tube than is the solid curve of Fig. 6.27(a) because the negative maximum of the induced wave is opposite the center of the electron bunch. The electrons in the bunch therefore encounter a negative or retarding field and as a result deliver energy to the wave on the helix, which therefore becomes larger than at the input, as shown in Fig. 6.27(b). Bunching action continues to take place in 2 in Fig. 6.27(c) in spite of the fact that electron e is no longer at the zero of the resultant wave on the helix. This is because although all the electrons near e in 2 are slowed down, those to the left of e are slowed down less than are the electrons to the right of e.

As the electrons travel further along the helix, the situation at 2 in Fig. 6.27(c) changes to that at 3. Here the bunching is more complete and the induced wave grows in amplitude. The shift in phase of the resultant wave relative to the electron bunch is also increased because of the fact that the induced wave is larger. Each electron in the bunch now encounters a stronger retarding field, and furthermore, there are more electrons in the bunch. A large and increasing amount of energy is thereby delivered by the electron bunch to the wave on the helix, which is now much larger than the original signal.

Analysis shows that the amplitude of the resultant wave traveling down the helix increases exponentially. The total interchange of energy between the electrons and the helix wave is such that large amounts of power amplification can be achieved, typically from 20 to 40 db in a single tube.

Suppression of Oscillations—Helix Attenuator : In order to prevent oscillations from being spontaneously generated in a traveling-wave tube, it is necessary to prevent internal; feedback arising from reflections due to slight impedance mismatches at the terminals. Thus energy reflected at the output terminals will travel back to the gun end of the tube, and upon reflection there provides a spurious or feedback signal that is further amplified along with the desired signal. It may also be necessary to prevent backward-wave oscillations from being generated.

This situation is controlled by introducing an **attenuator** some place moderately near the input end of the tube, as shown in Fig. (6.26), that absorbs any wave propagated along the helix. This attenuator can take a number of forms; a common arrangement is a conducting coating of Aquadag painted on the glass wall of the tube. The attenuator absorbs not only the undesired backward or feedback wave; it likewise absorbs the desired forward or growing wave that is present on the helix. However, the bunching of the electrons is to a first approximation unaffected by the presence of the attenuator. Hence as the bunches of electrons emerge from the attenuating region, they induce a new forward traveling wave on the helix on the output side of the attenuator. This wave then travels along the helix toward the output nearly synchronously with the electron bunches.

Bandwidth: The travelling-wave tube is inherently a non-resonant device. As a result it can be made to have bandwidths that are enormous compared with those obtainable from amplifiers involving resonant circuits. The amplification characteristic of a typical helix traveling-wave tube is shown in Fig. (6.28); in this case the amplification is constant to within ± 3 db from 2000 to 4000 MHz.

MICROWAVE TUBES

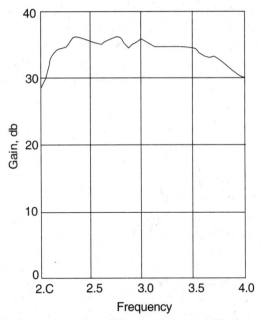

Fig. 6.28. Typical curve of amplification of a travelling-wave tube as a function of frequency.

The principal factors causing the gain of a traveling-wave tube to vary with frequency are:

(1) Variation in the velocity of the electric field along the axis of the tube (this causes the velocity of this wave and the electron velocity to be favorably related only over a limited range of frequencies).
(2) Variation of length of the tube in wavelengths.
(3) Variation in the strength of the axial electric field as a function of frequency,
(4) Failure to match the terminal impedance of the tube accurately at all frequencies.

The helix from of the travelling-wave tube is particularly suitable for achieving wideband operation. Over a wide frequency range the velocity of the axial field produced by the helix is substantially independent of frequency. Also the strength of the axial field in a helix increases at low frequencies in approximately the correct amount to take into account the fact that the length of the tube measured in wavelength becomes less as the frequency is reduced; thus factors 2 and 3 above tend to balance in the helix. It is also possible to obtain an impedance match to a helix over a very large frequency range.

Alternative Structures: Any arrangement that provides an axial component of electric field that advances with a velocity that is a small fraction of the velocity of light has the possibility of being used in a traveling-wave tube. The helix is only one of many 'slow-wave" structures that meet this requirement. Two other possibilities are illustrated in Fig. (6.29), both of these happen to be dispersive structures, i.e., they have a wave velocity that varies with frequency, and so have less bandwidth than the helix. However, their power dissipating ability is much greater.

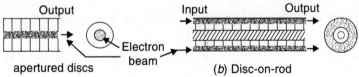

Fig. 6.29. Examples of slow-wave structures that can be used in travelling-wave tubes.

Operating Properties of Traveling-wave Tubes: Power gains of the order of 20 to 40 db combined with a bandwidth approaching 2 to 1 are readily realized when using a properly designed helix structure. Noise figures as low as 6 db at 3000 MHz and 11 db at 10,000 MHz have been achieved. Large amounts of continuous wave power can be generated in traveling-wave tubes, values of 1000 watts at range 500 to 1000 MHz and 100 watts at 3000 MHz having been realized in early experimental tubes. Pulsed powers of tens to hundreds of kilowatts have also been achieved in the laboratory. Efficiencies are from 10 to 40 per cent.

Traveling-wave tubes are best adapted for operation in the frequency range 500 to 10,000 MHz. They can, however, be constructed for still higher frequencies at the expense of extremely small helices; likewise, frequencies as low as 100 MHz can be realized by allowing the physical size of the tube to be relatively large.

Performance Characteristics of TWT:
(*i*) Frequency Range: 500 to 10000 MHz
(*ii*) Maximum Bandwidth: 2 : 1(0.8 GHz)
(*iii*) Efficiency: 10 to 40%
(*iv*) Power Output (CW): 1000 W (for the frequency range of 500-1000 MHz)
(*v*) Pulsed Power : 10 to 100 KW
(*vi*) Power gain: up to 60 db
(*vii*) Noise figure: 4 to 6 db

Applications of TWT:
(*i*) As a broad bond amplifier in microwave receiver and as a repeater amplifier in wide band communications linked.
(*ii*) Used as power output tubes in satellite communications because of in long life.
(*iii*) Used for CW radar and for radar jamming.
(*iv*) Pulsed high power tubes are used in airborne and ship borne radars.

Slow-Wave Structure: Slow wave structures are special circuits which are used in microwave tubes to reduce the velocity of the wave in a certain direction so that the electron beam and the signal wave can interact. We know that the phase velocity of a wave in waveguides is greater than the velocity of light in a vacuum. For the operation of TWT, the electron beam must keep in step with the microwave signal. The electron beam can be accelerated up to some extent which is about a fraction of the velocity of light; a slow wave structure must be incorporated in the microwave devices so that the phase velocity of the microwave signal can keep pace with that of the electron beam for effective interactions. There are several types of slow-wave structures shown in Fig. (6.30).

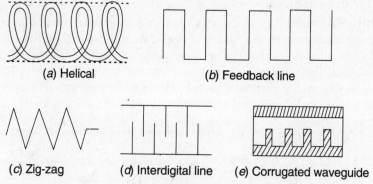

Fig. 6.30. Various slow wave structures.

MICROWAVE TUBES

The most commonly used slow wave structure is a helical coil with a concentric conducting cylinder as shown in fig. (6.31).

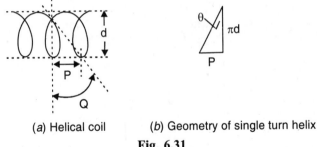

(a) Helical coil (b) Geometry of single turn helix

Fig. 6.31

If 'd' is the diameter of the helix and 'P' is the helix pitch, the time taken by the signal along the wire must be equal to that taken by the axial wave,

$$\frac{P}{v_P} = \frac{\sqrt{[P^2 + (\pi d)^2]}}{c}$$

or
$$v_P = \frac{cP}{\sqrt{[P^2 + (\pi d)^2]}} \approx \frac{cP}{\pi d} = \frac{\omega}{\beta} \qquad \ldots 6.45$$

when $P \ll \pi d$.

In general, the helical coil may be within a dielectric filled cylinder. For a very small pitch angle, the phase velocity along the coil in free space is approximately represented by above eq. (45).

Focusing: The beam of electrons moves axially without scattering by the influence of an axial magnetic field produced by the magnet around the helix. A permanent magnet is used for the low power tubes and a current carrying solenoid for high power tubes. The solenoids are suitable for high power tubes where *RF* power output is more than a few kW. But solenoid has main disadvantages such as it is bulky and consumes more power. Permanent magnet is used for focusing where weight and power consumption must be minimized such as in satellite communication.

Reflection of Waves : To attenuate the reflected wave propagating from any mismatch load at the output end to prevent from receiving the input end and resulting unwanted oscillation, an 'attenuator' is placed midway along the helix. The attenuation will attenuate both the forward and backward or reflected waves on the helix without affecting the electron beam.

Gain Characteristics of Helix TWT

The theory of TWT is first developed by Pierce. The analysis of TWT amplifier involves many complicated mathematical techniques. The output power gain is expressed as

$$G = 10 \log \left|\frac{\text{output voltage}}{\text{Input voltage}}\right|^2$$

or
$$G = -954 + 47.3 \, N \, C \, \text{dB} \qquad \ldots 6.46$$

where
G = The output power gain in dB,
N = Helix length in wavelength
l = Length of the slow wave structure in meters
$\lambda_S = v_p / f$

C = Gain parameter of the circuit
I_0 = DC beam current
Z_0 = Characteristics impedance of helix
V_0 = DC beam voltage
v_p = Axial phase velocity

The **gain will be maximum** when the beam velocity is approximately same as the velocity of the axial wave (v_p).

The high frequency of a TWT can be increased by decreasing the helix diameter. The lower and higher frequency limits are mainly due to the size limitation in TWT. The gain is linked by the length of helix at low frequencies.

The maximum **peak** power output of a single-helix type tube is about 3 kW because of the difficulty in removing heat due to ohmic loss from the helix conductor. The small signal gain is almost constant at low inputs. Now as we increase the RF power input, the RF power output does not increase proportionately but instead attains a maximum and start decreasing. The point at which the output power is maximum is known as **saturation point** and gain at this point is known as **saturation gain** as shown in Fig. (6.32).

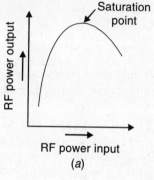
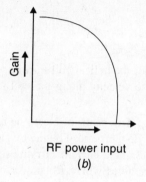

Fig. 6.32

Distortion

(1) At saturation level, the non linearity of the transfer characteristics gives rise to **inter-modulation distortion.**

(2) Another form of distortion occurs in a TWT amplifier as a result of **phase non-linearities.** At higher input levels, where more of the energy in the beam is converted to output power, the average beam velocity is reduced, and therefore the delay time is increased. Since the time delay in directly proportional to the phase delay, and its results in a phase shift at the output relative to the phase shift at saturation as shown in Fig. (6.33).

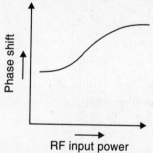

Fig. 6.33. Phase shift characteristic curve.

MICROWAVE TUBES

(3) **Harmonics distortions** occur at the saturation or near to the saturation point. Harmonics distortion occurs due to more intense electron bunching which produces sharp current peaks. These harmonics must be filtered out at the output end.

(4) Noise in TWT occurs mainly due to the **fluctuations in cathode** emission and in electron velocity which will contribute the current noise. The noise factor—for low power is less than 10 db, —for high power is 24 to 27 db, —for medium power is 18 to 30 db.

6.6 BACKWARD-WAVE OSCILLATORS

The **backward wave oscillator,** or the 'o' carcinotron, as it is sometimes called, is a development that has grown out of the traveling-wave idea. However, in contrast with the traveling-wave tube it is inherently an oscillator.

A typical backward-wave oscillator is illustrated in Fig. (6.34). The body of this tube consists of a folded transmission line, or alternately can be regarded as a waveguide operating in the TE_{10} mode, such that a wave traveling along the line winds itself back and forth, and in the process produces an axial component of electric field. The total path length from one end of the structure to the other is typically of the order of a dozen wavelengths under the usual operating conditions, and in a typical structure the line might cross the axis about fifty times.

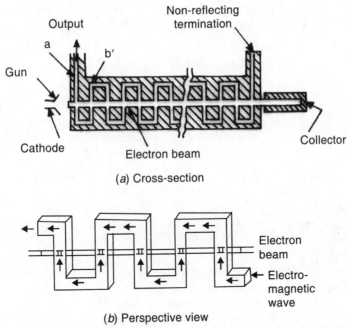

Fig. 6.34. Schematic diagram of backward-wave oscillator employing a folded-line structure.

An electron beam is directed along the axis through holes in the structures as shown. When this beam has a suitable velocity, there is an interaction between the electron stream and a wave traveling from right to left, *i.e.,* a *backward* wave, such that on the average energy is delivered to this wave by the electron beam. The collector end of the folded line is terminated with matched load impedance for the purpose of absorbing any power that might be reflected at the output or gun end of the structure as a result of an impedance mismatch at that point.

Mechanism of Operation: The principles involved in the operation of the backward-wave oscillator can be understood by assuming that a wave traveling from right to left in Fig. (6.34) (i.e., a backward wave) already exists on the folded-line structure, and then examining the interaction that results between this wave and the electron stream. First consider the situation existing at gap a. Here electrons enter from the cathode side at a uniform rate and are subjected to an alternating axial field that varies with time, as shown in Fig. 6.35(c). This situation is the same as that existing in the gap of the buncher of a multicavity klystron since the fact that the field happens to be produced by a backward wave instead of by a cavity does not make any difference to the electrons. The electrons passing through a thus experience velocity modulation, with a tendency for a bunch of electrons to be formed about the electron that passes through the gap when the alternating field at a is zero just turning decelerating {see Fig 6.35 (a) and (c)}. This reference electron also corresponds to the reference electron in the traveling-wave tube.

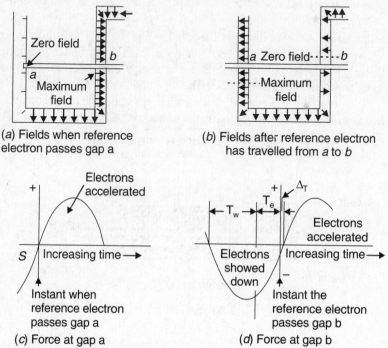

Fig. 6.35. Relationship of fields and electrons in the folded-line backward-wave tube of fig. (6.34)

The reference electron passes through gap b at a time T_e seconds after it passed through gap a, where T_e is the time it takes the electron to travel from a to b. This delay corresponds to $N_e = T_e f$ cycles, where f is the frequency. Now the field at a lags $N_w = T_w f$ cycles behind the field at b because of the time T_w it takes the wave to travel from gap b to a along the folded line. Thus when the electron crosses gap b the field it encounters is $N_e + N_w$ cycles behind the phase of the field it encountered at a. The geometry of the folded line is such, however, that a given flux line is directed oppositely at b from the direction of the same flux line when it reaches a (see Fig. 6.35). Therefore, if the electrons are given a velocity such that $N_w + N_e$ is just less than a half cycle, then the field that the reference electron encounters at gap b will be as shown in Fig. 6.35(b). This field at b will vary with time in the manner indicated by Fig. 6.35(d), where ΔT is a time such that $\Delta T \times f = N_\Delta$ is the fraction of a cycle by which $N_e + N_w$ fails to be exactly a half cycle. This field is substantially the same as that encountered at gap a. In fact, if $N_e + N_w$ is exactly a half cycle ($N_\Delta = 0$), then fields of identical phase are seen by the reference electron at gaps a and b. Thus the electrons passing through b experience further velocity modulation.

MICROWAVE TUBES

This situation is repeated over and over again as the electrons centered about the reference electron pass one gap after another, since the wave on the folded line always moves just far enough to the left during the transit time of the electrons from the last gap to the next gap to cause the field at the new gap to be of just the correct character to produce still further velocity modulation. Thus as the electrons travel toward the collector, the interaction with the backward wave on the folded line causes them gradually to group together in bunches, exactly as do the electrons in the traveling-wave tube. However, since the velocity of the electrons is so chosen that the quantity $N_e + N_w = f(T_e + T_w)$ is less than a half cycle by a small amount N_Δ cycles, these bunches of electrons advance in time with respect to the time of zero electric field at the gap. After passing n gaps the electron bunch has thus advanced $nN_\Delta = n\,\Delta T \times f$ cycles.

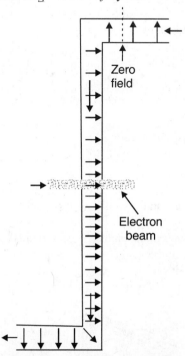

Fig. 6.36. Electron bunches passing through a gap in folded-line backward-wave oscillator under conditions such that the bunched electron beam delivers energy to the electric field at the gap.

By so choosing the electron velocity that at the collector and of the tube this total phase advance is approximately a half cycle, then over much of the length of the tube the electron bunches encounter relatively strong decelerating fields as they cross the gaps, as in Fig. (6.36). These fields slow down the electrons and cause energy to be delivered to the backward wave. This is very similar to the situation existing in a traveling-wave tube when the electron velocity is a little greater than the wave velocity. The result in the case of the backward-wave oscillator is that if the beam current is sufficiently great, self-sustaining oscillations are generated at a frequency such that $nN_\Delta \approx 0.5$.

The amplitude of the backward wave on the line builds up as shown in Fig. (6.37), this wave becomes larger as it progresses toward the gun end of the tube and passes more and more bunches of electrons from which it receives energy. At the same time, the alternating component of the beam current, i.e., the bunching becomes greater as the electrons progress toward the collector, as indicated in Fig. (6.37).

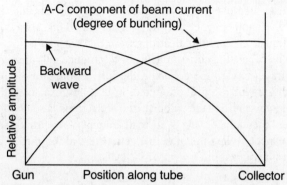

Fig. 6.37. Variation of amplitudes of the backward wave and of the a.c. component of beam current, as a function of position along a backward-wave oscillator.

It is seen that the backward-wave oscillator is a special kind of travelling-wave tube that possesses a built-in feedback mechanism whereby the power generated by a traveling-wave tube type of interaction is used to supply the signal required at the gun end of the tube to produce bunching of the electron stream.

Helix Form of Backward-wave Oscillator

The folded line of Fig. (6.34) is not the only arrangement that will support backward-wave operation. In particular, a helix structure forms the basis of an excellent backward-wave oscillator when the frequency is such that the circumference of a single helix turn lies between a quarter and a half wavelengths. Under these conditions the electric fields between successive turns have a phase difference similar to that existing between gaps a and b in Fig. (6.35). If now an electron beam of appropriate velocity travels along paths adjacent to the helix turns as shown in Fig. (6.38), it will encounter fields in gaps a and b qualitatively similar to those in gaps a and b of Fig. (6.35), when a backward wave is present on the helix. Although the exact details of the operation are a little more complicated than in the folded-line structure, the end result is the same.

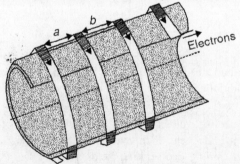

Fig. 6.38. Section of tape helix producing backward-wave action analogous to that existing in gaps a and b in figure (6.34).

In a helix backward-wave oscillator it is desirable to employ a hollow cylindrical electron beam, since only the electrons that pass adjacent to the wires are fully effective in delivering energy to the wave. In particular, electrons near the middle of the beam consume anode power but contribute little or nothing to the energy interchange that produces the backward-wave oscillations.

It is apparent that a given helix structure is able to function either as a traveling-wave amplifier in one frequency range, or as a backward-wave oscillator that operates at a somewhat higher frequency when using the same anode voltage. One may then wonder how a given tube knows whether it is supposed to be operating as a traveling-wave amplifier, or as a backward-wave

MICROWAVE TUBES

oscillator. The answer is determined at least in part by whether or not an attenuator is present that absorbs a wave traveling along the helix. With an attenuator as shown in Fig. (6.26), the backward wave cannot reach the gun end of the tube and produce the velocity modulation that is necessary to sustain the backward wave; if the electron velocity has a suitable value, traveling-wave action may then occur. However, if the attenuator is removed, or if it is not adequate, then Backward-wave oscillations may be generated spontaneously.

Frequency of Oscillation BWO

In the above discussion it was brought out that in order to generate oscillations the electron velocity must have a value such that $n\ N\Delta = n\Delta t\ f$ approximate one-half cycle, where n is the number of gaps and $N_\Delta = _\Delta T \times f$ represents the fraction of a cycle by which the field that the reference electron encounters at gap b differs in phase from a field passing through zero at this instant. Also $N_e + N_w$ is less than a half cycle by N_Δ. Hence

$$N_e + N_w + N_\Delta = T_e f + T_w f + N_\Delta = 0.5$$

and

$$N_\Delta \approx \frac{0.5}{n}$$

Hence

$$f \approx \frac{0.5[1-(1/n)]}{T_e + T_w} \qquad \ldots (47)$$

Here T_e, T_w, and f are defined as above.

Since T_w in Eq. (47) is determined by the geometry of the tube and so is nearly constant for any given tube, the frequency generated by the backward-wave tube is controlled by the transit time T_e, that is, by the velocity of the electron stream. The frequency hence depends upon the anode voltage, and is almost completely independent of the load impedance associated with the output terminals of the backward-wave oscillator, and the beam current. *Thus the backward-wave oscillator has the unique property of being voltage tuned.* Under practical conditions it is possible to achieve a 2-to-1 frequency range in a given tube by a voltage variation that is of the order of 10 to 1 or less.

Backward-wave oscillators are particularly suitable for use at microwave frequencies. Frequencies as large as 100 GHz have been achieved in structures that are physically large enough to be fabricated without undue difficulty. At the other extreme, frequencies below 200 MHz have been obtained from experimental tubes less than 2 ft long. In general, the maximum frequency that can be generated by backward-wave-oscillator action in a given structure is somewhat greater than the maximum frequency at which the same structure will amplify satisfactorily when used as a traveling-wave amplifier.

Power and Efficiency: Powers of the order of milliwatts are readily achieved in backward-wave oscillators at frequencies as large as 100 GHz. At lower frequencies relatively large powers can be developed, 100 watts at 3000 MHz having been reported in an experimental tube using a folded-line structure such as shown in Fig. (6.34).

Efficiencies obtainable in backward-wave oscillators are of the same order of magnitude as those realized with traveling-wave tubes. When powers are low the efficiencies will be of the order of a few per cent or less. With higher powers, efficiencies exceeding 10 per cent have been realized experimentally, and higher values appear to be possible.

6.7 Microwave Cross Field Tubes

Magnetron is a high power microwave oscillator tube invented by A.W. Hull in 1921. There are several types of magnetrons. However, all of them may be categorized in three groups as follows:

1. Cyclotron-frequency magnetrons
2. Traveling-wave magnetrons
3. Negative-resistance magnetrons

Magnetrons are basically vacuum diodes in which electrons are subjected to forces due to combined electric and magnetic fields that are mutually perpendicular to each other. In such conditions, it can be proved that electrons move in cycloidal paths; in these movements if they come across negative noise fields existing in the system, they deliver their power to these noise fields and make them to grow into powerful oscillations.

Magnetrons are M-Type devices or crossed field tubes in which the dc magnetic field and dc electric field are perpendicular to each other. The dc magnetic field plays a direct role in RF interaction.

Principle of Operation: the crossed field magnetron is a microwave generator device that uses electrical and magnetic fields crossed at right angles to each other in an interaction space between cathode and anode. The basis for operation of the device is the magnetron principle shown in Fig. 6.39.

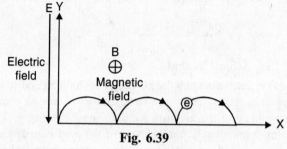

Fig. 6.39

Magnetron principle is derived from path of electron in crossed magnetic and electrical field. When an electron is injected into an electrical field it will accelerate from the cathode to the anode in a straight line. But if a perpendicular magnetic field is also present, then the electron moves in a curved cycloidal path. By properly arranging the cathode and anode structure, it is possible to keep electron cloud of the space charge moving in a curved path. This case is known as the **planer magnetron**.

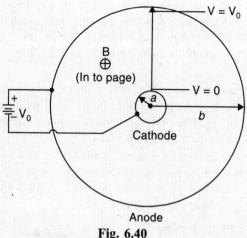

Fig. 6.40

MAGNETRON OSCILLATORS

The magnetron oscillator was the first device developed that was capable of generating large powers at microwave frequencies. It was the basis of the microwave radar transmitters of World

MICROWAVE TUBES

War II. Magnetrons are M-type devices or crossed field tubes in which the dc magnetic field and dc electric field are perpendicular to each other. In these tubes the dc magnetic field plays a direct role in RF interaction.

The essential elements of a typical magnetron oscillator are shown in Fig. (6.41). This consists of a cylindrical cathode surrounded by an anode structure that possesses cavities opening into the cathode-anode or *interaction* space by means of slots, as shown. Output power is withdrawn by means of a coupling loop, as illustrated, or alternatively a tapered waveguide can be employed.

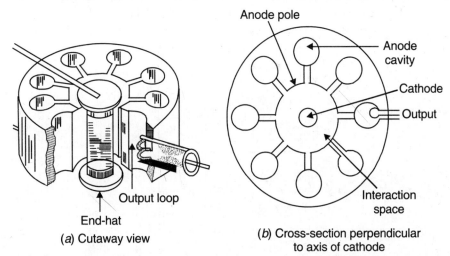

Fig. 6.41. Diagrams showing principal physical features of the cavity magnetron oscillator.

The magnetron requires an external magnetic field with flux lines parallel to the axis of the cathode. This field is usually provided by a permanent magnet, although an electromagnet can be employed.

Resonant Modes in Magnetrons and Their Separation: The anode cavities, together with the spaces at the top and bottom of the anode block, represent the resonant system of the oscillator. The fields associated with the cavities are of such a nature that the alternating magnetic flux lines pass through the cavities parallel to the cathode axis, while the alternating electric fields are confined largely to the slot and the region where the cavities open into the interaction space, and lie in planes perpendicular to the axis of the cathode (see Fig. 6.46).

The resonant system of a magnetron possesses a series of resonant frequencies, or *modes* as they are commonly called, equal in number to the number of cavities. This is because the resonant system can be regarded as consisting of a number of individual resonators, one for each cavity, which are all coupled together. It is known that when two resonant circuits are coupled together the result is to produce two resonant frequencies; similarly when n resonant cavities are coupled together, the result is n resonant frequencies or modes.

The mode employed in normal magnetron operation is that in which the phase difference between the adjacent anode poles is radians; this is called "mode." The other modes are characterized by some other value of phase difference between adjacent poles, but with the limitations that the total phase shift around the periphery of the interaction space must always be some multiple of 2π. Thus for the eight-cavity magnetron of Fig. (6.41), the π mode corresponds to a total phase shift of $\pm 8\pi$ radians around the periphery, while other modes correspond to total phase shifts of $\pm 6\pi, \pm 4\pi$, and $\pm 2\pi$ radians, corresponding to progressive phase differences between adjacent poles of $\pm 180, \pm 135, \pm 90$, and $\pm 45°$, respectively.

It will be noted that the electric fields existing in the interaction space of the magnetron correspond to the rotating fields in the air gap of a polyphase electrical machine. The mode corresponds to a single-phase system, while the other modes correspond to various polyphase arrangements. The phase differences that are characteristic of the various modes arise from the fact that each mode corresponds to a different frequency, and so is detuned differently from the resonant frequency of the cavities.

The relationship between the frequencies of the different modes in a typical magnetron is illustrated by the curve labeled "unstrapped" in Fig. (6.42). It will be noted that the desired mode differs very little in wavelength (or frequency) from the other modes.

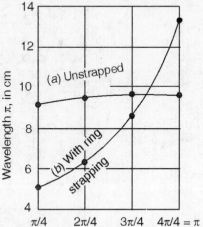

Fig. 6.42. Mode separation as observed experimentally in a particular magnetron oscillator when unstrapped and with ring strapping.

This situation introduces practical difficulties which make it important to separate the n mode from the other modes.

A method commonly used to achieve this result is called *strapping,* a typical example of which is shown schematically in Fig. (6.43). Here two rings are arranged in the end space as shown, with one ring connected to the even-numbered anode poles and the other ring connected to the odd-numbered anode poles. For the 7t mode all parts of each ring are at the same potential, but the two rings have opposite potential, as indicated by the + and − signs.

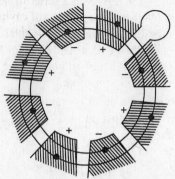

Fig. 6.43. Schematic representation of ring strapping.

The capacitance between the rings thus adds capacitive loading to the resonant cavities, thereby lowering the frequency of the π mode. For the other modes there is, however, a phase difference between the successive poles connected to a given ring. This causes current to flow

MICROWAVE TUBES

along the straps. Since the straps have inductance, this action places an inductive shunt in parallel with the equivalent resonant circuit of the cavity, thus raising the frequency for these modes. The net result is accordingly a separation in frequency of the π mode from other modes as illustrated by the curve in Fig. (6.42) labeled "ring strapping."

The frequency of the oscillations generated by a magnetron can be changed only by varying the resonant frequency of the magnetron circuits in some manner. One method of doing this consists in employing a "C ring" as illustrated in Fig. (6.44), which adds capacitance between the straps of a ring-strapped magnetron and thereby lowers the resonant frequency of the π mode by an amount depending upon the position of the C ring. The mechanism for adjusting the position of such a ring must transmit its effect through the vacuum-tight envelope of the magnetron by means of a flexible diaphragm.

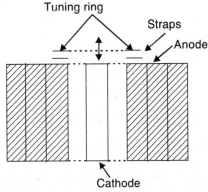

Fig. 6.44. Schematic representation of C-ring tuning of a magnetron.

Another method of controlling the resonant frequency of the anode system involves closely coupling a tunable high-Q resonant cavity to one of the anode cavities. It is possible in this way to couple sufficient reactance into the magnetron cavity system to modify the resonant frequency by a moderate amount.

Mechanism Oscillations in Magnetron

The first step in understanding the mechanism by which oscillations are generated in a magnetron consists in examining the behavior of the electrons emitted from the cathode when acted upon simultaneously by the d-c anode voltage and the axial magnetic field in the absence of radio-frequency oscillations. An electron emitted from the cathode under these conditions is accelerated toward the anode system by the radial electric field produced by the d-c anode voltage. If there were no magnetic field present, the electron would be drawn directly toward the anode in accordance with path a as shown in Fig. (6.45). However, as the electron gains velocity the axial magnetic field exerts a force on it.

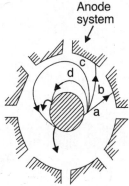

Fig. 6.45. Electron paths under the influence of a d.c. anode voltage for different values of axial magnetic field.

When the magnetic field is weak the electron path is deflected as shown by b in Fig. (6.45). However, when the intensity of the magnetic field is sufficiently great, the electrons are turned back toward the cathode without ever reaching the anode, as illustrated by paths c and d. The magnetic field which is just able to turn the electrons back to the cathode before reaching the anode is termed the *cutoff field;* when the magnetic field exceeds the cut-off value, and then in the absence of oscillations all the emitted electrons return to the cathode and the plate current is zero.

It will now be shown that if one postulates the existence of oscillations in the resonant structure, then when the magnetic field strength exceeds the cut-off there is an interaction between the electrons and the electric field that under favorable conditions causes the oscillations to receive energy from the electrons in the interaction space. Consider oscillations corresponding to the π mode; they produce radio-frequency fringing fields that extend into the interaction space in the manner illustrated in Fig. (6.46). If no radio-frequency field were present then electrons a and b would traverse paths shown by dotted lines a and b. However, the radio-frequency fields associated with the oscillations act on the electrons and modify the orbits. Thus electron a is so located with respect to these fields that its tangential velocity is opposed by the field. This electron is thus slowed down by the oscillations, and so delivers energy to the oscillations. Moreover, since electron a thereby loses velocity, the deflecting force that the magnetic field exerts on it is reduced; as a consequence this electron moves toward the anode, as shown by the solid path, instead of being turned back toward the cathode.

If the relationship between the d-c anode voltage and the magnetic field is now such that the tangential velocity of the electron makes the time required by electron a to travel from position 1 to 2 in Fig. (6.46) approximate a half cycle of the radio-frequency oscillations, then when electron a reaches point 2 in Fig. (6.46), it finds that the electric field has reversed its polarity from that shown.

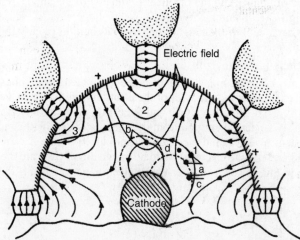

Fig. 6.46. Paths traversed by various electrons in a magnetron under oscillating conditions.

The result is that electron a continues to be slowed down, and continues to drift toward the anode. However, the velocity that electron a possesses does not change appreciably as it approaches the anode since the energy that this electron acquires from falling through the d-c anode-cathode voltage is in large measure delivered to the oscillations rather than being used to increase the velocity of the electron. Ultimately this working electron strikes the anode surface, after having delivered to the oscillations a large part of the energy represented by its fall through the d-c cathode-anode potential.

In contrast with this situation, consider next electron b in Fig. (6.46), which is emitted under circumstances such that it is accelerated by the radio-frequency field. Instead of being slowed down, this electron gains velocity and is therefore deflected more sharply by the magnetic field than if there were no oscillations present. As a result, this electron follows the solid line path *b* in Fig. (6.46), it is thus turned back toward the cathode even more quickly than indicated by the dotted path that would be followed in the absence of oscillations. This electron is harmful since it abstracts energy from the oscillations, but it is quickly removed from the scene of action and so does not have time to absorb very much energy. It does, however, bombard the cathode on its return with a velocity corresponding to whatever energy it has gained from the oscillating field; this causes "back heating" of the cathode. Typically about 5 per cent of the anode power of an operating magnetron is used in this way in heating the cathode.

It is very important to remember that there are three forces acting on electrons in the interaction region of the magnetron which are summarized as following:

1. Force due to electric field (-eE),
2. Force due to magnetic field (-e (v x B))
3. Centrifugal force (mv^2/r)

Associated with the action described above, there is also a **focusing mechanism** that tends to keep the working electrons in step with the fields in the interaction space in such a way that the working electrons deliver the maximum possible energy to the oscillations. For example, consider electron *c* in Fig. (6.46), which delivers some energy to the oscillators but was emitted a little too late to be in the correct position to make the maximum contribution. This electron is acted upon by a radial component of field from the oscillations as well as the tangential component. This radial field is in such a direction that it aids the d-c anode voltage. This increases the velocity with which electron *c* moves, and therefore assists it in catching up with electron *a* that is in the optimum position. Similarly, an electron *d* that is advanced beyond the optimum position encounters a radial field that opposes the anode voltage acting on it, and so attracted less strongly toward the anode. This causes electron *d* to be slowed down in its motion and thereby to fall back toward the optimum position. *This focusing action is equivalent to a velocity modulation that causes electrons such as c and d to form a bunch centered about electron a.*

The end result of these various actions that take place is to cause the orbit of the electrons to be confined to spokes, one for each two anodes, as illustrated in Fig. (6.47).

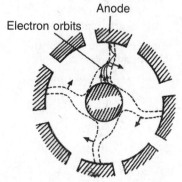

Fig. 6.47. Rotating space charge under oscillatory conditions and paths followed by electrons working their way through the space charge toward the anode.

In the case of the mode these spokes rotate at an angular velocity corresponding to two poles per cycle, and a certain fraction of the electrons emitted from the cathode travel out through the spokes, continuously delivering energy to the oscillations until these electrons reach the anode and disappear. Electrons emitted in the portions of the cathode between spokes are, however, returned very quickly to the cathode. Although these harmful electrons absorb some energy from the oscillations, this absorption is small in comparison with the energy that is delivered by the electrons in the spoke. The net result is that the oscillations receive a substantial net energy from the electrons.

Performance of Magnetrons

Magnetrons of the type described above are particularly suitable for generating relatively high powers at frequencies in the range 1000 to 25,000 MHz. Peak powers exceeding 1000 kw can be obtained in magnetrons designed for pulsed operation at 3000 MHz, while 1 kw continuous-wave power can be achieved at the same frequency. Efficiencies are typically 30 to 60 per cent, and in some cases may run even higher.

The performance of a magnetron is relatively sensitive to the strength of the magnetic field, the anode voltage, and the load impedance. Variations in any of these quantities will affect the output power, the efficiency, and the frequency to a marked extent.

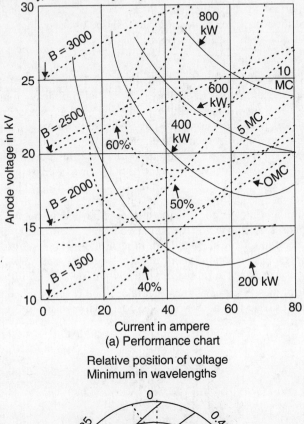

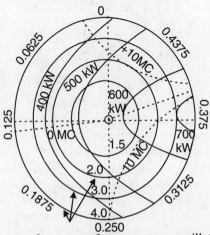

Fig. 6.48. Diagrams showing performance of a magnetron oscillator. The values of frequency indicated in the diagrams denote difference in frequency from an arbitrarily chosen reference value.

MICROWAVE TUBES

The two methods shown in Fig. (6.48) are in common use for portraying the relations involved.

1. The first is the *performance chart* which presents the magnetron performance for a given load impedance.
2. The second is termed the *Rieke diagram* and gives the magnetron performance as a function of load conditions for a given anode voltage and magnetic field strength. The Rieke diagram is a Smith chart, on which standing-wave circles and radial lines are drawn, but with the curvilinear impedance (or admittance) coordinates of the usual Smith chart are omitted.

It will be observed from Fig. 6.48(b) that the frequency of the magnetron oscillations is quite sensitive to changes in load impedance. This effect, termed **"pulling,"** is a source of considerable difficulty in many magnetron applications. It is similarly shown in Fig. 6.48(a) that for a given load impedance and magnetic field, the frequency also changes with variation of the d-c anode voltage; this effect is termed **"pushing."**

Mathematical Analysis of Magnetron Oscillators

Cavity magnetron which is most commonly used magnetron, we will deal with its mathematical analysis. Refer Fig. 6.49 for the analysis of cylindrical or cavity magnetron.

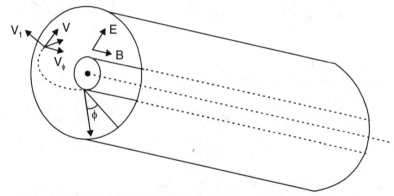

Fig. 6.49. Diagram for analysis of cylindrical.

Assume the radius of cathode is 'a' and anode is 'b' Let the angular displacement of the electron bends is ϕ. Since it is a cross field device magnetic and electric fields are perpendicular to each other and the path of the electrons is naturally parabolic in the presence of cross field.

A magnetic field is normal to the path of the electron; hence it creates a no work force. The force exerted by the magnetic field is given by the relation

$$F = (B \times v)e$$

The force component F_ϕ is given by

$$F = eBv_\rho$$

where F_ϕ = Force component in the direction

v_e = Velocity in the direction of the radial distance ρ from the center of the cathode cylinder
v = Velocity of electron
e = Charge of an electron
B = Magnetic field applied Torque in direction is given as

$$T_\phi = \rho F_\phi = e\rho v_\rho B \qquad \ldots 6.48$$

Angular momentum = angular velocity × moment of inertia

or Angular momentum $= \dfrac{d\varphi}{dt} \times m\rho^2$

The time rate of angular moments $= \dfrac{d}{dt}\left[\dfrac{d\varphi}{dt} \times m\rho^2\right]$ ...6.49

It gives the Torque in ϕ direction.

By equating eqn. (48 and 49), both the equation gives the value of Torque in ϕ direction, we get

$$\dfrac{d}{dt}\left[\left(\dfrac{d\varphi}{dt}\right)m\rho^2\right] = e\rho v_\rho B$$

$$2m\rho\dfrac{d\varphi}{dt} + m\rho^2\dfrac{d^2\varphi}{dt^2} = e.\rho.v_\rho.B \qquad ...6.50$$

$\therefore \quad v_\rho = \dfrac{d\rho}{dt}$

$\rho v_\rho = \rho \dfrac{d\rho}{dt}$

$\int \rho \dfrac{d\rho}{dt} = \dfrac{\rho^2}{2}$

By integrating the equation (50), w.r.t. 't', we get

$$2m\rho.\phi.m\rho^2.\dfrac{d\phi}{dt} = eB\dfrac{\rho^2}{2}$$

$m\rho\phi$ can be thought as a constant for a particular direction ϕ.

$$m\rho^2\dfrac{d\phi}{dt} + K = eB.\dfrac{\rho^2}{2} \qquad ...6.51$$

By applying boundary condition, at surface of the cathode $\rho = a$ and $\dfrac{d\phi}{dt} = 0$, we can find out the value of constant 'K'.

$$0 + Kd = \dfrac{e.B.a^2}{2}$$

or $\quad K = \dfrac{eBa^2}{2}$

Putting the value of K in equation (51), we get

$$m\rho^2\dfrac{d\phi}{dt} = \dfrac{eB}{2}(\rho^2 - a^2)$$

$$\dfrac{d\phi}{dt} = \dfrac{eB}{2m}\left[1 - \dfrac{a^2}{\rho^2}\right]$$

At cathode, when $\rho = a$ than $\dfrac{d\phi}{dt}$ tending to zero. And if $e \gg a$, $\dfrac{d\phi}{dt}$ approaches (ω_{max})

MICROWAVE TUBES

i.e.,
$$\left[\frac{d\phi}{dt}\right]_{max} = \omega_{max} = \frac{eB}{2m} = \frac{eBe}{2m} \qquad ...6.52$$

where B_e = Cut off magnetic flux density = B

We know that P.E. of electron = K.E. of electron

or
$$eV_0 = \frac{1}{2}mv^2$$

$$eV_0 = \frac{m}{2}(v_\rho^2 + v_\phi^2) \qquad ...6.53$$

(v_ρ and v_ϕ are components in ρ and ϕ directions in cylindrical coordinates)

$$v_\rho = \frac{d\rho}{dt}$$

and
$$v_\phi = \rho\frac{d\phi}{dt}$$

Substitute the values of v_ρ and v_ϕ in equation (53), we get

$$eV_0 = \frac{m}{2}\left[\left(\frac{d\rho}{dt}\right)^2 + \rho^2\left[\frac{d\phi}{dt}\right]^2\right]$$

From equation (52)

$$\left[\frac{d\phi}{dt}\right] = \omega_{max}\left[1 - \frac{a^2}{\rho^2}\right]$$

$$\therefore \quad eV_0 = \frac{m}{2}\left[\left(\frac{d\rho}{dt}\right)^2 + \rho^2.\omega_{max}^2\left[1 - \frac{a^2}{\rho^2}\right]^2\right] \qquad ...6.54$$

Now at anode $\rho = b\frac{d\rho}{dt} = 0$,

$$\frac{m}{2}\left[b^2.\omega_{max}^2\left(1 - \frac{a^2}{b^2}\right)^2\right] = eV_0 \qquad ...6.55$$

Also
$$\omega_{max}^2 = \left[\frac{eB_c}{2m}\right]^2 \qquad \text{(from eq. (52))}$$

B_c = cut off value of the magnetic flux density

$$\frac{m}{2}b^2\left[\frac{eB_c}{2m}\right]^2 \times \left[1 - \frac{a^2}{b^2}\right]^2 = eV_0$$

or
$$\frac{e^2 B_c^2 b^2}{8m}\left[1 - \frac{a^2}{b^2}\right]^2 = eV_0$$

or
$$B_c = \sqrt{\frac{8V_0 m}{e}} \Big/ b\left[1 - \frac{a^2}{b^2}\right]$$

or
$$B_c = \frac{1}{b}\sqrt{\frac{8mV_0}{e}} \qquad \ldots 6.56$$

(since $b \gg a$, so a^2/b^2 may be neglected)

Equation (56) is called the **Hull's cut off magnetic field equation.**
B_c is called the cut-off value of magnetic field.
If $B > B_c$ for a given V_0, the electron grazes the anode i.e., anode current is zero
The **cut off voltage** for a given B is

$$V_c = \frac{e}{8}mb^2.B^2\left(1 - \frac{a^2}{b^2}\right)^2 \qquad \ldots 6.57$$

Here it $V_0 < V_c$, for a given B, the electrons will not ready the anode. Eq. (57) is called the **Hull cut off voltage equation.**

Condition for maximum transfer of Energy: In the presence of the cross field, if we consider one such favorable electron, for the equilibrium state

$$\frac{mV^2}{r} = eVB$$

where r = radius of cycloidal path.

The Angular velocity, $\omega = \frac{v}{r} = \frac{eB}{m}$

$$T\frac{2\pi}{\omega} = \frac{2\pi m}{eB} \qquad (T = \text{Period for one revolution})$$

Now if magnetron has to oscillates, the feed back should be in phase or integral multiples of radius. If number of cavities are 'N' than the phaseshould be

$$\phi = \frac{2\pi n}{N}$$

where n is an integer indicating n^{th} mode of oscillations. It is already discussed that magnetron oscillators are operated in the n mode (i.e., $\phi = \pi$)

The angular velocity of RF field in the interaction space is expressed by

$$\frac{d\phi}{dt} = \frac{\omega}{\beta}$$

The maximum transfer of energy from the electrons to RF field takes place when the cyclotron frequency of electron is approximately equal to the angular velocity of the RF wave.

In other words,

$$\frac{\omega}{\beta} = \frac{d\phi}{dt}$$

or
$$\omega = \beta\frac{d\phi}{dt} = \frac{eB}{m} \qquad \ldots 6.58$$

Typical Characteristics of Magnetron Oscillator
(i) Frequency range : 500 MHz to 12 GHz
(ii) Power output: For pulsed mode - more than 250 kW, 10 mW at UHF band, 2mW at X-Band
(iii) Efficiency : 40% to 70%

MICROWAVE TUBES

Applications of Magnetron
(1) Radar Transmitter valve in pulse radar (mostly used)
(2) Voltage Tunable magnetrons (VTM's) are used as sweep oscillator) in missile application
(3) CW Radar Transmitter and fixed frequency magnetrons in industry as heating and in microwave oven.

SUMMARY

- Tubes that are efficient in the microwave range usually operate on the theory of VELOCITY MODULATION, a concept that avoids the problems encountered in conventional tubes.
- Conventional vacuum tubes such as Triode, Tetrodes and Pentode are less beneficial signal sources at frequency above 1 GHz because of following characteristics of ordinary vacuum tubes.
- These characteristics are:
  1. Interelectrode capacitance,
  2. Lead inductance,
  3. Electron transit time and
  4. Gain-Bandwidth product limitations.
- Transit Time is the time required for electrons to travel from the cathode to the plate.

$$\text{Transit Time } (t) = d / v$$

- Microwave tubes are generally categorized in two different groups one is linear beam tubes and other is cross field tubes.
- The basic principle of operation of the microwave tubes is the transfer of power from a dc voltage source to a source of ac voltage by means of a current density modulated electron beam.
- A Klystron is a vacuum tube that can be operated either as an oscillator or as an amplifier of power at microwave frequencies. Two basic configurations, of Klystron tubes such as one is called reflex klystron, used as a low power microwave oscillator and another is called multicavity klystron, used as a low power microwave amplifier.

$$v_1 = v_0 \left[1 + \frac{V_1}{2V_0} \cdot \sin \omega t_1 \right]$$

- **Equations of velocity modulation**

$$\omega t_1 = \omega t_0 + \theta_g / 2$$

$$L_{max} = 3.682 \frac{v_0 \cdot V_0}{\omega \beta V_1}$$

$$P_{out} = P_{max} = 0.58 I_0 V_2$$

- **Efficiency of Klystron**

$$\eta = \frac{0.58 I_0 V_2}{I_0 V_0} = 0.58 \frac{V_2}{V_0}$$

- This very considerably increases the voltage amplification of the tube, and likewise raises the efficiency that can be obtained. It is also possible to increase the bandwidth of a klystron amplifier by employing one or more intermediate cavities that are appropriately detuned;
- The *reflex klystron,* or *reflex oscillator* as it is sometimes called, is a form of klystron oscillator that requires only a single resonant cavity. The reflex klystron is particularly satisfactory for use in the frequency range 1000 to 25,000 MHz.

$$\frac{V_1}{2V_0}\theta'_0 = X' = \text{Bunching Parameter of Reflex Klystron Oscillator}$$

$$\frac{V_0}{(V_R - V_0)^2} = \frac{1}{8} \cdot \frac{1.e}{\omega^2 S^2 m} [2\pi n - \pi/2]^2$$

$$\frac{dV_R}{d\omega} = \frac{2\pi S}{(2\pi n - \pi/2)} \sqrt{\frac{8mV_0}{e}}$$

- **Maximum Theoretical Efficiency of Reflex Klystron**

  The maximum theoretical efficiency is given by

  $$\text{Efficiency } (\eta) = \frac{2X' \, J_1(X')}{(2n\pi - \pi/2)}$$

  $$\eta_{max} = \frac{2(2.408)(0.52)}{2\pi(2) - \pi/2} = 22.78\%$$

- The traveling-wave tube is an amplifier that makes use of distributed interaction between an electron beam and a traveling wave. It is particularly suitable for amplification of very high frequency, such as 3000 MHz and higher.

- The attenuator absorbs not only the undesired backward or feedback wave; it likewise absorbs the desired forward or growing wave that is present on the helix.

- The attenuator should also be so designed as to absorb the helix waves without introducing appreciable reflection.

- **Slow wave structures** are special circuits which are used in microwave tubes to reduce the velocity of the wave in a certain direction so that the electron beam and the signal wave can interact.

- **Gain Characteristics of Helix TWT**

  $$G = -9.54 + 47.3 \, N \text{CdB}$$

- The gain will be maximum when the beam velocity is approximately same as the velocity of the axial wave (v_p).

- **Frequency of Oscillation BWO**

  $$f \approx \frac{0.5 \, [1 - (1/n)]}{T_e + T_w}$$

- *Thus the backward-wave oscillator has the unique property of being voltage tuned.*

- The magnetron oscillator was the first device developed that was capable of generating large powers at microwave frequencies.

- The resonant system of a magnetron possesses a series of resonant frequencies, or *modes* as they are commonly called, equal in number to the number of cavities.

- The mode employed in normal magnetron operation is that in which the phase difference between the adjacent anode poles is radians; this is called "π **mode.**"

- The magnetic field which is just able to turn the electrons back to the cathode before reaching the anode is termed the "*cutoff field.*

- The frequency of the magnetron oscillations is quite sensitive to changes in load impedance. This effect, termed **"pulling,"** is a source of considerable difficulty in many magnetron applications. It is similarly that for a given load impedance and magnetic field, the frequency also changes with variation of the d-c anode voltage; this effect is termed **"pushing."**

MICROWAVE TUBES

$$B_c = \frac{1}{b}\sqrt{\frac{8mV_0}{e}}$$

- Hull's cut off magnetic field equation.
- The cut off voltage

$$V_c \frac{e}{8} mb^2 \cdot B^2 \left(1 - \frac{a^2}{b^2}\right)^2$$

$$\omega = \beta \frac{d\phi}{dt} = \frac{eB}{m}$$

SOLVED PROBLEMS

Problem 1. *A reflex klystron operates at the peak mode i.e., n = 2 with beam voltage is 300 V, beam current is 30 mA and signal voltage is 40 V. Calculate the (i) input and output power (ii) efficiency.*

Solution : Given V_0 = 300 V, I_0 = 30 mA

(i) Input power $(P_{d.c.}) = V_0 I_0$

$$P_{d.c.} = 300 \times 30 \times 10^{-3} = 9W$$

Output power $(P_{d.c.}) = \dfrac{2\ V_0 I_0 \times J_1(X')}{2n\pi - \pi/2}$

$$P_{d.c.} = \frac{2 \times 300 \times 30 \times 10^{-3} \times 1.25}{2 \times 2 \times \pi - \pi/2} = \frac{22.50}{10.99}$$

$$P_{d.c.} = 2.04\,W$$

(ii) Efficiency $(\eta) = \dfrac{P_{d.c.}}{P_{d.c.}} \times 100$

$$\eta = \frac{2.04}{9} = 0.22 \times 100 = 22\%$$

$$\eta = 22\%.$$

Problem 2. *A two cavity klystron amplifier having following parameters*
1. *Beam current (I_0) = 20 mA*
2. *Beam voltage (V_0) = 1000 V*
3. *Frequency (f) = 10 GHz*
4. *Spacing between centers of cavities (L) = 4 cm*
5. *Gap spacing in either cavity (d) = 1 mm*
6. *Shunt impedance (R_{sh}) = $50\,k\Omega$*

Calculate : (a) The electron velocity (b) The d.c. transit time of electrons (c) Maximum input voltage for maximum output voltage (d) Voltage gain.

Solution: (a) Electron velocity $(v_0) = 0.593 \times 10^6 \cdot \sqrt{V_0}$

$$= 0.593 \times 10^6 \cdot \sqrt{1000}$$

$$v_0 = 0.593 \times 10^6 \times 31.62$$

$$\boxed{v_0 = 18.75 \times 10^6\,m/\sec}$$

(b) DC transit time of electrons:

$$\theta_0 = \omega T_0 = \omega \frac{L}{v_0} \qquad \left(\therefore T_0 = \frac{L}{v_0}\right)$$

$$T_0 = \frac{L}{v_0} = \frac{4}{100} \times \frac{1}{18.75 \times 10^6}$$

$$T_0 = 0.21 \times 10^{-8} \text{ sec}$$

(c) Maximum input voltage

$$V_{in \text{ (max)}} = \frac{V_0 \times 3.68}{\beta_i \theta_0}$$

$$\theta_0 = \omega T_0 = 2\pi \times 10 \times 10^9 \times 0.21 \times 10^{-8}$$

$$\theta_0 = 1.31 \times 10^2 = 131 \text{ rad}$$

$$\beta_i = \sin\frac{\theta_g}{\theta_g/2}, \quad \theta_g = \omega\frac{d}{v_0} = \frac{2\pi \times 10 \times 10^9 \times 1 \times 10^{-3}}{18.75 \times 10^6} = 0.33 \times 10$$

$$\theta_g = 3.3 \text{ rad}$$

$$\theta_g/2 = \frac{3.3}{2} = 1.65 \text{ rad}, \quad \sin\theta_g/2 = \sin 1.65 \text{ rad} \approx 1$$

$$\therefore \quad \beta_0 = \beta_i = \frac{\sin(\theta_g/2)}{(\theta_g/2)} = \frac{1}{1.65} = 0.60$$

$$\therefore \quad V_{in \text{ (max)}} = \frac{1000 \times 3.68}{0.6 \times 1.31} = 40 \text{ volts}$$

$$\boxed{V_{in} \text{ (max)} = 40\text{V}}$$

(d) Voltage gain (A_v)

$$A_v = \frac{\beta_0^2 \theta_0 \left[\dfrac{J'(X)}{X}\right]}{R_0} \times R_{sh}$$

$$= \frac{(0.60)^2 (131)\left[\dfrac{0.582}{1.841}\right]}{20 \times 10^3} \times 50 \times 10^3$$

$$A_v = \frac{27.44}{20 \times 10^3 \times 1.841} \times 50 \times 10^3 = 37$$

$$\boxed{A_v = 37.}$$

Problem 3. *A pulsed cylindrical magnetron is operated with the following specifications:*
Anode voltage = 25 kV
Beam current = 25 A
Radius of cathode = 5 cm
Radius of anode = 10 cm
Magnetic density = 0.34 Wb/m^2
Calculate: (a) Angular frequency (b) Cut-off magnetic flux density (c) The cut-off voltage.

Solution: (a) Angular frequency $(\omega) = eB_0/m$

$$\omega = 1.759 \times 10^{11} \times 0.34 = 0.5981 \times 10^{11} \text{ rad}$$

$$\omega = 0.5981 \times 10^{11} \text{ rad}$$

(b) The cut-off voltage $(V_c) = \dfrac{eB_0^2 b^2}{8m} \cdot (1 - a^2/b^2)^2$

MICROWAVE TUBES

$$V_c = \frac{1}{8} \times 1.759 \times 10^{11} \times 0.34^2 \times (10 \times 10^{-2}) \times \left(1 - \frac{5^2}{10^2}\right)^2$$

$$\boxed{V_c = 142.97\, kV}$$

(c) The cut-off magnetic flux density (B_c)

$$= \frac{(8V_0\, m/e)^{1/2}}{b\left(1 - a^2/b^2\right)}$$

$$= \left[\frac{8 \times 25 \times 10^3 \times 1}{1.759 \times 10^{11}}\right]^{1/2} (10 \times 10^{-2})^{-1} \left(1 - 5^2/10^2\right)^{-1}$$

$$\boxed{B_c = 142.2\, mWb/m^2}$$

Problem 4. A helical TWT has diameter of 3 mm with 50 turns per cm
    (a) Calculate axial phase velocity
    (b) The anode voltage at which the TWT can be operated for useful gain.

Solution : (a) $V_P = c \times \dfrac{\text{Pitch}}{\text{Circumference}}$

$$\text{Pitch} = \frac{1}{50} = 0.02\, cm = 2 \times 10^{-4}\, m$$

$$\text{Circumference} = \pi \times D = 3.14 \times 3 \times 10^{-3}$$

$$= 9.42 \times 10^{-3}\, m$$

$$V_P = 3 \times 10^8 \times \frac{2 \times 10^{-4}}{9.42 \times 10^{-3}} = 0.63 \times 10^7\, m/s$$

$$\boxed{V_P = 0.63 \times 10^7\, m/s}$$

(b) $\quad eV_0 = \dfrac{1}{2} m V_P^2$

$$V_0 = \frac{1}{2} \cdot \frac{m}{e} \cdot V_P^2$$

$$V_0 = \frac{1}{2} \cdot \frac{9.1 \times 10^{-31}}{1.6 \times 10^{-19}} \times (0.63 \times 10^7)^2$$

$$= 2.84 \times 10^{-2} \times 0.39 \times 10^{14}$$

$$\boxed{V_0 = 1.10 \times 10^2\, V}$$

Problem 5. A cavity magnetron has the following specifications:
    Inner radius (a) = 0.15 m
    Outer radius (b) = 0.45 m
    Magnetic flux density = 1.2 mWb/m^2

Calculate:
(a) The null cut-off voltages (V_c)
(b) The cut-off magnetic flux density is the beam voltage is 6 kV.
(c) The cyclotron frequency.

Solution : (a) Cut-off voltage $(V_c) = \dfrac{eB_0^2 b^2}{8} \cdot m \left[1 - \dfrac{a^2}{b^2}\right]^2$

$= \dfrac{1.759 \times 10^{11}}{8} \times 1.2 \times 10^{-3} \times (0.45)^2 \left[1 - \left[\dfrac{0.15}{0.45}\right]^2\right]^2$

$\boxed{V_c = 50.666 \text{ kV}}$

(b) Cut-off magnetic flux density $(B_c) = \dfrac{(8V_0 m/e)^{1/2}}{b\left[1 - \dfrac{a^2}{b^2}\right]}$

$B_c = \dfrac{\sqrt{8 \times 6000}}{1.759 \times 10^{11}} \times \dfrac{1}{0.45 \left[1 - \left[\dfrac{0.15}{0.45}\right]^2\right]}$

$\boxed{B_c = 130.595 \, mWb/m^2}$

(c) Cyclotron frequency or angular frequency (ω_c)

$\omega_c = \dfrac{eB_0}{m} = 1.759 \times 10^{11} \times 1.2 \times \dfrac{10^{-3}}{2\pi}$

$\boxed{\omega_c = 0.0336 \, GHz}$

Problem 6. *A 250 kW pulsed cylindrical magnetron is operated with the following specifications:*

    Anode voltage = 25 kV
    Peak anode current = 25 A
    Magnetic field density = B_0 = 0.35 Wb/ m^2
    Radius of cathode cylinder = 4 cm
    Radius of anode cylinder = 8 cm

Calculate :
(a) The angular frequency
(b) Cut-off magnetic field
(c) Cut-off voltage
(d) Efficiency of the magnetron

Solution : (a) Angular frequency or cyclotron frequency

$\omega_c = \dfrac{eB_0}{m}$

$\boxed{\omega_c = 1.76 \times 10^{11} \times 0.35}$

$\omega_c = 0.616 \times 10^{11}$ rad

(b) Cut-off magnetic field B_c

$B_c = \dfrac{6.75 \times 10^{-6} \times 8 \times 10^{-2}}{(64-16) \times 10^{-4}} \sqrt{25 \times 10^3}$

$\boxed{B_c = 18 \ mWb/m^2}$

MICROWAVE TUBES

(c) Cut-off voltage (V_c)

$$V_c = \frac{1}{8} \times 1.76 \times 10^{11}(0.35)^2 \times (8 \times 10^{-2})^2 \left[1 - \frac{4^2}{8^2}\right]^2$$

$$\boxed{V_c = 151.58\,kV}$$

Problem 7. *A TWT has the following parameters.*
  Beam voltage V_0 = 2 kV
  Beam current I_0 = 4 mA
  Frequency f = 8 GHz
  Circuit length N = 50 in wavelength
  Characteristic impedance Z_0 = 20 Ω

Calculate :
(a) The gain parameter   (b) The power gain in db.

Solution : Given $V_0 = 2kV$
$$I_0 = 4\,mA$$
$$f = 8\,GHz$$
$$Z_0 = 20\Omega$$
$$N = 50\,\text{in wavelength}$$

(a) Gain parameter

$$C = \left[\frac{I_0 Z_0}{4V_0}\right]^{1/3} = \left[\frac{4 \times 10^{-3} \times 20}{4 \times 2 \times 10^3}\right]^{1/3}$$

$$\boxed{C = 2.16 \times 10^{-2}}$$

(b) Power gain

$$A_P = -9.54 + 47.3\ NC$$
$$= -9.54 + 47.3 \times 50 \times 2.16 \times 10^{-2}$$
$$= -9.54 + 51.0 = 41.46\,dB$$

$$\boxed{A_P = 41.46\,dB}$$

Problem 8. *A 400 kW cylindrical magnetron operating at X-band has the following specifications*

  Anode voltage $V_{d.c.}$ = 32 kV
  Beam current $I_{d.c}$ = 84 A
  Radius of cathode cylinder = 6 cm
  Radius of anode cylinder = 12 cm
  Magnetic field density = 0.01 Wb/ m^2

Calculate :
(e) Cyclotron angular frequency ω_c
(f) The cut-off magnetic flux density
(g) The cut-off voltage
(h) Efficiency (η).

Solution : (a) The cyclotron angular frequency

$$\omega_c = \frac{e}{m} B_0$$

$$\omega_c = 1.759 \times 10^{11} \times 0.01 = 1.759 \times 10^9\,rad/s$$

$$\omega_c = 1.759 \times 10^9\,rad/s$$

(b) The cut-off magnetic flux density B_c

$$B_c = \frac{1}{12} \cdot \sqrt{\frac{8 \times 32 \times 10^3}{1.759 \times 10^{11}}} \cdot \frac{1}{1-(6/12)^2}$$

$$\boxed{B_c = 0.0134 \, Wb/m^2}$$

(c) The cut-off voltage V_c

$$V_c = \frac{eB_0^2 b^2}{8 \, m} \left[1 - \frac{a^2}{b^2}\right]^2$$

$$V_c = \frac{1}{8} \times (0.01)^2 (12)^2 \times 1.759 \times 10^{11} \times \left[1 - \left(\frac{6}{12}\right)^2\right]^2$$

$$\boxed{V_c = 17.81 \, kV}$$

(d) Efficiency $(\eta) = \dfrac{P_{out}}{P_{in}} = \dfrac{400 \times 10^3}{32 \times 10^3 \times 84} \times 100$

$$\boxed{\eta = 14.8\%}$$

REVIEW QUESTIONS

1. Discuss the limitations of conventional tubes at microwave frequencies and explain the remedy for these.
2. How the microwaves tubes are classified?
3. Explain the working principle of two cavity klystron amplifier by giving the apple-gate diagram. **(MDU-2006)**
4. Explain the functioning of reflex klystron oscillators with the help of apple gate diagram and give the performance characteristics.
5. Differentiate between velocity and current modulation.
6. Explain the operation of TWT amplifier with the help of apple gate diagram.
7. What is the purpose of helix used in TWT and discuss various types of helix structures.
8. Differentiate linear and cross field devices.
9. Discuss the process of achieving bunching in magnetron.
10. What is the purpose of strapping in magnetron and how it is achieved?
11. Derive an expression for the cut-off magnetic flux density in the reference of cavity magnetron.
12. What is the effect of secondary emission at cathode in the magnetron?
13. Discuss the performance of magnetron and write down the typical application.
14. Differentiate between frequency pushing and frequency pulling.
15. What are slow wave structures and discuss the process of achieving amplification through the helix in TWT.
16. Write short notes:
    (i) Mode of operation in magnetron
    (ii) Slow wave structure
    (iii) Velocity modulation
    (iv) Current modulation
    (v) Strapping in magnetrons
    (vi) Mode jumping in magnetron
    (vii) Back heating
    (viii) BWO

MICROWAVE TUBES

    (ix) CFA   (x) Phase focusing effect.

17. Derive the expression for change in frequency due to repeller voltage variation in reflex klystron.
18. With suitable figure explain how oscillations are sustained in the cavity magnetron.
19. A reflex klystron has the following parameters:

    Beam voltage (V_{dc}) = 2.4 kV

    Repeller voltage (V_R) = – 150 V

    Resonant frequency f_r = 7.45 GHz

    Calculate : (i) the d.c. electron velocity (ii) the round trip d.c. transit time.

20. A magnetron is operating in the π-mode and has the following specifications

    $N = 8, f = 6$ GHz, $r_c = 0.3$ cm, $r_a = 0.9$ cm, $V_{dc} = 18$ kV, $B = 0.2$ Wb/m^2

    Determine:

    (a) Angular velocity of electrons

    (b) The radius at which the radial forces due to electric and magnetic fields are equal and opposite.

21. Why is magnetron also called 'extended interactions' tube? Derive the expression for the null cut-off magnetic flux density in a cylindrical magnetron. **(UPTU-2003, 2004)**
22. An identical two-cavity klystron amplifier operates at 4 GHz with $V_0 = 1$ kV, $I_0 = 22$ mA, cavity gap = 1 mm, drift space = 3 cm and catcher cavity total effective short conductance $G_{sh} = 0.3 \times 10^{-4}$ mhos. Calculate: (i) Ream coupling coefficient and the input cavity voltage magnitude for maximum output voltage. **(UPTU-2003, 2004)**
23. How is continuous interaction between the electron beams and RF field ensured in a TWT? Using suitable diagrams show that the favourable interactions are far more than the unfavourable interactions, resulting in amplification. **(UPTU-2003, 2004)**
24. What problems are encountered in extending the conventional multi-electrode tubes to microwave frequencies? Describe the principle of operations of reflex klystron oscillator. **(UPTU-2002, 2003)**
25. Draw the schematic diagram of a TWT amplifier and describe its principle of operation. Give the propagation characteristics of different waves generated in the amplifier. Explain how RF power output and gain vary with the change in RF power input. **(UPTU-2002, 2003)**
26. Draw the schematic diagram of a cylindrical multicavity magnetron and describe its principle of operation. **(UPTU-2002, 2003)**
27. Draw the cross-sectional of a cylindrical magnetron and derive the expression for cut-off voltage. **(MDU-2006)**
28. What is velocity modulation? How is it different from normal modulation? Explain how velocity modulation is utilized in klystron amplifier. **(MDU-2004)**
29. How is bunching achieved in a cavity magnetron? Explain the phase focusing effect. **(MDU-2004)**
30. Derive an expression for efficiency of a two cavity klystron amplifier. **(MDU-2003)**
31. A magnetron is operating in the µ mode and has the following specifications $N = 10$, $f = 3$ MHz, $a = 0.4$ cm, $b = 0.9$ cm and $l = 2.5$ cm (anode length), $V_0 = 18$ kV and $B = 0.2$ Wb/m^2. Determine:

(a) The angular velocity of electron?

(b) The radius of which the radial forces due to electric and magnetic fields are equal and opposite. **(MDU-2003)**

32. What is backward oscillator? **(MDU-2002)**

33. How for does the beam of 110 mA beam at 1.7 kV, travel before its radius becomes twice the initial value 0.2 cm.

34. A two-cavity klystron amplifier has the following parameters:

$V_0 = 1200V$, $I_0 = 25$ mA, $R_0 = 30$ kΩ, $f = 10$ GHz, $d = 1$ mm, $L = 4$ cm and $R_{SH} = 30$ kΩ

Calculate :

(i) The input voltage for maximum output voltage.

(ii) The voltage gain in decibels.

(iii) Efficiency.

35. A reflex klystron has the following parameters:

$V_0 = 800$ V, $L = 1.5$ mm, $R_{SH} = 15$kΩ, $f = 9$ GHz.

Calculate:

(i) The repeller voltage for which the tube can oscillate in 1¾ mode.

(ii) The direct current necessary to give a microwave gap voltage of 200 V and

(iii) Electron efficiency.

36. A two cavity, klystron has the following parameters:

Beam voltage $V_0 = 20$ kV

Operating frequency $f = 10$ GHz

Beam current $I_0 = 10$ GHz

Beam coupling coefficients $\beta_i = \beta_o = 1$

d.c. electron beam current density = 10^{-8} c/m^3

Signal voltage = $V_i = 10$ V (rms)

Shunt resistance of cavity = $R_{SH} = 15$ kΩ

Total shunt resistance including load = $R = 35$ kΩ

Calculate :

(i) Plasma frequency

(ii) The reduced plasma frequency for $R = 0.5$

(iii) The induced voltage in the output cavity

(iv) The output power delivered to the load

(v) The power gain

(vi) Electronic efficiency

CHAPTER 7

MICROWAVE SEMICONDUCTOR DEVICES

OBJECTIVES
- Introduction
- Tunnel Diode
- Varactor Diodes
- Step-Recovery Diode
- Parametric Amplifiers
- Gunn Diode
- LSA Diode
- Point-Contact Diode
- Crystal Diodes
- Schottky Barrier Diode
- PIN Diodes
- MASER
- LASER
- IMPATT
- TRAPATT
- BARITT
- BJT, Heterojunction Bipolar Transistor
- Junction Field Effect Transistors (JFETs)
- Metal Semiconductor Field Effect Transistors (MESFET)
- High Electron Mobility Transistors (HEMT)
- Metal Oxide Semiconductor Field Effect Transistors (MOSFETs)

7.1 INTRODUCTION

As with vacuum tubes, the special electronics effects encountered at microwave frequencies severely limit the usefulness of transistors in most circuit applications. The need for small-sized microwave devices has caused extensive research in this area. This research has produced solid-state devices with higher and higher frequency ranges. The new solid-state microwave devices are predominantly active, two-terminal diodes, such as Tunnel diodes, Varactors, Transferred-electron devices, and Avalanche transit-time diodes. Microwave solid-state devices are becoming increasingly important at microwave frequencies. These microwave semiconductor devices have been invented for various microwave applications such as detection, frequency multiplication, attenuation, amplitude limiting, generation of oscillations, phase shifting, switching and for low noise amplifiers. These microwave solid state devices have a great advantage of smaller size, lighter weight, and higher reliability of operation, lower cost and higher capability of being incorporated into Microwave Integrated Circuits (MICs), as compared to electron transit time devices. This section will describe the basic theory of operation and some of the applications of these relatively new solid-state devices.

These devices are broadly categorized in to four groups as shown in Fig. (7.1).
1. Transferred Electron Devices (TED)
2. Avalanche Transit Time Devices
3. Microwave Bipolar Junction Transistor (BJT)
4. Microwave Field Effect Transistor (FET's)

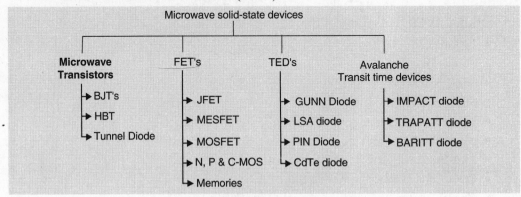

Fig. 7.1. Microwave solid-state devices.

7.2 MICROWAVE SEMICONDUCTOR DIODES

Microwave semiconductor diodes may be classified in accordance to the usage of these diodes in various microwave circuits and which are as following:

1. **Tunnel Diodes**: Tunnel diodes are used for generation of microwave oscillations and signal amplification.
2. **Varactor Diodes**: Varactor diodes are used for frequency multiplication, parametric amplification and frequency tuning.
3. **Crystal Diodes**: Crystal diodes are used for mixing of two microwave frequencies and also for detection of microwave signals.
4. **Schottky Diodes**: Schottky diodes are also used as microwave mixer and detector.
5. **PIN Diodes**: PIN diodes are used for microwave signals attenuation, switching, modulation, phase shifting, and amplitude limiting.
6. **Gunn Diodes**: Gunn diodes are used to generate microwave signals.
7. **Read Diodes**: Read diodes such as IMPATT, TRAPATT and BARITT diodes are used for amplification and generation of microwave signals.

In the following sections, we will be discussing the construction, structure, equivalent circuit and operation of the some of the semiconductor diodes.

7.3 TUNNEL DIODE DEVICES

Tunnel diodes, also known as **Esaki diodes** in honor of their Japanese inventor Leo Esaki, are able to transition between peak and valley current levels very quickly, "switching" between high and low states of conduction much faster than even Schottky diodes. Tunnel diode characteristics are also relatively unaffected by changes in temperature. In 1958, Leo Esaki, discovered that if a semiconductor junction diode is heavily doped with impurities, it will have a region of **negative**

MICROWAVE SEMICONDUCTOR DEVICES

resistance. The normal junction diode uses semiconductor materials that are lightly doped with one impurity atom for ten-million semiconductor atoms. This low doping level results in a relatively wide depletion region. Conduction occurs in the normal junction diode only if the voltage applied to it is large enough to overcome the potential barrier of the junction. In the Tunnel diode, the semiconductor materials used in forming a junction are doped to the extent of one-thousand impurity atoms for ten-million semiconductor atoms. This heavy doping produces an extremely narrow depletion zone similar to that in the Zener diode. Also because of the heavy doping, a tunnel diode exhibits an unusual current-voltage characteristic curve as compared with that of an ordinary junction diode. The characteristic curve for a tunnel diode is illustrated in figure (7.2).

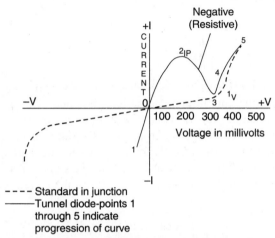

Fig. 7.2. Characteristic curve of a tunnel diode compared to that of a standard PN junction.

The three most important aspects of this characteristic curve are (1) the forward current increase to a peak (I_p) with a small applied forward bias, (2) the decreasing forward current with an increasing forward bias to a minimum valley current (I_v), and (3) the normal increasing forward current with further increases in the bias voltage. The portion of the characteristic curve between IP and IV is the region of negative resistance. Simply stated the theory known as quantum-mechanical tunneling is an electron crossing a PN- junction without having sufficient energy to do so otherwise. Because of the heavy doping the width of the depletion region is only one-millionth of an inch. You might think of the process simply as an arc- over between the N- and the P-side across the depletion region.

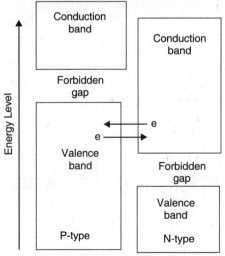

Fig. 7.3 (a). Tunnel diode energy diagram with no bias.

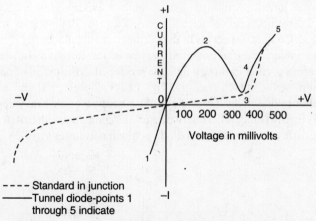

Fig. 7.3 (*b*). Tunnel diode energy diagram with no bias.

Figure (7.3) shows the equilibrium energy level diagram of a tunnel diode with no bias applied. Note in view A that the valence band of the P-material overlaps the conduction band of the N-material. The majority electrons and holes are at the same energy level in the equilibrium state. If there is any movement of current carriers across the depletion region due to thermal energy, the net current flow will be zero because equal numbers of current carriers flow in opposite directions. The zero net current flow is marked by a "0" on the current-voltage curve illustrated in view B.

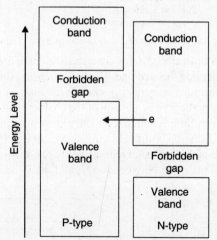

Fig. 7.4 (*a*). Tunnel diode energy diagram with 50 millivolts bias.

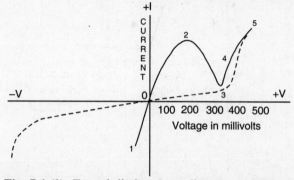

Fig. 7.4 (*b*). Tunnel diode energy diagram with 50 mv bias.

MICROWAVE SEMICONDUCTOR DEVICES

Figure (7.4), view A, shows the energy diagram of a tunnel diode with a small forward bias (50 mv) applied. The bias causes unequal energy levels between some of the majority carriers at the energy band overlap point, but not enough of a potential difference to cause the carriers to cross the forbidden gap in the normal manner. Since the valence band of the P-material and the conduction band of the N-material still overlap, current carriers tunnel across at the overlap and cause a substantial current flow. The amount of current flow is marked by point 2 on the curve in view B. Note in view A that the amount of overlap between the valence band and the conduction band decreased when forward bias was applied.

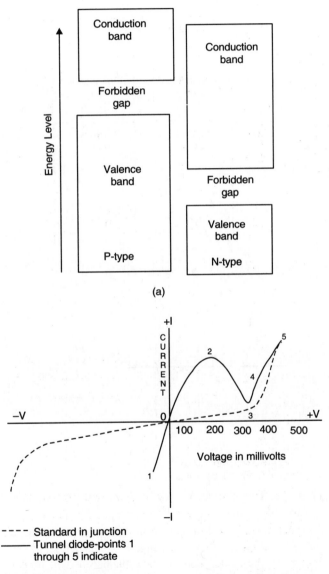

Fig. 7.5. Tunnel diode energy diagram with 450 mv bias

Figure (7.5), view A, is the energy diagram of a tunnel diode in which the forward bias has been increased to 450 mv. As you can see, the valence band and the conduction band no longer overlap at this point, and tunneling can no longer occur. The portion of the curve in view B from point 2 to point 3 shows the decreasing current that occurs as the bias is increased and the area of overlap becomes smaller. As the overlap between the two energy bands becomes smaller, fewer and fewer electrons can tunnel across the junction. The portion of the curve between point 2 and point 3 in which current decreases as the voltage increases is the negative resistance region of the tunnel diode.

Figure (7.6), view A, is the energy diagram of a tunnel diode in which the forward bias has been increased even further. The energy bands no longer overlap and the diode operates in the same manner as a normal PN junction, as shown by the portion of the curve in view (B) from point 3 to point 4.

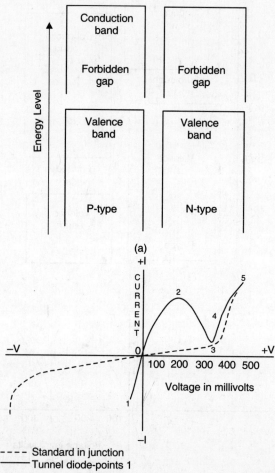

Fig. 7.6. Tunnel diode energy diagram with 600 mv bias.

The negative resistance region is the most important and most widely used characteristic of the tunnel diode. A tunnel diode biased to operate in the negative resistance region can be used as either an oscillator or an amplifier in a wide range of frequencies and applications. Very high frequency applications using the tunnel diode are possible because the tunneling action occurs so rapidly that there is no transit time effect and therefore no signal distortion. Tunnel diodes are also used extensively in high- speed switching circuits because of the speed of the tunneling action. Several schematic symbols are used to indicate a tunnel diode. Most commercial tunnel diodes are made of Ge or GaAs since high peak to valley current ratio Ip/Iv is difficult to obtain with Si. These symbols are illustrated in figure (7.7) (view A. view B, view C, and view D).

MICROWAVE SEMICONDUCTOR DEVICES

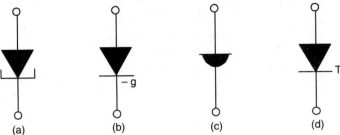

Fig. 7.7. Tunnel diode schematic symbols.

7.3.1 Tunnel Diode Applications

Tunneling causes the negative-resistance action and is so fast that no transit-time effects occur even at microwave frequencies. The lack of a transit-time effect permits the use of tunnel diodes in a wide variety of microwave circuits, such as:

1. Microwave amplifiers: Tunnel diode amplifiers are used in communication systems (such as satellite communication system) where low level, low noise RF amplification is needed. Frequency of operation range up to 80 GHz with Gallium Arsenide diodes.
2. Microwave oscillators: Tunnel diodes are used as microwave oscillator because of the negative resistance in the forward characteristics. These are low cost oscillators and power handling capability is low. These oscillators have very simple circuit with the lowest noise level among other microwave semiconductor devices.
3. Microwave frequency converters and mixers.
4. Microwave switching devices.

7.3.2 Constructions of Tunnel Diode

Normally Germanium and Gallium Arsenide materials are used. These are preferred to silicon as their forbidden energy gaps are smaller and ion mobility's are higher. Figure- 7.8 (a) shows the construction of the tunnel diode, which is quite simple. A very small dot of n-type sphere is alloyed into a heavily doped p-type wafer. The pellet is soldered to a base. An electrical connection is made from the top of the dot to the top of the package through a metal tape. The ceramic spacer and the lid (metallic) house the P-N device. Low inductance leads are taken out from the base and lid. And hence the tunnel diode is a two terminal circuit element which exhibits negative resistance property under certain bias and frequency ranges.

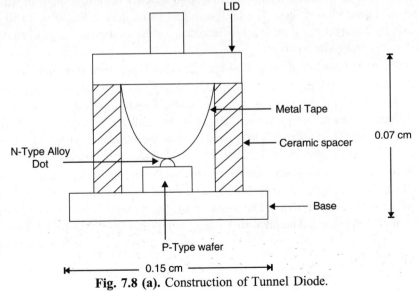

Fig. 7.8 (a). Construction of Tunnel Diode.

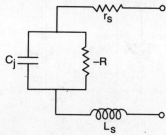

Fig. 7.8 (b). Tunnel diode equivalent circuit.

7.3.3 Equivalent Circuit of Tunnel Diodes

Figure 7.8 (b) shows the equivalent circuit of the tunnel diode, when it is biased in the negative-resistance region. The series resistance (R_s) and inductance (L_s) may be ignored at all except the higher frequencies. The resulting diode equivalent circuit is thus reduced to the parallel combinations of the junction capacitances C_j and the negative resistance R.

The typical values of components in the circuit are C_j = 0.6 PF, R = –75Ω, r_s = 6Ω, L_s, = 0.1 nH.

The total impedance, looking at the equivalent circuit of figure- , is found to be

$$Z = [R_s - \{R_j / 1+(\omega C_j R_j)^2\}] + j \, \omega C_j R_j [\{ L_s / C_j R_j\} - \{R_j / 1 + (\omega C_j R_j)^2\}] \quad \ldots 1$$

By equating the real part and imaginary part of Z to zero separately, two resonant frequencies can be obtained as given below,

Self Resonant frequency

$$f_r = \{1/ 2\pi \, C_j R_j\} \{\sqrt{} (R_j - R_s)/ R_s\} \quad \ldots 2$$

Resistive Resonant frequency

$$f_0 = 1/ 2\pi \sqrt{\{1/(L_s C_j) - 1/(R_j C_j)\}} \quad \ldots 3$$

7.3.4 Tunnel-Diode Amplifiers

Low-noise, tunnel-diode amplifiers represent an important microwave application of tunnel diodes. Tunnel-diode amplifiers with frequencies up to 85 GHz have been built in waveguides, coaxial lines, and transmission lines. The low-noise generation, gain ratios of up to 30 dB, high reliability, and light weight make these amplifiers ideal for use as the first stage of amplification in communications and radar receivers.

Most microwave tunnel-diode amplifiers are **reflection-type, circulator-coupled amplifiers**.

A circulator is a waveguide device that allows energy to travel in one direction only, as shown in figure (7.9). The tunnel diode is connected across a tuned-input circuit. This arrangement normally produces feedback that causes oscillations if the feedback is allowed to reflect back to the tuned-input circuit. The feedback is prevented because the circulator carries all excess energy to the absorptive load (R_L). In this configuration the tunnel diode cannot oscillate, but will amplify.

The desired frequency input signal is fed to port 1 of the circulator through a bandpass filter. The filter serves a dual purpose as a bandwidth selector and an impedance-matching device that improves the gain of the amplifiers. The input energy enters port 2 of the circulator and is amplified by the tunnel diode. The amplified energy is fed from port 2 to port 3 and on to the mixer. If any energy is reflected from port 3, it is passed to port 4, where it is absorbed by the matched load resistance.

MICROWAVE SEMICONDUCTOR DEVICES

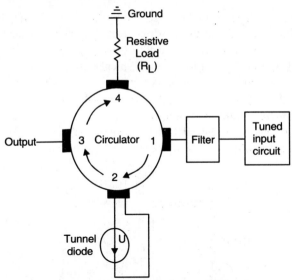

Fig. 7.9. Tunnel-diode amplifier.

7.3.5 Tunnel-Diode Oscillators

A tunnel diode, biased at the center point of the negative-resistance range (point B in figure (7.10)) and coupled to a tuned circuit or cavity, produces a very stable oscillator. The oscillation frequency is the same as the tuned circuit or cavity frequency.

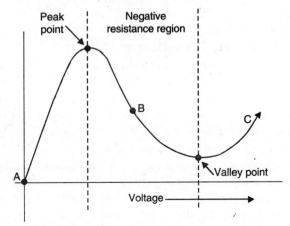

Fig. 7.10. Tunnel-diode characteristic curve.

Microwave tunnel-diode oscillators are useful in applications that require microwatts or, at most, a few mill watts of power, such as local oscillators for microwave superheterodyne receivers. Tunnel-diode oscillators can be mechanically or electronically tuned over frequency ranges of about one octave and have a top-end frequency limit of approximately 10 GHz.

Tunnel-diode oscillators that are designed to operate at microwave frequencies generally use some form of transmission line as a tuned circuit. Suitable tuned circuits can be built from coaxial lines, transmission lines, and waveguides.

An example of a highly stable tunnel-diode oscillator is shown in figure (7.11). A tunnel-diode is loosely coupled to a high-Q tunable cavity. Loose coupling is achieved by using a short; antenna feed probe placed off-center in the cavity. Loose coupling is used to increase the stability of the oscillations and the output power over a wider bandwidth.

The output power produced is in the range of a few hundred microwatts, sufficient for many microwave applications. The frequency at which the oscillator operates is determined by the physical positioning of the tuner screw in the cavity. Changing the output frequency by this method is called **Mechanical tuning**. In addition to mechanical tuning, tunnel-diode oscillators may be tuned electronically. One method is called **Bias tuning** and involves nothing more than changing the bias voltage to change the bias point on the characteristic curve of the tunnel-diode. Another method is called **Varactor tuning** and requires the addition of a varactor to the basic circuit. Varactors diodes are discussed in the next section. Tuning is achieved by changing the voltage applied across the varactor which alters the capacitance of the tuned circuit.

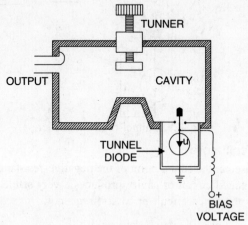

Fig. 7.11. Tunnel-diode oscillator.

7.3.6 Tunnel-Diode Frequency Converters and Mixers

Tunnel diodes make excellent mixers and frequency converters because their current-voltage characteristics are highly nonlinear. While other types of frequency converters usually have a conversion power loss, tunnel-diode converters can actually have a conversion power gain. A single tunnel diode can also be designed to act as both the nonlinear element in a converter and as the negative-resistance element in a local oscillator at the same time.

Practical tunnel-diode frequency converters usually have either a unity conversion gain or a small conversion loss. Conversion gains as high as 20 dB are possible if the tunnel diode is biased near or into the negative-resistance region. Although high gain is useful in some applications, it presents problems in stability. For example, the greatly increased sensitivity to variations in input impedance can cause high-gain converters to be unstable unless they are protected by isolation circuitry.

Performance Characteristics

1. Operating Frequency range - up to 100 GHz
2. Power - 750 Watts
3. Tuning range - 1 to 4-3 GHz

As with tunnel-diode amplifiers, low-noise generation is one of the more attractive characteristics of tunnel-diode frequency converters. Low-noise generation is a primary concern in the design of today's extremely sensitive communications and radar receivers. This is one reason tunnel-diode circuits are finding increasingly wide application in these fields

7.4 VARACTOR DIODE

The **VARACTOR** is another of the active two-terminal diodes that operates in the microwave range. The varactor is a semiconductor diode with the properties of a voltage-dependent capacitor.

MICROWAVE SEMICONDUCTOR DEVICES

Varactor diode has non-linearity of capacitance which is fast enough to follows microwaves. Specifically, it is a variable-capacitance, pn-junction diode that makes good use of the voltage dependency of the depletion-area capacitance of the diode. Losses in the non-linear element are almost negligible.

Varactor diodes are used as harmonic generators, up-conversion, low noise amplifier (Parametric amplifier), pulse shaping and pulse generation.

7.4.1 Material and Construction

Nowadays at microwave frequencies Gallium Arsenide (GaAs) varactor diodes are used instead of using Diffused-Junction mesa silicon diodes, were used earlier at microwave frequencies. The diode encapsulation contains electrical leads attached to the semiconductor wafer and a low loss ceramic case as shown in figure 7.12.

Fig (7.12) shows a varactor diode made of GaAs. GaAs varactor has many advantages due to the higher mobility of charge carrier exhibited by gallium arsenide such as:

1. It has a higher maximum operating frequency approximately up to 1000 GHz.
2. It has better functioning at the lowest temperature about –269 degree c.

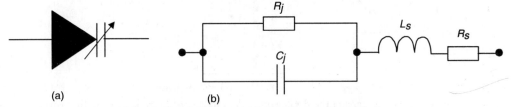

Fig. 7.12 (a). Varactor diode schematic symbols.
Fig. 7.12 (b). Varactor diode equivalent circuit.

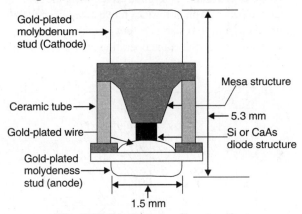

Fig. 7.12 (c). Varactor Construction

7.4.2 Principle and Working of Varactor Diode

Any semiconductor diode has a junction capacitance varying with reverse bias. If such a diode has microwave characteristics then it is referred as varactor diode.

The VARACTOR, or varicap (variable capacitance), as the schematic drawing in figure (7.13) suggests, is a diode that behaves like a variable capacitor, with the PN junction functioning like the dielectric and plates of a common capacitor.

Figure (7.13) shows a PN junction, surrounding the junction of the P and N materials, is a narrow region consists of both positively and negatively charged current carriers. This area is called the depletion region.

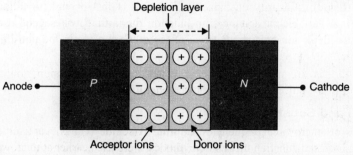

Fig. 7.13. *PN* junction.

The size of the depletion region in a varactor diode is directly related to the bias. Forward biasing makes the region smaller by repelling the current carriers toward the PN junction. If the applied voltage is large enough (about 0.5 volt for silicon material), the negative particles will cross the junction and join with the positive particles, as shown in figure (7.14).

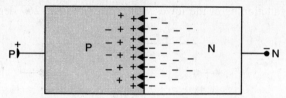

Fig. 7.14. Forward-biased *PN* junction

This forward biasing causes the depletion region to decrease, producing a low resistance at the PN junction and a large current flow across it. This is the condition for a forward-biased diode. On the other hand, if reverse-bias voltage is applied to the PN junction, the size of its depletion region increases as the charged particles on both sides move away from the junction. This condition, shown in figure (7.15), produces a high resistance between the terminals and allows little current flow (only in the microampere range). This is the operating condition for the varactor diode, which is nothing more than a special PN junction.

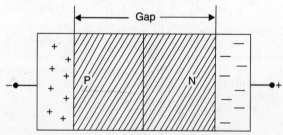

Fig. 7.15. Reverse-biased *PN* junction.

As the figure shows, the insulation gap formed by reverse biasing of the varactor is comparable to the layer of dielectric material between the plates of a common capacitor.

Furthermore, the formula used to calculate capacitance

$$C = \frac{AK}{d}$$

where A = plate area, K = a constant value and d = distance between plates can be applied to both the varactor and the capacitor. In this case, the size of the insulation gap of the varactor, or depletion region, is substituted for the distance between the plates of the capacitor. By varying the reverse-bias voltage applied to the varactor, the width of the "gap" may be varied. An increase in reverse bias increases the width of the gap (d) which reduces the capacitance (C) of the PN

MICROWAVE SEMICONDUCTOR DEVICES

junction. Therefore, the capacitance of the varactor is inversely proportional to the applied reverse bias. The ratio of varactor capacitance to reverse-bias voltage change may be as high as 10 to 1.

Figure (7.16) shows one example of the voltage-to-capacitance ratio. View (A) shows that a reverse bias of 3 volts produces a capacitance of 20 pf in the varactor.

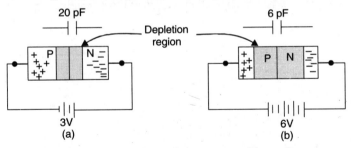

Fig. 7.16. Varactor capacitance versus bias voltage.

If the reverse bias is increased to 6 volts, as shown in view (b), the depletion region widens and capacitance drops to 5 pf. Each 1-volt increase in bias voltage causes a 5 pf decrease in the capacitance of the varactor; the ratio of change is therefore 5 to 1. Of course any decrease in applied bias voltage would cause a proportionate increase in capacitance, as the depletion region narrows. Notice that the value of the capacitance is small in the picofarad range.

The depletion region increases as reverse voltage across it increases; and since capacitance varies inversely as dielectric thickness, the junction capacitance will decrease as the voltage across the PN junction increases. So by varying the reverse voltage across a PN junction the junction capacitance can be varied. This is shown in the typical varactor voltage-capacitance curve below in figure (7.17).

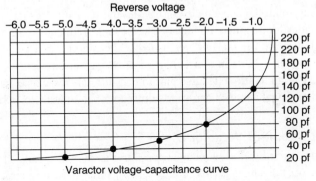

Fig. 7.17. The typical varactor voltage-capacitance curve.

Notice the nonlinear increase in capacitance as the reverse voltage is decreased. This nonlinearity allows the varactor to be used also as a harmonic generator.

Major varactor considerations are:
(a) Capacitance Value
(b) Voltage
(d) Maximum working voltage
(a) Capacitance value
(c) Variation in capacitance with voltage
(e) Leakage current

In general, varactors are used to replace the old style variable capacitor tuning. They are used in tuning circuits of more sophisticated communication equipment and in other circuits where variable capacitance is required. One **advantage** of the varactor is that it allows a dc voltage to be used to tune a circuit for simple remote control or automatic tuning functions. One such application of the varactor is as a variable tuning capacitor in a receiver or transmitter tank circuit like that shown in figure (7.18).

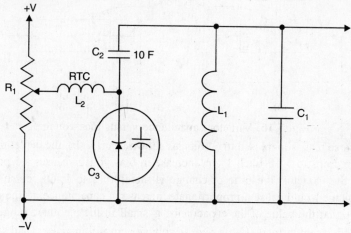

Fig. 7.18 (*a*). Varactor tuned resonant circuit.

Figure 7.18 (a) shows a dc voltage felt at the wiper of potentiometer R_1 which can be adjusted between +V and -V. The dc voltage, passed through the low resistance of RF choke L_2, acts to reverse bias varactor diode C_3. The capacitance of C_3 is in series with C_2 and the equivalent capacitance of C_2 and C_3 is in parallel with tank circuit L_1-C_1. Therefore, any variation in the dc voltage at R_1 will vary both the capacitance of C3 and the resonant frequency of the tank circuit. The RF choke provides high inductive reactance at the tank frequency to prevent tank loading by R_1. C_2 acts to block dc from the tank as well as to fix the tuning range of C_3. An ohmmeter can be used to check a varactor diode in a circuit. A high reverse-bias resistance and a low forward-bias resistance with a 10 to 1 ratio in reverse-bias to forward-bias resistance are considered normal. The capacitance of a typical varactor can vary from 2 to 50 picofarads for a bias variation of just 2 volts.

7.4.3 Equivalent Circuit of Varactor Diode

The equivalent circuit of the semiconductor wafer is shown in fig-7.18(b). It consisting of C_j (Junction capacitance), R_j (Junction resistance) and R_s (Series resistance including bulk resistance of the wafer and resistance of ohmic electrical leads). Junction resistance (R_j) is approximately 10 M ohm which is neglected as compared to the capacitive reactance at microwave frequency.

Encapsulation of the varactor adds parasitic resistances and reactance's to the semiconductor wafer.

L_s = Lead Capacitance
C_c = Capacitance of ceramic case
C_f = Fringing capacitance
$R_j \gg 1/\omega\, C_j$

The cut off frequency at a specified bias (*V*) is given by

$$F_{cv} = \{1/\, 2\pi\, R_s C_{jv}\}$$ ...7.4

MICROWAVE SEMICONDUCTOR DEVICES

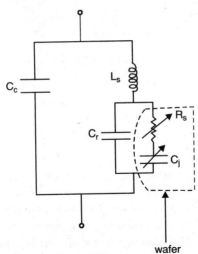

Fig. 7.18 (b). The equivalent circuit of the semiconductor wafer.

7.4.4 Varactor Diode Circuit Applications

The variable capacitance property of the varactor allows it to be used in circuit applications, such as

1. Amplifiers that produce much lower internal noise levels than circuits that depend upon resistance properties. Since noise is of primary concern in receivers, circuits using varactors are an important development in the field of low-noise amplification.
2. The most significant use of varactors has been as the basic component in parametric amplifiers.
3. Pulse generation and pulse shaping.
4. Microwave frequency converter.
5. As RF tuner in a Radio receiver.
6. Active Filters
7. Switching circuit
8. Modulation of a microwave signal.

7.5 STEP-RECOVERY DIODE

A step-recovery varactor diode is a P-N diode similar in construction to the varactor diode. Step-recovery diodes are mainly made-up of silicon because of its longer minority-carrier lifetimes, which are on the order of 0.5-5 µm. The step-recovery diode is operated under forward bias whereas the varactor diode which is operated under reverse-bias. The capacitance under consideration for its operation is the diffusion capacitance C_d.

Characteristics of Step-Recovery diode

The step-recovery diode exhibits low capacitance under applied reverse bias, and high capacitance under forward bias. The capacitance varies with applied bias according to the relation:

$$C_d \infty I_F \qquad \text{...7.5}$$

Equivalent Circuit of Step-Recovery Diode

The equivalent circuit of the step-recovery diode is exactly similar to the varactor diode, shown in figure 7.19.

However, the junction capacitance C_j is replaced by the diffusion capacitance C_d, and the junction resistance R_j is replaced by the junction resistance under forward bias R_d in this case.

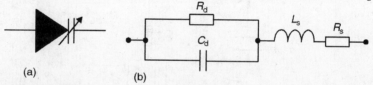

(a) (b)

Fig. 7.19. (a) Symbol and (b) equivalent circuit of Step-Recovery Diode.

7.6 Parametric Amplifiers

The parametric amplifier is named for the time-varying parameter, or value of capacitance, associated with the operation. Since the underlying principle of operation is based on reactance, the parametric amplifier is sometimes called a **Reactance amplifier**.

The conventional amplifier is essentially a variable resistance that uses energy from a dc source to increase ac energy. The parametric amplifier uses a nonlinear variable reactance to supply energy from an ac source to a load. Since reactance does not add thermal noise to a circuit, parametric amplifiers produce much less noise than most conventional amplifiers.

Because the most important feature of the parametric amplifier is the **low-noise characteristic**, the nature of electronic noise and the effect of this type of noise on receiver operation must first be considered. **Electronic noise** is the primary limitation on receiver sensitivity and is the name given to very small randomly fluctuating voltages that are always present in electronic circuits. The sensitivity limit of the receiver is reached when the incoming signal falls below the level of the noise generated by the receiver circuits. At this point the incoming signal is hidden by the noise, and further amplification has no effect because the noise is amplified at the same rate as the signal. The effects of noise can be reduced by careful circuit design and control of operating conditions, but it cannot be entirely eliminated. Therefore, circuits such as the parametric amplifier are important developments in the fields of communication and radar.

7.6.1 The Basic Theory of Parametric Amplifier

The basic theory of parametric amplification centers on a capacitance that varies with time. Consider the simple series circuit shown in figure (7.20).

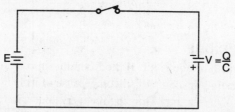

Fig. 7.20. Voltage amplification from a varying capacitor.

If the switch is closed, the capacitor charges to value charge, Q and if the switch is opened, the isolated capacitor has a voltage across the plates determined by the charge Q divided by the capacitance C.

$$V = \frac{Q}{C}$$

An increase in the charge Q or a decrease in the capacitance C causes an increase in the voltage across the plates. Thus, a voltage increase, or amplification, can be obtained by mechanically or electronically varying the amount of capacitance in the circuit. In practice a voltage-variable capacitance, such as a varactor, is used. The energy required to vary the capacitance is obtained from an electrical source called a **PUMP**.

Fig. (7.21) shows on equivalent circuit for a parametric amplifier.

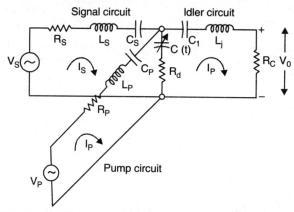

Fig. 7.21. Equivalent circuit for a parametric amplifier.

Figure (7.21), view (a), shows a circuit application using a voltage-variable capacitor and a pump circuit. The pump circuit decreases the capacitance each time the input signal (E) across the capacitor reaches maximum. The decreased capacitance causes a voltage buildup as shown by the dotted line in view (b). Therefore, each time the pump decreases capacitance (view (c)), energy transfers from the pump circuit to the input signal. The step-by-step buildup of the input-signal energy level is shown in view (d).

Proper phasing between the pump and the input signal is crucial in this circuit. The electrical pump action is simply a sine-wave voltage applied to a varactor located in a resonant cavity. For proper operation, the capacitance must be decreased when the input voltage is maximum and increased when the input voltage is minimum. In other words, the pump signal frequency must be exactly double the frequency of the input signal. This relationship can be seen when you compare views (b) and (c). A parametric amplifier of the type shown in figure (7.22) is quite phase-sensitive.

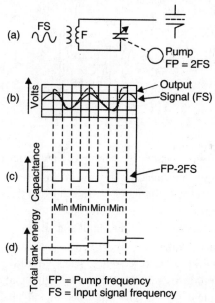

Fig. 7.22. Energy transfer from pump signal to input signal.

The input signal and the capacitor variation are often in the wrong phase for long periods of time. Parametric amplifier which is not phase-sensitive is referred as a **non degenerative**

parametric amplifier, uses a pump circuit with a frequency higher than the twice of the input signal. The higher-frequency pump signal mixes with the input signal and produces additional frequencies that represent both the sum and difference of the input signal and pump frequencies.

Figure (7.23), view (a), is a diagram of a typical non degenerative parametric amplifier with the equivalent circuit shown in view (b). The pump signal (fp) is applied to the varactor. The cavity on the left is resonant at the input frequency (fs), and the cavity on the right is resonant at the difference frequency (fp-fs).

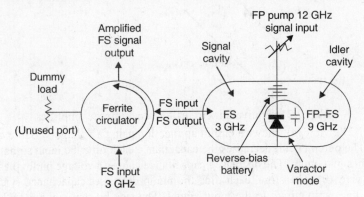

Fig. 7.23. (*a*) Non-degenerative parametric amplifier.

The difference frequency is called the **idler- or lower-sideband frequency**. The varactor is located at the high-voltage points of the two cavities and is reverse biased by a small battery. The pump signal varies the bias above and below the fixed-bias level.

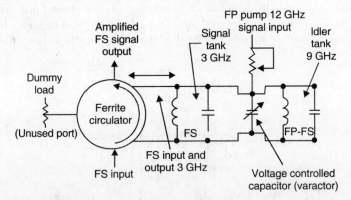

Fig. 7.23 (*b*). Non-degenerative parametric amplifier electrical equivalent.

The pump signal causes the capacitor in view (a) to vary at a 12 GHz rate. The 3 GHz input signal enters via a four-port ferrite circulator, is developed in the signal cavity, and applied across the varactor. The nonlinear action of the varactor produces a 9 GHz difference frequency (fp-fs) with an energy-level higher than the original input signal.

The difference (idler) frequency is reapplied to the varactor to increase the gain and to produce an output signal of the correct frequency. The 9 GHz idler frequency recombines with the 12 GHz pump signal and produces a 3 GHz difference signal that has much larger amplitude than the original 3-gigahertz input signal. The amplified signal is sent to the ferrite circulator for transfer to the next stage.

As with tunnel-diode amplifiers, the circulator improves stability by preventing reflection of the signal back into the amplifier. Reflections would be amplified and cause uncontrollable

MICROWAVE SEMICONDUCTOR DEVICES

oscillations. The ferrite circulator also serves as an isolator to prevent source and load impedance changes from affecting gain.

Typically, the gain of a parametric amplifier is about 20 dB. The gain can be controlled with a variable attenuator that changes the amount of pump power applied to the varactor.

Parametric amplifiers are relatively simple in construction. The only component is a varactor diode placed in an arrangement of cavities and waveguides. The most elaborate feature of the amplifier is the mechanical tuning mechanism. Figure (7.24) illustrates an actual parametric amplifier.

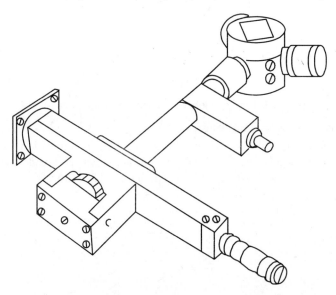

Fig. 7.24. Parametric amplifier.

7.6.2 Parametric Frequency Converters

Parametric frequency converters, using varactors, are of three basic types:

1. The upper-sideband parametric up-converter produces an output frequency that is the sum of the input frequency and the pump frequency.
2. The lower-sideband parametric down-converter produces an output frequency that is the difference between the pump frequency and the input frequency.
3. The double-sideband parametric up-converter produces an output, in which both the sum and the difference of the pump and input frequencies are available.

Parametric frequency converters are very similar to parametric amplifiers in both construction and operation. Figure (7.25) is a functional diagram of a parametric down-converter.

The parametric frequency converter operates in the same manner as the parametric amplifier except that the sideband frequencies are not reapplied to the varactor. Therefore, the output is one or both of the sideband frequencies and is not the same as the input frequency. The output frequency is determined by the cavity used as an output. For example, the idler cavity in figure (7.25) could be replaced by a cavity that is resonant at the upper-sideband frequency (22 GHz) to produce an upper-sideband parametric up-converter. Since input and output signals are at different frequencies, the parametric frequency converter does not require a ferrite circulator. However, a ferrite isolator is used to isolate the converter from changes in source impedance.

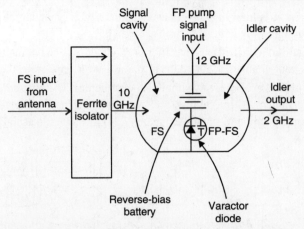

Fig. 7.25. Lower-sideband parametric down-converter.

7.6.3 Manley-Rowe Power Relations

Manley and Rowe have derived some set of energy relations regarding power flowing into and out of an ideal nonlinear reactance. Those relations are very useful while determining whether power gain is possible in a parametric amplifier. The figure (7.26) shows the equivalent circuit for Manley power derivation.

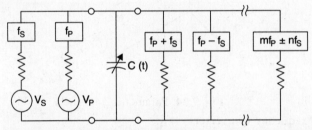

Fig. 7.26. Equivalent circuit for Manley-Rowe derivation.

A signal generator at frequency (f_s) and a pump generator at frequency (f_p), series resistance and band pass filter are applied to a nonlinear capacitance $(c(t))$. These resonant filter circuits are designed to stop power at all frequencies other than their respective signal frequencies. An infinite number of resonant frequencies of $m f_p \pm n f_p$ are being generated by the two frequencies f_s and f_p, where m and n are integers from 0 to ∞. It is assumed that resonant circuits are ideal so that power losses by the nonlinear susceptances are negligible. It means that the power entering the nonlinear capacitor at the pump frequency is equal to the power leaving the capacitor at the other frequencies through the non-linear interaction. Manley and Rowe established the power relations between the input power at the signal generator (f_s) and the pump generator frequency (f_p) and output power at the $m f_p \pm n f_s$.

The two independent equations give the Manley-Rowe relation for any single valued, nonlinear and loss-less reactance, which are:

1st equations:

$$\sum_{m=0}^{\infty} \sum_{n=-\infty}^{\infty} \frac{m P_{m,n}}{m f_p + n f_s} = 0 \qquad \ldots 7.6$$

2nd equations:

$$\sum_{m=-\infty}^{\infty} \sum_{n=0}^{\infty} \frac{n P_{m,n}}{m f_p + n f_s} = 0 \qquad \ldots 7.7$$

MICROWAVE SEMICONDUCTOR DEVICES

where $P_{m,n}$ = Average power flowing into the nonlinear reactance at frequencies $\pm(mf_p + nf_s)$

m and d = integers varying form 0 to ∞
f_s = Frequency of signal generator
f_p = Frequency of pump generator

The power gain is defined as the ratio of the power delivered buy the capacitor at a frequency of $f_p + f_s$ to the power absorbed by the capacitor at a frequency f_p as shown in equation

$$\text{Power gain} = \frac{f_p + f_s}{f_s} \quad \text{(In case of modulator)} \quad \ldots 7.8$$

When $f_p + f_s = f_0$ and $f_p + f_s > f_p > f_s$, than

$$\text{Gain} = \frac{f_0}{f_s} \quad \ldots 7.9$$

The maximum power gain is simply the ratio of the output frequency to the input frequency. This type of parametric amplifier is known as the **sum frequency parametric amplifier or up converter**.

If the signal frequency is the sum of output frequency and the pump frequency than the gain of parametric device is given by

$$\text{Power gain} = \frac{f_s}{f_p + f_s} \quad \text{(In case of demodulator)}$$

where $f_s = f_p + f_0$ and $f_0 = f_s - f_p$. This type of parametric device is known as **parametric down converter** and in this case gain is actually a loss.

If the signal frequency f_s, the pump frequency f_p and the output frequency is f_0, where $f_p = f_s + f_0$, the power P_{11} supplied at f_p is positive than the power $P_{1,0}$ and $P_{0,1}$ are negative. It means the capacitor delivers power to the signal generator f_s instead of absorbing it. The power gain will become infinite and leads to an unstable condition and circuit is oscillating both at f_s and f_0. This type of parametric devices is known as **negative-resistance parametric amplifier**.

7.6.4 Parametric Up-converter

The properties of up-convertor parametric amplifier are as following:

(i) The output frequency $f_0 = f_p + f_s$.

(ii) There is no power flow in the parametric amplifier other the signal, pumped output frequencies.

If the above conditions are satisfied in up-converter parametric amplifier the parameters such as power gain, bandwidth and noise figure.

(i) **Power gain**: The maximum power gain of a parametric up convertor is given by

$$\text{Power gain (maximum)} = \frac{f_0}{f_s} \frac{x}{\left(1 + \sqrt{1+x}\right)^2} \quad \ldots 7.10$$

where f_0 = output frequency = $f_p + f_s$

$$x = \frac{f_c}{f_0}(\gamma Q)^2 \text{ and } Q = \frac{1}{2\pi f_s C R_d}$$

R_d = series resistance of P-N junction.

γQ = figure of merit for the nonlinear capacitor.

The quantity of $x/(1+\sqrt{1+x})^2$ is a gain-degradation. As R_d approaches to zero, the figure of merit (γQ) tending to infinity and gain degradation factor becomes equal to unity. As a result, the power gain of parametric up-convertor for a loss less varactor diode is equal f_0/f_c (according to Manley-Rowe relation). For a typical microwave diode f_0/f_s = 15 and γQ is equal to 10, than the maximum gain is given by eq. (10) is 7.3 db.

(*ii*) **Bandwidth**: The bandwidth of a parametric up convertor is related to the gain degradation factor and the signal frequency and output frequency. It is given by

$$Bw = 2\gamma \sqrt{\frac{f_0}{f_s}} \qquad \qquad ...7.11$$

If f_0/f_s = 10 and γ = 0-2 than the bandwidth (*BW*) is equal to 1.264.

(*iii*) **Noise figure**: The noise figure F, for the parametric up convertor is given

$$F = 1 + | \frac{2T_d}{T_0} \left[\frac{1}{\gamma Q} + \frac{1}{(\gamma Q)^2} \right] \qquad \qquad ...7.12$$

where T_d = Diode temperature (degree Kelvin)

T_0 = Ambient temperature (300°K)

γQ = Figure of merit for the non linear capacitor

Parametric amplifier has the advantage over the transistor amplifier is its low noise figure because of pure reactance does not tribute thermal noise to the circuit.

The typical microwave diode has γQ = 10, f_0/f_s = 10 and T_d = 300K, the minimum noise figure is 0.90 dB.

7.6.5 Parametric Down convertor

In a parametric down convertor, the $f_s = f_p + f_0$, it means that the input power must feed in to the idler circuit and the output power must move out from the signal circuit.

The down-conversion gain (loss) is given by

$$\text{Gain} = \frac{f_s}{f_0} \frac{x}{f_0(1+\sqrt{1+x})^2} \qquad \qquad ...7.13$$

7.6.6 Negative-Resistance Parametric Amplifier

When a significant portion of power flows only at the f_s, f_p and the idler frequency f_i, than a regenerative condition with the possibility of oscillation at both the signal frequency and the idler frequency will occur. The idler frequency may be defined as the difference between the pump frequency and the signal frequency, *i.e.*, $f_i = f_p - f_s$. When the mode operates below the oscillation threshold the device behaves as a bilateral negative resistance parametric amplifier.

The various parametric of negative resistance parametric amplifier such as power gain, noise figure and the bandwidth are as following:

(*i*) **Power gain**: The output power is taken from the resistance R_i at a frequency and the conversion gain from f_s of negative-resistance parametric amplifier, is given by

MICROWAVE SEMICONDUCTOR DEVICES

$$\text{Power gain} = \frac{4f_i}{f_s} \cdot \frac{R_g R_i}{R_{TS} R_{Ti}} \cdot \frac{a}{(1-a)^2} \qquad \ldots 7.14$$

where, f_s = signal generator frequency
f_p = pump generator frequency
$fi = fp - fs$ idler frequency
R_g = Output resistance of the signal generator
R_i = Output resistance of the idler generator
R_{TS} = Total series resistance at f_s,
R_{Ti} = Total series resistance at f_i

$$R = \gamma^2 / \left(\omega_s \omega_i C^2 R_{Ti} \right) = \text{Equivalent negative resistance}$$

$$a = R/R_{Ts}$$

(*ii*) **Bandwidth**: The bandwidth of a negative-resistance parametric amplifier is given by

$$BW = \frac{\gamma}{2} \sqrt{\frac{f_i}{f_s \text{ gain}}} \qquad \ldots 7.15$$

For example, if gain = 20 db, $f_i = 4f_s$ and $\gamma = 0.3$, then the BW for single tuned circuit is 0.03.

(*iii*) **Noise figure**: The optimum noise figure is expressed

or

$$F = 1 + 2 \frac{T_d}{T_0} \left[\frac{1}{\gamma Q} + \frac{1}{(\gamma Q)^2} \right] \qquad \ldots 7.16$$

It is same as for the parametric up convertor.

7.6.7 Advantages of the up convertor over the negative resistance devices:

The up convertor parametric device has many advantages over the negative resistance parametric devices, which are as follows:

(*i*) Its input impedance is always greater than 0 (*i.e.*, positive).
(*ii*) Highly stable and unilateral.
(*iii*) Its power gain is independent of changes in its source impedance.
(*iv*) It does not require circulator.
(*v*) Its bandwidth is approximately 5% (large bandwidth).

7.6.8 Degenerate Parametric Amplifier

When the signal frequency is equal to the idler frequency in a negative-resistance amplifier then it is known as the **degenerate parametric amplifier** or oscillator. Since the idler frequency

$$f_i = f_p - f_s \text{ and } f_s = f_p - f_i$$

or

$$f_p = 2f_s \text{ or } f_s = \frac{1}{2} f_p \qquad \ldots 7.17$$

In degenerate parametric amplifier the condition of above eq. (17) must be satisfied and if it is not than it is known as **non-degenerate parametric amplifier**.

(*i*) **Power gain**: Its power gain is exactly same as parametric up convertor. With $f_s = f_i$ and $f_p = 2f_s$, the power delivered from pump to signal frequency is equal to the power transferred from pump to idler frequency. At high gain, the total power at the signal frequency (f_i). Hence the total power to have 3 dB more gain in the pass band.

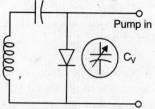

Fig. 7.27. A simple parametric amplifier without idler circuit.

(*ii*) **Bandwidth**: The bandwidth of degenerate parametric amplifier is the same is the parametric up converter.

(*iii*) **Noise figure:** The noise figure of a degenerate parametric amplifier is given

$$F_{SSB} = 2 + \frac{2T_d R_d}{T_0 R_g} \qquad \ldots 7.18$$

$$F_{DSB} = 2 + \frac{T_d R_d}{T_0 R_g} \qquad \ldots 7.19$$

where
F_d = Average diode temperature in K
T_0 = Ambient temperature (300°K)
R_d = Diode series resistance
R_g = External output resistance of the signal generator
F_{SSB} = Noise figure for single side band
F_{DSB} = Noise figure for double side band

F_{DSB} is 3 db less than that for F_{SSB} operation.

7.7 BULK-EFFECT SEMICONDUCTORS

Bulk-effect semiconductors are unlike normal pn-junction diodes in both construction and operation. Some types have no junctions and the processes necessary for operation occur in a solid block of semiconductor material. Other types have more than one junction but still use bulk-effect action. Bulk-effect devices are among the latest of developments in the field of microwave semiconductors and new applications are being developed rapidly. They seem destined to revolutionize the field of high-power, solid-state microwave generation because they can produce much larger microwave power outputs than any currently available pn-junction semi-conductors.

Bulk-effect semiconductors are of two basic types:
1. The transferred-electron devices and
2. The avalanche transit-time devices.

7.7.1 Transferred-Electron Semiconductors

The discovery that microwaves could be generated by applying a steady voltage across a chip of n-type gallium-arsenide (GaAs) crystal was made in 1963 by J.B. Gunn. The device is operated by raising electrons in the crystal to conduction-band energy levels that are higher than the level they normally occupy. The overall effect is called the **transferred-electron effect or Gunn Effect.**

MICROWAVE SEMICONDUCTOR DEVICES

In a gallium-arsenide semiconductor, empty electron conduction bands exist that are at a higher energy level than the conduction bands occupied by most of the electrons. Any electrons that do occupy the higher conduction band essentially have no mobility. If an electric field of sufficient intensity is applied to the semiconductor electrons, they will move from the low-energy conduction band to the high-energy conduction, band and become essentially immobile. The immobile electrons no longer contribute to the current flow and the applied voltage progressively increases the rate at which the electrons move from the low band to the high band. As the curve in figure (7.28) shows, the maximum current rate is reached and begins to decrease even though the applied voltage continues to increase. The point at which the current on the curve begins to decrease is called the Threshold. This point is the beginning of **the negative-resistance region**. Negative resistance is caused by electrons moving to the higher conduction band and becoming immobile.

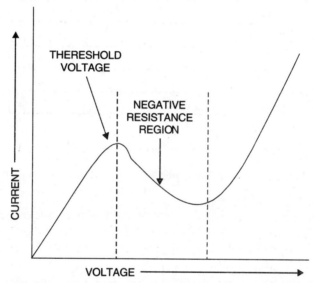

Fig. 7.28. Characteristic curve for a bulk-effect semiconductor.

If an increase in voltage is applied to a gallium-arsenide semiconductor, which is biased to operate in the negative-resistance region, it divides into regions of varying electric fields. A tiny region, known as a DOMAIN, is form that has an electric field of much greater intensity than the fields in the rest of the semiconductor. The applied voltage causes the domain to travel across the semiconductor chip from the cathode to the anode. The high field intensity of the domain is caused by the interaction of the slow electrons in the high-energy band and the faster electrons in the low-energy band. The electrons in the low-energy band travel faster than the moving domain and continually catch up during the transit from cathode to anode. When the fast electrons catch up to the domain, the high field intensity forces them into the higher band where they lose most of their mobility. This also causes them to fall behind the moving domain. Random scattering causes the electrons to lose some energy and drop back into the lower, faster, energy band and race again after the moving domain. The movement from the low-energy band to the high-energy band causes the electrons to bunch up at the back of the domain and to provide the electron-transfer energy that creates the high field intensity in the domain.

The domains form at or near the cathode and move across the semiconductor to the anode, as shown in figure (7.29). As the domain disappears at the anode, a new domain forms near the cathode and repeats the process.

Fig. 7.29. Gallium-arsenide semiconductor domain movement.

7.7.2 Two-Valley Model Theory

The **Gunn effect** can be explained on the basis of two valley theory or Ridley Watkins-Hilsum (RHW) theory or the transferred electron mechanism. The basic mechanism involved in the operation of bulk n-type GaAs devices is the transfer of electrons from lower conditions valley (*L*-valley) to upper subsidiary valley (*u*-valley).

According to the energy band theory of *n*-type GaAs, a high mobility lower valley is separated by energy of 0.36 eV from a low mobility upper valley as shown in figure (7.30).

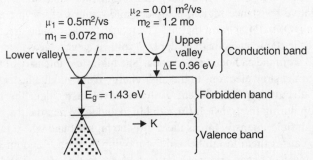

Fig. 7.30. Two-valley model of electron-energy for *n*-type GaAs.

In the figure (7.30) shows that in *L*-valley has a smaller effective mass ($m_1 = 0.072 m_0$) and u-valley has an effective mass ($m_2 = 1.2\ m_0$).

Accordingly mobilities for the *L*-valley in $\mu_1 = 0.5 m^2/vs$ and the u-valley mobility is $\mu_2 = 0.01 m^2/vs$.

Electron densities in the lower and upper valley have the ratio of about 60. In equilibrium condition the electron densities in the lower and upper valley remain the same. When the applied electric field is lower than the electric field of the lower valley then there is no transfer of electrons to the upper valley as shown in figure (7.31).

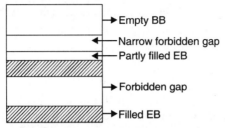

Fig. 7.31. GaAs energy bands.

As we increased the applied electric field, the electron gain energy from it and move upward in the upper valley. When the condition arrives that the applied electric field at lower valley is more than the upper valley, then all the electrons will transfer to the upper valley as shown in Fig. (7.32).

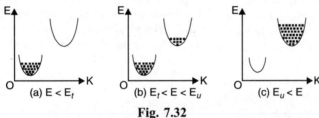

Fig. 7.32

As the electrons transfer to the upper valley, their mobility start decreasing and the effective mass is increasing, results in decreasing the current density and hence the negative differential conductivity. The inter valley transfer which is also known as **population inversion**, of charges from lower valley to upper valley will occur at a certain threshold field which is approximately 3.3 KV/cm and above. The inter valley transfer of charges is known as the Transfer Electron Effect. If the electron densities in the lower-valley and upper valley are n_1 and n_2 respectively, then the conductivity of the n-type GaAs is

$$\sigma = e(\mu_e.n_1 + \mu_u.n_2) \qquad \ldots 7.20$$

where
σ = conductivity,
μ_l = electron mobility in lower valley
$\mu\mu$ = electron mobility in upper valley
n_1 = electron density in lower valley
n_2 = electron density in upper valley

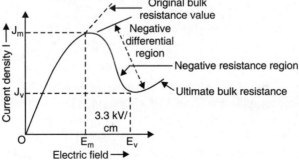

Fig. 7.33. V-I characteristics curve of Gunn diode.

J_m = maximum current density
J_v = current density at valley
E_m = maximum electric field required to have negative conduction region
E_v = electric field at J_v

The region between E_m and E_V in the characteristics curve; where current density decreases with increasing electric field is one of negative differential resistivity.

The condition for negative resistance (conductance) is

$$\frac{dJ}{dE} < 0 \quad \text{or} \quad 1 + \frac{E}{\sigma} \cdot \frac{d\sigma}{dE} < 0 \qquad \ldots 7.21$$

If $dE/d\sigma$ satisfies the above condition in eq. (21) that mean if differential conductivity is negative, then the slope of the V–I characteristics curve will be negative and the device will have a net negative resistance (conductance) to the external circuit.

7.7.3 Modes of Operation

The formation of strong space charge instability depends on the condition that enough charge is available in the crystal and the device is long enough so that a space charge of required strength can be built up within the transit time of the electron. This requirement set up a criterion for the various modes of operation of bulk negative-differential-resistance devices. Four basic modes of operation of uniformly doped bulk diodes with low-resistance contacts are proposed by the Copeland shown in Fig. (7.34).

(i) Gun oscillation mode
(ii) Stable amplification mode
(iii) LSA oscillation mode
(iv) Bias-circuit oscillation mode

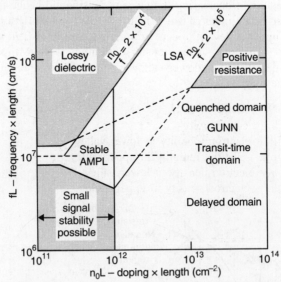

Fig. 7.34. Modes of operation for Gunn diodes.

Criterion for classifying the modes of operation: The Gunn-effect diodes are basically used an n-type GaAs, with the concentration of free electron ranging from 10^{14} to 10^{17} per cubic centimeter at room temperature. The typical dimensions are 150×150 μm in cross section and 30 μm long. The time rate of growth of the space charge layers, during the early stages of space-charge accumulation, is given by

$$Q(X,t) = Q(x-vt; 0) \exp\left(\frac{t}{t_d}\right) \qquad \ldots 7.22$$

where $t_d = \dfrac{\varepsilon_s}{\sigma} = \dfrac{\varepsilon_s}{en_0 |\mu_n|}$ is the magnitude of the negative dielectric relaxation time,

ε = dielectric permittivity of semiconductor
n_0 = doping concentration
μ_n = negative mobility
e = electron charge
σ = conductivity

The eq. (22) remain valid throughout the entire transit time of the space charge layer, and the factor of maximum growth is given by

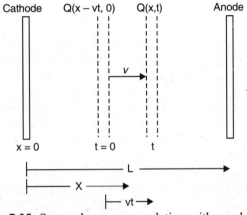

Fig. 7.35. Space charge accumulation with a velocity of v.

$$\text{Growth factor} = \frac{Q(L, L/v)}{Q(0, 0)} = \exp\left(\frac{L}{vt_d}\right)$$

$$= \exp\left(\frac{Ln_0 e |\mu_n|}{\varepsilon_s v}\right) \qquad \ldots 7.23$$

In eq. (23) the layer is assumed to start at the cathode at $t = 0$, $X = 0$ and arrive at the anode at $t = L/v$ and $X = L$. For a large space-charge growth, it should be larger than unity.

So that $\qquad n_0 L > \dfrac{\varepsilon_s v}{e |\mu_n|} \qquad \ldots 7.24$

Eq. (24) gives the criterion for classifying the modes of operation for the Gunn-effect diode. The value of $\varepsilon_s ve|\mu_n|$ n-type GaAs is about $10^{12}/cm^2$, where $|\mu_n|$ is assumed to be 150 cm²/vs.

For example, an n-type GaAs Gunn diode has the parameters given as
$v_d = 2.5 \times 10^5$ m/s

Negative Electron mobility $(|\mu_n|) = 0.015$ m²/v.s
E_r (Relative dielectric constant) = 13.1
The criterion for classifying the modes of operation is given by

$$\frac{\varepsilon_s v_d}{e |\mu_n|} = \frac{8.854 \times 10^{-12} \times 13.1 \times 2.5 \times 10^5}{1.6 \times 10^{-19} \times 0.015}$$

$$= 1.19 \times 10^{16} / m^2$$

$$= 1.19 \times 10^{12} / cm^2$$

It means that the product of the doping concentration and the device length must be $n_0 L > 1.19 \times 10^{12} / cm^2$

In this chapter we will discuss about the Gunn oscillation modes $(10^{12}/cm^2 < (n_0 L) < 10^{14}/cm^2)$ in detail.

Gunn Oscillation Mode

In the Gunn oscillation modes, the range of $n_0 L$ is given by $10^{12}/cm^2 < (n_0 L) < 10^{14}/cm^2$. The space charge perturbations in the specimen increases exponentially in space and time when $n_0 L > 10^{12} cm^2$ in GaAs. A high field domain is formed and moves from the cathode to anode and frequency of oscillation is given by

$$F = \frac{V_{dom}}{L_{eff}} \qquad \ldots 7.25$$

where V_{dom} = velocity of the electron in domain, L_{eff} = Effective length and the domain travels from the time it is formed till the time a new domain start forming.

Gunn described the behavior of Gunn oscillators under several circuits' configurations. When the circuit is mainly resistive or the voltage across the diode is constant, the period of oscillation is time required for the domain to drift from cathode to anode. This mode is not used in microwave oscillation. The negative conductivity devices are usually operated in resonant in with which have high-Q resonant microwave cavities. The frequency can be tuned to a range of about an octave without loss of efficiency.

In the normal Gunn oscillation mode is operated with the electric field greater than the threshold field ($E > E_{th}$). The high field domain drifts along the specimen until it reaches from cathode to the anode or until the low field value drops below the sustaining field E_s required maintaining v_s as shown in figure (7.36). For GaAs the sustaining drift velocity (v_s) is 10^7 cm/s. Since the electron drift velocity v varies with the electric field than there are four possible domain modes for the Gunn oscillation mode, which are as follows:

(i) Transit-Time domain mode
(ii) Delayed or Inhibited domain mode
(iii) Quenched domain mode
(iv) Limited space charge Accumulator mode (LSA mode)

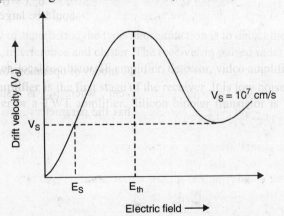

Fig. 7.36. Electron drifts velocity versus electric-field.

(i) Transit time Gunn domain mode : ($fL \approx 10^7$ cm/sec): In transit time domain mode, $fL \approx 10^7$ cm/s $= V_d$ when $V_d = V_s$ (drift velocity of electron V_d, is equal to the sustaining velocity V_s), the high field domain is stable. In other words, the electron drift velocity is written as

MICROWAVE SEMICONDUCTOR DEVICES

$$v_d = v_s = fL = 10^7 \text{ cm/s} \quad \ldots 7.26$$

In this condition the oscillation period is equal to the transit time as shown in fig. (7.37).

$$T_0 = T_e \quad \ldots 7.27$$

Its efficiency is less than 10% because the current is collected only when the domain arrives at the anode. The operating frequency f in this mode is quite sensitive to the applied voltage because the drift velocity of electron depends upon the bias voltage. The resonant frequency of the resonant circuit is equal to the transit time frequency for this mode. It is low power and low efficiency mode. It can be operated maximum frequency of 30 GHz because of device length which is ridiculously small.

(*ii*) **Delayed domain mode** (10^6 cm/s $< fL < 10^7$ cm/s) : In this mode the transit time is so chosen than the domain is collected while $E < E_m$ as shown in fig. (7.37). A new domain cannot form until the field rises above threshold again.

In this mode of operation the oscillation period is greater than the transit time (*i.e.*, $T_0 > T_e$) as shown in figure (7.37). The efficiency of this device is approximately 20%. This mode is also known as **inhibited mode** and the frequency of operation can be equal to or less than that in Gunn diode. This mode has the advantage that unlike the natural Gunn mode or the transit time mode, the frequency of oscillations can be varied by the resonant circuit.

(*iii*) **Quenched domain mode** ($fL > 2 \times 10^7$ cm/s): In this mode, if the bias field drops below the sustaining field E_s during the negative half cycle as shown in figure (7.37), the domain collapses before it reaches the anode. This is known as **Quenched domain mode**.

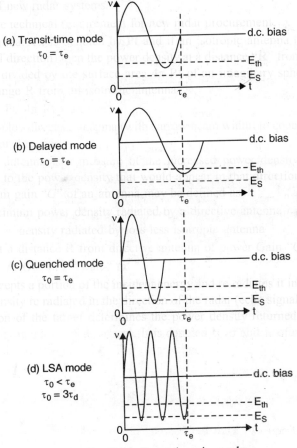

Fig. 7.37. Gunn domain modes.

A second domain is initiated only when the bias field goes above E_m, then the process is repeated. In this mode $T_0 < T_e$ and frequency of oscillation can be many times the transit time frequency. The oscillation occurs at the frequency of resonant circuit rather than at transit time frequency. The efficiency of this mode is around 13%.

(iv) **Limited Space-charge Accumulation (LSA) mode (*fL*>2 10^{7}cm/sec)**: This mode is very useful mode of operation for Gunn oscillator because of its high efficiency at more power. In this mode the domain is not formed at all. As the frequency in LSA mode of the field, never approach the peak value as in Gunn mode since the domain is not allowed to be formed. Hence this mode is suitable for the high operating voltage without breaking the avalanche breakdown. The operating frequency in this mode is about 0.5 to 50 times more than that for Gunn mode *i.e.*, approximately 100 GHz. The operation of LSA mode is mainly dependent on external circuit with high Q resonator.

7.7.4 LSA Diodes

As we have discussed LSA mode in the previous section that LSA stands for the limited space charge accumulation mode of the Gunn diode. In this mode, it is discussed that:

(i) The product $n_0 L$ is larger than $10^{12}/cm^2$,
(ii) The ratio of doping 'n_0' to frequency 'f' is within 2×10^5 to 2×10^4 s/cm^3 so that the high field domain and the space charge layer do not have sufficient time to build up.
(iii) The magnitude of the RF voltage must be large enough to drive the diode below threshold during each cycle in order to dissipate space charge,
(iv) The portion of each cycle during which the *RF* voltage is above threshold must be short enough to prevent the domain formations and the space charge accumulation V, only the primary accumulation layer forms near the cathode; thus with limited space-charge accumulation the remainder of the sample appears as a series negative resistance that increases the frequency of the oscillations in the resonant circuits is very high, the domain do not have sufficient time to form while the field is above threshold. As a result in this mode most of the domain is maintained in the negative conductance state during a large fraction of the voltage cycle. Accumulation of electrons near the cathode has time to collapse while the signal is below threshold. It consists of a uniformly doped semiconductor without any internal space charges. The current in the device is proportional to the drift velocity at this field level. The efficiency of this device is around 20%.

The LSA mode required that the period τ_0 of the resonant circuit is not more than a several times larger than the magnitude of the dielectric relaxation time in the negative conductance region T_d. This mode is shown in Fig. 7.37(d), in which $\tau_0 = 3T_d$. For the *n*-type GaAs, the product of doping (n_0) and length (L) is about $10^{12}/cm^2$ *i.e.*, $n_0 L = 10^{12}/cm^2$

At the low frequency limit the drift velocity is given by

$$v_e = fL = 5 \times 10^6 \, cm/s \qquad \ldots 7.29$$

The ratio of $(n_0 L)$ to fL is

$$\frac{n_0}{f} = 2 \times 10^5 \qquad \ldots 7.30$$

At the upper-frequency limit the drift velocity is

$$v_u = fL = 5 \times 10^7 \, cm/s$$

and ratio of $(n_0 L)$ to fL is

$$\frac{n_0}{f} = 2 \times 10^4 \qquad \qquad ...7.31$$

There is the lower and upper limit (boundaries) of the LSA mode.

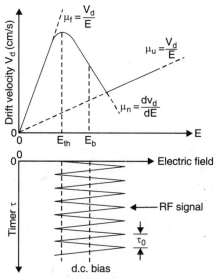

Fig. 7.38. LSA mode.

In the LSA mode the diode is placed in a resonator tuned to an oscillation frequency of and the device is biased to several times the threshold voltage as shown in figure (7.38).

$$f_0 = \frac{1}{\tau_0} \qquad \qquad ...7.32$$

As soon as the *RF* voltage swings beyond the threshold, the space charge starts building up at the cathode. Since the oscillation period (τ_0) of the *RF* signal is less than the T_g (domain-growth time constant), the total voltage swings below the threshold before the domain can form. Furthermore, since τ_0 is much greater than the dielectric relaxation time T_d, the accumulation space charge is drained in a very small part of the *RF* cycle. Therefore the device spends most of the *RF* cycle in the negative-resistance region and the space charge is not allowed to build up. In this mode the oscillation frequency is independent of the transit time of the carrier and is determined solely the external circuit of the device. The output power in the LSA mode can be greater than that in the other modes because the power impedance product not fall off as $1/f_0^2$.

The LSA diodes have some limitations such as :

(*i*) It is very sensitive to load conditions, temperatures and doping fluctuations,

(*ii*) The *RF* circuit must allow the field to build up quickly in order to prevent domain formation.

The power output of LSA diode can be written as

$$P = \eta V_0 I_0 = \eta (ME_{th}L)(n_0 ev_0 A) \qquad \qquad ...7.33$$

where P = power output of LSA oscillator

η = dc to *RF* conversion efficiency

V_0 and I_0 = operating voltage and operating current

M = multiple of the operating voltage above negative resistance threshold voltage

E_m = Threshold field (about 3400 V/cm)

L = Length of the device (10 to 200 μm)

n_0 = donor concentration (10^{15} e/cm^2)

e = electron charge (1.6×10^{-9} c)

v_0 = Average carrier drift velocity (about 10^7 cm/s)

A = Area of the diode (3×10^{-4} to 20×10^{-4} cm^2)

It power outputs vary from 6 KW (Pulse) at 1.75 GHz to 400 W (pulse) at 51 GHz. Practical Gunn oscillators are mounted in coaxial, waveguide and YIG-tuned strip line configurations.

7.7.5 Construction and Characteristics of Gunn Diodes

A simple construction of Gunn diode is shown in Fig. (7.54) below.

The typical characteristics of Gunn diodes are given as

(*a*) Power–CW operation = 25 mW to 250 mW at *X*-band

= 100 mW at 18–26.5 GHz.

= 40 mW at 26.5 – 40 GHz

Pulsed operation = 5 W (5-12 GHz)

(*b*) Efficiency: 2% to 12% at 1.5 W to 50 mW

Construction of Gunn diode

The construction of a Gunn diode is shown Fig. (7.39) (Construction details).

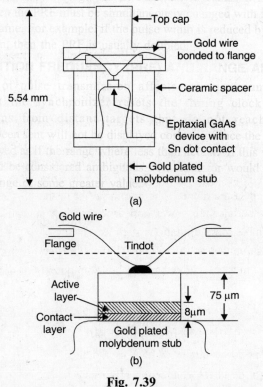

Fig. 7.39

MICROWAVE SEMICONDUCTOR DEVICES

Applications of Gunn Diode

1. Gunn diode is Amplifier—As Broad band amplifier
2. Gunn diode as oscillators used in transponders at ATC and in microwave receivers.
3. As pump source in parametric amplifiers.

7.7.6 Gunn Oscillator

The **Gunn oscillator** is a source of microwave energy that uses the bulk-effect, gallium-arsenide semiconductor. The basic frequency of a Gunn oscillator is inversely proportional to the transit time of a domain across the semiconductor.

Fig. 7.40. A Gunn diode type MA49156.

The transit time is proportional to the length of semiconductor material, and to some extent, the voltage applied. Each domain causes a pulse of current at the output; thus, the output is a frequency determined by the physical length of the semiconductor chip.

The Gunn oscillator can deliver continuous power up to about 65 milliwatts and pulsed outputs of up to about 200 watts peak. The power output of a solid chip is limited by the difficulty of removing heat from the small chip. Much higher power outputs have been achieved using wafers of gallium-arsenide as a single source.

7.8 THE POINT-CONTACT DIODE

Point-contact diodes, commonly called **crystals**, are the oldest microwave semiconductor devices. They were developed during World War II for use in microwave receivers and are still in widespread use as receiver mixers and detectors.

7.8.1 Principle and Working of Point-Contact Diode

Unlike the pn-junction diode, the point-contact diode depends on the pressure of contact between a point and a semiconductor crystal for its operation. Figure (7.41), views (a) and (b), illustrate a point-contact diode.

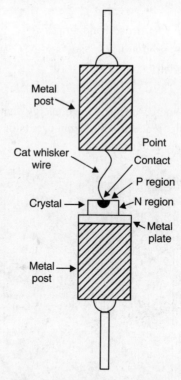

Fig. 7.41(a). Point-contact diode.

One section of the diode consists of a small rectangular crystal of n-type silicon. A fine beryllium-copper, bronze-phosphor, or tungsten wire called the cat whisker presses against the crystal and forms the other part of the diode. During the manufacture of the point contact diode, a relatively large current is passed from the cat whisker to the silicon crystal. The result of this large current is the formation of a small region of p-type material around the crystal in the vicinity of the point contact. Thus, a pn-junction is formed which behaves in the same way as a normal pn-junction.

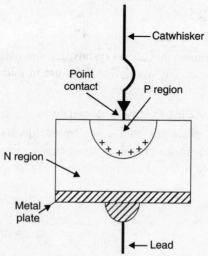

Fig. 7.41 (b). Point-contact diode, P region around point.

MICROWAVE SEMICONDUCTOR DEVICES

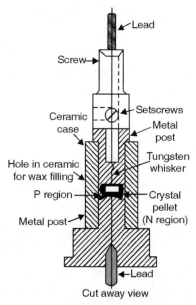

Fig. 7.41 (c). Point-contact diode cut away view.

Figure 7.41(d) shows the equivalent circuit of crystal diode consisting of series lead resistance R_s, series lead inductance L_s and ceramic case capacitance C_c.

C_j and R_j represent the effective junction capacitance and junction resistance respectively which vary with the applied bias. With reverse bias, the value of R_j is large, but with forward bias it is small. By using the combination of C_c, C_j and L_s can be tuned to any frequency for matching purpose.

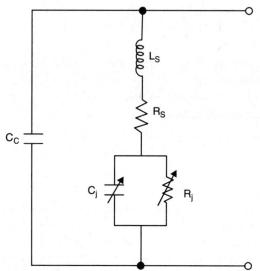

Fig. 7.41 (d). Equivalent circuit of Point-contact diode.

The pointed wire is used instead of a flat metal plate to produce a high-intensity electric field at the point contact without using a large external source voltage. It is not possible to apply large voltages across the average semiconductor because of the excessive heating.

The end of the cat whisker is one of the terminals of the diode. It has a low-resistance contact to the external circuit. A flat metal plate on which the crystal is mounted forms the lower contact of the diode with the external circuit. Both contacts with the external circuit are low-resistance contacts.

A typical example of the point contact diode is the OA 79 type of diode.

7.8.2 The Characteristics of the Point-Contact Diode

The characteristics of the point-contact diode under forward and reverse bias are somewhat different from those of the junction diode.

With forward bias, the resistance of the point-contact diode is higher than that of the junction diode. With reverse bias, the current flow through a point-contact diode is not as independent of the voltage applied to the crystal as it is in the junction diode. The point-contact diode has an advantage over the junction diode because the capacitance between the cat whisker and the crystal is less than the capacitance between the two sides of the junction diode. As such, the capacitive reactance existing across the point-contact diode is higher and the capacitive current that will flow in the circuit at high frequencies is smaller. A cutaway view of the entire point-contact diode is shown in figure (7.41), view (c). The schematic symbol of a point-contact diode is shown in view (e).

Fig. 7.41(e). Point-contact diode, schematic symbol.

Advantages of Point-Contact diodes:
1. The diode has an extremely narrow region junction. Therefore, the junction capacitance is extremely small. This makes it highly suitable as a high frequency diode, which can operate for higher microwave frequency range.
2. It does not affect by light radiation.
3. This is highly suitable as a high frequency rectifier and mixer diode because of its extremely small junction capacitance.
4. It has a lower cut-in voltage (0.2 v) than silicon.

Disadvantages of Point-Contact diodes:
1. The reverse leakage current is usually quite high.
2. Because of the very small junction, it cannot be used for applications exceeding a few mill amperes (mA).

7.9 CRYSTAL DIODES

Crystal diode is a special type of point-contact diode configured to operate in the microwave region. These diodes are special cases of the point-contact diode, and hence can be used in all applications where they are used. These crystal diodes are used as mixers and detectors in microwave circuits. These diodes have been used in applications involving frequencies in excess of 100 GHz.

Figure shows the structure of a crystal diode. The whisker is made up of gold-plated tungsten, and this is alloyed onto a silicon (GaAs or Ge) wafer in the form of pellet. This whisker-silicon pellet assembly is immersed in wax to prevent moisture and to have a protective covering. The complete structure is then housed in a ceramic enclosure, which has metal contact leads on the top and bottom. The top contact metal lead acts as the cathode and bottom metal contact lead acts as the anode.

MICROWAVE SEMICONDUCTOR DEVICES

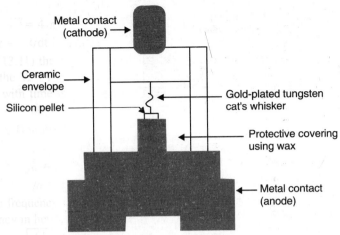

Fig. 7.42. Construction of Crystal Diode.

Fig. 7.43 shows the mounts in which crystal diodes are housed. These special types of crystal mounts are required to avoid mismatch in a microwave transmission path. In figure 7.43 the diode is mounted in a rectangular waveguide section, with coaxial lines used for taking the detected output. The input microwave signal is fed through the input opening of the waveguide.

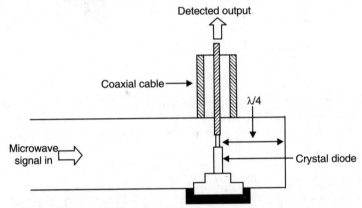

Fig. 7.43. Crystal Diode mounts for detector application.

7.10 SCHOTTKY BARRIER DIODE

The Schottky Barrier Diode is actually a variation of the point contact diode in which the metal semiconductor junction is a surface rather than a point contact. The advantage of schottky diode over point contact crystal diode is the elimination of minority carrier flow in the reverse-biased condition of the diode. Because of this elimination of holes there is no delay due to hole-electron recombination (which is present in junction diode) and hence the operation is faster. The large contact area, or barrier, between the metal and the semiconductor in the Schottky barrier diode provides some advantages over the point-contact diode. Lower forward resistance and lower noise generation are the most important advantages of the Schottky barrier diode. The applications of the Schottky barrier diode are the same as those of the point-contact diode. The low noise level generated by Schottky diodes makes them especially suitable as microwave receiver detectors and mixers.

The construction of schottky diode is illustrated in Fig. 7.44. The diode consists of n^+ silicon substrate upon which a thin layer of silicon of 2 to 3 micron thickness is epitixially grown. Then a thin insulating layer of silicon dioxide is grown thermally. After opening a window through masking process, a metal semiconductor junction is formed by depositing metal over SiO_2.

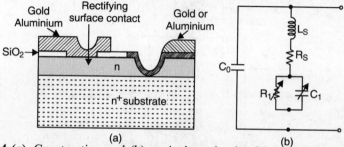

Fig. 7.44 (a). Construction and **(b)** equivalent circuit of Schottky-barrier diode.

The Schottky barrier diode is sometimes called the hot-electron or hot-carrier diode because the electrons flowing from the semiconductor to the metal have a higher energy level than the electrons in the metal. The effect is the same as it would be if the metal were heated to a higher temperature than normal.

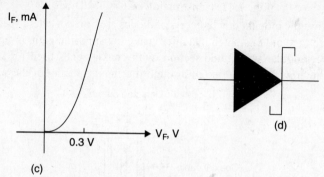

Fig. 7.44. (c) V-I characteristics **(d)** symbol of Schottky Barrier Diode.

Applications of Schottky Barrier Diode:
1. Schottky barrier diode can be used in fast switching applications because of the narrow junction and least charge at the junction.
2. They are used as mixers and detectors at microwaves.
3. They prevent transistors from going into saturation.

7.11 PIN DIODES

The PIN-diode consists of two narrow, but highly doped, semiconductor regions separated by a thicker, lightly-doped material called the intrinsic region, as shown in figure. As suggested in the name, P-I-N, one of the heavily doped regions is p-type material and the other is n-type. The same semiconductor material, usually silicon, is used for all three areas. Silicon is used most often for its power-handling capability and because it provides a highly resistive intrinsic (I) region. The pin diode acts as an ordinary diode at frequencies up to about 100 megahertz, but above this frequency the operational characteristics change.

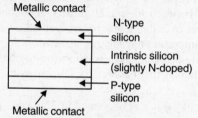

Fig. 7.45. P-I-N Diode construction.

The large intrinsic region increases the transit time of electrons crossing the region. Above 100 megahertz, electrons begin to accumulate in the intrinsic region. The carrier storage in the

MICROWAVE SEMICONDUCTOR DEVICES

intrinsic region causes the diode to stop acting as a rectifier and begin acting as a variable resistance.

The equivalent circuit of a pin diode at microwave frequencies is shown in figure (7.46), view (a). A resistance versus voltage characteristic curve is shown in view (b).

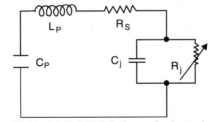

Fig. 7.46. (a) P-I-N Diode equivalent circuit.

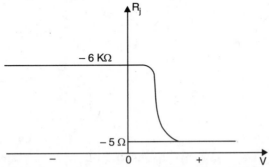

Fig. 7.46. (b) A resistance versus voltage characteristic curve.

In an equivalent circuit, L_p and C_p represent the package inductance and capacitance respectively. R_S is the bulk semiconductor layer and contact resistance. R_j and C_j represent the respective junction resistance and capacitance of the intrinsic layer.

When the bias on a pin diode is varied, its junction resistance, R_j, changes from a typical value of 6 kΩ under negative bias to about 5Ω when the bias is positive as shown in figure 7.46(b). Thus, if the diode is mounted across a transmission line or waveguide, the loading effect is insignificant while the diode is reverse-biased, and the diode presents no interference to power flow. When the diode is forward biased, the resistance drops to approximately 5 ohms and most power is reflected. In other words, the diode acts as a switch when mounted in parallel with a transmission line or waveguide. Several diodes in parallel can switch power in excess of 150 kilowatts peak. The upper power limit is determined by the ability of the diode to dissipate power. The upper frequency limit is determined by the shunt capacitance of the pn-junction, shown as C_j in figure 7.46, view (a). Pin diodes with upper limit frequencies in excess of 30 GHz are available.

7.11.1 Applications of PIN Diode

PIN diode may be used either as variable resistors or as electronic switches for RF signals. The PIN diode exhibits a characteristic, that the diode is basically a two valued resistor, with one value being very high and other being very low. Several applications makes use of this characteristic, are discussed in the following sections.

7.11.2 PIN Diode as Switch

PIN diode can be used to switch devices such as attenuators, filters and amplifiers in and out of the circuit. A PIN diode can be used either a series or a shunt configuration to form a single pole throw RF switch or in combination modes. These circuits are shown in figure 7.47, with bias networks.

1. Series switch

In the series configurations, as shown in figure 7.47, the diode (D_1) is placed in series with the signal line. The switch is ON when the diode is forward biased and OFF when it is reversed biased. When the diode is turned ON, the signal path has a low resistance and when the diode is turned off, it has a very high resistance (thus providing the switching action) when the switch S_1 is open, the diode is unbiased, so the circuit is open by virtue of the very high resistance. The ratio of OFF/ON resistances provides a measure of the isolation of the circuit.

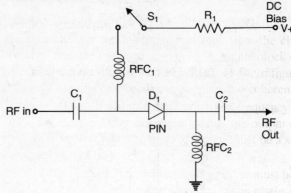

Fig. 7.47. Single pole switch – Series configuration.

2. Shunt configuration:

In the shunt configuration, as shown in figure 7.48, the diode is placed across the signal line. In the shunt configuration, forward biasing the diode "cuts-off" the transmission and reverse biasing the diode ensure transmissions from input to output. The DC blocks (capacitors) should have very low impedance at RF operating frequency and RF choke inductors should have very high impedance.

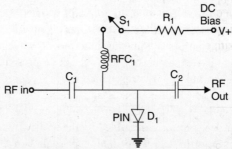

Fig 7.48. Shunt switch configuration.

3. Combinational switch

A combination series-shunt circuit is illustrated in fig- 7.49. In this configuration D_1 and D_2 are placed in series with the signal line, while D_3 is in parallel the signal line. D_1 and D_2 will turn ON with a positive potential applied, while D_3 turns ON when negative potential applied. When switch S_1 is in the ON position, a positive potential is applied to the junction of the three diodes. As a result D_1 and D_2, diodes are forward biased and thus take on a low resistance. At the same time D_3, diode is reverse biased and so has a very high resistance. Signal is passed through from input to output essentially unimpeded.

In the condition, when switch S_1 is in the OFF position, the opposite situation occurs. In this case, the applied potential is negative, so D_1 and D_2 are reverse biased (and take on a high series resistance), while D_3 is forward biased (and takes on a low series resistance). This circuit action exhibits a tremendous attenuation of the signal between input and output.

MICROWAVE SEMICONDUCTOR DEVICES

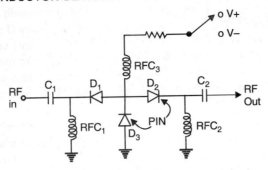

Fig. 7.49. PIN diode as combinational switch.

7.11.3 PIN Diode Attenuators

The forward resistance of the PIN diode decreases with increase of bias voltage as shown in the characteristics curve discussed earlier. This property of the PIN diode is utilized for the construction of attenuators. As it is known now that the PIN diode acts as a variable resistor only for frequencies above 100 MHz. Below this frequency, PIN diode acts as rectifier. To realize PIN diode attenuators, either series or shunt configuration can be used.

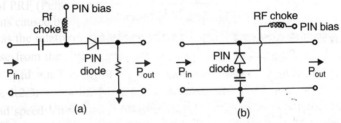

Fig. 7.50. PCIN attenuators (a) Series configuration (b) Shunt configuration.

In the shunt configuration of figure- 7.50 (b), as the PIN bias is increased, the resistance decreases and the attenuation increases. But in the series configuration of figure-(a), as the PIN bias is increased, the resistance decreases so also the attenuation introduced to the input power P_{in}.

7.11.4 PIN diode as Limiter

PIN diodes in shunt are used to limit the power, as shown in figure 7.51. A PIN limiter is a microwave switch that is controlled by self bias rather than external bias. Limiters are used for the protection of the microwave systems. The choke provides a return path for the self bias current. The power handling capability can be increased further by using more diodes in shunt. PIN limiters can withstand maximum power to 100 kW.

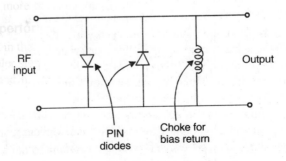

Fig. 7.51. PIN diode as a limiter.

7.11.5 PIN diode as Modulator

PIN diode may be employed as an amplitude modulator. In this application a PIN diode is connected across a transmission line or inserted into one end of piece of microwave waveguide. The audio modulating voltage is applied through an RF choke to the PIN diode. Suppose, when a CW signal is applied to the transmission line, the varying resistance of the PIN diode causes the signal to be amplitude modulated.

When a PIN attenuator is placed along a transmission line, sine wave, square wave and pulse modulation can be obtained by varying current through diode. Fig. 7.52 shows the arrangement of PIN diode as modulator.

A typical application of PIN modulator is in closed loop automatic leveling circuit.

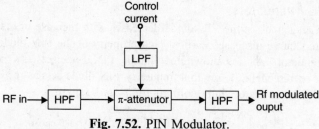

Fig. 7.52. PIN Modulator.

7.12 STIMULATED-EMISSION AND ASSOCIATED DEVICES

In 1954, Towens and his colleagues developed the first really low noise amplifier produced **Microwave Amplification** by **stimulated Emission of Radiation** (MASER) by a quantum-mechanical process. Another device, the LASER or optical MASER is a development of this idea, which permits the generation or amplification of coherent light. **LASER (Light Amplification by Stimulated Emission of Radiation)** and **MASER** are examples of, stimulated emission devices.

7.12.1 Fundamentals of Operations

As per atomic theory, electrons have various energy levels corresponding to different orbits. At a low temperature, most of the electrons exist in the lowest energy level but they may be raised by providing additional amount of energy. The electron can be stimulated from this energy level. As per Quantum theory, the specific amount of energy or quantum that can provide the necessary energy for raising the level of electron is given by the wave mechanical equation

$$E = hf \qquad \ldots (34)$$

where E = energy differences in joules

h = Planck's constant = 6.624×10^{-27} erg/sec

f = Photon frequency in Hz.

Having been excited by the absorption of a quantum, the atom may remain in the excited state (for a microsecond) or more likely remit the energy at the same frequency at which it was received, and the atom will thus return to its original or ground state. Re-emission of energy is stimulated at the expense of absorption by structures or cavities tuned to the frequency. It is also possible to supply energy to these atoms in such quantities and at such a frequency that they are raised to an energy level which is much higher from the ground state, rather than immediately above it. This is called **pumping the atom into the top energy level**. We can make the atom emit energy at a frequency corresponding to the difference between top level and an intermediate level, between the ground and the top energy levels. **Pumping** thus occurs at the frequency corresponding to the energy difference between the ground and the top energy levels. Re-emission of energy is stimulated at the desired frequency and signal at this frequency is amplified. It is to be noted that practically there is no noise is added to the amplified signal because there is no resistance involved

MICROWAVE SEMICONDUCTOR DEVICES

and no electron stream to produce shot noise. The energy re-emission is at a microwave frequency and reemission frequencies depend on the energy levels in the atom. Fig. (7.53) shows the energy levels in relevant to MASER operation.

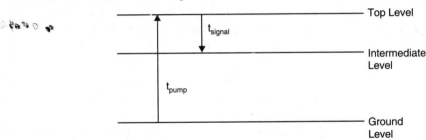

Fig. 7.53

Initially substance was the gas ammonia, while hydrogen and cesium used subsequently. The advantage of the gaseous substance is that it allowing absorbing atoms to be removed easily. Ammonia has only one frequency of re-emission, there was no method whatsoever of tuning the maser, so that signal at other frequencies just could not be amplified. So overcome these difficulties, the traveling-wave ruby maser was invented.

7.12-2 Ruby MASER

As gaseous materials is inconvenient in a maser amplifier, as can be appreciated, hence the search for more suitable materials revealed ruby, which is crystalline form of silica (Al_2O_3) with a slight natural doping of chromium. The presence of chromium makes the crystal paramagnetic or slightly magnetic. Ruby has the advantages of being solid, having suitably arranged energy levels, when a paramagnetic material is exposed in input signal under d.c. magnetic field, there will be Zeeman splitting. If the difference between the two levels be 10^{-5} ev, than by applying a suitable microwave signal of same frequency. We can observe the absorption of the signal.

Assume that there are two energy levels E_1 and E_3 in a paramagnetic material and the population on them be
N_1 and N_3 respectively. Then we can define Δ as :

$$\Delta \frac{E_3 - E_1}{KT} = \frac{hf_0}{KT} \qquad \ldots 7.35$$

where
K = Boltzman constant
T = Absolute Temperatures

We know that,

$$\frac{N_3}{N_1} = \exp.\left[-\frac{E_3 - E_1}{KT}\right] = e^{-\Delta} \qquad \ldots 7.36$$

and if $N_3 + N_1 = N$, then

$$N_1 = \frac{N}{1 + e^{-\Delta}} = \frac{1}{2} N \left(1 + \frac{1}{2}\Delta\right)$$

and
$$N_3 = \frac{N}{1 + e^{-\Delta}} \left(1 - \frac{1}{2}\Delta\right)$$

The signal strength in paramagnetic resonance is given by

$$\Delta N = N_3 - N_1 = N \tan h\left(\frac{\Delta}{2}\right) \qquad \ldots 7.37$$

$$\Delta N = \frac{1}{2}\Delta = \Delta \frac{1}{2} \cdot \frac{hf_0}{KT} \qquad \ldots 7.38$$

Thus at the operating frequency the greater signal strength is achieved at low temperature.

Fig. (7.53) shows the energy level situation in a three-level MASER. Energy at the correct pump frequency is added to the atom in the crystals lattice of ruby, biasing them to the uppermost of the levels. Normally, the number of electrons in the third energy level is smaller than the number in the ground level. Although as pumping is continued, the number of electron in level 3 increases until it is about equal to the number in the first level. At this point the crystal saturates and this is called **population inversion** has been accomplished.

Since the condition have been made suitable for re-radiation (not absorption) of this excess energy, electron in the third level may give off energy at the original pump frequency and thus return to ground level. At the other hand, they may give off smaller energy quanta at the frequency corresponding to the difference between the third and second levels and thus return to the intermediate level. A large number of electrons take the latter case, which is stimulated by the presence of the cavity surrounding the ruby, which is resonant at this frequency. This course is further aided by the presence of the input signal at this frequency. Since the amount of energy radiated or emitted by the excited ruby atoms at the signal frequency exceeds the energy applied at the input result amplifications.

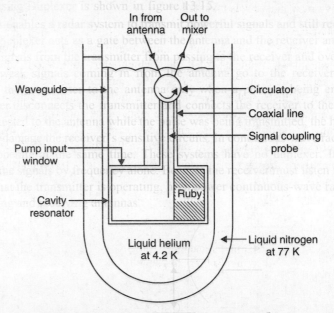

Fig. 7.54. Schematic diagram of ruby maser.

The sectional view of ruby cavity maser is shown in Fig. (7.54). Like some other microwave amplifier, it is also a single port device and it needed a circulator to separate the output power from the input. Like the parametric amplifier, a tuned circuit is must be provided for the pump signal as well as for the signal to be amplified.

The **disadvantage** of this type of maser is its narrow bandwidth and which is decided by the cavity itself. There may be compromising between gain and bandwidth but in that way one cannot increase the bandwidth to a very great extent. The solution for this is a TW maser (Traveling Wave), in this just like TWT a slow wave structure of ruby is made and the signal is allowed to pass through it and grows at the expenses of pump signal.

MICROWAVE SEMICONDUCTOR DEVICES

The **travelling wave maser**s have so many **advantages** such as
  (*i*) Greater Bandwidth
  (*ii*) Greater Stability
  (*iii*) Easy to tune ability
  (*iv*) No circulator is needed, hence losses due to circulator is eliminated.

Performance characteristics of typical TW maser:
  (*i*) Operating Frequency is 1.6 GHz
  (*ii*) Gain 25 dB
  (*iii*) Bandwidth is 25 MHz at 1.6 GHz
  (*iv*) Pump frequency is 48 GHz
  (*v*) Pump power is 140 mW (CW operation)
  (*vi*) Noise figure is better than 0.3 db.

Applications:
  (*i*) Low level low noise amplifier.
  (*ii*) Radio telescope and space probe receivers.
  (*iii*) Radio astronomy.

7.13 AVALANCHE TRANSIT TIME DEVICES

It is proposed by the W.T. Read in 1958, that the delay between voltage and current in an avalanche, together with transit time through the material, it is possible to make a microwave diode exhibit negative resistance. These types of devices are known as **Read Diodes**. At microwave frequencies they use carrier impact ionization and drift in the high field region of a semi-conductor junction to produce negative resistance.

They are of following three types:
  1. IMPATT (Impact Ionization Avalanche Transit Time device)
  2. TRAPATT (Trapped Plasma Avalanche Triggered Transit device)
  3. BARITT (Barrier Injected Transmit time device)

7.13.1 IMPATT Diode

The basic operating principle which is given by the 'Read' in 1958, of IMPATT diodes can easily be understood. A mode of the original Read diode with a doping profile and a dc electric field distribution that exist when a large reverse voltage is applied across the diode is shown in Fig. (7.55).

Any device which exhibit a dynamic negative resistance for direct current (d.c.) will also exhibit it for alternating current (a.c) also. Suppose we apply an ac voltage, the current will rise when voltage falls at an ac rate. The **negative resistance** may be defined as that property of a device which causes the current through it to be 180° out of phase with voltage across it. It can be shown that the negative resistance effect in IMPATT diode exhibit due to 180° phase difference between voltage and current.

An **IMPATT diode** is a combination of delay involved is generating avalanche current multiplication together with delay due to transit time through a drift space, which provides the necessary 180° phase difference between applied voltage and the resulting current. It is to note that IMPATT is a diode, the junction being between the P^+ and the n layers as shown in Fig. (7.55).

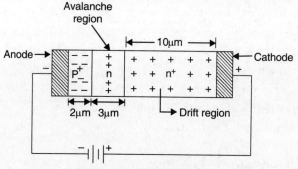

Fig. 7.55. IMPATT diode.

A very high voltage (400 kV/cm) is applied to the IMPATT diode, resulting in a very high current. A normal diode would easily breakdown under these conditions but the IMPATT diode is so constructed that it can withstand these conditions repeatedly. Its realization was delayed due to vast amount of heat dissipation. To ensure a satisfactorily low operating temperature for the IMPATT diode, so that it would not be damaged by melting. The typical operating temperatures are in the order of 250°C. Due to a high potential gradient causes a flow of minority carriers across the junction. It is now assumed oscillations exist and we may consider the effect of a positive swing of the RF voltage superimposed on top of the high dc voltage. The increased velocity of electrons and holes result additional holes and electrons by knocking them out of crystal structure by so called impact ionization. These additional carriers continue the process at the junction and it now snow balls into an avalanche. If the initially dc field was at the threshold of allowing this situation to develop this voltage will be exceeded during the whole of the positive RF cycle and avalanche current multiplication will be taking place during this entire time. Since it is multiplication process avalanche is not instantaneous. The process takes a time such that the current pulse maximum, at the junction, occurs at the instant when the RF voltage across the diode is zero and going negative. A phase difference of 90° between voltage and current has been obtained.

The current pulse in the IMPATT diode is situated at the junction. However it does not stay there and drift the current flows to the cathode as shown in Fig. (7.56). The time taken by the pulse to reach to the cathode depends upon the velocity and the thickness of the highly doped (n^+) layer. The time taken for current pulse to move from the position of $V = 0$ to the position of $V = -$ maximum of RF cycles is 90°. The voltage and current in the IMPATT diode are 180° out of phase and a dynamic RF negative resistance has been proved. Hence *IMPATT* diode is useful both as an oscillator and as an amplifier.

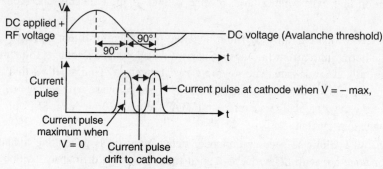

Fig. 7.56. IMPATT diode-V and I behavior.

The resonant frequency of IMPATT diode may be represented by

$$f_0 = \frac{V_d}{2L} \qquad \qquad \ldots 7.39$$

MICROWAVE SEMICONDUCTOR DEVICES

where
f_0 = Resonant frequency
V_d = Drift velocity of Carrier
L = Length of the drift space charge region

The IMPATT diode shown in Fig. (7.57) is a typical diode used below 50 GHz and house either of Ga As or Cu Si chip. Ga As is preferable since it gives low noise, higher efficiency and higher maximum operating frequency but it is difficult to fabricate. In a practical circuit, the IMPATT diode is generally embedded in the wall of a cavity, which then acts as an external heat sink.

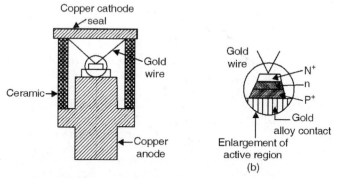

Fig. 7.57. Typical IMPATT diode construction details.

The maximum power in the IMPATT diode, that can be given to the mobile carriers decreases as $1/f_0$ the efficiency of IMPATT diode can be written as

$$\eta = \left[\frac{P_{ac}}{P_{dc}}\right] = \frac{V_{ac}}{V_{dc}} \cdot \frac{I_{ac}}{I_{dc}} \qquad \ldots (40)$$

where η = efficiency of IMPATT diode, $P_{a.c.}$ = a.c. power, $P_{d.c.}$ = d.c. power, $V_{a.c.}$ = a.c. voltage, $V_{d.c.}$ = d.c. voltage, $I_{a.c.}$ = a.c. current, $I_{d.c.}$ = d.c. current.

The impedance the IMPATT diode chip is capacitive because R_d (magnitude of negative resistance) is less than the reactance X_d.

An equivalent circuit of epitaxial IMPATT diode chip may be shown as in Fig. (7.58).

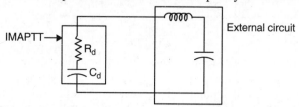

Fig. 7.58. Equivalent circuit of epitaxed IMPATT diode.

Performance Characteristics of IMPATT Diodes

(i) Efficiency $\eta < 30\%$
(ii) Operating frequency = 1 to 300 GHz
(iii) Pulsed Power = 4 KW
(iv) Maximum power output: (a) For single diode 5 W in X Band 0.5 W at 30 GHz (b) for several diodes cavities – 40 W at X Band.

IMPATT diodes are at present the most powerful CW solid state microwave power sources. These diodes have been fabricated from germanium, silicon and gallium arsenide and from other semiconductor as well. IMPATT diode provides potentially reliable, compact, inexpensive and moderately efficient microwave power sources.

Applications of IMPATT Diode

1. In the Transmission part of TV system
2. Used in the final stage of solid state microwave transmitter in RADARS.
3. It is used in TDM/FDM system
4. Microwave source for the experimental purpose
5. As a missile seeker head.

7.13.2 TRAPATT DIODES

TRAPATT stands for **Trapped Plasma Avalanche Triggered Transit mode**. It is a very high efficiency microwave generator capable, of operating from several hundred megahertz to several gigahertzes. The basic operation of the oscillator is a *PN* junction semiconductor diode reverse biased to current densities well in excess1 of those encountered in normal avalanche operation. They are typically silicon $n^+ - P - P^+$ or $P^+ - n - n^+$ structure high peak power diodes. Fig. (7.59), shows the schematic arrangement of TRAPATT diode.

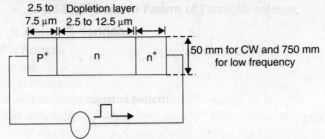

Fig. 7.59. The schematic diagram of TRAPATT diode.

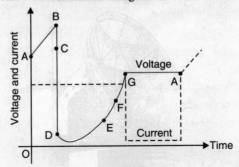

Fig. 7.60

Now the voltage decreases to point *D*. A long period is required to remove the plasma because the total plasma charge is large compared to the charge per unit time in the external current. The plasma is removed at point *E*, but a residual charge of electrons remains in one end of the depletion layer and a residual charge of holes in the other end. The voltage increases from point *E* to point *F* as soon as the residual charge is removed. At point *F* all the charge that was generated internally has been removed. This charge should be greater than or equal to the supplied by the external current, otherwise the voltage will exceed that at point *A*. The diode will charge up again like a fixed capacitor from point *F* to *G*. The diode current goes to zero at point *G* for a half period and the voltage remains constant at V_A until the current comes back on and the cycle repeats.

The electric field may be expressed as

$$E(x, t) = E_m - \frac{qN_A}{\varepsilon_s}x + \frac{J_t}{\varepsilon_s} \qquad \ldots 7.41$$

where, N_A is the doping concentration of the *n* region, *x* is the distance.

MICROWAVE SEMICONDUCTOR DEVICES

Thus the value of t at which the electric field reaches E_m at a given distance x into the depletion region is obtained by setting $E(x, t) = E_m$

$$We\ get\quad t = \frac{qN_A}{J} x \qquad \ldots 7.42$$

By differentiating the above eq. (41) with respect to time, we get

$$v_z = \frac{dx}{dt} = \frac{J}{qN_A} \qquad \ldots 7.42$$

where
v_z = Avalanche zero velocity
J = current density
q = Electron charge
N_A = Doping concentration of the n region.

The transit time of the carriers may be given by

$$I_S = \frac{L}{v_S} \qquad \ldots 7.44$$

where
T_s = transit time
v_s = standard carrier drift velocity
L = length of the specimen.

The transit time is also being used in this device (the time between injection and collection). The time delay of carriers in transit is utilized to obtain a current phase shift favorable for oscillation.

Performance Characteristics

(i) Frequency range from 3 to 50 GHz.
(ii) Operating voltage is 60 to 150 V.
(iii) Efficiency from 15 to 40%.
(iv) Noise figure is more than 30 db.
(v) Powers (a) CW power – 1 to 3 W between 8 GHz to 0.5 GHz.
(b) Pulse Power – 1.2 KW at 1.1 GHz.

Applications: It is used in 5-band pulse RADAR Transmitter of pulse array radar system.

7.13.3 BARITT Diodes

BARITT diodes mean **Barrier injected transit time diodes**, which are the latest addition to the family of active microwave diodes. BARITT diodes have long drift region similar to the IMPATT diodes. It is basically a back to back pair of diodes and has various structures including $P^+ n P^+$, $P^+ - n$-metal. When the BARITT diode is biased and voltage is applied across the diode the carriers are injected by crossing the metal semiconductor barrier and more through the long drift region with saturated drift velocity and result the transitions effect. The charge carriers are collected in the metal side at the other end of the drift layer (n-layer).

BARITT diodes are much-less noisy than IMPATT diodes. The noise figure is 15 db at C-band of frequencies with silicon BARITT amplifiers.

The fabrication of BARITT diodes is very simple. A crystal n-type silicon wafer with 11Ω-cm resistivity and of thickness 10 μm sandwiched between two P-type Si Schottky barrier contacts of about 0.1 μm thickness.

In the BARITT diode, the thermionic emission of hole injected from the forward biased contact is responsible for the increase of current, not the avalanche multiplication.

The microwave oscillation in BARITT diode depends upon the following two facts:
(*i*) The rapid increase of carrier injection process caused by the reducing potential barrier of the forward biased metal semiconductor contact.
(*ii*) An apparent transit angle ($3\pi/2$) of the injected carrier that traverses the semiconductor depletion region.

For a *m–n–m* BARITT diode P_S – *Si* Schottky barrier contacts metals with *n*-type Si wafer in between, a rapid increase in current with applied voltage, which is above 30 V, is due to the thermionic hole injection into the semiconductor. Fig. (7.61) shows the *V-I* characteristics of BARITT diode. The critical voltage is approximately given by

$$V_c = \frac{qNL^2}{2\varepsilon_s} \qquad \ldots 7.45$$

where
V_c = Critical voltage
N = Doping concentration
ε_s = Semiconductor dielectric permittivity

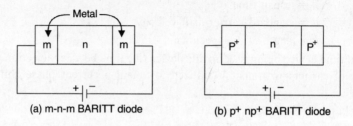

(a) m-n-m BARITT diode   (b) p+ np+ BARITT diode

Fig. 7.61

The breakdown electric field (E_{bd}) is given by

$$E_{bd} = \frac{V_{bd}}{L} = \frac{qNL}{\varepsilon_s} \qquad \ldots 7.46$$

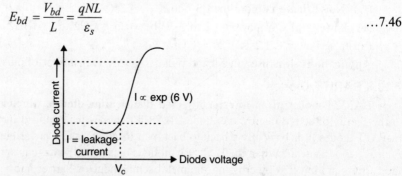

Fig. 7.62. Voltage and current characteristic curve of BARITT diode.

And the breakdown voltage field (V_{bd}),

$$V_{bd} = 2V_c$$

or
$$V_{bd} = \frac{2.qnC^2}{2\varepsilon_s} = \frac{qNL^2}{\varepsilon_s} \qquad \ldots 7.47$$

Typical Characteristics of a BARITT diode:
1. Frequency operation : 4 to 8 GHz
2. Maximum power output (CW) : 50 MW at 4.9 GHz
3. Conversion efficiency : 1.8%

4. Noise figure : 9 dB at 6.35 GHz

Applications:
1. BARITT diode is used as a local oscillator in radar receivers
2. Primarily BARITT diode is used as amplifier because of lower efficiencies.

7.14 MICROWAVE TRANSISTORS

A transistor is a semiconductor device that uses a small amount of voltage or electrical current to control a larger change in voltage or current. Because of its fast response and accuracy, it may be used in a wide variety of applications, including amplification, switching, voltage stabilization, signal modulation, and as an oscillator. The transistor is the fundamental building block of both digital and analog circuits -the circuitry that governs the operation of computers, cellular phones, and all other modern electronics. Transistors may be packaged individually or as part of an integrated circuit chip, which may hold thousands of transistors in a very small area.

Modern transistors are divided into two main categories:
1. Bipolar junction transistors (BJTs)
2. Field effect transistors (FETs).

Application of current in BJTs and voltage in FETs between the input and common terminals increases the conductivity between the common and output terminals, thereby controlling current flow between them.

The term "Transistor" originally referred to the point contact type, but these only saw very limited commercial application, being replaced by the much more practical bipolar junction types in the early 1950s.

Transistors are categorized by:
- *Semiconductor material* : Germanium, silicon, gallium arsenide, silicon carbide
- *Structure* : BJT, JFET, IGFET (MOSFET), IGBT, "other types"
- *Polarity* : NPN, PNP, N-channel, P-channel
- *Maximum power rating* : low, medium, high
- *Maximum operating frequency*: low, medium, high, radio frequency (RF), microwave (The maximum effective frequency of a transistor is denoted by the term f_T, an abbreviation for "frequency of transition". The frequency of transition is the frequency at which the transistor yields unity gain).
- *Physical packaging*: through hole metal, through hole plastic, surface mount, ball grid array.

7.14.1 Applications of Transistors

The BJT remains a device that excels in some applications, such as discrete circuit design, due to the very wide selection of BJT types available and because of knowledge about the bipolar transistor characteristics. The BJT is also the choice for demanding analog circuits, both integrated and discrete. This is especially true in very-high-frequency applications, such as radio-frequency circuits for wireless systems. The bipolar transistors can be combined with MOSFET's in an integrated circuit by using a BiCMOS process to create innovative circuits that take advantage of the best characteristics of both types of transistor.

7.14.2 Bipolar Junction Transistor (BJT)

The Bipolar junction transistor (BJT) was the first solid-state amplifier element and started the solid-state electronics revolution. Bardeen, Brattain and Shockley, while at Bell Laboratories, invented it in 1948 as part of a post-war effort to replace vacuum tubes with solid-state devices.

Solid-state rectifiers were already in use at the time and were preferred over vacuum diodes because of their smaller size, lower weight and higher reliability. A solid-state replacement for a vacuum triode was expected to yield similar advantages. The work at Bell Laboratories was highly successful and culminated in Bardeen, Brattain and Shockley receiving the Nobel Prize in 1956.

Nevertheless, bipolar transistors remain important devices for ultra-high-speed discreet logic circuits such as emitter coupled logic (ECL), power-switching applications and in microwave power amplifiers. Heterojunction bipolar transistors (HBTs) have emerged as the device of choice for cell phone amplifiers and other demanding applications.

Structure and Principle of Operation

Bipolar transistors, having 2 junctions, are 3 terminal semiconductor devices. The three terminals are emitter, collector, and base. A transistor can be either NPN or PNP.

A bipolar junction transistor consists of two back-to-back p-n junctions, who share a thin common region with width, wB. Contacts are made to all three regions, the two outer regions called the emitter and collector and the middle region called Since the device consists of two back-to-back diodes, there are depletion regions between the quasi-neutral regions. The width of the quasi neutral regions in the emitter, base and collector are indicated with the symbols wE', wB' and wC' and are calculated from the base. The structure of an npn bipolar transistor is shown in figure (7.63). The device is called "bipolar" since its operation involves both types of mobile carriers, electrons and holes.

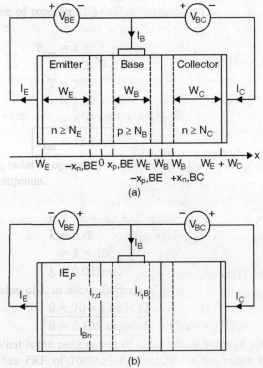

Fig. 7.63. (a) Structure and sign convention of a *npn* bipolar junction transistor.
(b) Electron and hole flow under forward active bias, $V_{BE} > 0$ and $V_{BC} = 0$.

$$w_E' = w_E - x_{n, BE}$$

$$w_B' = w_B - x_{p, BE} - x_{p, BC}$$

$$w_C' = w_C - w_{n, BC}$$

MICROWAVE SEMICONDUCTOR DEVICES

where the depletion region widths are given by:

$$x_{n,BE} = \sqrt{\frac{2\varepsilon_s(\phi_{i,BE} - V_{BE})}{q} \frac{N_B}{N_E}\left(\frac{1}{N_B + N_E}\right)}$$

$$x_{n,BE} = \sqrt{\frac{2\varepsilon_s(\phi_{i,BE} - V_{BE})}{q} \frac{N_E}{N_B}\left(\frac{1}{N_B + N_E}\right)}$$

$$x_{p,BC} = \sqrt{\frac{2\varepsilon_s(\phi_{i,BC} - V_{BC})}{q} \frac{N_C}{N_B}\left(\frac{1}{N_B + N_C}\right)}$$

$$\phi_{i,BE} = V_t \ln\frac{N_E N_B}{n_i^2}$$

$$\phi_{i,BC} = V_t \ln\frac{N_C N_B}{n_i^2}$$

The sign convention of the currents and voltage is indicated on figure 7.63(a). The base and collector current are positive if a positive current goes into the base or collector contact. The emitter current is positive for a current coming out of the emitter contact. This also implies that the emitter current, IE, equals the sum of the base current, I_B, and the collector current, I_C:

$$I_E = I_C + I_B$$

The base-emitter voltage and the base-collector voltage are positive if a positive voltage is applied to the base contact relative to the emitter and collector respectively.

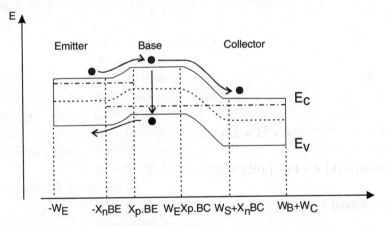

Fig. 7.64. Energy band diagram of a bipolar transistor biased in the forward active mode.

The operation of the device is illustrated with Figure 7.63(b). We consider here only the forward active bias mode of operation, obtained by forward biasing the base-emitter junction and reverse biasing the base-collector junction. To simplify the discussion further, we also set $V_{CE} = 0$. The corresponding energy band diagram is shown in Figure (7.64). Electrons diffuse from the emitter into the base and holes diffuse from the base into the emitter. This carrier diffusion is identical to that in a p-n junction. However, what is different is that the electrons can diffuse as minority carriers through the quasi-neutral base. Once the electrons arrive at the base-collector depletion region, they are swept through the depletion layer due to the electric field. These electrons contribute to the collector current. In addition, there are two more currents, the base recombination current, indicated on Figure (7.64) by the vertical arrow, and the base-emitter depletion layer recombination current, Ir,d, (not shown).

The total emitter current is the sum of the electron diffusion current, $I_{E,n}$, the hole diffusion current, $I_{E,p}$ and the base-emitter depletion layer recombination current, Ir,d.

$$I_E = I_{E,n} + I_{E,p} + I_{r,d}$$

The total collector current is the electron diffusion current, $I_{E,n}$, minus the base recombination current, $I_{r,B}$

$$I_C + I_{E,n} - I_{r,B}$$

The base current is the sum of the hole diffusion current, $I_{E,p}$, the base recombination current, $I_{r,B}$ and the base-emitter depletion layer recombination current, $I_{r,d}$.

$$I_B = I_{E,p} + I_{r,B} + I_{r,d}$$

The transport factor, a, is defined as the ratio of the collector and emitter current:

Using Kirchhoff's current law and the sign convention shown in Figure 7.63(a), we find that the base current equals the difference between the emitter and collector current. The current gain, a, is defined as the ratio of the collector and base current and equals

$$\beta = \frac{I_C}{I_B} = \frac{\alpha}{1-\alpha}$$

This explains how a bipolar junction transistor can provide current amplification. If the collector current is almost equal to the emitter current, the transport factor, a, approaches one. The current gain, a, can therefore become much larger than one.

See the schematic representations in Fig. (7.65) below:

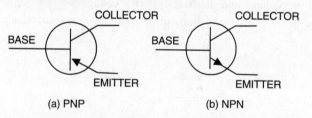

Fig. 7.65. The schematic representations of BJT's.

Note that the direction of the emitter arrow defines the type transistor. Biasing and power supply polarity is positive for NPN and negative for PNP transistors. The transistor is primarily used as a current amplifier. When a small current signal is applied to the base terminal, it is amplified in the collector circuit. This current amplification is referred to as H_{FE} or beta and equals I_c/I_b.

As with all semiconductors, breakdown voltage is a design limitation. There are breakdown voltages that must be taken into account for each combination of terminals, i.e., V_{ce}, V_{be} and V_{cb}. However, V_{ce}(collector-emitter voltage) with open base, designated as V_{ceo}, is usually of most concern and defines the maximum circuit voltage.

Also as with all semiconductors there are undesirable leakage currents, notably Icbo, collector junction leakage; and Iebo, emitter junction leakage.

A typical collector characteristic curve is shown in figure (7.66) below:

MICROWAVE SEMICONDUCTOR DEVICES

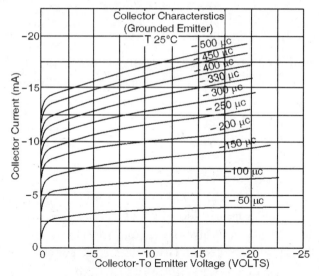

Fig. 7.66. A typical collector characteristic curve of BJT's.

Note that the negative collector-emitter voltage tells you that the transistor is PNP.

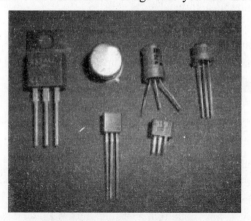

Fig. 7.67. Assorted transistors.

Performance Parameters: The behavior of microwave transistor is defined by f_T, the cut-off frequency and the maximum possible frequency of oscillation f_{max} rather than common base current (α), common emitter current gain (β) and cut-off frequencies (f_{ab} and f_{ac}).

$$\beta = \alpha/1-\alpha$$

$$f_{ac} = f_{ab}/\beta$$

The cut-off frequency $f_T = 2\pi\tau_{ec}$ is determined by the emitter-collector delay time τ_{ec}.

Where
$$\tau_{ec} = \tau_c + \tau_b + \tau_d + \tau_c$$

τ_e = the emitter base junction changing time,

τ_b = the base transit time

τ_d = the collector depletion layer transit time

τ_c = the collector depletion layer changing time

For microwave transistors, the cut-off frequency is

$$f_T = 1/=\tau_{ebc}$$

where $\tau_{ebc} = \tau_{bc} + \tau_{ec}$

Maximum Frequency of Oscillation

Maximum Frequency of Oscillation $(f_{max}) = \sqrt{f_T / 8\pi \ rb'C_c}$

f_{max} is higher than f_T because although β has fallen to unity at this frequency and power gain has not. That is β = 1 output impedance is higher than input impedance, voltage gain exits and hence both generation and oscillation are possible, where,

f_T = cut off frequency
rb' = base resistance
C_c = collector capacitance.

Performance Characteristics of Microwave Transistors
1. Noise figure : It is 3.3 dB to 14dB (4 GHz to 8GHz)
2. Power output : 20 W to 150 mW (1 GHz to 8 GHz)
3. Power gain = 31 ± 1.5 dB (4 to 6 GHz with average power out of 15mW.
4. Voltage Frequency limitation 2×10^{11} verses for silicon.

Primary considerations when selecting a transistor are:
(a) Voltage ratings of all three junctions
(b) Power rating and thermal resistance
(c) Current handling capability and the transistor case size
(d) Leakage currents, mainly Icbo and Iebo
(e) Frequency response and /or switching times.
(f) Current gain (H_{FE} and hfe)
(g) Temperature parameter variation,
(h) Saturation resistance
(i) h-parameters for linear applications

Semiconductor Material Used in BJT's

The first BJTs were made from germanium (Ge) and some high power types still are, Silicon (Si) types currently predominate but certain advanced microwave and high performance versions now employ the compound semiconductor material gallium arsenide (GaAs) and the semiconductor alloy silicon germanium (SiGe). Single element semiconductor material (Ge and Si) is described as elemental.

Characteristics of the most common semiconductor materials used to make transistors are given in the table (7.1) below:

Table 7.1 : Semiconductor material characteristics.

Semiconductor material	Junction forward voltage V @25°C	Electron mobility m^2/Vs @ 25°C	Hole mobility m^2/Vs @ 25°C	Max. Junction temp. °C
Ge	0.27	0.39	0.19	7. to 100
Si	0.71	0.14	0.05	150 to 200
GaAs	1.03	0.85	0.05	150 to 200
Al-Si junction	0.3			150 to 200

MICROWAVE SEMICONDUCTOR DEVICES

7.14.3 Heterojunction Bipolar Transistor

The heterojunction bipolar transistor (HBT) is an improvement of the BJT that can handle signals of very high frequencies up to several hundred GHz. It is common nowadays in ultrafast circuits, mostly RF systems.

Heterojunction transistors have different semiconductors for the elements of the transistor. Usually the emitter is composed of a larger bandgap material than the base. This helps reduce minority carrier injection from the base when the emitter-base junction is under forward bias and increases emitter injection efficiency. The improved injection of carriers into the base allows the base to have a higher doping level, resulting in lower resistance to access the base electrode. With a regular transistor, also referred to as homojunction, the efficiency of carrier injection from the emitter to the base is primarily determined by the doping ratio between the emitter and base. Because the base must be lightly doped to allow the high injection efficiency its resistance is relatively high. With a hererojunction the base can be highly doped' allowing a much lower base resistance and consequently higher frequency operation.

Two commonly used HBT's are:

(a) Silicon-germanium and
(b) Aluminum gallium arsenide.

Silicon-germanium is widely used because it is compatible with standard silicon digital processes, allowing integration of very high speed circuitry with complex lower speed digital circuitry.

Material Used in HBT's

The principal difference between the BJT and HBT is the use of differing semiconductor materials for the emitter and base regions, creating a heterojunction. The effect is to limit the injection of holes into the base region, since the potential barrier in the valence band is so large. Unlike BJT technology, this allows high doping to be used in the base, creating higher electron mobility while maintaining gain. The efficiency of the device is measured by the Kroemer factor, after Herbert Kroemer who received a Nobel Prize for his work in this field in 2000.

Materials used for the substrate include silicon-germanium alloys and gallium arsenide, while aluminum gallium arsenide, indium phosphide and indium gallium phosphide are used for the epitaxial layers. Wide-bandgap semiconductors are especially promising, eg. Gallium nitride and indium gallium nitride.

Fabrication of HBT's

Due to the need to manufacture HBT devices with extremely thin base layers, molecular beam epitaxy is principally employed. In addition to base, emitter and collector layers, highly doped layers are deposited either side of collector and emitter to facilitate an ohmic contact, which are placed on the contact layers after exposure by photolithography.

7.15 MICROWAVE FIELD-EFFECT TRANSISTORS

In conventional transistor both the majority and the minority carriers are involved and hence this type of transistors are referred to as bipolar transistor whereas in a field-effect transistor the current is carried by majority carriers only and it is referred to as a unipolar transistor. The bipolar transistors are controlled by a current whereas unipolar transistor or field effect transistors are controlled by voltage at the third terminal. The microwave field effect transistor has the capability of amplifying small signals up to the frequency range of X band with low noise figure.

There are various types of unipolar field-effect transistors devices such as:

1. Junction Field Effect Transistors (JFETs)
2. Metal Semiconductor Field Effect Transistors (MESFET)

3. High Electron Mobility Transistors (HEMT)
4. Metal Oxide Semiconductor Field Effect Transistors (MOSFETs)

Advantages over the BJTs
1. It efficiency is very high than bipolar transistor.
2. It has low noise figure.
3. It has voltage gain in addition to current gain.
4. It can operate up to X- band frequency range.

7.15.1 Junction FET

The J-FET (Junction Field Effect Transistor) and the MOS-FET (Metal-Oxide-Semiconductor FET) are voltage controlled devices and a small change in input voltage causes a large change in output current. FET operation involves an electric field which controls the flow of a charge (current) through the device. In contrast, a bipolar transistor employs a small input current to control a large output current. The source, drain, and gate terminal of the FET are analogous to the emitter, collector, and base of a bipolar transistor. The terms n-channel and p-channel refer to the material which the drain and source are connected. Fig. (7.68) shows the schematic symbol for the p-channel and n-channel JFET.

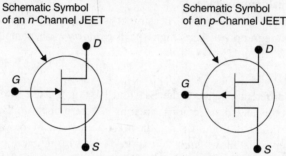

Fig. 7.68. The schematic symbol for the p-channel and n-channel JFET.

A simplified n-channel JFET construction is shown in Fig. (7.69). Note that the drain and source connections are made to the n-channel and the gate is connected to the p material. The n material provides a current path from the drain to the source. An n-channel JFET is biased so that the drain is positive in reference to the source. On the other hand, a p-channel JFET with n material gate would be biased in reverse.

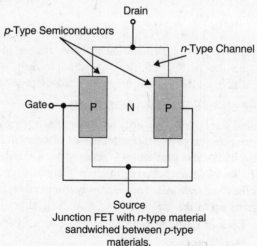

Junction FET with n-type material sandwiched between p-type materials.

Fig. 7.69. A simplified n-channel JFET.

MICROWAVE SEMICONDUCTOR DEVICES

As with any reversed biased PN junction, a depletion region is formed which increases as the reverse gate voltage is increased. This depletion region, being devoid of majority carriers, reduces the channel drain-source current.

As a result, the drain-source current is controlled by the gate voltage. Referring to the figure (7.70) below, a typical JFET characteristic curve; notice the effect of the gate-source voltage on the drain-source current, Ids. Notice the near linear relationship of drain current to drain voltage from zero to about one volt after that the JFET saturates. However, as the gate voltage is increased, the drain-source current is increased.

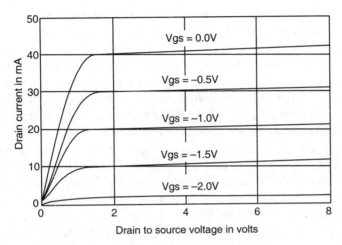

Fig. 7.70. A typical JFET characteristic curve.

The figure of merit, Gfs that is the ratio of drain-source current to gate voltage is the JFET Transconductance. As with the bipolar transistor, there can be many relevant specifications, depending upon the JFET circuit application; however the most common parameters are:

(a) Gate source breakdown voltage, BVgss
(b) Gate reverse leakage current, Igss
(c) Gate source cutoff voltage, Vgs(off)
(d) Drain current at zero gate voltage, Idss
(e) Forward transconductance, Gfs
(f) Input capacitance, Ciss
(g) Switching considerations
(h) Drain-source on resistance, Rds(on)

7.15.2 Metal-Semiconductor Field Effect Transistor (MESFETs)

Construction and operation: The Metal Semiconductor Field Effect Transistor (MESFET) consists of a conducting channel positioned between a source and drain contact region as shown in the figure (7.71), the carrier flow from source to drain is controlled by a Schottky metal gate. The control of the channel is obtained by varying the depletion layer width underneath the metal contact which modulates the thickness of the conducting channel and thereby the current between source and drain.

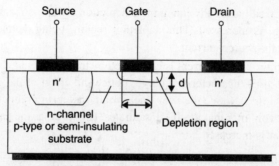

Fig. 7.71. Structure of MESFETs with gate length, L and channel thickness, d.

The **key advantage** of the MESFET is the higher mobility of the carriers in the channel as compared to the MOSFET. Since the carriers located in the inversion layer of a MOSFET have a wave function, which extends into the oxide, their mobility - also referred to as surface mobility is less than half of the mobility of bulk material. As the depletion region separates the carriers from the surface their mobility is close to that of bulk material. The higher mobility leads to a higher current, transconductance and transit frequency of the device.

The **disadvantage** of the MESFET structure is the presence of the Schottky metal gate. It limits the forward bias voltage on the gate to the turn-on voltage of the Schottky diode. This turn-on voltage is typically 0.7 V for GaAs Schottky diodes. The threshold voltage therefore must be lower than this turn-on voltage. As a result it is more difficult to fabricate circuits containing a large number of enhancement-mode MESFET.

The higher transit frequency of the MESFET makes it particularly of interest for microwave circuits.

While the advantage of the MESFET provides a superior microwave amplifier or circuit, the limitation by the diode turn-on is easily tolerated. Typically depletion-mode devices are used since they provide a larger current and larger transconductance and the circuits contain only a few transistors, so that threshold control is not a limiting factor. The buried channel also yields a better noise performance as trapping and release of carriers into and from surface states and defects is eliminated.

The use of GaAs rather than silicon MESFETs provides two more significant advantages: first, the electron mobility at room temperature is more than 5 times larger, while the peak electron velocity is about twice that of silicon. Second, it is possible to fabricate semi-insulating (SI) GaAs substrates, which eliminates the problem of absorbing microwave power in the substrate due to free carrier absorption.

The threshold voltage, V_T, of a MESFET is the voltage required to fully deplete the doped channel layer. This threshold voltage equals:

$$V_T = \phi_i - \frac{qN_d d^2}{e\varepsilon_s}$$

where f_i is the built-in potential and d is the thickness of the doped region. This threshold voltage can also be written as a function of the pinch-off voltage V_p:

$$V_T = \phi_i - V_P$$

where the pinch-off voltage equals

$$V_P = \frac{qN_d d^2}{2\varepsilon_s}$$

MICROWAVE SEMICONDUCTOR DEVICES

The derivation of the current in a MESFET starts by considering a small section of the device between y and $y + dy$. The current density at that point can be written as a function of the gradient of the channel voltage:

$$J = qn\dot{v} = qN_d\mu_s E = -qN_d\mu_s \frac{dV_C(y)}{dy}$$

The drain current is related to the current density and the part of the MESFET channel that is not depleted.

$$I_D = -JW(d - x_n(y))$$

where the depletion layer width at position y is related to the channel voltage, $V_C(y)$, by:

$$x_n(y) = \sqrt{\frac{2\varepsilon_s\left((\phi_i - V_G + V_C(y))\right)}{qN_d}}$$

The equation for the current can now be integrated from source to drain, yielding:

$$\int_0^L I_D dy = qN_d\mu_n dW \int_0^{V_D}\left(1 - \sqrt{\frac{\phi_i - V_G + V_C}{V_P}}\right) dV_C$$

Since the steady-state current in the device is independent of position, the left hand term: equals I_D times L so that

$$I_D = qN_d\mu_n d\frac{W}{L}\left(V_C\bigg|_0^{V_D} - \frac{(\phi_i - V_G + V_C)^{3/2}}{\sqrt{V_P}}\bigg|_0^{V_D}\right)$$

This result is valid as long as the width of the un-depleted channel $(d - xn(y))$ is positive, namely for:

$$V_D \le V_G - V_T$$

This condition also defines the quadratic region of a MESFET. For larger drain voltage, the current saturates and equals that at

$$V_D = V_G - V_T = V_{D,sat}$$

The corresponding current is the saturation current, $I_{D,sat}$:

$$I_{D,sat} = q\mu_n N_d d \frac{W}{L}\left[V_G = V_T - \frac{2}{3}\left(V_P - \frac{(\phi_i - V_G)^{3/2}}{\sqrt{V_P}}\right)\right]$$

GaAs is primary material for MESFETs. It has high electron mobility (8000 cm²/Vs).
Generally if frequency > 2GHz, GaAs transistors are used and if frequency <2GHz than — Si transistors are used.

MESFET Cutoff Frequency

The cutoff frequency of MESFET in a circuit depends on the way in which the transistor is being made. In a wide band lumped circuit the cutoff frequency (f_{co}) is given as

$$f_{co} = \frac{g_m}{2\pi C_{gs}} \quad \ldots (1)$$

or

$$f_{co} = \frac{v_s}{4\pi L} Hz \quad \ldots (2)$$

where
f_{co} = cutoff frequency
gm = transconductance
C_{gs} = gate-source capacitance $\left(\frac{dQ}{dV_{gs}}\bigg|_{V_{gd} = constant}\right)$

L = gate length

v_s = saturation drift velocity

Maximum Frequency of Oscillation in MESFET

In MESFET, the maximum frequency of oscillation depends on the device transconductance and the drain resistance in a distributed circuit. It is given as

$$f_{max} = \frac{f_{co}}{2}(g_m R_d)^{1/2} \qquad \ldots (3)$$

or

$$f_{max} = \frac{f_{co}}{2}\left[\frac{\mu E_p (u_m - e)}{v_s (1 - u_m)}\right]^{1/2} \qquad \ldots (4)$$

where,
f_{max} = maximum frequency of oscillation
R_d = drain resistance
g_m = device transconductance
E_p = electric field at the pinch-off region in the channel
u_m = saturation normalization of u
v_s = saturation drift velocity
μ = electron mobility

$\rho = \left(\frac{|V_g|}{V_P}\right)^{1/2}$ is the normalized gate voltage with respect to the saturation drift velocity.

It has been found that the maximum frequency of oscillation for a gallium arsenide FET with the gate length less than 10 μm is

$$f_{max} = \frac{33 \times 10^3}{L} Hz \qquad \ldots (5)$$

Where, L = gate length in meters (Typical value of D = 0-5 μm).

Applications of MESFETs

1. It is used in cellular devices, satellite receiver, radars and high frequency devices.
2. It is used in microwave receiver as Low noise amplifier.
3. As power amplifiers for output stage in microwave links
4. As power oscillator and driver amplifiers for high power transmitters.

7.16 MOS Field-Effect-Transistors

The *n*-type Metal-Oxide-Semiconductor Field-Effect-Transistor (nMOSFET) consists of a source and a drain, two highly conducting *n*-type semiconductor regions, which are isolated from the *p*-type substrate by reversed-biased *p-n* diodes. A metal or poly-crystalline gate covers the region between source and drain. The gate is separated from the semiconductor by the gate oxide. The basic structure of an *n*-type MOSFET and the corresponding circuit symbol are shown in figure (7.11).

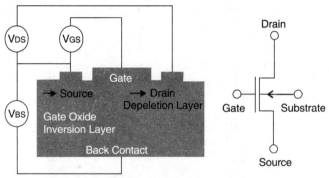

Fig. 7.72. Cross-section and circuit symbol of an n-type metal-oxide-semiconductor-field-effect-transistor (MOSFET).

As can be seen on the figure the source and drain regions are identical. It is the applied voltages, which determine which n-type region provides the electrons and becomes the source, while the other n-type region receives the electrons and becomes the drain. The voltages applied to the drain and gate electrode as well as to the substrate, by means of a back contact, are referred to the source potential, as also indicated figure (7.72). Initially, it was only possible to deplete an existing n-type channel by applying a negative voltage to the gate. Such device has a conducting channel between source and drain even when no gate voltage is applied. They are called **"depletion-mode"** devices.

A reduction of the surface states enabled the fabrication of devices, which do not have a conducting channel unless a positive voltage is applied. Such devices are referred to as "enhancement-mode" devices. The electrons at the oxide-semiconductor interface are concentrated in a thin (~10 nm thick) "inversion" layer. By now, most MOSFETs are "enhancement-mode" devices.

While a minimum requirement for amplification of electrical signals is power gain, one finds that a device with both voltage and current gain is a highly desirable circuit element. The MOSFET provides current and voltage gain yielding an output current into an external load, which exceeds the input current, and an output voltage across that external load which exceeds the input voltage.

The current gain capability of a Field-Effect-Transistor (FET) is easily explained by the fact that no gate current is required to maintain the inversion layer and the resulting current between drain and source. The device has therefore an infinite current gain in dc. The current gain is inversely proportional to the signal frequency, reaching unity current gain at the transit frequency. The voltage gain of the MOSFET is caused by the current saturation at higher drain-source voltages, so that a small drain-current variation can cause a large drain voltage variation.

Structure and Principle of Operation

A top view of the same MOSFET is shown in figure (7.73). Where the gate length, L, and gate width, W, are identified. Note that the gate length does not equal the physical dimension of the gate, but rather the distance between the sources and drain regions underneath the gate. The overlap between the gate and the source/drain region is required to ensure that the inversion layer forms a continuous conducting path between the sources and drain region. Typically this overlap is made as small as possible in order to minimize its parasitic capacitance.

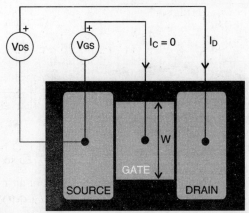

Fig. 7.73. Top view of an n-type metal-oxide-semiconductor-field-effect-transistor (MOSFET).

The voltage applied to the gate controls the flow of electrons from the source to the drain. A positive voltage applied to the gate attracts electrons to the interface between the gate dielectric and the semiconductor. These electrons form a conducting channel called the inversion layer. No gate current is required to maintain the inversion layer at the interface since the gate oxide blocks any carrier flow. The net result is that the applied gate voltage controls the current between drain and source.

The typical current versus voltage (I-V) characteristics of a MOSFET are shown in figure (7.74).

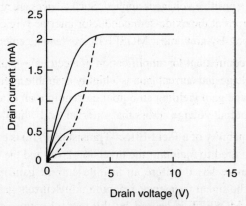

Fig. 7.74. I-V characteristics of an n-type MOSFET with $V_G = 5$ V (top curve), 4 V, 3 V and 2 V (bottom curve).

We will primarily consider the n-type or n-channel MOSFET in this chapter. This type of MOSFET is fabricated on a p-type semiconductor substrate. The complementary MOSFET is the p-type or p-channel MOSFET. The p-type MOSFET contains p-type source and drain regions in an n-type substrate. The inversion layer is formed when holes are attracted to the interface by a negative gate voltage. While the holes still flow from source to drain, they result in a negative drain current. CMOS circuits require both n-type and p-type MOSFETs.

MOSFET Circuits and Technology

The MOSFET circuit technology has dramatically changed over the last three decades. Starting with a ten-micron P-MOS process with an aluminum gate and a single metallization layer around 1970, the technology has evolved into a tenth-micron self-aligned-gate CMOS process with up to five metallization levels. The transition from doping diffusion to ion implantation, from thermal oxidation to oxide deposition, from a metal gate to a poly-silicon gate, from wet chemical etching to dry etching and more recently from aluminum (with 2% copper) wiring to copper wiring has provided vastly superior analog and digital CMOS circuits.

MOSFET Fabrication Process

A quick look at the history of the MOSFET fabrication process reveals that it has evolved significantly over the years. Around 1970, P-MOS circuits with aluminum gate metal and wiring were dominant. The corresponding steps of a typical P-MOSFET fabrication process steps are listed in Table (7.2).

Table 7.2 : P-MOS process steps.

Lithography step	Process step	Process
1	Field oxide growth	Thermal oxidation
	Oxide etch	HF etch
	Source-drain diffusion	Boron diffusion
2	Oxide etch	HF etch
	Gate oxide growth	Thermal oxidation
3	Via hole etch	HF etch
	Aluminium metal deposition	
4	Aluminium etch	Evaporation
	Contact anneal and surface	Wet chemical etch
	state reduction	Furnace anneal in H2/N2

The primary problem at the time was threshold voltage control. Positively charged ions in the oxide decreased the threshold voltage of the devices. P-type MOSFETs were therefore the device of choice despite the lower hole mobility, since they would still be enhancement-type devices even when charge was present. Circuits were still operational at somewhat higher power supply voltages despite the presence of some residual charge in the oxide.

Thermal oxidation of the silicon in an oxygen or water vapor atmosphere provided a quality gate oxide with easily controlled thickness. The same process was also used to provide a high-temperature mask for the diffusion process and a passivation and isolation layer. Some people claim that the quality and versatility of silicon's oxide made silicon the preferred semiconductor over germanium. The oxide was easily removed in hydrofluoric acid (HF), without removing the underlying silicon. Aluminum was evaporated over the whole wafer and then etched yielding both the gate metal and the metal wiring connecting the devices. A small amount of copper (~2%) was added to make the aluminum more resistant to electro migration. Electro migration is the movement of atoms due to the impact with the electrons carrying the current through the wire. This effect can cause open circuits and is therefore a well-known reliability problem. It typically occurs at points where the local current density is very high, in narrow wires, at corners or when crossing an oxide step. The addition of a small amount of copper provides a more rigid structure within the aluminum and eliminates the effect.

Annealing the metal in a nitrogen/hydrogen (N_2/H_2) ambient was used to improve the metal-semiconductor contact and to reduce the surface state density at the semiconductor/gate-oxide interface.

Since then the fabrication process has changed as illustrated with Table 7.3. Most changes were introduced to provide superior performance, better reliability and higher yield. The most important change has been the reduction of the gate length. A gate length reduction provides a shorter transit time and hence a faster device. In addition, a gate length reduction is typically linked to a reduction of the minimum feature size and therefore yields smaller transistors as well as a larger number of transistors on a chip with a given size. As the technology improved, it was also possible to make larger chips, so that the number of transistors per chip increased even faster. At the same time, the wafer size was increased to accommodate the larger chips while reducing the loss due to partial chips at the wafer periphery. Larger wafers further reduce the cost per chip as more chips can be accommodated on a single wafer.

The other changes can be split into process improvements and circuit improvements. The distinction is at times artificial, as circuit improvements typically require new processes.

The key circuit improvement is the use of CMOS circuits, containing both nMOS and pMOS transistors. Early on, the pMOS devices were replaced with nMOS transistors because of the better electron mobility. Enhancement-mode loads were replaced for first by resistor loads and then by depletion-mode loads yielding faster logic circuits with larger operating margins. Analog circuits benefited in similar ways. The use of complementary circuits was first introduced by RCA but did not immediately catch on since the logic circuits were somewhat slower and larger than the then-dominant nMOS depletion logic. It was only when the number of transistors per chip became much larger that the inherent advantages of CMOS circuits became clear. CMOS circuits have a lower power dissipation and larger operating margin. They became the technology of choice as thousands of devices we integrated on a single chip. Today, the CMOS technology is the dominant technology in the IC industry as the ten-fold reduction of power dissipation largely outweighs the 30%-50% speed reduction and size increase.

The process improvements can in turn be split into those aimed at improving the circuit performance and those improving the manufacturability and reliability. Again the split is somewhat artificial but it is beneficial to understand what factors affect the process changes. The latter group includes CVD deposition, ion implantation, RIE etching, sputtering, planarization and deuterium annealing. The process changes, which improve the circuit performance, are the self-aligned poly-silicon gate process, the silicide gate cap, LOCOS isolation, multilevel wiring and copper wiring.

Table 7.3 : **MOS process changes and improvements.**

Initial process and process parameters	*Current process and process parameters*
10 μm gate length	0.1 μm gate length
1 inch wafers	300 mm wafers
2 × 2 mm chips	1 × 2 cm chips
Thermal oxidation	CVD deposition
Field oxide isolation	LOCOS isolation, trench isolation
Wet chemical etching	Reactive ion etching (RIE)
Diffusion	Ion implantation
PMOS	nMOS, CMOS
Enhancement load, resistor load	Depletion load, complementary load
Aluminium gate	Poly-silicon/Silicide self-aligned gate
Evaporated aluminium wiring with 2% copper	Sputtered copper
One or two metal wiring levels without planarization	Up to six wiring levels with planarization and tungsten plugs
Metal evaporation	Sputtering
Hydrogen anneal	Deuterium anneal

The self-aligned poly-silicon gate process was introduced before CMOS and marked the beginning of modern day MOSFETs. The self-aligned structure is obtained by using the gate as the mask for the source-drain implant. Since the crystal damage caused by the high-energy ions must be annealed at high temperature (~800° C), an aluminum gate could no longer be used. Doped poly-silicon was found to be a very convenient gate material since it withstands the high anneal temperature and can be oxidized just like silicon. The self-aligned process lowers the parasitic capacitance between gate and drain and therefore improves the high-frequency performance and switching time. The addition of a silicide layer on top of the gate reduces the gate resistance while still providing a quality implant mask. The self-aligned process also reduced the transistor size and hence increased the density. The field oxide was replaced by a local oxidation isolation structure (LOCOS), where a Si_3N_4 layer is used to prevent the oxidation in the MOSFET region. This oxide provides an implant mask and contact hole mask yielding an even more compact device.

Multilevel wiring is a necessity when one increases the number of transistors per chip. After all, the number of wires increases with the square of the number of transistors and the average wire length increase linearly with the chip size. While multilevel wiring simply consists of a series of metal wiring levels separated by insulators, the multilevel wiring has increasingly become a bottleneck in the fabrication of high-performance circuits. Planarization techniques, as discussed below and the introduction of copper instead of aluminum-based metals have further increased the wiring density and lowered the wiring resistance.

Chemical vapor deposition (CVD) of insulating layers is now used instead of thermal oxidation since it does not consume the underlying silicon. Also, because there is no limit to the obtainable thickness and since materials other than SiO_2 (for instance $Si3N_4$) can be deposited. CVD deposition is also frequently used to deposit refractory metals such as tungsten.

Ion implantation has replaced diffusion because of its superior control and uniformity. Dry etching including plasma etching, reactive ion etching (REE) and ion beam etching has replaced wet chemical etching. These etch processes provide better etch rate uniformity and control as well as very pronounced anisotropic etching. The high etch rate selectivity of wet chemical etching is not obtained with these dry etch techniques, but are well compensated by the better uniformity.

Sputtering of metals has completely replaced evaporation. Sputtering typically provides better adhesion and thickness control. It is also better suited to deposit refractory metals and silicides.

Planarization is the process by which the top surface of the wafer is planarized after each step. The purpose of this planarization process is to provide a flat surface, so that tine-line lithography can be performed at all stages of the fabrication process. The planarization enables high-density multi-layer wiring levels.

Deuterium anneal is a recent modification of the standard hydrogen anneal, which passivates the surface states. The change to deuterium was prompted because it is a heavier isotope of hydrogen. It chemically acts the same way but is less likely to be knocked out of place by the energetic carriers in the inversion layer. The use of deuterium therefore reduces the increase of the surface state density due to hot-electron impact.

Poly-Silicon Gate Technology

An early improvement of the technology was obtained by using a poly-silicon gate. Such gate yields a compact self-aligned structure with better performance. The poly-silicon gate is used as a mask during the implantation so that the source and drain regions are self-aligned with respect to the gate. This self-alignment structure reduces the device size. In addition, it eliminates the large overlap capacitance between gate and drain, while maintaining a continuous inversion layer between source and drain.

A further improvement of this technique is the use of a low-doped drain (LDD) structure. As an example we consider the structure shown in Figure (7.75).

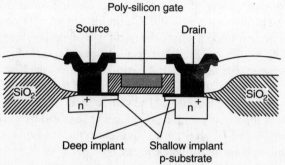

Fig. 7.75. Cross-sectional view of a self-aligned poly-silicon gate transistor with LOCOS isolation.

Here a first shallow implant is used to contact the inversion layer underneath the gate. The shallow implant causes only a small overlap between the gate and source/drain regions. After adding a sidewall to the gate a second deep implant is added to the first one. This deep implant has a low sheet resistance and adds a minimal series resistance. The combination of the two implants therefore yields a minimal overlap capacitance and low access resistance.

Shown is also the **local oxidation isolation (LOCOS)**. Typically, there would also be an additional field and channel implant. The field implant increases the doping density under the oxide and thereby increases the threshold voltage of the parasitic transistor formed by the metal wiring on top of the isolation oxide. The channel implant provides an adjustment of the threshold voltage of the transistors. The use of a poly-silicon gate has the disadvantage that the sheet resistance of the gate is much larger than that of a metal gate. This leads to high RC time-constants of long poly-silicon lines. Silicides (WSi, TaSi, CoSi etc.) or a combination of silicides and poly-silicon are used to reduce these RC delays. Also by using the poly-silicon only as gate material and not as a wiring level one can further eliminated such RC time delays.

Maximum Operating Frequency of MOSFET

An equivalent circuit of a MOSFET is show in Fig. (7.76). The maximum operating frequency of a MOSFET is determined by in circuit parameters.

In Fig. (7.76),

g_{in} = Input conductance due to the leakage current (g_{in} is negligible)

C_{fb}, = feedback capacitance

C_{in} = the input capacitance

g_{out} = output conductance

$C_{out} = C_i C_s/(C_i + C_s)$ is the sum of the P-N junction capacitances in series with the semiconductor capacitance per unit area.

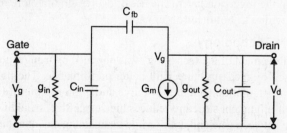

Fig. 7.76. An equivalent circuit of a common source MOSFET.

The maximum operating frequency of a MOSFET is given by following expression.

$$f_m = \frac{w_m}{2\pi} = \frac{g_m}{2\pi C_{in}} = \frac{\mu_n V_d}{2\pi L^2} \qquad \text{(In the linear region)} \qquad ...(6)$$

The transconductance g_m is reduced to

$$g_m \text{ sat} = \frac{Z}{L} \pi_n C_r V_g = Z C_i V_s \qquad \text{(In the saturation region)} \qquad ...(7)$$

Application of MOSFET's

The MOSFET's are generally used as power amplifiers because they offer same advantages over JFETs and MOSFET's.

(i) In the active region of an E-mode MOSFET, the input capacitance and the transconductance are almost independent of gate voltage and the output capacitance is independent of the drain voltage. This leads to very linear power amplification (class A),

(ii) The active gate voltage range can be larger because n-channel depletion type MOSFET's can be operated from the depletion made region ($-V_g$) to the enhancement mode region ($+V_g$).

7.17 CMOS

Complementary Metal-Oxide-Silicon circuits require an nMOS and pMOS transistor technology on the same substrate. To this end, an n-type well is provided in the p-type substrate. Alternatively one can use a p-well or both an n-type and p-type well in a low-doped substrate. The gate oxide, poly-silicon gate and source-drain contact metal are typically shared between the pMOS and nMOS technology, while the source-drain implants must be done separately. Since CMOS circuits contain pMOS devices, which are affected by the lower hole mobility, CMOS circuits are not faster than their all-nMOS counter parts. Even when scaling the size of the pMOS devices so that they provide the same current, the larger pMOS device has a higher capacitance.

The CMOS advantage is that the output of a CMOS inverter can be as high as the power supply voltage and as low as ground. This large voltage swing and the steep transition between logic levels yield large operation margins and therefore also a high circuit yield. In addition, there is no power dissipation in either. The CMOS advantage is that the output of a CMOS inverter can be as high as the power supply voltage and as low as ground. This large voltage swing and the sleep transition between logic levels yield large operation margins and therefore also a high circuit yield. In addition, there is no power dissipation in either logic state. Instead the power dissipation occurs only when a transition is made between logic states. CMOS circuits are therefore not faster than N-MOS circuits but are more suited for very/ultra large-scale integration (VLSI/ULSI).

CMOS circuits have one property, which is very undesirable, namely latch up. Latch up occurs when four alternating p-type and n-type regions are brought in close proximity. Together they form two bipolar transistors, one npn and one pnp transistor. The base of each transistor is connected to the collector of the other, forming a cross-coupled thyristor-like combination. As a current is applied to the base of one transistor, the current is amplified by the transistor and provided as the base current of the other one. If the product of the current gain of both transistors is larger than unity, the current through both devices increases until the series resistances of the circuit limits the current. Latch up therefore results in excessive power dissipation and faulty logic levels in the gates affected. In principle, this effect can be eliminated by separating the n-type and p-type device. A more effective and less space-consuming solution is the use of trenches, which block the minority carrier flow. A deep and narrow trench is etched between all n-type and p-type wells, passivated and refilled with an insulating layer.

7.18 MOSFET MEMORY

MOSFET memory is an important application of MOSFETs. Memory chips contain the largest number of devices per unit area since the transistors are arranged in a very dense regular structure. The generic structure of a memory chip is shown in figure (7.77).

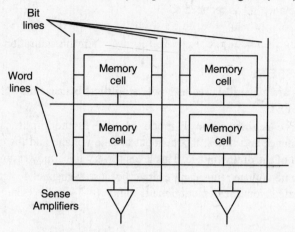

Fig. 7.77. Arrangement of memory cells into an array.

A two dimensional array of memory cells, each containing a single bit, are connected through a series of word lines and bit lines. One row of cells is activated by changing the voltage on the corresponding word line. The information is then stored in the cell by applying voltages to the bit lines. During a read operation, the information is retrieved by sensing the voltage on the bit lines with a sense amplifier. A possible implementation of **a static random access memory (SRAM)** is shown in figure 7.78(a).

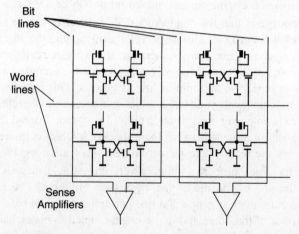

Fig. 7.78. (*a*) Static random access memory (SRAM) using a six-transistor cell.

Each memory cell consists of a flip-flop and the cells are accessed through two pass transistors connected to the bit lines and controlled by the word line. Depletion mode transistors are shown here as load devices. A common alternate load is an amorphous silicon resistor. A simpler cell leading to denser memory chips is the dynamic random access memory (DRAM) shown in figure 7.78(b).

MICROWAVE SEMICONDUCTOR DEVICES

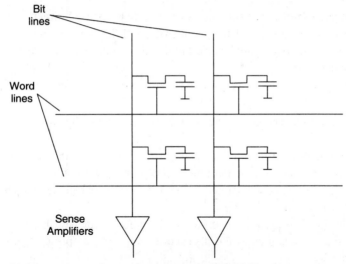

Fig. 7.78. (b) Dynamic random access memory (DRAM) using a one-transistor cell.

SUMMARY

- Microwave solid-state devices are becoming increasingly important at microwave frequencies.
- **Classification of Microwave Solid-state Devices**
   1. Microwave Bipolar Junction Transistor (BJT)
   2. Microwave Field Effect Transistor (FET's)
   3. Transferred Electron Devices (TED)
   4. Avalanche Transit Time Devices
      - **Tunnel Diodes**: Tunnel diodes are used for generation of microwave oscillations and signal amplification.
      - **Varactor Diodes**: Varactor diodes are used for frequency multiplication, parametric amplification and frequency tuning.
      - **Crystal Diodes**: Crystal diodes are used for mixing of two microwave frequencies and also for detection of microwave signals.
      - **Schottky Diodes**: Schottky diodes are also used as microwave mixer and detector.
      - **PIN Diodes**: PIN diodes are used for microwave signals attenuation, switching, modulation, phase shifting, and amplitude limiting.
      - **Gunn Diodes**: Gunn diodes are used to generate microwave signals.
- **Read Diodes**: Read diodes such as IMPATT, TRAPATT and BARITT diodes are used for amplification and generation of microwave signals.
- Tunnel Diode Applications
   1. Microwave amplifiers,
   2. Microwave oscillators,
   3. Microwave frequency converters and mixers,
   4. Microwave switching devices.
- Most microwave tunnel-diode amplifiers are reflection-**type, circulator- coupled amplifiers**.
- The varactor is a semiconductor diode with the properties of a voltage-dependent capacitor.
- The parametric amplifier is named for the time-varying parameter, or value of capacitance, associated with the operation.

- The parametric amplifier is sometimes called a Reactance amplifier.
- The most important feature of the parametric amplifier is the low-noise characteristic, the nature of electronic noise and the effect of this type of noise on receiver operation must first be considered.
- The properties of up-convertor parametric amplifier are as following :
  (i) The output frequency $f_0 = f_p + f_s$.
  (ii) There is no power flow in the parametric amplifier other the signal, pumped output frequencies.
- In a parametric down converter, the $f_s = f_p + f_0$ it means that the input power must feed in to the idler circuit and the output power must move out from the signal circuit.
- The idler frequency may be defined as the difference between the pump frequency and the signal frequency, i.e., $f_i = f_p + f_s$.
- When the signal frequency is equal to the idler frequency in a negative- resistance amplifier than it is known as the degenerate parametric amplifier or oscillator. Since the idler frequency
  $$f_i = f_{-p} + f_s \text{ and } f_s = f_p + f_i$$
- Bulk-effect semiconductors are of two basic types :
  1. The transferred-electron devices and
  2. The avalanche transit-time devices.
- A tiny region, known as a DOMAIN, is form that has an electric field of much greater intensity than the fields in the rest of the semiconductor.
- Four basic modes of operation of uniformly doped bulk diodes with low-resistance contacts are proposed by the Copeland.
  (i) Gun oscillation mode.
  (ii) Stable amplification mode.
  (iii) LSA oscillation mode.
  (iv) Bias-circuit oscillation mode.
- In the Gunn oscillation modes, the range of n_0L is given by $10^{12}/cm^2 < (n_0L) < 10^{14}/cm^2$.
- Four possible domain modes for the Gunn oscillation mode,
  (i) Transit-Time domain mode.
  (ii) Delayed or Inhibited domain mode.
  (iii) Quenched domain mode.
  (iv) Limited space charge Accumulator mode (LSA mode).
- **The Gunn oscillator** is a source of microwave energy that uses the bulk-effect, gallium-arsenide semiconductor.
- At microwave frequencies they use carrier impact ionization and drift in the high field region of a semi-conductor junction to produce negative resistance.
- They are of following three types :
  1. IMPATT   2. TRAPATT   3. BARITT
- The heterojunction bipolar transistor (HBT) is an improvement of the BJT that can handle signals of very high frequencies up to several hundred GHz. It is common nowadays in ultrafast circuits, mostly **RF** systems.
- The microwave field effect transistor has the capability of amplifying small signals up to the frequency range of X band with low noise figure.

MICROWAVE SEMICONDUCTOR DEVICES

- There are various types of unipolar field-effect transistors devices such as :
  1. Junction Field Effect Transistors (JFETs)
  2. Metal Semiconductor Field Effect Transistors (MESFET)
  3. High Electron Mobility Transistors (HEMT)
  4. Metal Oxide Semiconductor Field Effect Transistors (MOSFETs)
- **Advantages over the BJTs**
  1. It efficiency is very high than bipolar transistor.
  2. It has low noise figure.
  3. It has voltage gain in addition to current gain.
  4. It can operate up to X- band frequency range.
- The Metal Semiconductor Field Effect Transistor (MESFET) consists of a conducting channel positioned between a source and drain contact region.
- Applications of MESFETs
  1. It is used in cellular devices, satellite receiver, radars and high frequency devices.
  2. It is used in microwave receiver as Low noise amplifier.
  3. As power amplifiers for output stage in microwave links.
  4. As power oscillator and driver amplifiers for high power transmitters.

REVIEW QUESTIONS

1. Explain the operation of a varactor diode. Discuss the constructional details, equivalent circuit and figure of merit.
2. What is parametric amplifier? How it is different from a normal amplifier.
3. Discuss the advantages and list the applications of a parametric amplifier.
4. What is PIN diode? Explain the constructional detail and characteristics of PIN diode.
5. Explain the tunnel diode characteristics with the help of energy band diagram.
6. Discuss gun effect using the two valley theory.
7. Explain negative differential resistivity. Discuss the J-E characteristics of a gun diode.
8. Discuss the several domain formation modes of a gun diode.
9. What are avalanche transit time devices?
10. Discuss the principle of MASERS.
11. Write short notes :
    (*i*) IMPATT   (*ii*) TRAPATT   (*iii*) BARRIT
    (*iv*) MASERS   (*v*) VARACTOR diode   (*vi*) Gun effect
    (*vii*) Step Recovery Diode
12. Discuss the difference between transferred electron devices and avalanche transit time device.
13. How can PIN diode be used as a microwave switch? Describe a single PIN switch in shunt and series mounting configurations. **(UPTU-2003, 2004)**
14. How is plasma trapped in a TRAPATT diode? Why is the operating frequency of this diode lower than the IMPATT? Give its major merits and demerits. **(UPTU-2003, 2004)**
15. Discuss the formation and growth of the high field domain in a TED. **(UPTU-2003, 2004)**
16. What are the conditions to be satisfied by the semiconductor in order to exhibit transferred electron effect? What are the limitations of GaAs diode? **(UPTU-2003, 2004)**
17. Discuss the differences between tunnel diode and ordinary PN junction diode. Show that an amplifier with finite gain can be built by connecting the tunnel diode to a circulator.
    (UPTU-2003, 2004)

18. What does acronym IMPATT stand for? Why does the device show a differential negative resistance? Give the physical structure doping profile and electric field distribution of a double drift region IMPATT diode. What is the advantage of doubles drift region over single drift region? **(UPTU-2003, 2004)**
19. Why Gun diodes called is transferred electron devices? Give the mechanism of negative differential resistance in GaAs semiconductor. Describe the behaviour of different modes of operation of a Gun diode oscillator and give one of its applications. **(UPTU-2002, 2003)**
20. What is importance of PIN diodes at microwave frequency? Describe the behaviour of a PIN diode mounted in a transmission line. **(UPTU-2002, 2003)**
21. What is an IMPATT diode? Draw the schematic diagram and equivalent circuit of the IMPATT diode. **(UPTU-2002,2003)**
22. Give the constructional details of a crystal diode and list the requirements that a diode mount must fulfill if this diode is to be used as a detector. **(UPTU-2002, 2003)**
23. Explain the following terms in context with a gun diode.
    (*i*) Gunn effect   (*ii*) High field domain theory   (*iii*) Two valley theory **(MDU-2006)**
24. Discuss the applications of tunnel diode. **(MDU-2006)**
25. Discuss various modes of oscillation in a Gunn diode oscillator. **(MDU-2006)**
26. Write short notes on :
    (*i*) MASER   (*ii*) IMPATT   (*iii*) TRAPATT
    (*iv*) Parametric amplifier   (*v*) PIN diode **(MDU-2004)**
27. What is tunnel diode? Give volt-amp. Characteristics of a tunnel diode, what is negative resistance amplifier theory applied to tunnel diode. **(MDU-2004)**
28. What are MESFETS? Explain the constructions, operation, performance characteristics and their applications.
29. Explain operation, construction and applications of TRAPATT diode. **(MDU-2002, 2003)**
30. Write a note on parametric amplifier. **(MDU-2002)**
31. Explain the operation of a varactor diode and compare it with a step recovery diode with regard to frequency range, power output, efficiency and harmonic components.
    (MDU-2002)

CHAPTER 8

MICROWAVE MEASUREMENTS

OBJECTIVE

- Introduction
- Detection of Microwave Signal
- Frequency And Wavelength Measurements
- Measurement of VSWR
- Measurement of Impedance
- Insertion Loss And Attenuation Measurements
- Low & High Power measurements
- Radiation Pattern
- Network Analyzer (Scattering Parameters)

8.1 INTRODUCTION

Microwave measurements are concerned with distributed circuit elements whereas the basic measurement parameters in low frequency ac circuit containing lumped elements such as voltage, current, frequency and power. At low frequency from these parameters, the values of the power factor, the impedance and the phase angle can be calculated. But at microwave frequencies, the amplitude of the voltage and currents on a transmission line are functions of distance and are not easily measurable. Unlike low frequency measurements, many quantities measured at microwave frequencies are relative and it is not necessary to know their absolute values. At microwave frequencies it is more convenient to measure power instead of voltage and current. Most of the properties of devices and circuits at microwave frequencies are obtained from the measurement of S-parameter, power, frequency, phase shift, noise figure and *VSWR*.

At microwave frequencies most important parameters to be measured are the frequency, power, attenuation, *VSWR,* phase, impedance, insertion loss, dielectric constant and noise factor. It is found appropriate to include the descriptions of some important measurements devices in brief, in this chapter.

8.2 DETECTION OF MICROWAVE SIGNAL

As at low frequency diode detect, the signals and generates a proportional d.c. current signal at its output. Same way, the microwave frequency signals are being detected by the microwave semiconductor diode. These detector diode rectifies the received microwave signal for the measurement of power, impedance etc., and generates a proportional d.c. current signal at its output. Hence they are also known as square-law detectors.

Mainly two types of diodes are used for microwave detection:

(*i*) Schottky barrier diode   (*ii*) Point Contact Diode

8.2.1 Schottky Barrier Diode

In a Schottky barrier a junction is made between n-type semiconductor and metal to form the diode. A planar technology is used to fabricate it. Over the cleaned surface of n^+ type silicon, a thin active layer of n-type silicon is grown epitaxially. Over this layer, a thin layer of SiO_2 is grown and through which windows are created to make metal semiconductor junction.

Schottky barrier diode is more reliable than the point contact diode and has low noise.

Under normal operation condition these diodes should be mounted across the waveguide and the other end of the wave guide is shorted. The distance between the diode and shorted end of the waveguide can be varied by adjusting the plunger and tunes the detector at a desired frequency. The detectors performance is expressed with the help of current sensitivity and it is the ratio of charge in short circuit current (Δi) resulting from an available input power P is given by

$$\beta = \frac{\Delta i}{P}$$

8.2.2 Point Contact Diode

The constructional detail of point contact diode is shown in fig. (5.1). This diodes consist of a thin tungsten 'whisker' touching a silicon chip. On the cleaned and polished surface of polycrystalline P-type silicon a thin layer of oxide is grown. During oxidation diffuse out of P-type impurities from the substrate into oxide makes the layer highly resistive. The tungsten whisker is etched electrolytically to obtain a tip diameter of few microns.

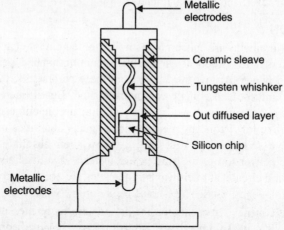

Fig. 8.1. Point contact diode.

8.2.3 TUNABLE DETECTOR

The non-reciprocal detector diode mounted in the microwave transmission line is used to detect the low frequency square wave modulated microwave signal. Normally, the non-reciprocal detector is a crystal diode or Schottky barrier diode. The matching between the detector and the transmission line is obtained by using a tunable stub as shown in figure (8.2). It is possible to match accurately waveguide sections to any other instrument through co-axial coupling with single tuning.

MICROWAVE MEASUREMENTS

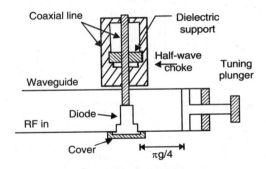

(a) Waveguide Tunable detector

(b) Tunable Crystal detector mount

Fig. 8.2. *a & b*

The crystal diode can also be inserted into the coaxial cable as illustrated in figure-8.3. The detected output which is a square wave, taken out of another coaxial cable connected to a BNC connector.

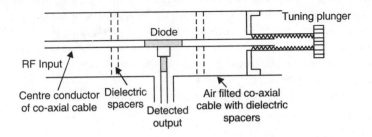

Fig. 8.3

A tunable probe detector is used for the detection of microwave signals which is made to propagate through a slotted line as shown in figure 8.4(a). A probe which is acting as a short antenna is used to sense the voltage at any point on the standing waves created inside the slotted line by an unmatched load resistance. This detector probe is mounted on carriage plate.

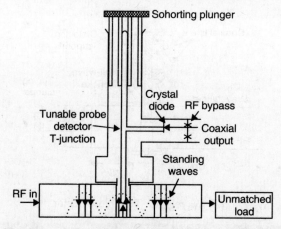

Fig. 8.4. (*a*) Tunable probe detector for slotted line.

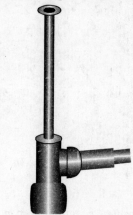

Fig. 8.4. (*b*) Probe detector mount.

8.3 FREQUENCY AND WAVELENGTH MEASUREMENTS

At microwave frequencies, measurement of frequency is made through a measurement of the wavelength of the given wavelength using **wavemeters**. It is easier to measure the wavelengths, rather than measuring frequencies, at these very high frequencies. At optical frequencies, there is no way of measuring the amplitude or frequency of the radiation involved.

Microwave frequency can be measured by either mechanical or electronic techniques. Frequency counter and cavity wave meter are commercially available to measure the microwave frequency. There are three methods for the measurement of frequency:

(*i*) Slotted Line Technique
(*ii*) Wave meter or calibrated resonant cavity technique
(*iii*) Electronic Technique

8.3.1 SLOTTED LINE TECHNIQUE

Slotted Line Technique is a useful method for measuring frequency since the distance between two successive voltage minima is equal to half the wavelength. It is required that the minima of voltage standing wave should be as sharp as possible. It can be achieved by putting a short circuit on the end of the line to have complete reflection. It is better to read as much minimum voltage position as possible and take the average spacing between them. By using following relations, free space wavelength (λ_0) can be calculated from the measured guided wavelength (λ_g)

MICROWAVE MEASUREMENTS

$$\frac{1}{\lambda_0^2} = \frac{1}{\lambda_g^2} - \frac{1}{4a^2} \quad (\lambda_g = 2d_{min}) \qquad \ldots 8.1$$

Where, a = width of the waveguide.
The frequency can be calculated from the following equation,

$$f = \frac{c}{\lambda_0} \quad \text{or} \quad f(GHz) = 30/\lambda_0(cm) \qquad \ldots 8.2$$

Where, c = speed of the light

8.3.2 WAVE METER OR CALIBRATED RESONANT CAVITY TECHNIQUE

Wavemeter technique is the better and more accurate method of measurement of frequency. A typical wavemeter is a cylindrical cavity with a variable short circuit termination which varies the resonance frequency of the cavity by adjusting the cavity length. In other words, a calibrated resonant cavity is used as a wave meter to measure the frequency of the microwave sources. There are two types of wave meters such as (a) Transmission type and (b) Absorption type.

(*a*) **Transmission Type:** In this type of wave meter the desired tuned frequency signal is made to pass through.

(*b*) **Absorption type:** In this type of wave meter the desired tuned frequency signal is being absorbs or attenuates. Since the power is absorbed in the wave meter at resonance this is called **absorption type wave meter.** This method is more preferred for the laboratory purpose; a typical set up with the absorption type cavity is shown in fig. (8.5)

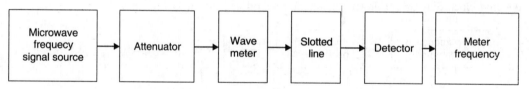

Fig. 8.5. Set-up for frequency measurement.

The unknown frequency signal should pass unaltered by determining the cavity and indicates its value at the meter. Now with the help of the shorting plunger which is attached with the cavity, the size of the cavity is varied to tune it with the incoming frequency. When the cavity is tuned, the cavity absorbs the power and which causes a dip in the output power level. The frequency is determined by reading the micrometer, which is calibrated directly in the terms of frequency.

Working of Wave meter

A typical wave meter is a cylindrical cavity, which is already discussed, with a variable short circuit plunger which changes the resonance frequency of the cavity by varying the cavity length as shown in Fig. (8.6). TM_{010} mode is used practically in the wave meter applications. The wave meter axis is placed perpendicular to the broad wall of the wave guide and it is coupled by means of hole in the narrow wall of the wave guide as shown in Fig. (8.6). Due to magnetic field coupling TM_{010} mode is excited. A block of polytron (as absorbing material) placed at the back of the tuning plunger which prevents oscillation on top of it. The cavity resonates at different frequencies for different position of the plunger. The tuning can be calibrated in terms of frequency by known frequency input signals and observes the dip in the display unit i.e., power meter connected at the outside of wave guide. Wave meter has very high 'Q' typically in the range of 1000-50,000 and accuracy in the range of 1 to 0.005% respectively.

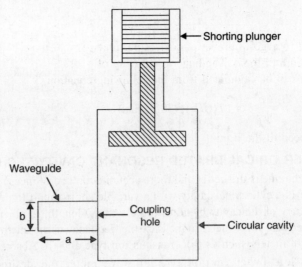

Fig. 8.6 Schematics of wave meter method of frequency measurements.

8.3.3 Electronic Techniques

Electronic techniques generally are more accurate than the mechanical techniques but they are expensive. Frequency counter or a heterodyne converter system can be used for the measurements of higher frequencies. In this system the unknown frequency is compared with the harmonics of a known lower frequency. A simple block diagram of this method is shown in Fig. (8.7) and a detailed block diagram of down conversion method are shown in Fig. (8.8)

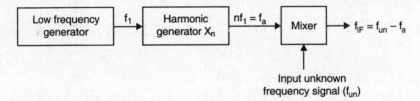

Fig. 8.7. Simple block diagram of frequency measurement using electronic method.

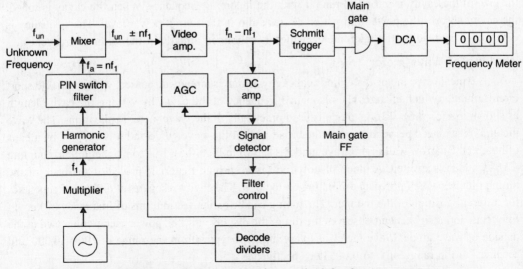

Fig. 8.8. Block diagram of down conversion method for frequency measurement.

MICROWAVE MEASUREMENTS

A heterodyne converter is used to down convert the unknown frequency (f_{un}) by mixing it with the known frequency (f_a). The difference of the frequencies i.e.

$f_{un} - f_a = F_{IF}$ is amplified then measured by the counter and display the frequency in display meter. The frequency f_a which is selected by first multiplying local oscillator known frequency to a convenient frequency f_1 than a harmonic generator produces a series of harmonics of f_1. The suitable harmonic $nf_1 = f_a$ is selected by tuning cavity such that f_a can be added with f_{IF} and display f_{un}. The system starts from $n = 1$ and the filter frequency is selected by a feed back circuit from IF stage until an IF frequency in the proper range in present. For the unknown frequency range up to 20 GHz, the typical value of $f_1 = 100$ to 500 MHz. To have better accuracy a low noise local oscillator and noise less multiplier are to be used.

8.4 MEASUREMENTS OF VSWR

Measurements of VSWR express the degree of mismatch between the load and transmission line. Any mismatched load leads to reflected waves resulting in standing waves along the transmission line. The ratios of maximum to minimum voltage determine the VSWR (voltage standing wave ratio) as shown in Fig. (8.9).

$$S = \frac{V_{max}}{V_{min}} = \frac{1+\rho}{1-\rho}$$

where 'S' varies from 1 to ∞

as ρ varies from 0 to ∞

ρ = Reflection coefficient

$$= \frac{P_{reflection}}{P_{incident}}$$

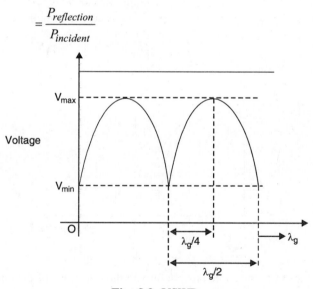

Fig. 8.9. VSWR.

8.4.1 LOW VSWR (S < 10)

The value of VSWR is not exceeding 10 than it is low VSWR and it can be directly measured of VSWR meter, the setup of low VSWR measurement is shown in Fig. (8.10). When the line is terminated with unmatched impedance of the characteristics impedance of the line, the incident and reflected wave sets a standing wave on the line. The probe is first moved to voltage minima in the slotted line and the detector is tuned to give the maximum output. The attenuation is now adjusted to get full scale reading. This full scale reading is noted down. Now adjust the probe on the slotted line is adjusted to get minimum reading on the meter

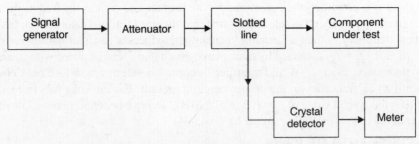

Fig. 8.10. Low VSWR measurements setup.

The ratio of V_{max} to the V_{min} gives the VSWR. The meter then gives the VSWR which can be directly read on the VSWR scale of the meter. The full scale deflection corresponds to VSWR of 1. For example, if a FSD of 20 mV corresponds to a VSWR. The travelling probe is adjusted to get minimum reading on the meter. If minimum reading corresponds to 10 mV, then

$$\text{VSWR} = \frac{20\,mV}{10\,mV} = 2.$$

High VSWR means, if S is more than 10. In this case the reflection is very high then the probe insertion at the minimum should be high and which may lead disturbance to the field. For high VSWR, we use double minimum method. In this method the probe depth does not introduce an error the probe is inserted to a depth where the minimum can be read without difficulty. Further the probe is then moved to a point where the power is double the minimum. Let this position is indicated as 'd_1'.

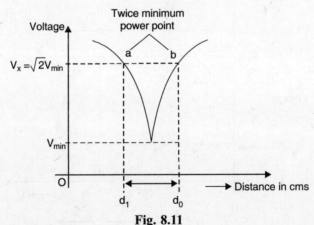

Fig. 8.11

The probe is moved to the twice power point on the other side of the minimum the position is denoted by d_0 as shown in Fig. (8.11).

We get
$$P_{min} \propto V_{min}^2$$
$$2P_{min} \propto V_x^2$$

$\therefore$
$$\frac{1}{2} = \frac{V_{min}^2}{V_x^2}$$

or
$$V_x^2 = 2(V_{min})^2$$

MICROWAVE MEASUREMENTS

or $\quad V_x = \sqrt{2}\, V_{min}$

For TE_{10} mode, $\lambda_c = 2a$

$$\lambda_0 = c/f$$

$$\lambda_g = \frac{\lambda_0}{\sqrt{(1-\lambda_0/\lambda_c)^2}},$$

The VSWR can be determined by using the empirical relation

$$VSWR = \frac{\lambda_0}{\pi\,(d_0 - d_1)} \qquad \ldots (8.3)$$

8.5 IMPEDANCE MEASUREMENT

To measure the impedance, both amplitude and phase of the test signal are required to be determined. Following methods are commonly used to measure the Impedance.

(i) Slotted Line Technique
(ii) Using Reflectometer
(iii) Impedance measurement of Reactive discontinuity
(iv) Using Magic-T

8.5.1 SLOTTED LINE TECHNIQUE

Impedance at microwave frequency is found by determining the VSWR of the unknown load connected in a slotted line. The complex impedance of load can be measured by measuring the phase angle ϕ_L of complex reflection coefficient (Γ_L) from the distance of first voltage standing wave minimum a_{min} and the magnitude of the same from the VSWR. To calculate the Z_L (Load Impedance) the following relations are important.

$$Z_L = Z_0 \frac{1+\Gamma_L}{1-\Gamma_L} \qquad \ldots 8.4$$

$$\Gamma_L = \rho_L\, e^{j\phi L} \qquad \ldots 8.5$$

$$S = (1+\rho_L)/(1-\rho_L) \qquad \ldots 8.6$$

$$\phi_L = 2\beta d_{min} - \pi \qquad \ldots 8.7$$

$\lambda_g = 2 \times$ Distance between two successive minima.

Where, Z_L = Load Impedance

Z_0 = Known impedance
ρ = Reflection Coefficient
Γ_L = Complex reflection coefficient
ϕ_L = Phase angle of the complex reflection coefficient

Fig. (8.12), shows the setup of determination of load impedance using slotted line.

First, measure the load VSWR to compute ρ_L using equation (8.6). Measure the distance d to find out the λ_g ($\lambda_g = 2d$) and $\beta = 2\pi/\lambda_g$. Now measure the distance d_{min} of the first voltage minimum from the load plane. Calculate the phase angle ϕ_L of the load by using equation (8.5). Finally the unknown impedance Z_L is than calculated from equation (8.4).

To ease the calculation smith chart may be used to determine Z_L from the measurement of VSWR and d_{min}.

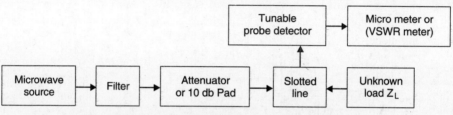

Fig. 8.12. Impedance measurement using slotted line.

8.5.2 Measurement of Impedance Using Reflectometer

In the reflectometer technique only the magnitude of the impedance is calculated whereas a slotted line wave guide measurement gives both magnitude and phase angle of load impedance. Fig. (8.13) shows the typical setup for reflectometer technique.

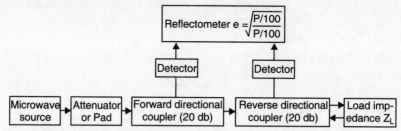

Fig. 8.13. Setup for measuring impedance using reflectometer.

In this technique we used two identical directional couplers to sample the incident power P_i and the reflected power P_r from the load. The magnitude of the reflection coefficient (ρ) can be measured directly from the reflectometer. By knowing the ρ, impedance can be calculated by using following relation.

$$\rho = \sqrt{\frac{P_r}{P_i}} \qquad \ldots (8.8)$$

$$S = \frac{1+\rho}{1-\rho} \text{ and } \frac{Z-Z_g}{Z+Z_g} = \rho$$

Where, $\rho =$ reflection coefficient

$S =$ VSWR

$Z_g =$ known wave impedance

$Z =$ Unknown wave impedance

Because of directional couplers property, there will be no interference between reverse and forward waves.

Attenuation of 10 db pad is used to keep the input power of low level. The accuracy of reflectometer is greatest at low VSWR.

8.5.3 IMPEDANCE MEASUREMENT OF REACTIVE DISCONTINUITY

The impedance of a shunt reactive discontinuity, such as windows or a post or a step in a microwave transmission line, can be measured with the help of the slotted line method from the measurement of line VSWR and the distance of first voltage minimum from the discontinuity plane. Let the reactance of the discontinuity at load distance $d = 0$. The line is shorted by a matched load R_0 at $d = 0$.

MICROWAVE MEASUREMENTS

The total impedance of the combination is

$$Z_L = \frac{R_0 \cdot jX}{R_0 + jX} \qquad \ldots 5.9$$

Where, Z_L = unknown impedance

R_0 = Matched load

jX = Reactance of the discontinuity

Fig. (8.14) shows the experimental setup for the measurement of impedance of a discontinuity.

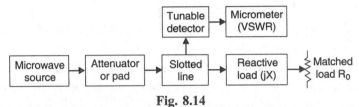

Fig. 8.14

8.5.4 IMPEDANCE MEASUREMENT BY USING MAGIC TEE

A magic Tee can be used to measure the unknown impedance by forming of a bridge as shown in Fig. (8.15).

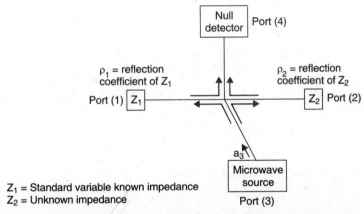

Fig. 8.15. Magic Tee used for impedance measurement.

The unknown impedance is connected at port -3 and standard variable known impedance at port-1. At port-3 Microwave source is connected at the port-4 a null detector is connected.

By using the properties of Magic Tee, the power from microwave source (a_3) gets divided equally between two ports (1) and (2) i.e., $\frac{a_3}{\sqrt{2}}$. As these impedances are not equal to characteristics impedance (Z_0) so there has to be reflections from port 1 and port 2.

If powers $\frac{\rho_1 a_3}{\sqrt{2}}$ and $\frac{\rho_2 a_3}{\sqrt{2}}$ enters the magic Tee junction from port 1 and 2 respectively as shown in Fig. (8.15). where ρ_1 and ρ_2 = reflection coefficient of impedance at port 1 and 2.

The resultant wave of the null detector can be given as follows:

$$= \frac{1}{\sqrt{2}}\left[\frac{1}{\sqrt{2}}a_3\rho_1\right] - \frac{1}{2}\left[\frac{1}{\sqrt{2}}a_3\rho_2\right]$$

$$= \frac{1}{2}a_3(\rho_1 - \rho_2) \qquad \ldots 8.10$$

$$= \frac{1}{2}a_3(\rho_1 - \rho_2) = 0 \qquad \text{(for perfect balancing of the bridge)}$$

or $\quad \rho_1 - \rho_2 = 0 \text{ or } \rho_1 = \rho_2$

or $\quad \dfrac{Z_1 - Z_z}{Z_1 + Z_z} = \dfrac{Z_2 - Z_z}{Z_2 + Z_z}$

$\therefore \quad Z_1 = Z_2$

i.e., $\quad R_1 + jX_1 = R_2 + jX_2$

$R_1 = R_2 \quad \text{and} \quad X_1 = X_2$

By adjusting the standard variable impedance till the bridge is balanced and both impedances become equal, the unknown impedance can be measured.

8.6 INSERTION LOSS AND ATTENUATION MEASUREMENTS

If input power is P_i, then insertion loss is given by the sum of power reflected (P_r) by the device to mismatch and the power attenuated i.e., $P_i - P_r$. The output signal power (P_0) is therefore less than input power(P_i). Insertion loss may be defined in another way such as the difference in the power reaching at the terminating load with and without the network in the circuit.

Insertion loss = Reflection loss + Attenuation loss

Since, $\qquad \dfrac{P_0}{P_i} = \dfrac{P_i - P_r}{P_i} \times \dfrac{P_0}{P_i - P_r} \qquad \qquad \ldots 8.11$

or $\qquad 10\log\dfrac{P_0}{P_i} = 10\ \log\left(1 - \dfrac{P_r}{P_i}\right) + 10\ \log\left(\dfrac{P_0}{P_i - P_r}\right) \qquad \ldots 8.12$

We can write,

$$\text{Insertion loss (dB)} = 10 \log \dfrac{P_0}{P_i} \qquad \ldots 8.13$$

$$\text{Reflection loss (dB)} = 10\log\left[1 - \dfrac{P_r}{P_i}\right] \qquad \ldots 8.14$$

or $\qquad \qquad \qquad = 10\log\dfrac{4S}{(1+S)^r} \qquad \ldots 8.15$

where, $\qquad S = VSWR = \dfrac{1 - |\Gamma|}{1 + |\Gamma|}$

$$\text{Attenuation loss (dB)} = 10\log\left[\dfrac{P_0}{P_i - P_r}\right] \qquad \ldots 8.16$$

$$\text{Return loss (dB)} = 10\log\dfrac{P_i}{P_r} = 20\log|\Gamma| \qquad \ldots 8.17$$

In case of perfect matching of impedance, the $P_r = 0$, that mean no reflection loss and the insertion loss is the same as the attenuation loss.

Attenuation is the ratio of output power (P_0) to the input power (P_i) of a transmission network terminated with the matched load.

Attenuation measurements are made mainly by using two methods, which are:

(i) Power Ratio method
(ii) Substitution method

MICROWAVE MEASUREMENTS

8.6.1 POWER RATIO METHOD

Power ratio method is the simplest form, which involves connecting an unmodulated microwave source to a bolometer mount and a power meter. Initially power level is measured and then the attenuating device is inserted between the source and the bolometer. The new power level reading is taken. The ratio of these two power level will be the measure of the attenuation of the device. With this method we can measure the attenuation up to 20 dB only.

For determining the attenuation, measurement of absolute power is not essential, since we are mostly concerned with the relative power levels. A detector having a square law characteristic can be used to measure the power ratio without knowing the absolute power levels.

Fig. (8.16) shows the attenuation measurements setup.

The steps required to follow to measure the attenuation by using power ratio method are as follows:

(*i*) Adjust and set the variable precision attenuator to approximately 20 dB and the flap attenuator to 0 dB.

(*ii*) Modulate the microwave signal from the source with a 1 KHz square wave. Adjust the modulation for a peak level on the SWR meter.

(*iii*) For the crystal detector to operate within its square law characteristics the value of power level at the input should be approximately 0.1 mW.

(*iv*) Adjust the control of the flap attenuator to provide the unknown loss. The level of loss can be read directly in the decibel scale of the SWR meter.

In case both the power levels with and without the unknown attenuation are not operating the crystal detector within its square law characteristics, than there will be an error in the measurement. By using substitution method the greater accuracy can be provided.

8.6.2 SUBSTITUTION METHOD

In this method the network, whose attenuation is to be measured, is replaced by a calibrated attenuator.

By changing the attenuation it is adjusted in such a way that it gives the same power as that was measured at the output of the network. In this situation, the attenuation of the two is the same and can be read off from the calibrated attenuator. Experiment setup is same as shown in Fig. (8.17)

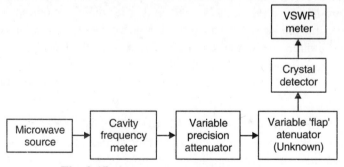

Fig. 8.17. Attenuator measurement setup.

(*i*) Adjust and set the precision variable attenuator to approximately 40 db higher than the expected attenuation of the component to be tested.

(*ii*) Modulate the microwave from the source with a 1 KHz square wave and adjust the modulation for a peak level or the SWR.

(*iii*) For the crystal detector to operate within its square law characteristics, the microwave power level at the crystal should be around 0.1 mW.

(iv) Rotate the knob of the flap attenuator to provide the 'unknown' loss. The needle of the SWR meter will fall back but should be restored to the 0 dB mark by reading of the precision variable attenuator.

(v) The difference between the new reading of the precision variable attenuator and its earlier reading is equal to the unknown attenuation.

The advantage of the substitution method over the power ratio method is that the power level at the crystal detector is the same when taking both reading of the precision attenuator.

8.7 POWER MEASUREMENT

Microwave power is one of the few parameters that can be measured directly. It is helpful in assessing the proper functioning of the microwave circuits and devices. Power is defined as the quantity of energy dissipated or stored per unit of time. The microwave power is divided in to three categories such as:

(i) Low power measurements (less than 10 mW)

(ii) Medium power measurements (10 mW to 10 W)

(iii) High power measurements (more than 10 W)

The average power is measured while propagations in the transmissions line is defined by

$$P_{av} = \frac{1}{nT} \int_0^{nT} V(t)i(t)\,dt. \qquad \ldots (8.18)$$

where, P_{av} = Average power

T = Time period of the lowest frequency

n = cycles are considered.

The most convenient unit of power at microwave frequency is dB_m.

$$P(dB_m) = 10 \log \frac{P(mW)}{1mW} \qquad \ldots (8.19)$$

The microwave power measurement devices consists of a power sensor, which converts the microwave power in to heat energy and the rise in temperature provides a change in the electrical parameters which is output current and indicates the power. Generally the sensors used for power measurements are the detectors, such as Schottky barrier diode, Bolometer and thermocouple. High power is generally measured using microwave calorimeter in which the temperature rises of the load provides a direct measure of the power absorbed by the load.

During the measurements of power, the power source should be isolated from the load with the help of an isolator, since microwave power sources are very sensitive to load impedance variation.

8.7.1 LOW POWER MEASUREMENTS (LESS THAN 10 mW)

The devices such as bolometer and thermocouple are used for measuring of the low power. The resistances of these devices changes with the applied power. Bolometers are mostly used for the measurement of low power.

Bolometer measurements are based on the dissipation of the RF power in a small temperature sensitive resistive component called Bolometer. There are two types of bolometer such as:

(i) Barretters (ii) Thermistors

(i) **Barretters** : Barretters are positive temperature coefficient and their resistance is increases with an increase in temperature as shown in Fig. (8.18a).

MICROWAVE MEASUREMENTS

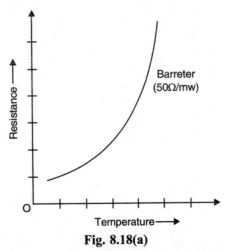

Fig. 8.18(a)

Barretters are consisting of a short thin metallic (Platinum) wire whose diameter is less than 0.0001 inch, mounted in a cartridge similar as ordinary fuse as shown in figure 8.18(b). The barretters are biased to have optimum resistance approximately around 200Ω.

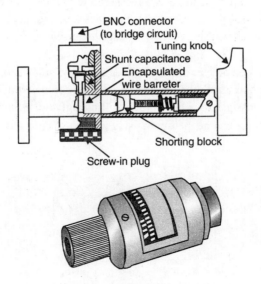

Fig. 8.18(b). Barretters.

(*ii*) **Thermistors :** The thermistor is a semiconductor sensor which has a negative temperature coefficient of resistance, their resistances decreases as the temperature increases as shown in Fig. (8.19). The impedances of these bolometer are in the range of 100-200 Ω.

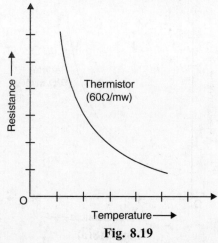

Fig. 8.19

The microwave power to be measured, heats the bolometer and resulting a change in its electrical resistances, which serves as an indication of magnitude of power. Bolometers are a square law device such as crystal diode and its produces a current that is proportional to the input applied power. Barretters are more delicate than thermistors, hence they are used only for power of few mW. For medium and high power are measured with a low power thermistor sensor after precisely attenuating the signal. The thermistor provides low loss, good impedance match, better isolation and physical shock and less energy leakage. The sensitivity of thermistor is limited to approximately 20 dBm.

Fig. (8.20) shows, a bolometer mounted inside the wave guide and acts as a load. When low microwave power is applied which is to be measured, than some part of the power is absorbed in the bolometer load and dissipated as heat which will change the resistance value.

The microwave power can be measured using a bridge, is proportional to the change in the resistance $(R_1 - R_0)$.

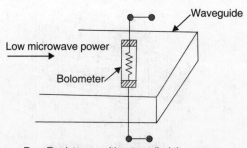

R_0 = Resistance without applied the power
R_1 = Resistance of the applied power

Fig. 8.20. A bolometer mounted inside the wave guide.

Balanced Bolometer Bridge Technique

Fig. (8.21) shows the setup of balanced bolometer bridge technique, in this the bolometer is used as one of the arms of the bridge. The microwave power is applied and this power is get dissipated in the bolometer, the bolometer heats up and changes its resistance causing an unbalance in the bridge from its initial balance position under no power is incident. The non-zero output is recorded on a voltmeter which is calibrated to measure the level of the input power.

MICROWAVE MEASUREMENTS

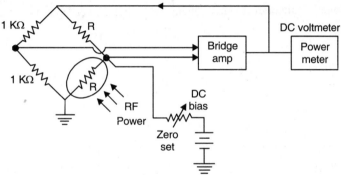

Fig. 8.21. Balanced Bolometer Bridge.

A single bridge has many disadvantages such as

(i) The change of resistance due to mismatch at the microwave input port.

(ii) Effect of ambient temperature on the thermistor resulting in false reading.

The above mentioned problem can be eliminated by using double identical bridges as shown in Fig. (8.22). The upper bridge measures the micro power and another bridge circuit compensates the effect of ambient temperature variation. The added microwave power due to mismatch is compensated by a self balancing circuit by decreasing the dc power V_2 carried by the sensing thermistor until bridge becomes balance or net change in the thermistor resistance is zero due to negative dc feedback.

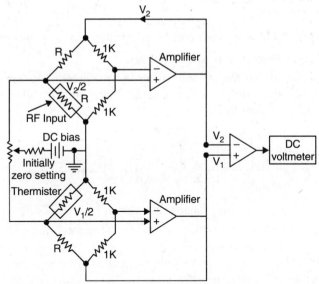

Fig. 8.22. Power meter using double bridge for compensation.

When there is no input microwave signal applied than the initial zero setting of the bridge is done by adjusting $V_0 = V_1 = V_2$. R is the resistance of the thermistor at balance.

At balance the dc voltages across the sensor are $\dfrac{V_1}{2}$ and $\dfrac{V_2}{2}$, without and with microwave present, respectively.

Average input power (P_{av}) is given by

$$P_{av} = \frac{V_1^2}{4R} - \frac{V_2^2}{4R} = \frac{(V_1 - V_2)(V_1 + V_2)}{4R} \qquad \ldots 8.19$$

If any change is temperature due to change in voltage be ΔV than the change in RF power is

$$P_{av} = \frac{(V_1 + \Delta V)^2}{4R} - \frac{(V_2 + \Delta V)^2}{4R} \qquad \ldots 8.20$$

$$P_{av} = \frac{(V_1 - V_2)\ (V_1 - V_2 + 2\Delta V)}{4R} \qquad \ldots 8.21$$

$(V_1 + V_2 \gg 2\Delta V,$ hence $\Delta P \approx 0)$

The meter responds as per equation (8.21) to determine microwave power P_{av}.

By using directional coupler and attenuator, the power handling capacity of Barretters and Thermistors can be increased many times in range of 1000 times even. For example if 20 db directional coupler and 10 db attenuator is used, the power can be increased by 1000 times as shown in Fig. (8.23).

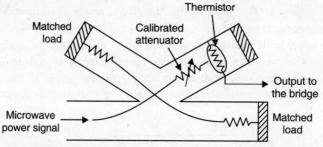

Fig. 8.23. Directional coupler using attenuator as power meter.

8.7.2 Medium Power Meters (10 mw to 10 w)

The medium power which is ranging from 10 mw to 10w can be measured by using calorimetric technique. Its principle is quite simple, here rise in temperature in the load is monitored which is responsible for the rise as shown in Fig. (8.24).

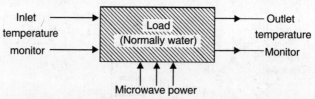

Fig. 8.24. Principle of measurement of medium power.

Normally water is used as load because of its high specific heat. By knowing the mass, specific heat and temperature rise at a constant rate of fluid flow, the power can be measured. The normally adopted method is the self balancing bridge technique, which is shown in the Fig. (8.25).

The self balancing bridge technique consists of two identical temperature sensitive resistors in two of its arms. The input load resistor senses the unknown input microwave power. The comparison load is related with the comparison power. Both, the input temperature resistor and input load power resistor are placed in the close vicinity to each other so that the temperature increased in the input load resistor, raises the temperature of the gauge (resistor). Now the bridge will become unbalance and the signal due to this effect is amplified and then applied to the comparison load resistor which is placed near to the comparison resistor. Hence the heat generated in the comparison load resistor is transferred to comparison resistor and make the bridge balance again. The meter measures the amount of power that is supplied to the comparison load to make the bridge balance.

MICROWAVE MEASUREMENTS

The meter can be calibrated in terms of input microwave power. To avoid the errors, it is necessary that the two gauges or resistors should be matched and indentical. For fast balancing of the bridge and to have efficient transfer of heat from loads to the gauges, the components are immersed in an oil stream.

The accuracy of measuring the power is balancing bridge technique is approximately ± 5%. Just to have constant temperature, the streams are passed through a parallel flow heat exchanger before they enters the heads. An amplifier is used to amplify the error signal and, the 1.2 KHz source and meter are separated by means of a transformer which form the other arms of the bridge.

8.7.3 High Power Measurement (10 w and more)

High power microwave energy is usually measured by a calorimeter-wattmeter. These meters may be only dry type or flow type. The calorimeter-wattmeter method involves conversion of the microwave energy in to heat, absorbing this heat in a fluid (usually water) and then measuring the temperature rise of the fluid as discussed in the medium power measurement method. A dry type calorimeter consists of a 50 ohm coaxial cable which is filled by a dielectric load with a high hysteresis loss. In case of flow type calorimeter, by knowing the rate of the fluid flow the exact value of power can be calculated by using the equation

$$P_{av} = \frac{R \; C\rho \; (T_2 - T_1)}{4.18} \text{ watts} \quad \ldots 8.22 \text{ (for circulating calorimeter)}$$

or

$$P_{av} = \frac{4.187 \, m C \rho T}{t} \text{ watts} \quad \ldots 8.23 \text{ (for static calorimeter)}$$

where
P = average power
R = Rate of flow in cm³/s
C = Specific heat of liquid in cal/gm
$T_2 - T_1 = T$ = Temperature difference in °C
t = Time sec
m = Mass of the liquid flow
ρ = Specific gravity in g/cm²

The flow type calorimeter is also known as circulating calorimeter shown in fig. (8.21) and the dry type calorimeter is also known as static calorimeter.

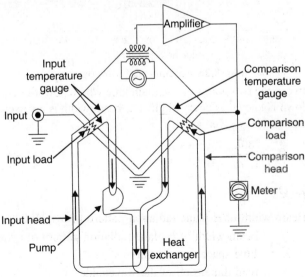

Fig. 8.25. Microwave calorimeter.

The average power can be converted into peak power by using following relations

$$P_P = \frac{P_{av}}{\text{duty cycle}} \qquad \ldots 8.24$$

The disadvantages of calorimetric-wattmeter system are as follows:
1. It has low accuracy approximately 5% only.
2. Inherent inertia of the system.
3. This method only can be used for high power measurement of microwave power.

8.8 RADIATION PATTERN MEASUREMENTS

The important parameters required to be measured to determine the performance characteristics of microwave antennas are radiation pattern, gain, directivity, radiation efficiency, beam width, input impedance, bandwidth and polarization. We will be discussing about the radiation pattern of an antenna system.

A short dipole radiates equal power in all the direction or in omni directions is called an isotropic antenna. But any practical antenna does not have the property of isotropic antenna i.e., omni directional radiations. The radiation pattern is a representation of the radiation characteristics of the antenna as a function of azimuth angle 'ϕ' and elevation angle 'θ' for a constant radical distance and frequency. The field strength or power intensity as a function of aspect angle at a constant distance from the radiating antenna is known as the radiation pattern of the transmitting antenna. The radiation pattern in the H- and E-plane will depend upon the fact, as which field is parallel to the ground. When the receiving and transmitting antennas are identical and have same polarization, then the transmitting and receiving radiation patterns are also same or identical. Fig. (8.26) shows the typical radiation pattern, shows a main lobe, side to be and a back to be. The major portion of the radiated power is concentrated in the main lobe (major lobe).

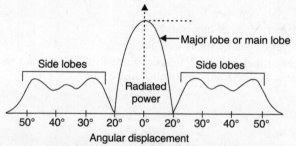

Fig. 8.26 Antenna radiation patterns.

The angle between the two points on the main lobe where power intensity falls to half of the maximum or 3 db down is known as beam width. The beam width is inversely proportional to their respective dimensions and it is given as

$$Q_H = \frac{80\lambda_0}{a} \qquad \ldots 8.25$$

$$Q_E = \frac{53\lambda_0}{b} \qquad \ldots 8.26$$

where Q_H = Beam width of H-plane radiation pattern of pyramidal horn
$\quad Q_E$ = Beam width of E-plane radiation pattern of pyramidal horn
$\quad \lambda_0$ = Free space wavelength
$\quad a$ = Broad dimension of the waveguide
$\quad b$ = Small dimension of the wave guide

MICROWAVE MEASUREMENTS

Two identical horns are placed in line and in the same orientation for measuring Q_H and Q_E experimentally. The microwave power is being fed to one of the antenna and other horn antenna (mounted on a rotary mount) the power received at various angles. The radiation pattern may be plotted and Q_E and Q_H may be calculated.

The gain of the horn antenna is calculated from the measured Q_E and Q_H value, by the following relation

$$G = \frac{133.25.71}{Q_H Q_E} \qquad \ldots 8.27$$

A setup for measuring gain of a horn antenna is shown in Fig. (8.23).

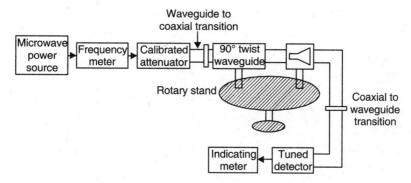

Fig. 8.27. Setup for measuring radiation pattern and gain of a horn antenna.

8.9. NETWORK ANALYSERS

A network analyser makes measurements of complex reflection coefficients on 2-port microwave networks. In addition, it can make measurements of the complex amplitude ratio between the outgoing wave on one port and the incoming wave on the other. There are thus four possible complex amplitude ratios which can be measured. If we designate the two ports 1 and 2 respectively, these ratios may be written s_{11} s_{12} s_{21} s_{22}. These are the four "s-parameters" or "scattering parameters" for the network. Together they may be assembled into a matrix called the "s-matrix" or "scattering matrix".

The network analyser works on a different principle to the slotted line. It forms sums and differences of the port currents and voltages, by using a cunning bridge arrangement. The phase angles are found by using synchronous detection having in-phase and quadrature components. From the measured voltage and currents it determines the incoming and outgoing wave amplitudes.

$$V+ = (V + ZoI)/2 \text{ and } V-- = (V - ZoI)/2.$$

Network analysers can be automated and controlled by computer, and make measurements at a series of different frequencies derived from a computer controlled master oscillator. They then plot the s-parameters against frequency, either on a SMITH chart or directly.

The important experimental technique to the use of a network analyser lies in the calibration procedure. It is usual to present the analyser with known scattering events, from matched terminations and short circuits at known places. It can then adjust its presentation of s-parameters for imperfections in the transmission lines connecting the analyser to the network, so that the user never has to consider the errors directly. It is even possible to calibrate out the effects of intervening transmission components, such as chip packages, and measure the "bare" s-parameters of a chip at reference planes on-chip.

SUMMARY

- Microwave measurements are concerned with distributed circuit elements.
- Most of the properties of devices and circuits at microwave frequencies are obtained from the measurement of S-parameter, power, frequency, phase shift, noise figure and VSWR.
- Mainly two types of diodes are used for microwave detection :
  (i) Schottky barrier diode (ii) Point Contact Diode
- Microwave frequency can be measured by either mechanical or electronic techniques.
- There are three methods for the measurement of frequency :
  (i) Slotted Line Technique
  (ii) Wave meter or calibrated resonant cavity technique (iii) Electronic Technique
- Measurement of VSWR express the degree of mismatch between the load and transmission line.
- To measure the impedance, both amplitude and phase of the test signal are required to be determined.
- In the reflectometer technique only the magnitude of the impedance is calculated whereas a slotted line wave guide measurement gives both magnitude and phase angle of load impedance.
- A magic Tee can be used to measure the unknown impedance by forming of a bridge.
- Insertion loss = Reflection loss + Attenuation loss
- Attenuation loss (dB) $= 10 \log \left[\dfrac{P_0}{P_i - P_r} \right]$
- Return loss (dB) $= 10 \log \dfrac{P_r}{P_i} = 20 \log |\Gamma|$
- Microwave power is one of the few parameters that can be measured directly.
- The microwave power is divided in to three categories such as :
  (i) Low power measurements (less than 10 mW)
  (ii) Medium power measurements (10 mW to 10 W)
  (iii) High power measurements (more than 10 W)
- The devices such as bolometer and thermocouple are used for measuring of the low power.
- Bolometer measurements are based on the dissipation of the RF power in a small temperature sensitive resistive component called Bolometer. There are two types of bolometers such as :
  (i) Barretters (ii) Thermistors
  (i) **Barretters**: Barretters are positive temperature coefficient and their resistance is increases with an increase in temperature as shown.
  (ii) **Thermistors** : The thermistor is a semiconductor sensor which has a negative temperature coefficient of resistance, their resistances decreases as the temperature increases.
- The medium power which is ranging from 10 mw to 10 w, can be measured by using calorimetric technique
- High power microwave energy is usually measured by a calorimeter-wattmeter.
- A network analyser makes measurements of complex reflection coefficients on 2-port microwave networks.

MICROWAVE MEASUREMENTS

REVIEW QUESTIONS

1. How are microwave measurements different from measurements of low frequencies?
2. Explain the power measurements techniques in brief with the diagram.
3. Discuss the detection of microwaves.
4. Explain the working principle of Bolometer for microwave power measurements.
5. Discuss the range of power measurements and how it can be extended.
6. Explain the self balancing bridge technique used for measuring medium power.
7. Discuss the various techniques used for microwave frequency measurements.
8. Discuss the measurements of **VSWR** and explain the double minimum method of measuring of VSWR.
9. Explain the methods of measuring impedance of a terminating load in microwave systems.
10. Discuss the measurements of attenuation in microwave.
11. Discuss the measurement of radiations pattern of an antenna.
12. Write short notes on:
    (*i*) Microwave measurements
    (*ii*) Power measurements
    (*iii*) Calorimeter technique
    (*iv*) Wavemeter
    (*v*) Measurement of frequency
    (*vi*) Measurements of VSWR
    (*vii*) Network analyser
13. Two identical directional couplers are used in a waveguide to provide samples of the incident and reflected powers. The outputs of the directional couplers are 2-5 mW and 015mW. Calculate the VSWR. **[Ans. 1.64]**
14. How is slotted line used for measurement of impedance of an unknown load? **(UPTU-2003, 2004)**
15. A 50 Ω loss-less line terminated in unknown impedance. The VSWR of the load is 2. When the load is replaced by a short, the minima shifts 1.5 cm towards the load and the successive voltage minima are 5 cm apart. What is the load impedance? **(UPTU-2003, 2004)**
16. What is Bolometer? Give the construction of a thermistor mount and explain the method of microwave power measurement using dual bolometer bridge. What is the function of second bridge? **(UPTU-2003,2004)**
17. What is wavemeter? Differentiate between transmission and reactions type wavemeters. Which type is generally used in the microwave test bench in the laboratory? Explain its construction and working. **(UPTU-2003, 2004)**
18. State how to measure the frequency of the source without using a wavemeter in the microwave test bench. **(UPTU-2003, 2004)**
19. Describe the standing wave detector method for measuring the impedance. Derive the formula used. Why it is a common practice to use minima rather than the maxima in measuring distances in connection with the standing wave pattern. List the sources of error in measurement. **(UPTU-2003,2004)**
20. Distinguish between insertion loss and attenuation. Describe a method to measure insertions loss of a network over a load of microwave frequencies. How can the loss due to reflection at the input port of the network be taken into account in the measurement process? **(UPTU-2002, 2003)**

21. Explain the method of measurement of unknown impedance through VSWR measurement. **(MDU-2006)**
22. Discuss the importance of VSWR in characterization of microwave components. What is the difference between measurements of low and high VSWRs? **(MDU-2006)**
23. Explain bolometers, bolometer mounts and bolometer bridges. **(MDU-2004)**
24. Discuss the return loss measurements by reflectometer. **(MDU-2004)**
25. Describe various techniques of measuring unknown frequency pf a microwave generator. **(MDU-2004)**
26. A slotted line is used to measure the frequency and it was found that distance between nulls is 1.85 cm. Given the guide dimensions as 3 × 1.5 cm, calculate the value of frequency. **(MDU-2004)**
27. How are microwave measurements different from low frequency measurements? **(MDU-2003)**
28. How can power of a microwave generator be measured using bolometer? **(MDU-2003)**

9

CHAPTER

MICROWAVES ANTENNAS

OBJECTIVE
- Introduction
- Types of Microwaves Antennas
- Parabolic Reflector
- Cassegrain Antenna
- Horn Antenna
- Lens Antennas
- Slot Antennas
- Microstrip Antennas

9.1 INTRODUCTION

It is to understand that the system operating in the range of 1 GHz and extended up to 100 GHz, is known as microwave system. The antenna used in these systems may be called as microwave antennas, tend to be directive *i.e.*, very high gain and narrow beam width in both the planes such as horizontal and vertical. Theoretically all the antennas could be used for all the frequencies but practically the actual shape of antenna depend upon the frequency band for which it is designed. It is known as the frequency increases, the wavelength decreases and thus it becomes easier to construct an antenna system that is large in terms of wavelengths.

Microwaves are normally used for communication purpose not for broadcast purpose. As they are used for communication, so we need highly directional and also higher antenna gains are required to take care of induced and shot noise effects while generation and reception at microwaves frequencies. Since wavelength is small at higher frequencies, a small height of antenna is adequate to have (l/λ) ratio.

9.2 TYPES OF MICROWAVE ANTENNAS

At microwave frequency, following types of antennas are used,
1. Parabolic reflector   2. Cassegrain antenna
3. Horn antenna   4. Slot antenna
5. Lens antenna   6. Microstrip antenna

9.3 PARABOLIC REFLECTOR

A parabolic reflector can be used to concentrate the radiation from an antenna located at the focus in the same way that a search light reflector produces a sharply defined beam of light. The parabola is a plane curve, defined as the locus of a point which moves so that its distance from another point (called the focus) plus its distance from a straight line (directrix) is constant. These geometric properties yield an excellent microwave reflector.

The parabolic reflector does this by converting the spherical waves originated by the radiator at the focus of the parabola in to a plane wave of uniforms phase across the mouth or aperture of the parabola as shown in Fig. (9.1).

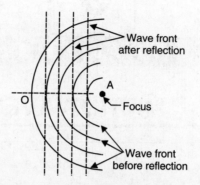

Fig. 9.1. Wavefronts in parabolic reflection system.

The parabolic antenna is the form, which is most frequently used as the radar antenna. Fig. (9.2), illustrates the parabolic antenna. A feed horn as a radiation source Parabolic surface is placed at the local point 'A' that is known as feed. The field leaves this feed horn with a spherical wave front.

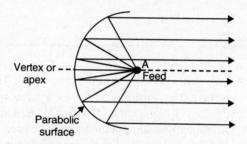

Fig. 9.2. Parabolic reflector antenna.

As each part of the wave front reaches the reflecting surface, it is shifted 180° in phase and sent outward and at angles that cause all parts of the field to travel in parallel paths. Because of the shape of a parabolic reflector surface, all the parts from 'A' to the reflector and back are the same length. These characteristics makes parabolic reflector is most suitable for the microwave antenna.

The radiation pattern of a parabolic antenna contains a major lobe (or main lobe) which is directed along the axis of propagation and several small minor lobes as shown in fig. (9.3).

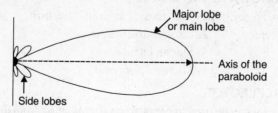

Fig. 9.3. Parabolic radiation pattern.

The shape and width of the main lobe of the radiation pattern of a parabolic reflector depend upon the size and shape of the mouth of the parabola and the variation of field intensity over the aperture defined by the mouth.

MICROWAVES ANTENNAS

The directive gain of antenna system using parabolic reflection is given as:

Directive gain
$$(G_D) = \frac{4\pi A}{\lambda^2} \qquad \ldots 9.1$$

Since $A = KA_0$, then
$$G_D = \frac{4\pi K A_0}{\lambda^2} \qquad \ldots 9.2$$

where A = Capture area
A_0 = Actual area of mouth or aperture
K = Constant that's depends on the type of antenna feed used.
G_D = The directive gain or power gain of circular aperture with diameter D.

Since we know,
$$A = \pi D^2 / 4 \qquad \ldots 9.3$$

By putting the value of 'A' from eq. 9.3 in eq. 9.2, we get
$$G_D = \frac{(4\pi K)(\pi D^2)}{\lambda^2 \times 4} \qquad \ldots 9.4$$

The value of K constant for half wave dipole as approximately 0.65. By putting the value of K in the above equation 9.4,

$$G_D = \frac{0.65 \times 4\pi \times \pi \times D^2}{4 \times \lambda^2}$$

$$= 0.65 \times \frac{\pi^2 D^2}{\lambda^2}$$

$$= 6.389 (D/\lambda)^2$$

$$G_D = 6\left(\frac{D}{\lambda}\right)^2 \qquad \ldots 9.5$$

If the feed used is an isotropic than the paraboloid produces a beam of radiation. If the circular mouth is assumed to be large than the beam width between half power points (HPBW) is given by

$$HPBW = 70° \lambda / D$$

where D is the diameter of circular aperture.

9.3.1 Various Types of Parabolic Reflectors

There are various types of parabolic reflectors, which can be used for various applications:
1. Truncated or cut parabola
2. Cylindrical parabola
3. Cheese antenna or pill box antenna

1. Truncated or cut parabola: A certain amount of control of beam can be achieved by making use of a section of parabolic as shown in Fig. (9.4), with large horizontal plane and a small vertical aperture, this will give a narrow beam is horizontal plane and a wider beam in vertical plane and vice-versa.

The tapering illumination has an important advantage to decrease the amplitude of side lobes through, there is consequent decrease in gain a beam width is increases.

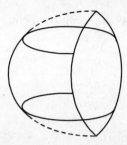

Fig. 9.4. Truncated parabola.

2. Cylindrical Parabolic: Another means of producing a symmetrical or an asymmetrical antenna pattern is with the parabolic cylinder is generated by moving the parabolic contour parallel to itself. A line source such as linear array or planner array is used to the parabolic cylinder. A cylindrical parabolic reflector is shown in the Fig. (9.5). The gain for the rectangular aperture $= 7.7 \ (L/\lambda)^2$

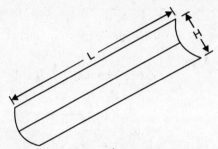

Fig. 9.5. A cylindrical parabolic reflector.

3. Cheese Reflector: In the Cheese reflector the aperture is perfect rectangular. It is commonly used in a short parabolic cylinder enclosed by parallel plates. The reflector surface is illuminated by a line source preferable, linear array. It generates a fan shaped beam. If the Cheese reflector placed horizontally the beam width in the azimuth is small and in the vertical plane is large. Similarly, for a vertical Cheese, the vertical beam width is smaller than a horizontal beam width. Fig. (9.6), shows the Cheese paraboloid. All the three parabolic reflectors have advantages that they are smaller in size.

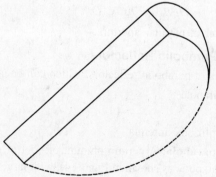

Fig. 9.6. Cheese paraboloid.

9.3.2 Feeds for Parabolic

The ideal feed for a parabolic consists of a point source of illumination with a pattern of proper shape to achieve the desired aperture distribution efficiency. The antenna placed at the focus of a parabolic is known as **feed or primary radiation** and its radiation pattern is called

MICROWAVES ANTENNAS

Primary pattern; the radiation pattern of aperture where illuminated by the feed is called the **secondary pattern**.

The entire parabolic reflector antenna consists of two basic components:
1. Reflector and
2. Source of primary radiations.

Practice feeds for parabolic reflector are only approximate the ideal. An ideal feed is that in which the radiations which radiates towards the reflector in such a way that no energy will radiates in any other directions except to the entire surface of the reflectors. Such an ideal radiator is not available in practice. There are several types of feed that can be used. Some are as follows:
1. Dipole antenna (end fire feed)
2. Horn feed
3. Parabola with Yagi antenna

An isotropic antenna as feed is not a better choice. For the primary feed a dipole antenna was used in the earlier days but due to some limitation, they were not used in a dipole with parabolic reflector such as Yagi- uda or a plane sheet or a half cylinder or a hemi sphere, which is feed with a conical line. Various configurations of feeds are shown in Fig. (9.7).

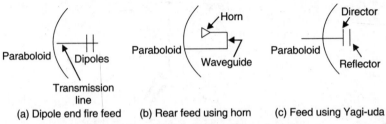

Fig. 9.7. Various configurations of feeds.

A better feed than the half wave dipole is the open-ended wave-guide. The total energy is directed in the forward directions and the sphere characteristics are very good, if it is radiating in suitable mode. The wave guide horn is more popular method of feeding a paraboloid for radar applications if circular polarization is required than conical horn antenna or helix antenna can be used as feed.

9.3.3 Feed Support

Two types of feed arrangement are used to feed the paraboloid. One is rear feed support system and the other is the front feed arrangement. The wave guide rear feed as shown in Fig. 9.7(b) produce as asymmetrical pattern because the transmission line is not in the center of the feed, dual-aperture rear feed in which the wave-guide is in the center of the dish and the energy is made to bend 190° at the end of the wave guide by a properly designed reflecting plate. The advantage of rear feed is that it utilizes a minimum length of transmission line and its compactness.

The other way of arranging the feed is front feed which is shown in Fig. (9.8). It is well suited for supporting horn feeds but it obstructs the aperture.

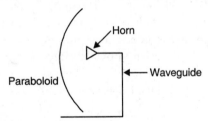

Fig. 9.8. Front feed using long horn.

There are two limitations in the above-mentioned feed arranged configurations.
1. Aperture Blocking and
2. Impedance Mismatching in feed.

The feed, transmission line, and supporting structure intercept a portion of the radiated and alter antenna pattern. The energy reflected by the paraboloid enters the feed and act as a reverse direction wave.

Standing wave is produced along the line, causing an impedance mismatch and degradation of Transmitter performance.

9.3.4 Offset Feed

Using the offset feeds parabolic antenna can eliminate the aperture blocking and mismatching of feeds. The center of the feed is placed at the focus of parabola but the horn is tipped with respect to the parabolas axis. For all practice purpose the feed is out of the path of the reflected energy. With this arrangement as shown in Fig. (9.9), there is no aperture blocking and no impedance-mismatching can occur.

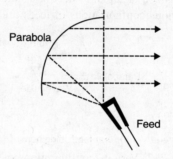

Fig. 9.9. Reflector with off-set feed.

The offset parabola eliminates two of the main draw backs of rear and front feed configurations. However, it has problem of its own.

The cross polarizations lobes are produced in the offset configurations, which may deteriorate the radar system performance. It is more difficult to scan and to properly support the offset antenna than a circular paraboloid.

Drawbacks of Parabolic Reflector:
1. The side lobes are strong enough causes false echoes.
2. The making of large dishes geometrically exact is quite difficult and any deviations may cause reception from undesired sources.
3. The wind loading is more problematic area in the large size dish antenna, however using of wire meshed reflector may be quite safe from stormy wind.
4. The primary feed is not at the center of at the focus point reducing the directivity.

9.4 CASSEGRAIN ANTENNA

The cassegrain antenna is named after the name of seventeenth century astronomer, William Cassegrain. The cassegrain principle is the adaptive to the microwave region of an optical technique, which is widely used in telescope design to obtain high magnification with a physically short telescope. Its application in microwave reflector antenna allows the reduction in the aerial dimensions of the antenna. It also allows reducing the length of the transmission lines and the flexibility in the design of feed system.

MICROWAVES ANTENNAS

Fig. (9.10) shows the arrangement of cassegrain antenna, in this the primary feed radiator is positioned around an opening near the vertex of the parabolic instead of at focus and a sub reflector is located in front of the parabola between the vertex and focus. The rays from the primary horn reflector get reflected from the sub-reflector or hyperbola surface on to the parabolic reflector, which reflects the electromagnetic wave into a parallel plane wave emerging from the antenna. The hyperbolic reflector images the feed so that it appears as a virtual image at the focal point of the parabola.

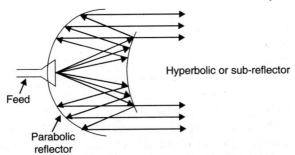

Fig. 9.10. Cassegrain antenna showing the hyperbolic sub-reflector and the feed at vertex of the main parabolic reflector.

9.5 HORN ANTENNAS

A typical horn antenna consists of an open ended waveguide flared so that the wave inside the guide expands in an orderly manner. When a waveguide is terminated by a horn, the abrupt discontinuity that existed is replaced by a gradual transformation such as shown in Fig. (9.11). All the energy travelling forward in the wave guide will be radiated properly if the impedance matching is correct. With this arrangement the directivity is improved and diffraction is minimized. Relative to the parabolic reflector, the horn is thus an alternative means of producing a field distribution across an aperture. The amplitude and the phase of the fields in the plane of the horn mouth depend upon the type of wave fed into horn from the waveguide, and upon the horn proportions.

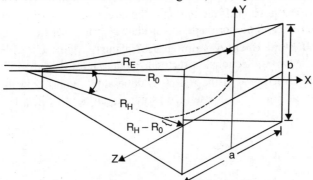

Fig. 9.11. Pyramidal horn excited from a waveguide carrying a TE_{10} wave.

In normal condition where the wave is of the TE_{10} type, the intensity of the electric field at the mouth is constant along the axis (F-axis) and is sinusoidal distributed along the other axis of the horn (Z-axis) as shown in the Fig. (9.11).

In the figure (9.11)

a = mouth dimension in Z

b = mouth dimension in Y

l = horn length from mouth to apex (R_0)

However, the field across the mouth of the horn, unlike the field across the mouth of the parabola, is a section of spherical wave front having its center at the apex of the horn.

Accordingly the field near the rim of the horn lags in phase behind the field at the center because the distances R_H and R_E from the equivalent apex of the horn to the corresponding edges are greater than the distance R_0 to the center of the plane defined by the mouth. The extent of the resulting variation in phase across the plane of the mouth depends upon the flare angles of the horn, and the horn length is wavelengths.

In case when the relation of flare angles and horn length is such that $R_H - R_0$ and also $R_E - R_0$, is a small fraction of a wavelength (*i.e.*, less than 0.15 λ), then the field distribution across the horn mouth has everywhere nearly the same phase, and the action of the horn is almost exactly the same as that of a parabola having the same shape of mouth and the same field distribution across this mouth. However if the flare angles of the horn are too large, *i.e.*, the horn length is to small in proportion to the mouth dimensions, than the distance that the wave must travel to reach the rim of the mouth is enough greater than the distance it must travel to reach the center so that the field across the mouth can no longer be considered to be an equi-phase field. This increases the beam width and reduces the power gain. In addition if the difference in distance is appreciably more than a half wave length ($\lambda/2$), large minor lobes develop and ultimately the major lobe splits into two lobes.

By analyzing the above discussion, it is apparent that very small flare angles given the sharpest beam and highest power gain for a given mouth size. However, small flare angles correspond to a horn of excessive length.

Typically in practice, Here angles, ϕ varies from 40° when $L/\lambda = 6$, at which the beam width in the plane of the horn in 66° and the maximum directive gain is 40, to 15° when $L/\lambda = 50$. For which beam width is 23° and gain is 120°.

9.5.1 Various Types of Horns

(*i*) Sectional E-plane horn
(*ii*) Sectional //-plane horn
(*iii*) Pyramidal horn
(*iv*) Conical horn

(*i*) **Sectional E-Plane Horn:** If the flaring is the direction of *E*-field vector than it is known as **sectional or sectoral E-plane** horn antenna as shown in Fig. 9.12(c).

(*ii*) **Sectional H-Plane Horn**: It is similar as *E*-plane horn, if the flaring is in the direction of *H*-field then it is known as sectional is sectoral *H*-plane horn as shown in Fig. 9.12(b).

(*iii*) **Pyramidal Horn**: Now if the flaring is done along both the walls of the rectangular wave guide as shown in Fig. 9.12(c) is termed as pyramidal horn. It has the shape of truncated pyramid.

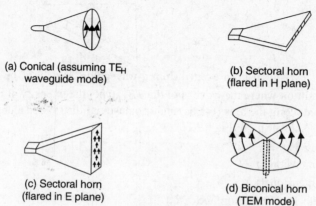

(a) Conical (assuming TE_H waveguide mode)

(b) Sectoral horn (flared in H plane)

(c) Sectoral horn (flared in E plane)

(d) Biconical horn (TEM mode)

Fig. 9.12. Various forms of horn antennas.

MICROWAVES ANTENNAS

(iv) Conical Horn: When open end of a circular wave guide is flared, it is termed as conical horn as shown in Fig. 9.12(a), Fig. 9.12(d) shown the diagram of Biconical horn, it produces a practice shaped beam, thin in the vertical direction but uniform in the horizontal plane.

The pyramidal horn and the conical horn give pencil like beams that have pronounced directivity in both horizontal and vertical planes. For shaped beams result when the dimension of the horn mouth is much smaller than the other; the sectoral horns formed by flaring is only one dimension, have this behavior.

The biconical horn produces a pancake-shaped beam, thin in the vertical direction but uniform in the horizontal plane. It is generally excited by mean of coaxial transmission line.

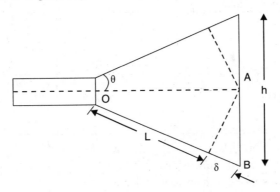

Fig. 9.13

By referring fig. (9.13), a = aperture, Q = 1/2 of flare angle and L = axial length. We have

$$\cos\theta = \frac{L}{L+\delta} \quad \text{and} \quad \tan\theta = \frac{n/2}{L}$$

or

$$\theta = \cos^{-1}\left[\frac{L}{L+\delta}\right] = \tan^{-1}\left[\frac{h}{2L}\right] \qquad \ldots 9.6$$

From the right-angled triangle AOB

$$(L+g)^2 = L^2 + \left[\frac{h}{2}\right]^2$$

or

$$L^2 + f^2 + 2K\delta = L^2 + \left[\frac{h}{2}\right]^2$$

As δ is so small that δ^2 may be neglected.

$$2L\delta = \frac{h^2}{4}$$

or

$$L = \frac{h^2}{\delta} \qquad \ldots 9.7$$

Both the equation (9.6 and 9.7) may be used to calculate the value of L and θ, to design a horn antenna. Table 9.1, gives the formulas for optimum horns in terms of beam width and gain.

Table 9.1

Types of horn	Property which is optimized for given length	Optimum proportions	Half power beam width in degrees	
			H-plane	E-plane
Sectoral-E-plane	Beam width in E-plane	$b = \sqrt{2/\lambda}$	$\dfrac{68}{(a/\lambda)}$	$\dfrac{53}{(b/\lambda)}$
Sectoral-H-plane	Beam width in H-plane	$a = \sqrt{3/\lambda}$	$\dfrac{68}{(a/\lambda)}$	$\dfrac{53}{(b/\lambda)}$
Pyramidal	Gain	$a = \sqrt{3/\lambda}$ $b = 0.910$ Gain $= 15.3\, l/\lambda$	$\dfrac{80}{(a/\lambda)}$	$\dfrac{53}{(b/\lambda)}$
Conical (TE_{11} mode)	Gain	$d = \sqrt{2.8/\lambda}$	$\dfrac{70}{(d/\lambda)}$	$\dfrac{60}{(d/\lambda)}$

Notation used in Table 9.1

a = mouth dimensions in Z-direction
b = mouth dimensions in Y direction
d = mouth diameter in conical horn
l = horn length from mouth to apex.

Horn antennas find extensive use at microwave frequencies under condition where the power gain required is only moderate. For high power gains the horn must be so long that for such application it is at a disadvantage compared with the corresponding parabola or lens antenna.

9.5.2 Lens Antennas

It is another type of microwave antenna based on the optical principle. Lens antenna may be used in the higher frequency in excess of 3 GHz and its works in the same way as a glass lens used in optics. At lower frequencies, the lens antenna becomes bulky and heavy.

When the source is located at the focal point of the lens then after refraction the beam forms parallel collimated beam or on the other side when parallel collimated beam refracted through the lens and converges at the focus as shown in the Fig. (9.14).

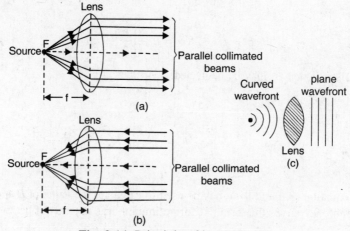

Fig. 9.14. Principle of lens antenna.

MICROWAVES ANTENNAS

The same principle of refraction of optics is applicable for electromagnetic wave. Instead of using glass lens, we use dielectric lens which is made of polystyrene.

Using an electromagnetic wave approach, we see that a curved wave front is present on the source side of the lens. It is known that a **plane wavefront** is required on the opposite side of the lens; to ensure a correct phase relationship. The function of the lens is to straighten out the wavefront. This can be achieved through the lens, by greatly slowing down the portion of the wave in the center. The parts of the wavefront near the edges of the lens are slowed slightly, since the thickness of the dielectric lens at the edge is less, in which velocity is reduced.

There are mainly two types of lens antenna used such as:

(*i*) Dielectric lens or *H*-plane metal plane lens

(*ii*) *E*-plane metal plate lens.

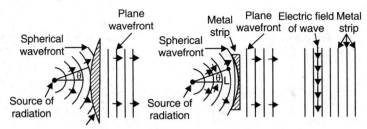

Fig. 9.15. Metallic conducting-strip lens.

In dielectric lens or *H*-plane metal plate are those in which the travelling wave front are delayed by lens medium where as in E-plane metal plate lens the travelling wavefronts are speeded up by the lens medium as shown in Fig. (9.15a & b).

Fig. (9.16) shows the stepping or zoning, of dielectric lenses which is often used to eliminate the problem of great thickness required of lenses used at lower microwave frequencies. The thickness of the dielectric lens can be reduced by means of stepping. The thickness of this type of lens is a function of wavelength, which is given by

$$t = \frac{\lambda}{\mu} - 1$$

where μ is the refractive index of the dielectric material.

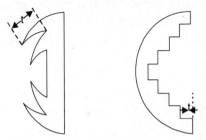

Fig. 9.16. Stepped dielectric lens.

Unstepped dielectric lens antennas are wide band and its bandwidth is around 12%, whereas that of a stepped dielectric lens antenna is about 5%.

The advantages and disadvantages of dielectric lens are as follows:

Advantages:

(*i*) More design flexibility

(*ii*) No aperture blocking

Disadvantages:
(*i*) They are bulky (*ii*) Design complexity (*iii*) High cost

9.6 SLOT ANTENNAS

The slot antenna makes use of the fact that energy is radiated when a RF field exists across a narrow slot in conducting plane. A typical slot antenna is shown in Fig. 9.17(a) here the fields are excited by a twin wire transmission line. The electric field across the slot is maximum at the center and takes off towards the edge as indicated, while at the same time current flow in the conducting plane in the general manner indicated. When the slot is exactly half wavelength ($\lambda/2$) long, the electrical field distribution is sinusoidal and the impedance offered by the slot to the two wire line is resistance of approximately 363Ω.

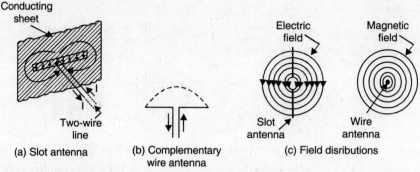

Fig. 9.17. Slot antenna and it complimentary wire antenna.

It is suggested that the radiation produced by a narrow slot in a conducting plane has exactly the same directional pattern as the radiations from a thin, flat, wire type antenna having a shape corresponding to the slot, and a current distribution that is the same as the distribution of electric field across the slot as shown in Fig. 9.17(b).

The difference between the slot and its complementary antenna are as follows:

(*i*) The polarizations are differed since the electric field associated with the slot is identical with the magnetic field of the complementary wire antenna.

(*ii*) The radiations from the back side of the conducting phase has the opposite polarity from that of the complementary antenna because of the way in which the fields are directed shows in Fig. 9.17(c).

There are some more possibilities of arrangements of slot antennas as shown in Fig. (9.18).

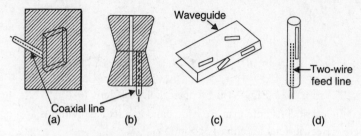

Fig. 9.18. Various slot antennas.

A broad side array of slots is shown in Fig. (9.19). Here, the centers of the successive slots are placed a half guide wavelength a part and placed on opposite sides of the center line as shown in Fig. (9.19).

MICROWAVES ANTENNAS

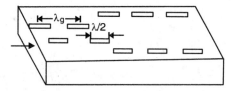

Fig. 9.19. Slotted waveguide antennas designed to act as a broad side array of slots.

In this way all slots radiate in the same phase, since the several of polarity of the field inside the guide at half wavelength interval is compensated by the fact that alternate slots are on opposite sides of the center lines. The radiation from all the slots can be made the same inspite of the fact that the fields inside the guide become weaker as they progress down the guide, by the expedient of placing. The slots are progressively farther from the center line.

Slots provide a particularly desirable form of high frequency antenna for high speed aircraft. By closing the aperture with insulating material, such an antenna does not affect the streamlining of the plane. Slots are however, limited to use at relatively high frequencies because to radiate efficiently the slot length must be of the order of a half wavelength or more, and this is a reasonable dimension only at very high frequencies.

9.7 MICROSTRIP ANTENNA

Generally **microstrip** or **patch antennas** are used in space craft or aircraft applications. In the aircraft, where size, weight, cost, performance and aerodynamic profiles are constraints such low profile antennas are used.

These antennas can be flash mounted to metal or even other existing surfaces; they require the space only for the feed line which is normally placed behind the ground plane. The disadvantages of microstrip or patch antennas are as follows:

(*i*) Very narrow band frequency (*ii*) Low efficiency

These antennas are more popular for low profile application at frequencies above 100 MHz. They are fabricated normally over the dielectric-coated ground plane printed circuit board. Normally the rectangular metal patch ($t << \lambda$) placed a small fraction up wavelength ($n << \lambda$) above a ground plane. Any dielectric substrate sheet referred as substrate is used to separate the strip or patch from the ground plane. The feed lines and radiating elements is normally photo etched on the dielectric substrate as shown in Fig. (9.20). The shape of patch may be square, rectangular, circular, elliptical or any other shape, mostly square, rectangular and circular are preferred because of ease of analysis and fabrication. The feed line is also a conducting strip of smaller width. The feed line which is used normally is coaxial lines where the inner conductor of the coaxial line is attached to the radiating patch.

Linear and circular polarization can be achieved with the microstrip antennas. To obtain greater directivity arrays of microstrip elements with single or multiple feeds may be used. These antennas are considered to be low efficient and they behaves more like a cavity rather than a radiator because of the thickness of the microstrip which is very small, the wave generated within the dielectric substrate undergo reflection to some extent when they arrive at the edge of the strip that results radiation of only small fraction of the incident energy.

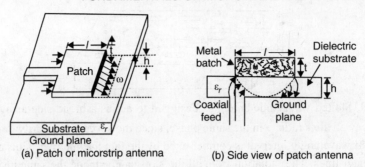

(a) Patch or microstrip antenna (b) Side view of patch antenna

Fig. 9.20. Microstrip or patch antenna.

These patch antennas acts as resonant half wavelength ($\lambda/2$) parallel plate microstrip transmission line with characteristics impedance which is equal to the reciprocal of the number n of parallel field cell transmission lines. The characteristics impedance (Z_0) of each field transmission line is equal to the intrinsic impedance of the medium.

It may be written as

$$Z_0 = \eta_i = \sqrt{\frac{\mu}{\varepsilon}} = \sqrt{\frac{\mu_0}{\varepsilon_0}}\sqrt{\frac{\mu_r}{\varepsilon_r}} \qquad \ldots 9.9$$

$(\because \mu = \mu_0 \mu_r \text{ and } \varepsilon = \varepsilon_0 \varepsilon_r)$

or $$Z_0 = 120\pi\sqrt{\frac{\mu_r}{\varepsilon_r}} \qquad \ldots 9.9$$

Refer figure (9.21), it is obvious from the plates, that the cross-ection has total 10 field cells transmission lines, hence for $\varepsilon_r = 2$ the characteristics impedance (Z_c) of pattern antenna is expressed by

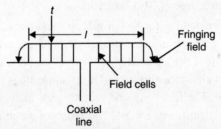

Fig. 9.21. Microstrip antenna with gap or slot divided into n-square field cells.

$$Z_c = \frac{Z_0}{n\sqrt{\varepsilon_r}} \qquad \ldots 9.10$$

or $$\frac{476.7}{10\sqrt{2}} = 26.63\,\Omega$$

$$Z_c = 26.63\,\Omega$$

Eq. 9 may written as

$$Z_c \frac{Z_0 l}{l\sqrt{\varepsilon_r}} \qquad \ldots 9.11$$

where $$n = \frac{l}{r}.$$

The radiation pattern of the patch antenna is broad. The typical value of beam area Ω_A is half of a half space or about n steradian. The directivity (D) of the patch antenna is given by

MICROWAVES ANTENNAS

$$D = \frac{4\pi}{\Omega_A} = \frac{4\pi}{\pi} = 4 \qquad \ldots 9.12$$

or it may be expressed in *dBs*, as follows :

$$D_{dB} = 10\log_{10} 4 = 6.021 dB$$

$$D_{dB} = 6.021 dB \qquad \ldots 9.13$$

The main limitation of the patch antenna is its narrow bandwidth which is normally controlled by the characteristics of the parallel plate transmission lines. By using, following measures, the bandwidth can be increased:

(i) Increased the thickness n of the parallel plate transmission line.
(ii) Use of high dielectric constant or substrate, so that the physical dimensions of the parallel plate are decreased.
(iii) By increasing inductance of the patch by cutting holes or slot in it.
(iv) By reducing VSWR using reactive component.

SUMMARY

- It understands that the system operating in the range of 1 GHz and extended up to 100 GHz, is known as microwave systems. The antenna used in these systems may be called as microwave antennas, tend to be directive *i.e.*, very high gain and narrow beam width in both the planes such as horizontal and vertical.
- **Types of Microwave Antennas**
  1. Parabolic reflector   2. Cassegrain antenna
  3. Horn antenna   4. Slot antenna
  5. Lens antenna   6. Microstrip antenna
- A parabolic reflector can be used to concentrate the radiation from an antenna located at the focus in the same way that a search light reflector produces a sharply defined beam of light.
- The parabolic reflector does this by converting the spherical waves originated by the radiator at the focus of the parabola in to a plane wave of uniforms phase across the mouth or aperture of the parabola.
- There are various types of parabolic reflectors, which can be used for various applications:
  1. Truncated or cut parabola
  2. Cylindrical parabola
  3. Cheese antenna or pill box antenna.
- The antenna placed at the focus of a parabolic is known as feed or primary radiation and its radiation pattern is called Primary pattern; the radiation pattern of aperture where illuminated by the feed is called the secondary pattern.
  There are several types of feed that can be used. Some are as follows :
  1. Dipole antenna (end fire feed).
  2. Horn feed,
  3. Parabola with Yagi antenna.
- Two types of feed arrangement are used to feed the paraboloid. One is rear feed support system and the other is the front feed arrangement.
- Using the offset feeds parabolic antenna can eliminate the aperture blocking and mismatching of feeds.
- The cassegrain principle is the adaptive to the microwave region of an optical technique, which is widely used in telescope design to obtain high magnification with a physically

short telescope. Its application in microwave reflector antenna allows the reduction in the aerial dimensions of the antenna.
- A typical horn antenna, consists of an open ended waveguide flared so that the wave inside the guide expands in an orderly manner
    (i) Sectional E-plane horn,
    (ii) Sectional H-plane horn,
    (iii) Pyramidal horn,
    (iv) Conical horn.
- It is another type of microwave antenna based on the optical principle. Lens antenna may be used in the higher frequency in excess of 3 GHz and its works in the same way as a glass lens used in optics. At lower frequencies, the lens antenna becomes bulky and heavy.
- There are mainly two types of lens antenna used such as:
    (i) Dielectric lens or W-plane metal plane lens
    (ii) E-plane metal plate lens.
- **Stepping** or zoning, of dielectric lenses which is often used to eliminate the problem of great thickness required of lenses used at lower microwave frequencies.
- The slot antenna makes use of the fact that energy is radiated when a RF field exists across a narrow slot in conducting plane.
- Generally **microstrip** or **patch antennas** are used in space craft or aircraft applications. In the aircraft, where size, weight, cost, performance and aerodynamic profiles are constraints such low profile antennas are used.
- The disadvantages of microstrip or patch antennas are as follows:
    (i) Very narrow band frequency,
    (ii) Low efficiency.
- Linear and circular polarization can be achieved with the microstrip antennas.

REVIEW QUESTIONS

1. What do you understand by term 'microwave antennas'?
2. What are the various types of microwave antennas and explain horn antenna?
3. Explain the principle of parabolic reflector and give the merits and demerits of its.
4. What is antenna feed? Discuss various types of feeds used in parabolic reflector antenna system.
5. What are the various type of parabolic reflector and explain them.
6. Enumerate the advantage and disadvantages of horn antenna.
7. Discuss the working principle of lens antenna and explain various types of lens antenna.
8. Discuss the advantages and disadvantages of lens antenna.
9. What is horn antenna and discuss various types of horn antennas.
10. Explain the working of cassegrain antenna and discuss its merits and demerits.
11. Explain the microwave strip antenna.
12. Write down the advantages and disadvantages of microstrip antenna.

CHAPTER 10

RADIO WAVE PROPAGATION – AN OVERVIEW

OBJECTIVE

- Introduction
- Radio Links
- EM Wave Propagation : Introduction
- Mechanism/Modes of Propagation
- Vertical and Oblique Incidence – Critical Frequency and Critical Angle
- Skip Distance, Skip Zone and Multiple Hop Transmission
- MUF, LUF and OUF (Usable Frequencies)
- Fading

10.1 INTRODUCTION

This chapter is tended to provide an overview of the area of Radio-wave propagation so that the reader can better appreciate the text that follows.

We have already studied the propagation of radio waves directed by any man-made structure such as transmission lines, coaxial lines wave guides, and strip or single-wire lines. In this chapter we shall discuss the unguided propagation of radio waves, that is, in terrestrial atmosphere, over and through the earth and in outer space. Unguided wave propagation is also used in geophysics, in the study of the upper atmosphere, radio astronomy, the study of the sun, stars and nebular inside and outside our Galaxy.

In all the above applications (despite the differences) of unguided radio wave propagation, the common aspect is that all use a radio link, consisting of transmitter, a receiver and a propagation medium. In radio transmission, the atmosphere around the earth or outer space is the natural propagation medium.

10.2 RADIO LINKS

There are three commonly used radio links. Figure 10.1 shows the simplest one in which transmitter and receiver located at its terminal points. In Fig. (10.1), the transmitted waves reach the receiver after reflection or scattering in the ionosphere. in other cases radio waves can reach the receiver through diffraction round the globe (Fig. 10.11), scattering the troposphere (Fig. 10.13) or some other way (space wave Fig. 10.12), figure (10.2) shows a line-or-sight radio relay system (a microwave link). In which there are number of repeaters. At each repeater, the R.F. signal is converted back to the I.F. band, amplified, translated to a microwave frequency, and restrained in the direction of the next repeater at a frequency different from the received one. Figure (10.3) shows a scatter radio link. In which the transmitted signal dies not reach the receiver R_X directly. Instead transmitted signal from T_X illuminated a man-made or natural body, C, which differs from the surrounding medium in electrical characteristics.

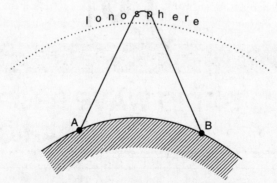

Fig. 10.1. A simple radio link (T- transmitter, R-receiver).

As a result, the body C reflects or scatters the incident radiation, and the reflected and forward-scattered signal is picked up at R_X (tropospheric scatter, ionospheric scatter, reflection from moon or artificial earth satellite used as passive reflectors). In case of Radar transmitted and receiver system is located together, C in the target or targets.

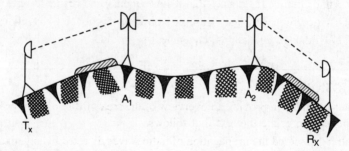

Fig. 10.2. A radio relay system (microwave link) (A_1 and A_2 are repeaters).

10.3 EM WAVE PROPAGATION: INTRODUCTION

An electromagnetic wave consists of two primary components – an electric field and a magnetic field. The electric field results from the force of voltage and the magnetic fields results from the flow of current. Although electromagnetic fields that are radiated are commonly considered to be waves, under certain circumstances their behavior makes them appear to have some of the properties of particles. In general, however, it is easier to picture electromagnetic radiation in space as horizontal and vertical lines of force oriented at right angles to each other. These lines of forces are made up of an electric field (E) and a magnetic field (H), which together makes up the electromagnetic field in space. The electric and magnetic fields radiated from an antenna from the electromagnetic field. This field is responsible for the transmission and reception or electromagnetic energy through free space. An antenna, however, is also part of the electrical circuit of a transmitter or a receiver and is equivalent to a circuit containing inductance, capacitance and resistance. Therefore, the antenna can be expected to display definite voltage and current relationships with respect to a given input. A current through the antenna produces a magnetic field and a charge on the antenna produces an electric field. These two fields combine to form the induction field.

Polarization: For maximum absorption of energy form the electromagnetic fields, the receiving antenna must be located in the plane of polarization. This places the conductor of the antenna at

RADIO WAVE PROPAGATION – AN OVERVIEW

right angles to the magnetic lines of force moving through the antenna and parallel to the electric lines, causing maximum induction. Normally, the plane of polarization of a radio wave is the plane in which the E field propagates with respect to the earth. If the E field component of the radiated wave travels in a plane perpendicular to the Earth's surface (vertical), the radiation is said to be vertically polarized, as shown in figure (10.3). If the E field propagates in a plane parallel to the Earth's surface the position of the antenna in space is important because it affects the polarization of the electromagnetic wave.

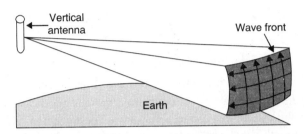

Fig. 10.3. Vertical polarization.

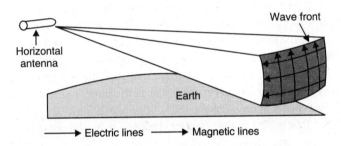

Fig. 10.4. Horizontal polarization.

When the transmitting antenna is close to the ground, vertically polarized waves cause greater signal strength along the Earth's surface. On the other hand, antennas high above the ground should be horizontally polarized to get the greatest possible signal strength to the Earth's surface. The radiated energy from an antenna is in the form of an expanding sphere. Any small section of this sphere is perpendicular to the direction the energy travels and is called a wave front. All energy on a wave front is in the phase. Usually all points on the wave front are at equal distances from the antenna. The farther the wave front is form the antenna, the less spherical the wave appears. At a considerable distance the wave front is from the antenna, the less spherical the wave appears. At a considerable distance the wave front can be considered as a plane surface at a right angle to the direction of propagation. If you know the directions of the E and H components, you can use the "right-hand rule" as shown in figure (10.5) to determine the direction of wave propagation. This rule states that if the thumb, forefinger and middle finger of the right and are extended so they are mutually perpendicular. The middle finger will point in the direction of wave propagation if the thumb points in the direction of the E field and the forefinger points in the direction of the H field. Since both the E and H fields reverse direction simultaneously, propagation of a particular wave front is always in the same direction (away from the antenna). With the atmosphere, radio waves can be reflected, refracted and diffracted like light and heat waves.

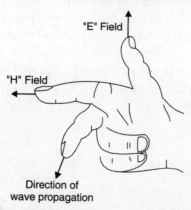

Fig. 10.5. Right-hand rule for propagation.

Reflection: Radio waves may be reflected from various substances or objects they meet during travel between the transmitting and receiving sites. The amount of reflection depends on the reflecting material. Smooth metal surfaces of good electrical conductivity are efficient reflectors of radio waves. The surface of the Earth itself is a fairly good reflector. The radio wave is not reflected form a single point on the reflector, but rather from an area on its surface. The size of the area required form reflection to take place depends on the wavelength of the radio wave and the angle at which the wave strikes the reflecting substances. When radio waves are reflected from flat surfaces, a phase shift in the wave occurs. Figure (10.6) shows two radio waves being reflected from the Earth's surface. Notice that the positive and negative alternations of radio waves (A) and (B) are in phase with each other in their paths towards the Earth's surface. After reflection takes place, however, the waves are approximately 180° out of phase from their initial relationship. The amount of phase shift that occurs is not constant. It depends on the polarization of the wave and the angle at which the wave strikes the reflecting surface. Radio waves that keep their phase relationships after reflection normally produce a stronger signal at the receiving site. Those that are received out of the phase produce a weak or fading signal. The shifting in the phase relationships of reflected radio waves is one of the major reasons for fading.

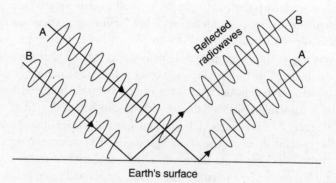

Fig. 10.6. Reflection.

Refraction: Another phenomenon common to most radio waves is the bending of the waves as they move from one medium into another in which the velocity of propagation is different. This bending of the waves is called refraction. This simple principle applies to radio waves as changes occur in the medium through which they are passing. As an example, the radio wave shown in figure (10.7) is traveling through the Earth's atmosphere at a constant speed.

RADIO WAVE PROPAGATION – AN OVERVIEW

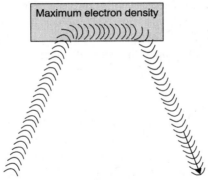

Fig. 10.7. Radiowave refraction.

As the wave enters the dense layer of electrically charged ions, the part of the wave that enters the new medium first travels faster than the parts of the wave that have not yet entered the new medium. This abrupt increase in velocity of the upper part of the wave causes the wave to bend back toward the Earth. This bending, or change of direction, is always toward the medium that has the lower velocity of propagation.

Radio waves passing through the atmosphere are affected by certain factors, such as temperature, pressure, humidity and density. These factors can cause the radio waves to be refracted.

Diffraction: A radio wave that meets an obstacle has a natural tendency to bend around the obstacle as illustrated in figure (10.8). The bending, called **diffraction**, results in a change of direction of part of the wave energy form the normal line-of-sight path. This change makes it possible to receive energy around the edges of an obstacle as shown in figure (view *A*) or at some distance below the highest point of an obstruction, as shown in figure (view *B*). Although diffracted R.F. energy usually is weak, it can still be detected by a suitable receiver.

The principal effect of diffraction extends the radio range beyond the visible horizon. In certain cases, by using high power and very low frequencies, radio waves can be made to encircle the Earth by diffraction.

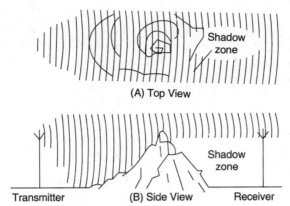

Fig. 10.8. Diffraction around an object.

The Effect of the Earth's Atmosphere on Radio Waves

This discussion of electromagnetic wave propagation is concerned mainly with the properties and effects of the medium located between the transmitting antenna and the receiving antenna. While radio waves traveling in free space have little outside influence affecting them, radio waves traveling within the Earth's atmosphere are affected by varying conditions. The influence exerted on radio waves by the Earth's atmosphere adds many new factors to complicate what at first seems to be a relatively simple problem. These complications are because of a lack of uniformity within the Earth's atmosphere. Atmospheric conditions vary with changes in height, geographical location, and even with changes in time (day, night, season, and year). Knowledge of the composition of the Earth's atmosphere is extremely important for understanding wave propagation. The Earth's atmosphere is divided into three separate regions, or layers.

Stratosphere: The stratosphere is located between the troposphere and the ionosphere. The temperature through this region is considered to be almost constant there are little water vapors present. One stratosphere has relatively little effect on radio waves because it is a relatively calm region with little or no temperature changes.

Ionosphere: The ionosphere extends upward from about 31.1 miles (50 km) to a height of about 250 miles (402 km). It contains four cloud-like layers (D, E, $F1$ and F_2) of electrically charged ions, which enable radio waves to be propagated to great distances around the Earth. This is the most important region of the atmosphere for long distance point-to-point communication.

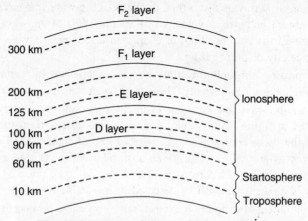

Fig. 10.9. Layers of the earth's atmosphere.

Troposphere: The troposphere is the portion of the earth's atmosphere that extends from the surface of the earth to a height of about 3.7 miles (6 km) at the north pole or the south pole and 11.2 miles 918 km) at the equator. Virtually all weather phenomena take place in the troposphere. The temperature in this region decreases rapidly with altitude, clouds from and there may be much turbulence because of variations in temperature, density and pressure. These conditions have a great effect on the propagation of radio waves.

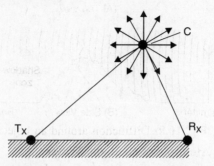

Fig. 10.10. Scatter radio link (T_X-transmitter, R_X-receiver).

It has the following properties which influence the propagation of radio waves:
1. The percentage of the gas components does not vary with height, remaining practically the same at it is at the surface of the earth.

RADIO WAVE PROPAGATION – AN OVERVIEW

2. The water vapor content sharply decreases with height and strongly dependent on the weather conditions.
3. Its temperature decreases with height. The average vertical temperature gradient of the troposphere in 6 degree per kilometer.
4. The troposphere is an inhomogeneous dielectric whose refractive index varies with height.

10.4 MECHANISM/MODES OF PROPAGATION

When an electromagnetic wave is radiated form a transmitting antenna, part of the radiated energy travels along or near the earth's surface, and another part travels from the antenna upward into space. The energy that stays close to the surface of the earth is called the **ground wave**. The energy that travels upward into space is called the **sky wave**. Energy that travels directly from the transmitting antenna to the receiving antenna, without following the surface of the earth or radiating toward the sky, is called **space wave** or **direct wave**.

10.4.1 GROUND WAVES

As just stated, the ground wave is a radio wave that travels along the surface of the earth. Ground waves are the primary mode of propagation in the LF and MF bands. The longer wavelengths in these bands tend to follow the curvature of the earth due to diffraction and actually travel beyond the horizon, as shown in Fig. (10.11). The radio wave propagation due to diffraction cover a distance not exceeding 3000 km to 4000 km. However, as the frequency increases, the ground wave is more effectively absorbed by the irregularities on the earth's surface, because hills, mountains, trees and buildings become significant relative to the transmitted wavelength. For example, at 30 kHz, the wavelength is 10 km and even mountains are relatively insignificant compared to this value. But, at 3 MHz, the wavelength is only about 100m, short enough that hills, trees, and large buildings break up and absorb the ground wave.

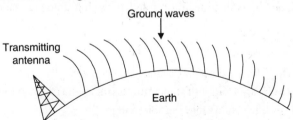

Fig. 10.11. Ground wave propagation.

One way of greatly improving ground wave coverage is to use vertical polarization. When horizontal polarization is used the electric field is parallel to the earth's surface, and any ground wave is effectively short-circuited by the conductivity of the earth. There are few communications services in the LF band. Most ground wave communications are in the MF where antenna size is more practical. In LF band full-size quarter wave antenna does present problem.

10.4.2 SPACE WAVES

Figure (10.12) illustrates the concept of space wave in which energy radiated form the transmitter travels directly in a straight line to receiving antenna. When the transmitted signal is increased above 4 or 5 MHz, the usable ground wave signal is limited to a few kilometers. Therefore, at these frequencies and above, signals can be transmitted farther using the pace or direct wave. Space wave propagation is used primarily in the VHF, UHF, and higher frequency bands (microwaves).

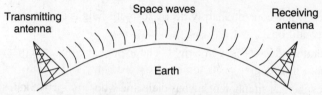

Fig. 10.12. Space wave propagation.

In general, space wave propagation is limited to line-of-sight distance. Most of the energy in radio waves at frequencies above 30 MHz moves through space in straight lines like light waves, although some of the energy at these higher frequencies follows the curvature of the earth, and good VHF reception is often obtained considerably beyond line-of-sight distances. At still higher frequencies, such as those in the UHF and SHF ranges, normal transmissions from conventional antenna are limited to line-of-sight.

When the transmitting and receiving antennas are within sight of each other, the signal is considered to be a space wave. It follows, then, that to increase the range of space wave propagation, you need only to increase the height of either or both the transmitting and receiving antennas. Example of increased antenna height is satellites and aircraft antennas. In these circumstances, propagation ranges are greatly extended.

10.4.3 RADIO HORIZON

In space wave transmission, there is a phenomenon called the radio horizon, the distance of which is about one third greater than that of the optical horizon. This phenomenon is called by refraction in the earth's lower atmosphere. The refraction occurs because the density of the earth's atmosphere decreases linearly as height increase. As a result, the top of the wave travels slightly faster than the bottom of the wave, effectively bending the wave slightly downward, where is follows the curvature of the earth beyond the optical horizon.

The radio horizon for both transmitting and receiving antennas can be calculated by the following equation:

$$D_T = 4.12\sqrt{H_T} \text{ or } D_R = 4.12\sqrt{H_R} \qquad \ldots (1)$$

Where, D_T and D_R = Radio horizon distance in kilometer

$H_T (H_R)$ = Height of transmitting (receiving) antenna in meter

When measured in miles, substitute the following:

$$D = \sqrt{2H} \qquad \ldots (2)$$

Where, D = Horizon distance in mile

H = Height of antenna in feet

The maximum space wave communications distance is the sum of the numbers obtained by using Eq. (1) for both antennas.

$$D_{max} = 4.12\sqrt{H_T} + 4.12\sqrt{H_R} \qquad \ldots (3a)$$

$$D_{max} = D_T + D_R \qquad \ldots (3b)$$

Example 1. *A receiver is located 76 km from the transmitter. The transmitting antenna is 81 km high. Find the height of the receiving antenna.*

Solution: Algebraic manipulation of Eq. (3a) gives

$$H_R = \left(\frac{D_{max} - 4.12\sqrt{H_T}}{4}\right)^2$$

$$= \left(\frac{76 - 4.12\sqrt{81}}{4}\right)^2$$

$$H_R \approx 100\,m$$

10.4.4 SKY WAVES

Ionization is the loss or gain of electrons by an atom. There are different degrees of ionization, forming several recognizable ionized layers (Fig. 10.9), and known collectively as the ionosphere. These ionized layers of the atmosphere between 50 and 400 km above the surface of the earth make sky wave reception possible. All of the energy radiated upward in the sky wave would be wasted, so far as earthbound radio communication is concerned, if it continued to travel in a straight line off into space. However, at certain frequencies and radiation angle, the ionosphere reflects radio waves much the same way a mirror reflects light. Also, radio waves at other frequencies and angles are refracted (bend) in such a manner that they return to earth because of decreasing refractive index. These type of wave propagation is known as **sky wave** or ionospheric wave, as shown if Fig. (10.13).

The amount of refraction depends on three factors:

(1) The frequency of the wave f,

(2) The density of the ionized layer (refractive index $n = \sqrt{1 - \frac{81N}{f^2}}$) and

(3) The angle at which the wave enters the ionosphere.

When conditions are just right, the wave will be refracted enough to return to earth (Fig. 10.13). The refracted radio waves may return to earth very far from the point when they originated.

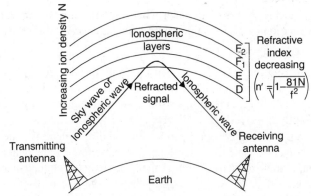

Fig. 10.13. Sky wave ionospheric wave propagation.

Long distance communications often makes use of this characteristic by sing carrier frequencies in the MF and HF bands, because waves radiated at these frequencies can be refracted back to earth with comparative ease. Almost all HF propagation, and night-time long distance MF propagation, is by ionospheric wave. However, waves at frequencies above 30 MHz are much more likely to penetrate the ionosphere and continue moving out into space. In the UHF and SHF bands, a very small percentage to the wave's energy is refracted back to earth.

10.4.5 TROPOSPHERIC SCATTER

Tropospheric scatter, commonly called troposcatter, is a special case of sky wave propagation used for frequencies higher than those in standard sky wave propagation techniques. These higher frequencies are directed to the troposphere instead of to the ionosphere.

The troposphere is the upper region of the earth's atmosphere, ranging from about 10 km to 16 km above the surface of the earth. Propagation by tropospheric scatter uses the properties of the troposphere as a reflector of UHF signals in much the same manner as the ionosphere is used to reflect lower-frequency signals. However, the troposphere only partially reflects the signal, scattering the energy, some extremely small portion of which is scattered in the forward direction or in the direction of the receiver. Local in homogeneities in the troposphere cause scattering of radio wave, and scattered waves can cover ranges up to 1000 km away from the transmitter. It is a very reliable technique for extending UHF signal communications paths beyond the horizon; however, it is not cost effective.

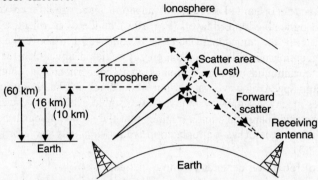

Fig. 10.14. Tropospheric scatter propagation.

The scattering process, illustrated in figure (10.14) shows directional transmitting and receiving antennas aimed so that their beams intercept in the troposphere. Note that much of the energy is scattered in undesired directions and is lost.

10.5 VERTICAL AND OBLIQUE INCIDENCE – CRITICAL FREQUENCY AND CRITICAL ANGLE

The highest frequency returned to earth when radiated upward in the vertical direction is called the **critical frequency** f_c. Its value could be calculated by the relation

$$f_c = 9\sqrt{N_{max}},$$

where f_c is expressed in MHz and N_{max} in the ionic density in m^3. The value of the critical frequency depends on the condition of the ionosphere. Since ionospheric conditions change form hour to hour, day to day, month to month, season to season, and year to year, the critical frequency also changes constantly.

Lowering the radiation angle from the exact vertical direction allows the wave to travel longer through the ionized layer, causing a greater degree of refraction. In practical terms, this means that, with one limitation, signal frequencies higher than the critical frequency can be refracted back to earth. The limitation is that, for any given frequency, there is a **critical angle** above which the signal will not be refracted enough to return to earth. Figure (10.15) shows how signals above the critical angle penetrate the ionospheric layer while signals below the critical angle return to earth.

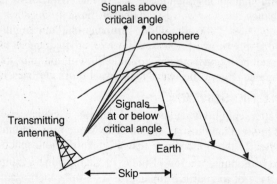

Fig. 10.15. Signals above the critical angle penetrate the ionosphere.

RADIO WAVE PROPAGATION – AN OVERVIEW

10.6 SKIP DISTANCE, SKIP ZONE AND MULTIPLE HOP TRANSMISSION

In Fig. (10.15) note that as the radiation angle decreases the distance signal wave travels over the earth increase. This distance is known as the **skip distance** and is an important consideration in long distance communications. At high frequencies, there may be long distances between the end of the usable ground wave signal and the reappearance of the reflected sky wave, as illustrated in Fig. (10.16). This is a no-signal zone known as **skip zone.**

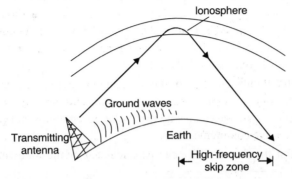

Fig. 10.16. Skip Zone.

The skip distance can be maximized by using the lowest radiation angle possible and the highest frequency that will be refracted at that angle. If the sky wave returns to earth and strikes a good conducting surface such as salt water, it can be reflected back into the ionosphere and take a double hop. If the signal strength is sufficient, the wave will be reflected again and take another hop, as shown in figure (10.17). This concept is known as **multiple-hop transmission** and is used extensively for long-distance communications.

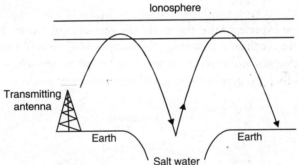

Fig. 10.17. Multiple-hop transmission.

Under the best conditions, the maximum distance of a single hop is about 3200 km. Limiting factors are the frequency used and the radiation angle, and since the radiation angle cannot be reduced below the horizon, multiple-hop transmission is used. Attenuation of the signal from the ionosphere and the ground reflection points are the primary determining factors for the distance that, multiple-hop sky wave signals can travel.

10.7 MUF, LUF & OUF (USABLE FREQUENCIES)

The maximum usable frequency (MUF) is the highest frequency that can be used for transmission between two points via sky wave (ionospheric wave) that is, the highest frequency at which a signal radiated into the ionosphere will be refracted back to earth with **usable strength.** The MUF varies between 8 MHz and 30 MHz with time of day, distance, direction, season, and solar activity.

The MUF (f_{muf}) can be calculated by the SECANT Law ($f_{muf} = f_c \sec i$) where, i, is angle of incidence and f_c is critical frequency.

A frequency lower than the MUF can actually only be used, because it will also be refracted. However, as the signal frequency decreases, the signal energy absorbed by the ionosphere increases, drastically reducing the usable signal strength. Below a certain frequency, sometimes called the **lowest usable frequency** (LUF), the RF signal will be totally absorbed by the ionosphere. Therefore, operating at the MUF will produce the maximum received signal strength.

The MUF changes constantly with atmospheric conditions, and the LUF is border-line for refraction. Therefore, to obtain the most reliable ionospheric propagation, the **optimum usable frequency** (OUF) is used. The OUF is well above the LUF but far enough below the MUF that the signal is not appreciably affected by the constant atmospheric changes, thus providing relatively reliable ionospheric wave communications.

Because the usable frequencies are constantly changing, charts are available that predict them for every hour of the day, over any path on the earth, and for any month of the year. The predictions on these charts are based on extensive solar observations and can be used to optimize sky wave/ionospheric wave communications.

When earth is flat the skip distance can be calculated by the following equation

$$D_{skip} = 2H\sqrt{\frac{f_{muf}^2}{f_c^2} - 1}$$

where, H = height of ionosphere layer,
f_{muf} = maximum usable frequency and
f_c = critical frequency for the layer under consideration.

10.8 FADING

It is our common experience that the received signal strength varies. If the changes in signal strength are small, the receiver's automatic gain control (AGC on AVC) may compensate. On the other hand, if the changes are large, the signal may be lost completely. This variation in signal strength is called **fading** and may be the result of multipath reception as shown in Fig. (10.18) and Fig. (10.19).

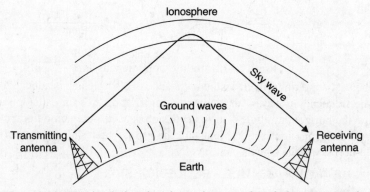

Fig. 10.18. Received signal arrives via ground and sky waves paths.

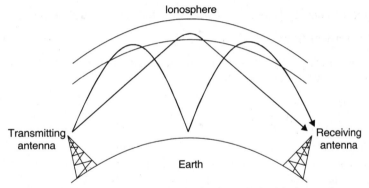

Fig. 10.19. Received signal arrives via both single and double hope sky waves paths.

Multipath fading exists in more or less in all type of communication systems.

Several methods can be used to reduce fading such as installing highly directive receiving antenna, stagger stacking etc.

SUMMARY

- Radio waves may be reflected form various substances or objects they meet during travel between the transmitting and receiving sites.
- Another phenomenon common to most radio waves is the bending of the waves as they move from one medium into another in which the velocity of propagation is different. This bending of the waves is called refraction.
- Radio waves passing through the atmosphere are affected by certain factors, such as temperature, pressure, humidity and density.
- A radio wave that meets an obstacle has a natural tendency to bend around the obstacle. The bending, called **diffraction**.
- The Earth's atmosphere is divided into three separate regions, or layers.
- **Stratosphere:** The ionosphere extends upward from about 31.1 miles (50 km) to a height of about 250 miles (402 km).
- **Troposphere:** The troposphere is the portion of the earth's atmosphere that extends from the surface of the earth to a height of about 3.7 miles (6 km) at the North Pole or the South Pole and 11.2 miles (18 km) at the equator.
- When an electromagnetic wave is radiated from a transmitting antenna, part of the radiated energy travels along or near the earth's surface.
- The energy that stays close to the surface of the earth is called the **ground wave**. The energy that travels upward into space is called the **sky wave**. Energy that travels directly from the transmitting antenna to the receiving antenna, without following the surface of the earth or radiating toward the sky, is called the **space wave** or **direct wave**.
- One way of greatly improving ground wave coverage is to use vertical polarization.
- In space wave transmission, there is phenomenon called the radio horizon, the distance of which is about one third greater than that of the optical horizon.

$$D_{max} = 4.12\sqrt{H_T} + 4.12\sqrt{H_R}$$

- Refractive index $n = \sqrt{1 - \dfrac{81N}{f^2}}$

- Tropospheric scatter, commonly called troposcatter, is special case of sky wave propagation used for frequencies higher than those standard sky wave propagation techniques.

- The highest frequency returned to earth when radiated upward in the vertical direction is called the **critical frequency** f_c. $f_c = 9\sqrt{N_{max}}$
- For any given frequency, there is a **critical angle.** Above which the signal will not be refracted enough to return to earth.
- The radiation angle decrease, the distance the signal wave travels over the earth increases. This distance is known as the **skip distance** and is an important consideration in long distance communications.
- If the signal strength is sufficient, the wave will be reflected again and take another hop. This concept is known as **multiple-hop transmission** and is used extensively for long-distance communications.
- The maximum usable frequency (MUF) is the highest frequency that can be used for transmission between two points via sky wave (ionospheric wave) that is, the highest frequency at which a signal radiated into the ionosphere will be refracted back to earth with **usable strength.** $\left(f_{muf} = f_c \sec i\right)$
- This variation in signal strength is called fading and may be the result of multipath reception.

IMPORTANT FORMULAS

- $$D_{max} = 4.12\sqrt{H_T} + 4.12\sqrt{H_R}$$
- Refractive index $$n = \sqrt{1 - \frac{81N}{f^2}}$$
- $$f_c = 9\sqrt{N_{max}}$$
- $$\left(f_{muf} = f_c \sec i\right)$$

REVIEW QUESTIONS

1. What do you understand by radio wave propagation?
2. Describe the structure of atmosphere with the help of neat diagram.
3. What are the various layers in atmosphere? Explain them.
4. Explain various modes of propagation.
5. Write short notes on :
   - (i) Modes of propagation
   - (ii) Ground wave propagation
   - (iii) Space wave propagation
   - (iv) Sky wave propagation
   - (v) Reflection and Refraction of wave
   - (vii) Radio horizon
   - (viii) Tropospheric scatter
   - (ix) Fading
6. Differentiate between critical frequency and critical angle.
7. Explain in brief skip distance; skip zone and multiple hop transmission.
8. Explain if brief MUF, LUF and optimum usable frequencies.
9. A receiver is located 76 km from the transmitter. The transmitting antenna is 81 m high. Calculate the required height of the receiving antenna.

   [Hint: $H_R = \left[\dfrac{D_{max} - 4.12\sqrt{H_T}}{4}\right]^2$] (Ans. $H_R \cong 100\,m$)

10. What do you understand by 'radio horizon' and give the formula to calculate the height of receiver?

CHAPTER 11

GROUND WAVE PROPAGATION

OBJECTIVE

- Introduction
- Friss Power Transmission Equation and Free Space Path Loss
- Space Wave Propagation
- Surface Wave Propagation

11.1 INTRODUCTION

In Chapter-10, we presented an overview of radio wave propagation, where we introduced most of the sky concepts and basics of propagation. We have seen that there are two major classification of wave propagation, namely, ground and sky wave propagation.

Energy propagation over other paths near the earth's surface is considered to be ground wave. Ground wave may again be subdivided into space wave and surface wave. The space wave is made of the direct wave, the signal that propagates from the transmitter to receiver in straight-line, and the ground reflected wave, which is the wave arriving at the receiver after being reflected from the surface of the earth. The wave propagation in the troposphere is known as tropospheric wave, it is a space wave. The surface wave is a wave which is guided along the earth's surface.

In this chapter we shall be discussing ground wave propagation, considering ground effects, atmospheric effects, ducting, diffraction, etc.

11.2 FRISS POWER TRANSMISSION EQUATION AND FREE SPACE PATH LOSS

Consider an isotropic antenna (radiator) which is radiating average power P_T (watt) in all directions, is placed in free-space (*i.e.*, homogeneous and loss-free medium of dielectric constant unity). A receiving antenna may be positioned so that it collects maximum power from the incoming wave. Let P_R be the received power at the receiver in Watts, one can derive (using the concept of power flux density or Poynting vector). $\dfrac{P_R}{P_T} = G_T \ G_R \left(\dfrac{\lambda}{4\pi D}\right)^2$

or $$P_R = \dfrac{P_T G_T G_R}{\left(\dfrac{4\pi D}{\lambda}\right)^2} \qquad \ldots 11.1$$

where
G_T = maximum directivity gain of transmitting, antenna
G_R = maximum directivity gain of the receiving, antenna
λ = wave length of the signal radiated in meters = c/f and
D = distance between the antennas.

The Eq. (1) is known as **Friis transmission equation**. It is apparent from the Eq. (1) that the received power is proportional to the gain of both antennas and decrease as $1/D^2$. Thus one can note from the Eq. (1) that larger the denominator, smaller the value of received power P_R and hence we may write

$$L_p = \left(\frac{4\pi D}{\lambda}\right)^2 \qquad \ldots 11.2$$

and is known as **free-space path loss** or **spatial attenuation**. It is defined as the loss incurred by an EM wave as it propagates in a straight line (line-of-sight) through free-space (i.e., vacuum) with no absorption or reflection of EM energy from near by objects.

Converting to *dB* yields

$$L_{p(dB)} = 10\log_{10}\left(\frac{4\pi fD}{c}\right)^2 \quad ; \quad \left(\frac{1}{\lambda} = \frac{f}{c}\right)$$

$$= 20\log_{10}\left(\frac{4\pi fD}{c}\right)$$

$$= 20\log_{10}\frac{4\pi}{10C} + 20\log_{10} f + 20\log_{10} D \qquad \ldots 11.3a$$

when the frequency is given in MHz and the distance in km, then Eq. (3a) will take the form

$$L_{p(dB)} = 32.4 + 20\ \log_{10} f_{(MHz)} + 20\ \log_{10} D \qquad \ldots 11.3b$$

and when the frequency is given in GHz and the distance in km, then Eq. (3a) will take the form

$$L_{p(bB)} = 92.4 + 20\log_{10} f_{GHz} + 20\ lof_{10}D_{(un)} \qquad \ldots 11.3c$$

It is clear from Eq. (3) that the EM wave (energy) gets attenuated in strength even while travelling through ideal space. The first term of Eq. (3) is standard unit distance loss (it is 32.5 dB for 1 km from transmitter when frequency is in MHz and it is 92.4 dB when frequency is in GHz). The second term of Eq. (3) shows the dependence of free-space path loss on operating frequency. It increases with increase in frequency. One can see from Eq. (3b) that for the same distance an EM wave at 100 MHz will be attenuated by 40 dB more than an EM wave at 1 MHz, The third term of Eq. 3 shows the dependence of free-space path loss on distance from the transmitter antenna. An EM wave propagating from a transmitting antenna the free-space path loss at 10 km distance will be more by 6 dB as compared to the attenuation at 5 km from the transmitter. If the distance is increased 10 times from the reference level, the free-space path loss will increase by 20 dB.

11.3 SPACE WAVE PROPAGATION

Friis power transmission equation and free-space path loss equations were derived assuming that radiated EM energy is propagating through free-space (loss free medium) in straight line (line-of-sight) from the transmitting antenna to the receiving antenna (receiver). The waves which travel in a straight line from the transmitter to receiver are known as **direct waves or space waves**. These waves, when travels in troposphere, are called **tropospheric waves**. Free space, however, is an idealization, in the real world; the earth's surface and atmosphere have major effects on the space wave propagation.

GROUND WAVE PROPAGATION

11.3.1 GROUND EFFECT

The most obvious effect of the presence of the ground on radio wave propagation is reflection from the earth surface (land or sea). For elevated transmitting and receiving antennas within line-of-sight of each other (Fig. 11.1), radiated signal from the transmitting antenna arrives at the receiving antenna via two separate waves (paths).

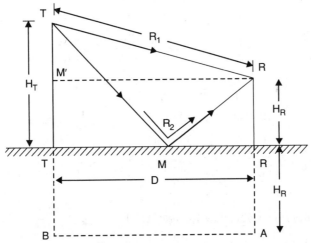

Fig. 11.1. Geometry illustrating radio-wave propagation over, a plane reflecting surface.

One is the direct wave (direct path TR) from transmitting antenna to receiving antenna; the other is reflected wave (path TMR) from the ground, which is smooth, plane and perfectly reflecting flat earth.

The reflected wave travels a longer distance than the direct wave and suffers a significant phase difference relative to the wave before reflection. The reflection from the earth's surface also affects the amplitude and phase of the reflected wave relative to the direct wave. The nature of reflection dependents in a complicated way constitution of earth's reflecting layers, the angle of incidence of the wave and whether the wave is horizontally or vertically polarized. **It has been experimentally found for a wide range of condition that the amplitude change in negligible and the phase change is 180°.**

The received signal at the receiving antenna will be vector sum of the two wave components and, depending on the relative phases of the two waves, may be greater or less than the direct wave alone. When earth is rough the reflected wave tends to be scattered and may be much reduced in amplitude compared with smooth flat earth reflection.

Field Strength

Now we calculate the field strength received at the receiving antenna (Fig. 1) assuming a plane, smooth, perfectly reflecting plane earth. The modification of the field strength of (measured in Volts/meter) caused by the presence of the earth's surface may be expressed by the ratio

$$\eta = \frac{\text{Field strength at receiving point in presence of earth's surface}}{\text{Field strength at receiving point if in free space}} \qquad \ldots 11.4$$

The geometry is shown in Fig. (11.1). The transmitting antenna (T) is located at a height H_T above the plane surface and receiving antenna (R) is at height H_R. The ground distance between transmitting antenna and receiving antenna is D. R_1 direct ray path TR and R_2 is surface reflected path TMR (Fig. 11.1). E_0 is field strength at R due to direct wave.

It is assumed in this analysis that the path lengths of the direct and surface-reflected signals are approximately equal so that the amplitudes of the two signals are essentially the same, except for any loss of signal suffered on reflection from the earth's surface. If reflection coefficient at the ground is Γ, then the magnitude of ground reflected wave is ΓE_0. The two waves combine at the receiving point (R) vectorially. Thus the resultant field at receiving point (R) is given by

$$E_R = E_0\left(1+\Gamma e^{-j\theta}\right) \qquad \ldots 11.5$$

where θ is total phase difference which is the sum of the phase difference caused due to path difference ($R_2 - R_1$) and phase difference of π due to reflection from the ground, i.e.,

$$\theta = (R_2 - R_1)\frac{2\pi}{\lambda} + \pi \qquad \ldots 11.6$$

Considering right angle ΔTMR (Fig. 1) we can write path length R_1

$$R_1 = \left[D^2 + (H_T - H_R)^2\right]^{1/2} \qquad \ldots 11.7$$

$$= D + \frac{(H_T - H_R)^2}{2D}$$

and similarly from the right angle triangle ΔTAB

$$R_2 = \left[D^2 + (H_T - H_R)^2\right]^{1/2} \qquad \ldots 11.8$$

$$= D + \frac{(H_T + H_R)^2}{2D}$$

$\therefore$ Path difference between direct wave and surface reflected wave is Path difference

$$= R_{@} - R_1 = D + \frac{(H_T + H_R)}{2D} - D - \frac{(H_T - H_R)}{2D}$$

$$= \frac{2H_T H_R}{D} \qquad \ldots 11.9$$

The phase difference (α) caused due to path difference is

$$\alpha = \frac{2\pi}{\lambda} \text{ (path difference)}$$

$$= \frac{2\pi}{\lambda}\left(\frac{2H_T H_R}{D}\right) = 4\pi\frac{H_T H_R}{D} \qquad \ldots 11.10$$

$\therefore$ To a phase difference

$$\theta = \frac{4\pi H_T H_R}{D} + \pi \qquad \ldots 11.11$$

The Eq. (5) for resultant field at receiving point R is expanded as

$$E_R = E_0\left[1+\Gamma(\cos\theta\, 0 - i\sin\theta)\right]$$

$$|E_R| = E_0\sqrt{(1+\Gamma\cos\theta)^2 - (i\,\Gamma\sin\theta)^2}$$

$$= E_0\sqrt{1+\Gamma^2 + \cos^2\theta + 2\Gamma\cos\theta + \Gamma^2\sin^2\theta}$$

$$= E_0\sqrt{1+\Gamma^2 + 2\Gamma\cos\theta} \qquad \ldots 11.12$$

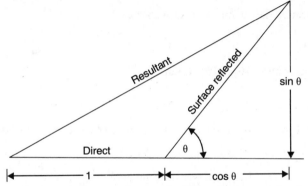

Fig. 11.2. Vector addition of direct and ground reflected wave, each of unity amplitude, with a phase difference θ.

If the earth assume to be perfect reflector $\Gamma = 1$, in that case

$$|E_R| = E_0\sqrt{2 + 2\left(2\cos^2\frac{\theta}{2} - 1\right)}, \left[\cos\theta = 2\cos^2\frac{\theta}{2} - 1\right]$$

$$= 2E_0 \cos\frac{\theta}{2}$$

$$= 2E_0 \cos\frac{(\alpha + \pi)}{2} = 2E_0 \sin\frac{\alpha}{2}$$

$$= 2E_0 \sin\left(\frac{4\pi H_T H_R}{2\lambda D}\right) \qquad \ldots 11.14$$

$$\therefore \quad \eta^2 = \left(\frac{E_R}{E_0}\right)^2 = 4\sin^2\frac{2\pi H_T H_R}{\lambda D} \qquad \ldots 11.15$$

Here the value of η^2 is the signal power density (W/m²) at the receiving antenna compared to the power density that would have been at the receiving point, if it were in free space.

When $D >> H_T, H_R$

$$\sin\frac{2\pi H_T H_R}{2D\lambda} = \frac{4\pi H_T H_R}{2D\lambda}$$

$$\therefore \quad |E_R| = 2E_0 \frac{4\pi H_T H_R}{2D\lambda} = E_0 \frac{4\pi H_T H_R}{D\lambda} \qquad \ldots 11.16$$

If E_f = field strength of the direct wave, i.e. free space field strength at unit distance

$$E_0 = \frac{E_f}{D}$$

$$E_f = 7\sqrt{P} \text{ volt / metre}$$

where P is effective power radiated, in watt

$$\therefore \quad E_0 = \frac{7\sqrt{P}}{D} \qquad \ldots 11.17$$

Substituting this value is Eq. (16), we obtain

$$|E_R| = \frac{7\sqrt{P}}{D} \cdot \frac{H_T H_R}{D\lambda} = 88\sqrt{P} \frac{H_T H_R}{\lambda D^2} \; volt/metre \qquad \ldots 11.18$$

Thus, one see from Eq. (18), for a propagation of space wave the effect of earth

$E_p \propto \sqrt{P}$, effective power radiated

$\propto H_T$, height of transmitting antenna

$\propto H_R$, height of receiving antenna

$\propto \dfrac{1}{\lambda}$, inversely with λ

$\propto \dfrac{1}{D^2}$, inversely with square of the distance from the transmitting antenna

Limitations

The above analysis is based on the assumption that both the direct and surface reflected rays (waves) are propagated in straight lines at constant velocity. In reality, none of these conditions hold. Due to atmospheric refraction (section 11.3.2) the paths of the direct and surface refracted waves are curved toward the earth, as shown in Fig. (11.3), which also shows the ray paths (the broken lines) without refraction (straight line propagation). Therefore, the path difference will not be the same for straight and curved paths.

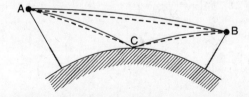

Fig. 11.3. A direct and surface reflected wave in real atmosphere.

The effect a fluctuations in the troposphere on space wave/Tropospheric wave propagation/Fading

In actual, the state of troposphere is varying continually, affecting the conditions of propagation, and as a consequence, the received field. In Fig. (11.4), the received resultant signal is due to three waves

(i) Direct

(ii) Ground reflected and

(iii) Reflected wave from reflection or scattering from local irregularities in the troposphere. The local irregularities also affect the velocities of these waves, as a consequence, the resultant phase difference between two rays. Phase fluctuations lead to fluctuations in the resultant field known as **fading**.

GROUND WAVE PROPAGATION

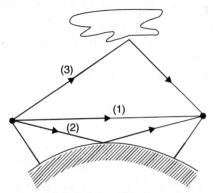

Fig. 11.4. Origin of Fading.

11.3.2 ATMOSPHERIC EFFECT (NORMAL REFRACTIVE CONDITIONS)

When EM wave (radiated by transmitting antenna) travels in troposphere (earth's atmosphere closest to the earth's surface), wave propagates beyond the line-of-sight (geometrical horizon, Fig. 11.5. This phenomenon is caused by refraction in earth's lower atmosphere, i.e., troposphere,

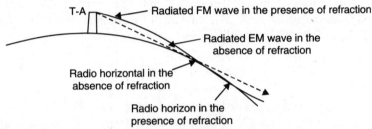

Fig. 11.5. Extension of radio horizon due to refraction of EM waves by the atmosphere/troposphere.

which is inhomogeneous and its density decreases with altitude. According to the International Telecommunications Union (ITU), the average value of the earth's surface refractive index (μ) is 1.000315. Since we are dealing with very small changes in (μ) it is more convenient to use a modified parameter called the **refractivity** (N), which is defined as

$$\boxed{N = (\mu - 1) \times 10^6 \text{ Units}} \qquad \ldots 11.19$$

Thus, as index of refraction $\mu = 1.000315$ corresponds to a refractivity $N = 315$. The refractivity can be calculated through the relation

$$\boxed{N = (\mu - 1) \times 10^6 = \frac{77.6}{T}\left(P + 4810\frac{e}{T}\right)} \qquad \ldots 11.20$$

when p = pressure in mill bar, mbar (1 mm Hg = 1.3332 m bar)
              T = absolute temperature, K
              e = vapour pressure of water vapour, m bar

It is clear from Eq. (20) that atmospheric refractivity depends on the pressure, temperature and amount of water vapour present.

The refractivity index (N) decreases with height (around 40 N units for every increase of 1 km height for standard atmosphere). The decrease of N means that the velocity of EM wave propagation increases with height, causing the EM waves to bent or refracted, and propagates beyond line-of-sight, i.e., produces a virtual increase of radio-horizon distance (Fig.11.5). Since, there is small change of refractivity with height, the magnitude of the atmospheric refractivity at some particular height is not as important, it is the **gradient of refractivity that causes the waves to bent or refracted**.

The gradient of the refractivity (N), within the troposphere remains constant

$$\frac{dN}{dH} = -4.3 \times 10^{-2} \text{ N per metre}(-43N \text{ per km}) \qquad \ldots 11.21a$$

The gradient of the refractivity (N), within the standard atmosphere is given by

$$\frac{dN}{dH} = -4.3 \times 10^{-2} \text{ N per metre}(-43 N \text{ per km}) \qquad \ldots 11.22b$$

Thus, we find that gradient of N for troposphere and standard atmosphere are nearly same. However, for practical purposes it is often taken as $\frac{dN}{dH} \approx -4 \times 10^{-2} m^{-1} = -40 N$ per km for standard atmosphere ...11.21c

From Eq. (20) it follows that N tends to zero to the monotonic decrease in pressure and humidity with height. Figure (11.6) shows idealized profile of refractive (N) with altitude, under actual condition, there may be marked departure from the idealized profile, since troposphere is greatly affected by the weathers conditions. In case of temperature inversions, changes in N are significantly different from the standard propagation, which results into formation of ducts.

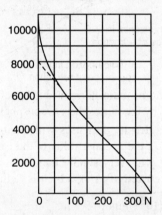

Fig. 11.6. An idealized profile of refractivity N with altitude.

11.3.3 EXPRESSION FOR THE RADIUS OF CURVATURE OF WAVE PATH IN THE TROPOSPHERE (LOWER ATMOSPHERE)

Due to atmospheric refraction (refraction in troposphere) the wave path is curve (Figs. 11. 4 & 11. 5). For simplicity, we shall neglect the effect of curvature of the earth's surface and assume that the troposphere consists of strata or planes each with a constant value of refractivity N and parallel to the flat earth. We consider two such planes (Fig. 11.7) spaced a distance dH apart. The refractive index of the lower plane or surface is μ and that of the upper surface, $\mu + d$.

Suppose a wave, incident upon the lower surface at an angle and refracted along the distance dH, it will fall upon the upper surface at an angle $\psi + d\psi$. Since at point B the element of the wave path is turned through an angle d relative to the element of the wave path at point A. The same angle is obtained between the normals to these path elements, that is, the angle at the centre O of the curvature (Fig. 11.7).

GROUND WAVE PROPAGATION

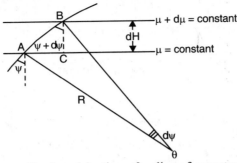

Fig. 11.7. Geometry for determination of radius of curvature of curved path due to atmospheric refraction.

The required radius of curvature, R, is given by

$$R = \frac{AB}{d\psi} \text{ metre}(m) \qquad \ldots 11.22a$$

From the $\triangle ABC$, we get

$$AB = \frac{dH}{\cos(\psi + d\psi)} = \frac{dH}{\cos\psi} m \qquad \ldots 11.22b$$

Hence

$$R = \frac{dH}{\cos\psi \, d\psi} m \qquad \ldots 11.22c$$

The law of refraction must hold at any point along the propagation path in media with a gradually varying refractive index. Thus, we may write

$$\mu \sin\psi = (\mu + d\mu)\sin(\psi + d\psi) \qquad \ldots 11.23$$

Expanding the right hand term and neglecting the second order terms (infinitesimals), we get

$$\mu \sin\psi = \mu \sin\psi + \mu \cos\psi \, d\psi + \sin\psi \, d\mu \qquad \ldots 11.24a$$

$$\cos\psi \, d\psi = -\sin\psi \, \frac{d\mu}{\mu} \qquad \ldots 11.24b$$

Substituting it in Eq. (22c) gives

$$R = \frac{\mu}{\sin\psi \left(\frac{-d\mu}{dH}\right)} m \qquad \ldots 11.25$$

We can set $\mu = 1$, without any determinant to the accuracy of calculation ($\mu = 1.000315 \approx 1$) and also we are more interested in waves propagated at small elevation angles for which $\sin\psi \approx 1$ (very nearly). As a result, the expression of Eq. (25) is further simplified as follows

$$R = \frac{1}{-\frac{d\mu}{dH}} = \frac{10^6}{-dH/dH} m \qquad \ldots 11.26$$

$$= -\frac{dH}{d\mu} = -\frac{dH}{dN} \times 10^6 \, m$$

It is apparent from Eq. (26) that the radius of curvature of the wave path is decided by the gradient of refractive index (refractivity) with height, and not by its absolute value. The minus sign of the derivative implies that the radius of curvature will be positive, that is, the propagation path will be convex only when the refractive index decreases with altitude.

If we take $\dfrac{dN}{dH} = -4 \times 10^{-2}$, We find (Standard atmosphere)

$$R = \dfrac{10^6}{-4 \times 10^{-2}} = -2.5 \times 10^7 \, m = -25{,}000 \, km.$$

In the standard atmosphere (troposphere) for which gradient $\dfrac{dN}{dH}$ is fixed, $\left(-4 \times 10^{-12} \, m^{-1}\right)$ radio waves propagate at small elevation angles will trend in arcs of a circle whose radius is 25,000 km.

11.3.4 EFFECTIVE EARTH'S RADIUS CONCEPT

To account for atmospheric refraction to tropospheric waves is to assume an effective radius of curvature of the earth which is bit greater than the actual radius and a uniform atmosphere in which waves propagate in straight lines rather than along curved path (Fig. 11.8).

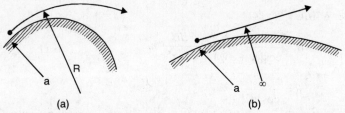

Fig. 11.8. (a) Actual path over the actual earth (b) Path over the effective earth.

In analytical geometry, the relative curvature is defined as the difference $\dfrac{1}{a} - \dfrac{1}{R}$. The relative curvatures for the situations shown in Fig. 11.8 (a) and in Fig. 11.8 (b) are equal. As such, we get

$$\dfrac{1}{a} - \dfrac{1}{R} = \dfrac{1}{a'} - \dfrac{1}{\infty} \qquad \ldots 11.27$$

Hence the effective radius of the earth

$$a' = \dfrac{a}{1 - \dfrac{a}{R}} \, m \qquad \ldots 11.28$$

Substituting the value of R form the Eq. (26) in Eq. (28), we get

$$a' = \dfrac{a}{1 + a \dfrac{dN}{dH} \times 10^{-6}} \, m \qquad \ldots 11.29$$

and

$$k' = \dfrac{a'}{a} = \dfrac{1}{1 + a \dfrac{dN}{dH} \times 10^{-6}} \qquad \ldots 11.30a$$

$$k' = \dfrac{1}{\left(1 + \dfrac{ad\mu}{dH}\right)} \qquad \ldots 11.30b$$

where k' is the ratio of the effective to the true earth's radius and is known as the **effective earth's radius factor**.

Using the value of $\dfrac{dN}{dH} = -4 \times 1^{-2} \, m^{-1}$ (for standard atmospheric refraction) and the true earth's radius $a = 6.37 \times 10^{-6} \, m$ in Eq. (30a), we get

GROUND WAVE PROPAGATION

$$k = \frac{4}{3}$$

and $a' = 8,500 \text{ km}$

Thus for standard atmospheric refraction the effective earth's radius is 4/3 times the actual earth's radius. In other words the term standard refraction is used to signify a value of $k = 4/3$ with the refractivity decreasing uniformly with altitude with a gradient of

$$\frac{dN}{dH} = -4 \times 10^2 \, m \quad \text{or} \quad \frac{dN}{dH} = -40 \text{ per km}$$

or

$$\frac{d\mu}{dH} = -4 \times 10^{-9} \, m \quad \text{or} \quad \frac{d\mu}{dH} = -40 \times 10^{-6} \text{ per km}$$

The $\frac{4}{3}$ rd earth radius represents an average and average and should not be used where precision is important.

11.3.5 RADIO HORIZON DISTANCE

In the line-of-sight/space wave propagation, the radio horizon distance depends upon the heights of the transmitting and receiving antennas. When a transmitting antenna is at height, H_T above the earth surface and receiving antenna is at height H_R above the earth as shown in Fig. (11.4), the actual wave paths, from the transmitting antenna to the receiving antenna, are curved paths due to atmospheric refraction (Fig. 11.5). The concept of effective earth radius, allow us to take the wave path as straight lines instead of being curved as they are actually. From simple geometrical consideration of Fig. 11.9, the radio-horizon distance D is

$$\boxed{D = \sqrt{2ka} \left(\sqrt{H_T} + \sqrt{H_R} \right)} \quad \ldots 11.31$$

where ka is the effective earth's radius and heights of both antennas above the earth are assumed to be much small compared to the real earth's radius a. Substituting the values of $k = \frac{4}{3}$ for standard atmospheric refraction and $a = 6.37 \, 10^6$ m in Eq. 31, yields

$$D(km) = 4.12 \left(\sqrt{H_{T(m)}} + \sqrt{H_{R(m)}} \right) \quad \ldots 11.32$$

For example if H_T and H_R are 100 meter each the value of radio-horizon D is 82.4 km.

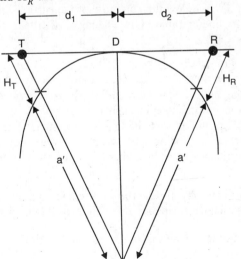

Fig. 11.9. Determining radio horizon distance with both antennas elevated with effective earth radius $a' = ka$.

11.3.6 FORMS OF ATMOSPHERIC REFRACTION AND FORMATION OF DUCTS

Some times weather conditions may lead to wide changes in refractivity gradient with altitude from the standard one described earlier. Obviously it can cause significant changes in radio-wave propagation. Such conditions are known as **anomalous or nonstandard propagation**.

In this section we shall consider several forms of refraction in the troposphere. Our classification is based on the value of $\frac{dN}{dH}$, which effects radio wave propagation. Broadly, there are seven different refractive conditions depending on the value of dN/dH, and which are listed in Table 11.1.

Sub-refraction

Sometimes weather conditions may be such that refractivity gradient may increase with height (positive gradient) instead of the more usual decrease, the propagating rays (waves) would curve upward (Table 11.1 where actual path is also shown) and the radio-range would decrease as compared to normal conditions. This type of refraction is known as sub refraction. Sub refraction can occur when warm, moist air flows over a cool ocean surface or coyer a cooler air mass is just above the ocean.

No refraction

Sometimes weather conditions may be such that N will remain fixed in an interval of height (uniform atmosphere). This will be zero refraction (second line of Table 11.1).

Normal and Standard Refraction

When the refractivity gradient $\frac{dN}{dH}$ is between 0 and -79 N/km normal refraction occurs, *i.e.*, wave paths are curved toward the earth and there is an increase in the range of propagation. When its value is -40 N/km, standard refraction occurs as defined earlier.

Critical Refraction

When the refractivity gradient $\left(\frac{dN}{dH}\right)$ equals -157 N/km, the effective earth's radius as given by Eq. (29) takes infinitely large values, that is, earth reduces to a plane. Under such conditions, a horizontally directed ray (wave) will be propagated at a fixed height above the earth's surface. That is the wave will follow the curvature of the earth. This type of refraction is known as critical refraction (fifth line of Table 11.1). The range of propagation is significantly increased.

Super-refraction

When refractivity gradient lies between -79 and -157 N/km, result in what is called **super-refraction**, in which the bending of rays is more pronounced than in critical refraction (Table 1). With super-refraction, the radius of curvature of the ray path is smaller than the earth's radius, and the waves leaving the transmitting antenna at small angles of elevation will undergo total internal reflection in the troposphere and return to earth at some distance from the transmitter (Table-1) on reaching the earth's surface and being reflected it, the waves can skip large distances, thereby giving abnormally large ranges beyond the line-of-sight due to multiple reflections.

Trapping or ducting

Weather conditions can some times produce a temperature inversion, where the temperature increases with height, which cause an extremely high lapse rate of refractivity with altitude, that is $\frac{dN}{dH}$ decreases much faster than normal, with increasing altitude. When the refractivity gradient $\frac{dN}{dH}$ exceeds -157 N/km, the curvature of the propagating wave exceeds the curvature of the earth

GROUND WAVE PROPAGATION

and ducts can form by the layer of air along the temperature inversion that trap the radio wave energy. Such ducts can range in height from 15-150 m or more.

Modified Refractivity

We have used the refractivity denoted by N to describe the property of the atmosphere to bend radio waves. However, the propagation engineers employ a **modified refractivity** (M), in determining effects of refraction, defined as

$$M = N + (H/a) \times 10^6 = \left(\mu - 1 + \frac{H}{a}\right) \times 10^6 \qquad \ldots 11.33$$

where H = height above the earth's surface and
         a = earth's radius (in the same units as H).

The modified refractivity M takes account of the curvature of the earth. It is useful in identifying ducting, since the trapping of radio-waves occurs for all negative gradient of M that is M decreases with height.

11.3.7 DUCTS AND DUCT PROPAGATION

Surface-Based Ducts: If the base (lower side) of the duct is at the earth's surface, it is known as **surface duct** (Fig. 11.10). These ducts are formed when the upper air in exceptionally warm and dry, compared with the air at the surface. This leads to a temperature inversion at the ground and sharp decrease in moisture, as a result refractivity gradient $\frac{dN}{dH}$ decrease much faster than normal with increasing height. Surface based ducting in more noticeable at night and usually disappears during most part of the day.

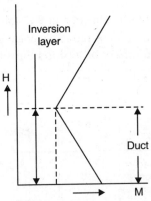

Fig. 11.10. Surface Duct.

Elevated Ducts: When the base (lower side) of the duct lies above the earth's surface, it is known as elevated duct (Fig. 11.11). These ducts can occur in the trade wind belt.

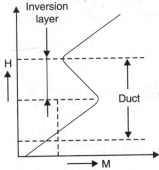

Fig. 11.11. Elevated duct.

The surface and elevated ducts can occur over land as well as over water (sea).

The duct propagation is very similar to wave propagation in a dielectric wave guide or leaky wave guide. As in Wave-guide signals of wavelengths above certain critical value (maximum value) can not propagate through guide, similarly duct propagation will not occur for the signals for which the wavelengths are above the critical value. For the most commonly encountered cases, the following relationship exist $\lambda_c \approx 0.084 H^{3/2}$

Where, λ_c = wavelength in cm, and H = duct height in meters. The duct propagation is mostly important at UHF and microwave frequencies. The UHF or microwave signals undergo specular reflections in stages/hopes within duct (Fig. 11.12).

To propagate wave within the duct, the angle the wave makes with the duct should be small, usually less than one half degree. Thus, only those waves launched nearly parallel to the duct are trapped and propagate over abnormally large ranges (as large as 1000 km).

When the receiving antenna is inside the duct, the received signal may be large. When, it is outside the duct, either below or above it, the received signal will be very small.

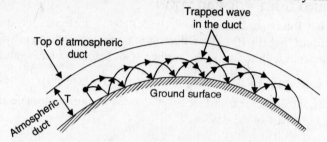

Fig. 11.12. Super-refraction in the atmospheric duct.

Table 1 Summarizes the various refractive conditions/forms of refraction.

Table 11.1. Summary of Various Forms of Refraction

S.No	Refraction	Number per km	Actual path
1	Sub-refraction	0 to 40	
2	No refraction (Uniform-atmosphere)	0	
3	Standard refraction	−40	
4	Normal refraction	0 to −79	
5	Critcal refraction	−157	
6	Super refraction	−79 ∞ −157	
7	Trapping refraction	Exceeding −157	Elevated duct / Surface duct

GROUND WAVE PROPAGATION

The main condition for the formation of a duct is a temperature inversion which may be caused by advection processes, cooling the earth's surface through radiation and compression of the air masses.

11.3.8 TROPOSPHERIC SCATTERING

The range of tropospheric propagation in the VHF, UHF and higher frequency bands can be extended to a much longer distance (around 160 km to 1600 km or so as against tropospheric space wave propagation of about 100 km) through the forward scattering in the tropospheric irregularities. The scattering process, illustrated in figure (11.13), show directional transmitting and receiving antennas aimed so that their beams intercept in the troposphere. Note that much of the energy is scattered from the irregularities within the common volume (Fig. 11.13) of the troposphere in undesired directions and is lost.

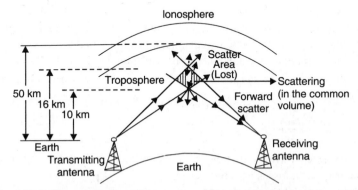

Fig. 11.13. Tropospheric scatter propagation (scattering in the common volume of transmitting and receiving antenna beam).

However, some extremely small portion of energy, which is scattered in the forward direction or in the direction of the receiver, is reaching the receiving antenna. It is a very reliable technique for extending UHF or microwave signal communication paths beyond the horizon. The scattering process also creates fading problems, caused by multipath transmission within the scattering path and by atmospheric changes.

It is inferred from experimental results that beyond the horizon the received power decreases at about the 7th or 8th power of the distance, *i.e.*, strength of the field $\alpha \dfrac{1}{d^{7-8}}$. The signal level has seasonal variations of ±10 dB which seem to be proportional to variation is k the effective earth radius factor. Another observation or rule of thumb is that at about 160 km range the field strength is approximately 57 dB below free space value, and a further loss of about 0.12 dB per 1.6 km occurs at great distances.

11.4 SURFACE WAVE PROPAGATION: INTRODUCTION

As mentioned earlier, the surface Wave is a radio wave that travels close to the earth's surface and follows the curvature of the earth due to diffraction and actually travels beyond the line-of-sight horizon. For such propagation both transmitting and receiving antennas are located almost at the earth surface. Since surface waves are propagated close to the earth's surface, the attenuation of the wave directly depends upon the surface irregularities, permittivity (dielectric constant) and conductivity of the earth. Surface wave propagation is a primary mode of propagation in the LF and MF bands. At microwave frequencies there is very little energy diffracted by the earth surface, so that microwave coverage cannot be significantly extended beyond the line-of-sight and in such frequencies, space wave propagation predominates.

Received Field Strength

The field strength at a distance (d) from the transmitting antenna due to surface wave has been obtained from the Maxwell Equations as

$$E = \frac{120\pi H_T \cdot H_R I_S}{\lambda d} \text{ Volt/meter} \qquad \ldots 11.34$$

where $120\pi = 377 =$ Intrinsic impedance of free space.

$H_T, H_R =$ Effective heights of transmitting and receiving antennas.
$I_S =$ Antenna currents.
$\lambda =$ wavelength
$d =$ distance between transmitting and receiving points.

If, however, the distance d is fairly large, the reduction in the field strength due to ground attenuation and atmospheric absorption increases and thus the actual voltage received at receiving point decreases. This results in less field strength than that shown by Eq. (34).

According to Sommerfield, the field strength for ground or surface wave propagation for a flat earth is given by

$$E_g = \frac{E_0 A}{d} \qquad \ldots 11.35$$

where $E_0 =$ Surface wave field strength at the surface of earth, at unit distance from the transmitting antenna. Earth losses not considered.

$E_g =$ Surface wave field strength at receiving point.
$A =$ Factor accounting for earth losses called attenuation factor.

Unit distance field strength E_0 depends upon
(i) Power radiated by transmitting antenna,
(ii) Directivity in vertical and horizontal planes.

If the antenna is non-directional in the horizontal plane, producing a radiated field which is proportional to the cosine of the angle of elevation (as in case of short vertical antenna), then the field at unit distance (i.e., 1 km) for a radiated power of 1 kW is given by the general formula

$$E_0 = \frac{300\sqrt{P}}{d} V/m = \frac{300\sqrt{1}}{1000} V/m = 300\, mV/m \qquad \ldots 11.36$$

where $P =$ radiated power in kilo-Watts and $d =$ distance in kilometers.

This is because, for a short vertical unipole antenna (grounded antenna), the field strength E_0 at a distance of d on a hypothetical flat perfectly conducting earth is

$$E_0 = \frac{\sqrt{90 P}}{d} \text{ Volt/meter} \qquad \ldots 11.37a$$

where $P =$ radiated power in watt and
$d =$ distance in meter.

When P is expressed in kilowatt (kW) and d in kilometer (km), then Eq. (37a) reduced to

$$E_0 = \frac{\sqrt{90 \times P \times 1000}}{d} V/m$$

Since $P = 1 kW = 1000\, watt$
and $d = 1 km = 1000\, m$

GROUND WAVE PROPAGATION

$$E_0 = \frac{300\sqrt{P}}{d} V/m$$

If d is expressed in miles, then

$$E_0 = \frac{300\sqrt{P}}{1.609} mV/mile = 186.45 \, mV/mile \qquad \ldots 11.37b$$

Thus for radiated power of 1 kW, E_0 = 300 mV/m at a distance of 1 km and E_0 = 186.45 mV/m at a distance of 1 mile. For other values of radiated power, E_0 will be proportional to the square root of the power P and will accordingly be modified in accordance with the directivity in horizontal plane, and for any added directivity when the field is not proportional to the cosine of the angle of elevation.

Surface wave Attenuation Factor A

P being effective power radiated in kilo-Watts and d, the distance in kilometers. The reduction factor A in Eq. (35), accounting for earth losses too, depends on:

(i) frequency   (ii) dielectric constant   (iii) conductivity of the earth

A is a complicated function of above factors, expressed in terms of two auxiliary variables, the numerical distance p and phase constant b.

For all value of b, the phase constant

$$A = \frac{2+0.3p}{2 = p + 0.6p^2} - \sin b \sqrt{\frac{p}{e}} \cdot e^{\frac{5}{g^p}} \qquad \ldots 11.38$$

For $b < 5°$

$$A = \frac{2+0.3p}{2+p+0.6p^2} \qquad \ldots 11.39$$

where

$$p = \frac{0.582 \, d_{km} f^2 (MHz)}{\sigma \, (mS/m)} \qquad \ldots 11.40$$

The two constant p and b are determined by the frequency, distance and dielectric characteristics of ground considered as a conductor of radio frequency currents and are given as follows.

(i) For Vertically Polarized Wave: The reduction factor A is expressed in terms of the auxiliary parameters p and b.

The parameter p and b are related as

$$p = \frac{\pi}{x} \cdot \frac{d}{\lambda} \cos b \qquad \ldots 11.41a$$

and

$$b = (2b_2 - b_1) = \tan^{-1}\left(\frac{\varepsilon_r + 1}{x}\right) \qquad \ldots 11.41b$$

(ii) For Horizontally Polarized Wave: The parameters p and b are given by

$$p = \frac{\pi d}{\lambda} \cdot \frac{x}{\cos b} \qquad \ldots 11.42a$$

$$b = 180° - b_1 \qquad \ldots 11.42b$$

where $x = \dfrac{1.8 \times 10^{12}\,\sigma}{f(Hz)}$ mhos/cm $= \dfrac{1.8 \times 10^{4}\,\sigma}{f\,(MHz)}$ mhos/m

$$b_1 = \tan^{-1}\left(\dfrac{\varepsilon_r - 1}{x}\right),\ b_2 = \tan^{-1}\dfrac{\varepsilon_r}{x}$$

b_2 = Power factor angle of the impedance offered by the earth to the flow of current.

f = frequency, in Hz

σ = conductivity of the earth, in mhos/cm.

ε_r = dielectric constant of the earth relative to air.

λ = wavelength in same unit as d

The relation between numerical distance p and phase constant b is shown in Fig. (11.14). The numerical distance p depends upon the frequency and the ground constants and the actual distance to the transmitter. It is proportional to the distance and the square of the frequency and varies almost inversely with the ground conductivity. We see that low values of p all curves tend to $A = 1.41$ and for $p > 25$, the curves also merge together (Fig. 11.14). The phase constant b is a measure of the power-factor angle of the earth.

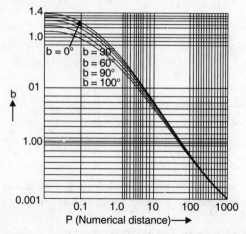

Fig. 11.14. Variation of ground attenuation factor A with numerical distance p for different values of b.

Earth offers a resistive impedance to the flow of radio frequency currents when $b = 0$ for vertical polarization and $b = 180°$ for horizontal polarization and offers a capacitive impedance when $b = 90°$ for either polarization. The study of Fig. **(11.14)** shows that

  (*i*) **For $p < 1$:** The ground attenuation factor A differs only slightly from unity and reduces slowly with the increase of p. The ground losses are then not significant. From Eq. (35) it can be seen that the field strength of the ground wave varies inversely with the distance.

 (*ii*) **For $p > 1$:** As the numerical distance p becomes greater than unity, the attenuation factor A decreases rapidly.

(*iii*) **For $p > 10$:** The attenuation factor A is almost exactly inversely proportional to actual physical distance. Hence for $p > 10$, the field strength of the ground wave is inversely proportional to the square of the distance.

GROUND WAVE PROPAGATION

The value of numerical distance p of the plane earth [Eq. (40)] surface wave attenuation factor ignores the effect of diffraction and ground permittivity. The variation of p and ground attenuation factor A is shown in Fig. (11.15). This give realistic answers for the distance less than d_{max} where

$$d_{max} = -\frac{100}{f^{1/3}} = \frac{100}{3\sqrt{f}} = km \qquad \ldots 11.46$$

where d_{max} is in km and f is in MHz. Typical value of the maximum distance is 125 km. to 90 km corresponding to a frequency range of $f = 0.5$ MHz to 1.5 MHz.

A slightly more accurate solution is obtained within the same limit of distance by incorporating the relative permittivity of the ground path. This is achieved through yet another auxiliary parameter b defined by

$$\tan b° = \frac{(\varepsilon_r - 1) f}{18\sigma} \qquad \ldots 11.47$$

with the auxiliary parameter for numerical distance reduced to

$$p = \frac{0.582 \, d \cdot f^2 \cos b°}{\sigma} \qquad \ldots 11.48$$

where d is in km, f is in MHz and σ is in mS/m.

According to Sommerfield space wave predominates at a larger distance above the earth, where as, the surface wave is the larger near the surface of the earth.

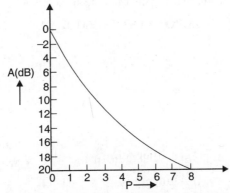

Fig. 11.15. Shows graph of the approximate values of ground wave attenuation factor A against numerical distance p based on Eq. (39).

SUMMARY

- $P_R = \dfrac{P_T G_T G_R}{\left(\dfrac{4\pi D}{\lambda}\right)^2}$

- $L_p = \left(\dfrac{4\pi D}{\lambda}\right)^2$

- $L_{p(dB)} = 32.4 + 20 \log_{10} f_{(MHz)} + 20 \log_{10} D_{(km)}$

- $L_{p(dB)} = 92.4 + 20 \log_{10} f_{GHz} + 20 \log_{10} D_{(un)}$

- Friis power transmission equation and free-space path loss equations were derived assuming that radiated EM energy is propagating through free-space (loss free medium) in straight line (line-of-sight) from the transmitting antenna to the receiving antenna (receiver). The waves which travel in a straight line from the transmitter to receiver are known as **direct waves or space waves**.

- The most obvious effect of the presence of the ground on radio wave propagation is reflection from the earth surface (land or sea).

- $|E_R| = E_0 \sqrt{2 + 2\left(2\cos^2\frac{\theta}{2} - 1\right)}, \quad \left[\cos\theta = 2\cos^2\frac{\theta}{2} - 1\right]$

- $|E_R| = 2E_0 \dfrac{2\pi H_T H_R}{2D\lambda} = E_0 \dfrac{4\pi H_T H_R}{D\lambda}$

- It is the gradient of refractivity that causes the waves to bent or refracted.
- When refractivity gradient lies between −79 and −157 N/km, result in what is called **super-refraction**, in which the bending of rays is more pronounced than in critical refraction.
- **Surface-Based Ducts**: If the base (lower side) of the duct is at the earth's surface, it is known as **surface duct**.
- When the base (lower side) of the duct lies above the earth's surface, it is known as **elevated duct**.
- $\lambda_e = 0.084\, H^{3/2}$

where = wavelength in cm, and H = duct height in meters.

IMPORTANT FORMULAS

$\Rightarrow P_R = \dfrac{P_T G_T G_R}{\left(\dfrac{4\pi D}{\lambda}\right)^2}$

$\Rightarrow L_p = \left(\dfrac{4\pi D}{\lambda}\right)^2$

$\Rightarrow L_{p(dB)} = 32.4 + 20\log_{10} f_{(MHz)} + 20\log_{10} D_{(km)}$

$\Rightarrow L_{p(dB)} = 92.4 + 20\log_{10} f_{GHz} + 20\log_{10} D_{(un)}$

$\Rightarrow |E_R| = \dfrac{7\sqrt{P}}{D} \cdot \dfrac{H_T H_R}{D\lambda} = 88\sqrt{P}\, \dfrac{H_T\, H_R}{\lambda D}\ \text{volt/metre}$

$\Rightarrow N = (\mu - 1) \times 10^6 = \dfrac{77.6}{T}\left(P + 4810\dfrac{e}{T}\right)$

where p = pressure in millibar, mbar (1 mm Hg = 1.3332 mbar)
T = absolute temperature, K
e = vapour pressure of water vapor, mbar

$\Rightarrow R = \dfrac{\mu}{\sin\psi\left(\dfrac{-d\mu}{dH}\right)}\, m$

$\Rightarrow D(km) = 4.12\left(\sqrt{H_{T(m)}} + \sqrt{H_{R(m)}}\right)$

$\Rightarrow \lambda_e = 0.084\, H^{3/2}$

GROUND WAVE PROPAGATION

$\Rightarrow E = \dfrac{120\pi H_T \cdot H_R I_S}{\lambda d}$ Volt/meter

$\Rightarrow E_g = \dfrac{E_0 A}{d}$

$\Rightarrow E_0 = \dfrac{300\sqrt{P}}{d} V/m$

$\Rightarrow E_0 = \dfrac{\sqrt{90}\,P}{d}$ volt/meter

$\Rightarrow E_0 = \dfrac{300\sqrt{P}}{1.609} mV/mile = 186.45\,mV/mile$

$\Rightarrow A = \dfrac{2+0.3p}{2 = p + 0.6p^2} - \sin b \sqrt{\dfrac{p}{e}} \cdot e^{-\tfrac{5}{g^p}}$

$\Rightarrow p = \dfrac{0.582\, d_{km} f^2 (MHz)}{\sigma\,(mS/m)}$

$\Rightarrow p = \dfrac{\pi}{x} \cdot \dfrac{d}{\lambda} \cos b$

$\Rightarrow b = (2b_2 - b_1) = \tan^{-1}\left(\dfrac{\varepsilon_r + 1}{x}\right)$

$\Rightarrow p = \dfrac{\pi d}{\lambda} \cdot \dfrac{x}{\cos b}$

$\Rightarrow b = 180° - b_1$

$\Rightarrow d_{max} = -\dfrac{100}{f^{1/3}} = \dfrac{100}{3\sqrt{f}} = km$

$\Rightarrow \tan b° = \dfrac{(\varepsilon_r - 1) f}{18\sigma}$

$\Rightarrow p = \dfrac{0.582 d . f^2 \cos b°}{\sigma}$

SOLVED PROBLEMS

Problem 1. *A television transmitting antenna mounted at height of 50 m and radiates 20 kW of power at a frequency of 60 MHz. Assuming the power is radiated in all the direction uniformly. Determine:*
  (i) *Maximum line-of-sight range*
  (ii) *The field strength at receiving antenna at height of 10 m at a distance of 10 km.*
  (iii) *Calculate the distance at which the field strength reduces to 1 mV/m.*

Solution
The distance to horizon

$$d = 3.55\left(\sqrt{H_t} + \sqrt{H_R}\right) km$$

Given,

$$f = 60\ MHz \qquad \therefore \lambda = \frac{300}{60(MHz)} = 5\ m$$

$$H_T = 50\ m$$
$$H_R = 10\ m$$
$$P = 10\ kW$$

(i)
$$d = 3.55\left(\sqrt{H_T} + \sqrt{H_R}\right) km$$
$$d = 3.55\left(\sqrt{50} + \sqrt{10}\right) km$$
$$d = 3.55\ (7.07 + 3.16) = 3.55 \times 10.23\ km$$
$$d = 36.32\ km \quad \textbf{Ans.}$$

(ii) At 10 km, the field strength is given by

$$E_t = \frac{88\sqrt{P}\ H_T H_R}{\lambda D^2} = \frac{88\sqrt{10^4} \times 50 \times 10}{5 \times (10^4)^2}$$

$$E_t = \frac{88 \times 100 \times 50 \times 10}{5 \times 10^8} = \frac{44 \times 10^5}{5 \times 10^8}$$

$$8.8 \times 10^{-3} V/m \quad \textbf{Ans.}$$

(iii) Now if the field strength is reduced to 1 mV/m at a distance D metre.

Then
$$10^{-3} = \frac{88\sqrt{10^4} \times 50 \times 10}{5 \times D^2}$$

$$D^2 = \frac{88 \times 100 \times 50 \times 10}{5 \times 10^{-3}} = \frac{44 \times 10^5 \times 10^3}{5}$$

$$D^2 = 8.8 \times 10^8\ m$$

$$D = \sqrt{8.8 \times 10^8}\ m$$

$$D = 29664.79\ km$$

$$D = 29.664\ km \quad \textbf{Ans.}$$

Problem 2: *Calculate the field strength at a distance of 20 km form a 200 kW, medium wave broadcast transmitter using short vertical antenna. Assume that field strength value is 300 mV/m at a distances of 1 km from the transmitter for a radiated power of 1 kW.*

Solution :
$$E = \frac{300\sqrt{P}}{D} mV/m$$

$$E = \frac{300\ \sqrt{200}}{20} = \frac{14.142 \times 300}{20} = 212.13\ mV/m$$

$$E = 212.13\ mV/m \quad \textbf{Ans.}$$

Problem 3: *A VHF mobile radio system, the base station transmits 200 W at 150 MHz and the antenna is 10m above ground. The transmitting antenna is a $\lambda/2$ dipole for which the gain is 1.64. Calculate the field strength at a receiving antenna of height 3m at distance of 50 km.*

GROUND WAVE PROPAGATION

Solution: $E_R = \dfrac{88\sqrt{PH_T H_R}}{\lambda D^2}$

Given, $P = 200$ W
$H_T = 10$ m
$H_R = 3$ m
$D = 50$ km
$f = 150$ MHz

So $\lambda = \dfrac{300}{150(MHz)} = 2\,m$

By putting the values in above equation, we get

$$E_R = \dfrac{88\sqrt{200} \times 10 \times 3}{2 \times (50 \times 10^3)^2} = \dfrac{37335.30}{5 \times 10^9} = \dfrac{7.46 \times 10^3}{10^9}$$

$E_R = 7.46 \times 10^{-6}$

$= 7.46\,\mu V/m$ **Ans.**

Problem 4: *What is the radio horizon of a TV antenna placed at a height of 100 m? If the signal is to be received at a distance of 60 km, what should be the height of the receiving antenna?*

Solution : $d = 4.12\left[\sqrt{H_T} + \sqrt{H_R}\right]$ km

Given, $H_T = 100\,m$

$d = 60\,km$

$60 = 4.12\left[\sqrt{100} + \sqrt{H_R}\right]$ km

$\dfrac{60}{4.12} = 10 + \sqrt{H_R}$

$\sqrt{H_R} = 14.56 - 10 = 4.56$

$H_R = (4.56)^2$

$H_R = 20.79\,m$

Problem 5: *Two aircrafts are flying at altitudes of 3000 m and 6000 m respectively. What is the maximum possible distance along the surface of the earth over which they can have effective point to point microwave communication? Assume radius of earth = $6.37\,10^6$ m.*

Solution: After accounting the earth radius of earth, the maximum possible distance on the earth surface is,

$$d = 4.12\left(\sqrt{H_T} + \sqrt{H_R}\right) km$$

Given, $H_T = 300\,m$

$H_R = 6000\,m$

$d = 4.12 \times \left(\sqrt{3000} + \sqrt{6000}\right) km$

$$d = 4.12 \times (54.77 + 77.45)$$

$$d = 544.79 \, km \quad \textbf{Ans.}$$

Problem 6: *Find the basic loss for a communication form the moon to the earth operating at 3000 MHz, assure distance between moon and earth is 384000 km.*

Solution: Path loss $= 32.45 + 20 \log_{10} f_{MHz} + 20 \log_{10} d_{km}$

$$= 32.45 + 20 \log_{10} 3000 + 20 \log_{10} 384000$$

Path loss $= 213.68$ dB

Problem 7: *Calculate the basic transmission loss in free space if*
(i) $D = 10$ km and $\lambda = 10{,}000$ m
(ii) $D = 10^6$ km and $\lambda = 0.3$ cm

Solution: Path loss $L_P = \left(\dfrac{4\pi D}{\lambda}\right)^2$

(i) $$L_P = \left[\dfrac{4\pi \times 10^3}{10 \times 10^3}\right]^2$$

$$L_P = \left[\dfrac{12.56}{10}\right]^2 = (1.256)^2$$

$$L_P = 1.58$$

(ii) $$L_P = \left[\dfrac{4\pi \times 10^6}{10 \times 10^{-2}}\right]^2$$

$$L_P = \left[\dfrac{12.56 \times 10^4}{3}\right]^2 = \left[4.19 \times 10^4\right]^2$$

$$L_P = 1755610000 = 1.7 \times 10^9$$

$$L_P = 1.7 \times 10^9 \quad \textbf{Ans.}$$

Problem 8: *Find the range of LOS system when the receiving and transmitting antenna height are 5 m respectively. Take the effective earth's radius into consideration.*

Solution: $$d = 4.12\left(\sqrt{H_T} + \sqrt{H_R}\right) \, km$$

$$d = 4.12\left(\sqrt{50} + \sqrt{5}\right) km$$

$$d = 4.12 \times 9.31 = 38.36 \, km$$

$$d = 38.36 \quad \textbf{Ans.}$$

Problem 9: *A communication link is to be established between two stations using half wave length antenna for maximum directive gain. Transmitter power is 5 kW and distance between transmitter and receiver is 100 km. What is the maximum power received by receiver? The frequency of operation is 300 MHz.*

Solution: The maximum power received power by antenna is

$$\dfrac{P_R}{R_T} = G_T G_r \left[\dfrac{\lambda}{4 \, \lambda \, D}\right]^2$$

GROUND WAVE PROPAGATION

where
P_R = Received power by antenna
P_T = Transmitted power
G_R = Directivity of receiver = 1.64
G_T = Directivity of transmitter = 1.64
λ = Wavelength = $\dfrac{300}{f_{MHz}} = \dfrac{300}{300} = 1\,m$

$P_T = 5 \times 10^3\,W$

$D = 100 \times 10^3\,m$

Hence, $P_{r(max)} = P_T\,G_T\,G_R \left[\dfrac{\lambda}{4\pi D}\right]^2$

$P_{r(max)} = 5 \times 10^3 \times 1.64 \times 1.64 \times \left[\dfrac{1}{4\pi \times 100 \times 10^3}\right]^2$

$= 13.45 \times 10^3 \times \dfrac{1}{16 \times (3.14)^2 \times (100)^2 \times 10^6}$

$= \dfrac{13.45 \times 10^3}{157.76 \times 10^3 \times 10^6} = 85.26 \times 10^{-9}\,W$

$P_{r(max)} = 85.26 \times 10^{-9}\,W$ **Ans.**

REVIEW QUESTIONS

1. Briefly describe the three classification of radio wave propagation and list the primary frequency bands for each.
2. How can ground wave coverage be increased?
3. How can the range of space wave transmission be increased?
4. Explain the mode of wave propagation.
5. Discuss the phenomenon- of wave propagation.
6. Explain the tropospheric scatters fading.
7. Derive an expression for range of LOS microwave system.
8. Explain the concept of effective earth radius.
9. Explain the effect of earth's curvature on troposphere propagation.
10. Explain the duct propagation.
11. Derive an expression for field strength at receiving point due to space wave and ground reflected wave.
12. Show that the radius of curvature of the wave path is decided by the gradient of refraction index (refractivity with height and not by its absolute value).
13. Describe different forms of atmospheric refraction.
14. Calculate the value of the factor by which the horizon distance of a transmitter will be modified, if the gradient of refraction index of air near ground is (= 0.05 x 10^{-6} per meter). The radius of the earth is given by 6370 km.

[Hint : $K = \dfrac{a'}{a} = \dfrac{1}{\left[1 + a\dfrac{d\mu}{dH}\right]}$]   (**Ans.** $a' = \dfrac{25}{17}a$)

400   FUNDAMENTALS OF MICROWAVE AND RADAR ENGINEERING

15. A TV transmitting antenna mounted at height of 100 m radiates 10 kW of power operating at 60 MHz. Power is radiated equally in all directions. Calculate
    (*i*) Maximum line-of-sight range.
    (*ii*) The field strength at receiving antenna at height of 9 m at a distance of 10 km.
    (*iii*) The distance at which the field strength reduces to 1 mV/m.

    [Hint : $d = 3.54\left(\sqrt{H_T} + \sqrt{H_R}\right) km$, $E_t = \dfrac{88\sqrt{PH_T H_R}}{\lambda d^2}$]

    (**Ans.** $d = 4615 km$, $E_t = 15.84 mV/m$, $D = 39.8 \times 10^3 m$)

16. 16. Calculate the field strength at a distance of 10 km from a 100 kW, medium wave broadcast transmitter employing a short vertical antenna.

    [Hint: $E = \dfrac{300\sqrt{P}}{D} mV/m$]   (**Ans.** 300 mV/m)

17. An VHF mobile radio system, the base station transmits 100 watts at 150 MHz (λ = 2 m) and the antenna is 20 m above ground. The transmitting antenna is /2 dipole for which the gain is 1.64. Calculate the field strength at a receiving antenna of height 2 m at distance of 40 km.

    [Hint: $E_R = \dfrac{88\sqrt{PH_T H_R}}{\lambda D^2}$]   (**Ans.** 11 μ V/m)

18. Two aircrafts are flying at altitudes of 3000 m and 5000 m respectively. What is the maximum possible distance along the surface of the earth over which they can have effective point to point microwave communication? Assume radius of earth is 6.37 x 10^6 m.

    [Hint: $D = 4.12\left[\sqrt{H_T} + \sqrt{H_R}\right] km$]   (**Ans.** 516.985 km)

19. Find the range of LOS system when the receiving and transmitting antenna height are 10 m and 100 m respectively. Assume effective earth's radius into consideration.

    [Hint: $D = 4.12\left[\sqrt{H_T} + \sqrt{H_R}\right] km$]   (**Ans.** D = 54.23 km)

20. A communication link is to be established between two stations using half length antenna for maximum directive gain. Transmitter power is 2 kW, d = 200 km, what is the maximum power received by receiver. The operating frequency is 150 MHz.

    [Hint: $P_R = P_T G_T G_R \left(\dfrac{\lambda}{4 \pi D}\right)^2$, $G_T = G_R = 1.64$]   (**Ans.** P_r = 3.41 10^{-9} W)

21. Obtain the expression of effective radius of earth for space- wave propagation through troposphere and prove the following relation :
    The range of transmission is

    $$d = 4.126\sqrt{H_T} + \sqrt{H_R} \;\; km$$

    where H_T and H_R are heights of transmitter and receiver antenna in meters.

12

CHAPTER

SKY WAVE PROPAGATION – THE IONOSPHERE WAVES

OBJECTIVES
- Introduction: Ionosphere
- Characteristics of Different ionized Regions
- Ionospheric Variations
- Effect of Ionosphere on Radio Waves

12.1 INTRODUCTION: IONOSPHERE

Experimental work by Appleton showed that the atmosphere receives sufficient energy from the sun and its molecules dissociate into positive and negative ions. They remain ionized for a long period. There are several layers of ionization at different heights (Fig. 12.1) which reflect back the high frequency wave to the earth which otherwise would have escaped into space. The various layers of the ionosphere have different effects on the propagation of radio waves.

The ionosphere is the upper portion of the atmosphere, which continually absorbs large quantities of radiant energy from the sun, gets heated and ionized. The physical properties of the atmosphere, such as temperature, density and decomposition, vary. Owing to this and the different types of radiation received, the ionosphere is not regular in distribution. The most important ionizing agents are ultraviolet and α, β and γ radiations from the sun as well as cosmic rays and meteors. The overall result (see Fig. 12.1) is a range of, four main layers D, E, F_1 and F_2 in ascending order.

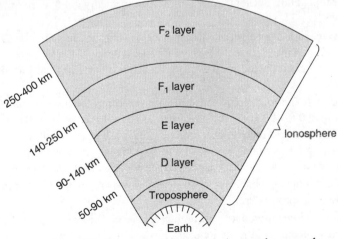

Fig. 12.1. Four main ionospheric layers above earth.

Figure (12.2) shows approximate distributions of electron density among the various regions and layers of the atmosphere by day (when the atmosphere is exposed to solar radiations) and at night. The last two layers merge in a single layer at night time.

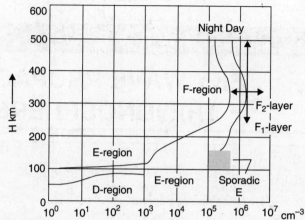

Fig. 12.2. Approximate distribution of electron density with height by day and at night. The arrows show the limits of variations for the height and electron density for the F_2 layer.

12.2 CHARACTERISTICS OF DIFFERENT IONIZED REGIONS

1. The **D layer** is the lowest layer existing at height of 50 to 90 km and an average thickness of 10 km. The degree of its ionization depends on the altitude of the sun and hence it disappears at night. It is the least important layer for HF propagation. It reflects VLF and LF waves absorbing MF and HF waves to a certain extent.

2. The next higher layer is the **E layer** which appears at the height of 100 km and has a thickness of 25 km. It also disappears at night like the D layer due to the deionization of ions in these layers. It is because of the absence of the sun (at night) when radiation is, consequently, not received. The main function of this layer is to aid MF surface-wave propagation a little and to reflect some HF waves in day time.

3. The E_2 **layer** is a thin layer of very high ionization density. It is also called the **sporadic E layer**, and often persists also at night. It normally does not play an important part in long-distance propagation but sometimes permits unexpectedly good reception over large distance, even of VHF or TV signals.

4. The F_1 **layer** exists at a height of 180 km in the day time and has a thickness of 20 km. At night, it combines with the F_2 layer. Although it reflects some of the HF waves, most of them pass through it and get reflected from the F_2 layer. Thus, it provides some absorption of HF waves both ways.

5. The F_2 layer is the most important reflecting medium for high-frequency radio waves. Its height from 200 to 400 km in the day time and has a thickness of 200 km. At night, its height falls to 300 km where it combines with the F_1 layer. Its height and ionization density change enormously and depend on the time of day, the average ambient temperature and the sunspot cycle. Unlike other layers, the F layer persists at night. The reason are as follows:

    (a) Since it is the topmost layer, it is highly ionized and in spite of recombination ionization persists to some extent.

    (b) Although ionization density is high in this layer, the actual air density is not, hence most of the molecules in its, are ionized.

    (c) Because of low density, the molecules have a large mean free path. The molecular collision rate is low; hence ionization does not disappear as soon as the sun sets.

SKY WAVE PROPAGATION – THE IONOSPHERE WAVES

The reasons for better HF reception at night are the combination of the F_1 and F_2 layers into one F layer, and the virtual disappearance of the other two layers, which had caused noticeable absorption during the day.

The key characteristics of the various regions and layers in the ionosphere are summarized in Table 12.1.

Table 12.1 : Characteristics of the various regions in the ionosphere

Particulars	D layer	E layer	F_1 layer	F_2 layer
Likely origin	(a) Ionization of NO with Lyman-alpha radiation (b) Ionization of all gases by soft X-rays	Ionization of all gases by soft X-rays	Ionization of O with fast decrease of recombination coefficient with height	Ionization of O by UV, X-rays and probably corpuscular radiation (α, β, γ rays)
Height, km	50–90 by day; disappears at night	90–140	140–250 by day; disappears at night	250–400
Molecular density, per cubic centimeter	10^{14}–10^{16}	5×10^{10} – 10^{13}	About 10^{11}	About 10^{10}
Electron or ion density, per cubic centimeter	100–10^3 for electrons; 10^6 –10^8 for ions max. at noon	Up to 10^5 to 4.5×10^5 by day; fixed at about 5×10^3 to 10^4 at night max. at 110 km approx.	2×10^5 –4.5×10^5 max. at 220 km approx.	Max 2×10^6 by day in winter; max. 2×10^5 by day in summer, 3×10^5 at night in winter
Collisions, per second	10^7 at lower edge	10^5	10^4	10^3 –10^4
Recombination coefficient, cu. cm per sec	10^{-5} – 10^{-7}	10^{-8}	4×10^{-9}	8×10^{-11} by day; 3×10^{-11} at night
Critical frequency	100 kHz	3 MHz to 5 MHz		5 MHz to 12 MHz

12.3 IONOSPHERIC VARIATIONS

As we have seen that the existence of the ionosphere depends on solar radiation; therefore, any variations in the radiation must influence the ionosphere. The amount of solar radiation reaching the earth and thus influencing the ionosphere is determined, in part, by the earth's rotation and its revolution around the sun. The variations of the earth, sun, and ionosphere occur regularly and are predictable. They are divided into four categories:

1. **Diurnal variations:** the hour-to-hour changes in the various ionospheric layers caused by the rotation of the earth around its axis.

2. **Seasonal variations:** caused by the constantly changing position of any point on earth relative to the sun as the earth orbits the sun.
3. **Geographical variations:** caused by the varying intensity of the solar radiation striking the ionosphere at different latitudes.
4. **Cyclical variations**: caused by sunspot activity over an 11-years cycle.

The 11-years sunspot cycle has the most influence on the ionosphere. The greater the number of sunspots, the intensity of solar radiation is greater. The critical frequency during a sunspot maximum, which occurs every 11 years, is about twice that of a sunspot minimum and results in greatly increased communications range and reliability during years of high sunspot activity.

Time of day may also affect propagation. Ground waves/surface waves remain the same during the day and the night, as do space waves. However, there can be radical changes in the sky wave during these periods. At nightfall, the sun can no longer ionize the atmosphere above the darkened part of the earth, producing thinner ionized layers in these parts of the world. The thinly ionized layers refract sky waves back to earth over a wider area; thus, sky waves generally return to earth farther away at night than during the day. However, different frequency bands can be affected differently.

In the VLF and LF bands, there is little difference between day and night propagation because signals are received primarily by ground/surface wave. In general, however, at distances of several thousands kilometers the signals at night will be stronger.

Beyond the ground wave range, signals in the MF band improve significantly at night. The distance these signals travel may increase from hundreds of kilometers in the daytime to thousands of kilometers at night.

During the daytime, frequencies at the lower end of the HF band may return to earth 35-800 km away, whereas at night, these same frequencies may be returned to earth 320 km to many thousands of kilometers away. Frequencies at the high end of the band may return to earth 320-8000 km away during the day, but at night they may penetrate the ionosphere and not return to earth at all. For these reasons, to maintain signals at usable strengths, HF communications systems may have to shift from one frequency during the day to another during the night.

Frequencies at the lower end of the VHF band may sometimes be refracted during the day, but rarely at night. Although unreliable for long distance communications, these lower-end frequencies are useful for communications upto about 80 km, Frequencies above 100 MHz are rarely refracted by the ionosphere. The UHF, SHF, and EHF bands are used primarily for line-of-sight communications both day and night, so there is rarely any change in travel distance between day and night.

Table 12.2 shows the frequency bands and the major systems they service. Note the overlapping services in some of the bands.

Table 12.2 : Frequency Bands and Major Services

Band	Range	Major Services
VLFLF	10–30 kHz (30–10 km) 30-300 kHz (10-1 km)	Radio navigation; time and frequency broadcasts; maritime mobile communications; aeronautical communications.
MF	300 kHz–3 kHz (1 km–100 m)	AM broadcasting; amateur communications; time and frequency broadcasts; fixed and mobile communications; maritime and aeronautical aids and communications.
HF	3–30 MHz (100–10 m)	Shortwave broadcasting; time and frequency broadcasts; point-to-point communication; amateur communications; land, maritime and aeronautical communications.

VHF	30–300 MHz (10–1 m)	Land and aeronautical mobile communications; industrial and amateur communications; FM and TV broadcasting; space and meteorological communications; radio navigation.
UHF	300 MHz–3 GHz* (1 m – 10 cm)	TV broadcasting; aeronautical and land mobile communications; radio astronomy; telemetry; satellite communications; amateur communications.
SHF EHF	3–30 GHz* (10-1 cm) 30–300 GHz* (1 cm–1 mm)	Microwave relay; satellite and exploratory communications; amateur communications.

* Frequencies above about 900 MHz are considered to be microwaves.

12.4 EFFECT OF IONOSPHERE ON RADIO WAVES

The presence of electrons and ions in the ionosphere affects the propagation of a radio wave passing through it. The ions, due to their heavier mass, have little effect on the propagation and are ignored in the following analysis.

Let, at any height in the ionosphere, there are N free electrons per cubic meter. Each electron has a charge $-e$ and mass m. Also, let an electric field of value

$$E = E_m \sin \omega t \text{ volt/meter} \qquad \ldots 12.1$$

is acting across *a cubic meter of space in the ionosphere*, where ω is the angular frequency and E_m, the maximum amplitude. Force exerted by electric field on each electron is given by

$$F = -eE \text{ Newton} \qquad \ldots 12.2a$$

where $\quad e$ = charge of an electron in Coulomb.

Let us assume that there is no collision, then the electron will have an instantaneous velocity v metre/sec in the direction opposite to the field.

Forced = Mass × Acceleration

$$-Ee = m \frac{dv}{dt} \qquad \ldots 12.2b$$

where m = mass of an electron in kg; $\frac{dv}{dt}$ = Acceleration

or $\qquad \dfrac{dv}{dt} = -\dfrac{Ee}{m} \quad$ or $\quad dv = -\dfrac{Ee}{m} dt$

Integrating both sides

$$\int dv = -\int \frac{eE}{m} dt; \quad v = -\frac{e}{m} \int E_m \sin \omega t \, dt, \quad (\text{Using } E = E_m \sin \omega t)$$

$$v = +\frac{eE_m \cos \omega t}{m\omega}$$

$$v = \left(\frac{e}{m\omega}\right) E_m \cos \omega t \qquad \ldots 12.3$$

Constant of integration is set to zero.

If N be the number of electron per cubic meter, then instantaneous electronic current density constituted by these N electrons moving with instantaneous velocity v is

$$J_e = -Nevamp/m^2 = -Ne\left(\frac{e}{m\omega}\right) e_m \cos \omega t \qquad \ldots 12.4$$

or $$J_e = -\left(\frac{Ne^2}{m\omega}\right) E_m \cos\omega t \qquad \ldots 12.5$$

which shows current J lags behind the electric field $E = E_m \sin \omega t$ by $90°$.

Ignoring the presence of the electrons, the ionosphere can be considered to be a free space, and in that case the displacement current density is

$$J_D = \frac{dD}{dt} = \frac{d}{dt}(\varepsilon_0 E) = \varepsilon_0 \frac{d}{dt}(E_m \sin \omega t)$$

Since $D = \varepsilon_0 D$ and where ε_0 is the permittivity of free space.

$$J_D = \varepsilon_0 E_m \omega \cos\omega t \qquad \ldots 12.6$$

thus, the displacement current leads the applied electric field by $90°$.

The total current density J_T that flows through a cubic meter of ionized medium is

$$J_T = J_D + J_e = \varepsilon_0 E_m \omega \cdot \cos\omega t - \frac{Ne^2}{m\omega} E_m \cos\omega t$$

$$= E_m \omega \cos\omega t \left[\varepsilon_0 - \frac{Ne^2}{m\,\omega^2}\right] \qquad \ldots 12.7$$

Comparing Eq. (7) and (6), the effective dielectric constant ε of the ionosphere (*i.e.*, ionized space)

$$\varepsilon = \varepsilon_0 - \frac{Ne^2}{m\omega^2} = \varepsilon_0\left[1 - \frac{Ne^2}{m\omega^2 \varepsilon_0}\right]$$

Hence, the relative dielectric constant w.r.t. vacuum of air

$$\varepsilon_r = \frac{\varepsilon}{\varepsilon_0} = 1 - \frac{Ne^2}{m\omega^2 \varepsilon_0}$$

$$\varepsilon = \varepsilon_0\left(1 - \frac{Ne^2}{m\omega^2 \varepsilon_0}\right) \qquad \ldots 12.8$$

We have used SI units in the above analysis. Substituting $e = 1.6 \times 10^{-19} C$, $m = 9.1 \times 10^{-31} kg$, and $\varepsilon_0 = 8.854 \times 10^{-12} F/m$ in Eq. (8) we get relative dielectric constant as

$$\varepsilon_r = \frac{\varepsilon}{\varepsilon_0} = \left(1 - \frac{81N}{f^2}\right) \qquad \ldots 12.9$$

Where $f = \frac{\omega}{2\pi}$ = frequency of the wave. Equation (9) suggests that the effective permittivity of the ionized space containing free electrons is less then the free space value ε_0.

12.4.1 REFRACTIVE INDEX OF IONOSPHERE AND MECHANISM OF RADIO WAVE BENDING BY THE IONOSPHERE

The phase velocity of a plane wave in a medium of permittivity ε and permeability μ, is given by

$$v_p = \frac{1}{\sqrt{\varepsilon\mu}} \qquad \ldots 12.10$$

In free space $\varepsilon = \varepsilon_0$ and $\mu_0 = \mu_0$; and the phase velocity is equal to c, the speed of light. In an ionized medium $\mu_0 = \mu_0$ but ε is less than ε_0 (Eq. 9). Therefore, the phase velocity in the ionosphere is greater than the speed of light. The group velocity cannot, however, exceed c.

The refractive index of an ionized medium with respect to free space is given by

$$n = \frac{c}{v_p} = \sqrt{\frac{\varepsilon}{\varepsilon_0}} \qquad \ldots 12.11$$

Suppose that a radio wave enters into the ionosphere from the underlying ionized medium. As the phase velocity in the ionosphere is larger, the refractive index is smaller. That is, the ionosphere will behave like a rarer medium. As a consequence at the boundary of the media, the incident ray will deviate from the straight line path and will move in the ionosphere in a direction away from the normal drawn at the point of incidence. This is shown in Fig. (12.3). The angle of incidence, i and the angle of refraction, r, shown in Fig. (12.3) are related to the refractive index, n, of the ionosphere through *Snell's* law of refraction

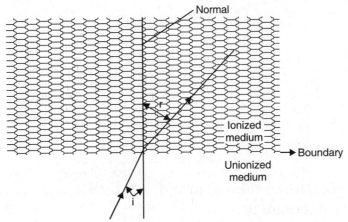

Fig. 12.3. Refraction of a radio wave at the boundary between the ionosphere and free space.

$$n = \frac{\sin i}{\sin r} = \sqrt{\frac{\varepsilon}{\varepsilon_0}} = \left(1 - \frac{81N}{f^2}\right)^{1/2} \qquad \ldots 12.12$$

As the wave penetrates more and more into the layer where electron density N gradually increases n decreases more and more. Hence r gradually increases and the bending of the ray becomes more prominent, as shown in Fig. (12.4). At P_m, the top of the ray trajectory, $r = 90°$, and at this point if the

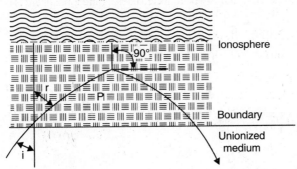

Fig. 12.4. Bending and reflection of a radio wave in the ionosphere.

electron density is denoted by N_1, we get from Eq. (12).

$$n = \sin i = \left(1 - \frac{81N}{f^2}\right)^{1/2} \qquad \ldots 12.13$$

The point, P_m, is referred to as the point of reflection. At this point, the ray becomes parallel to the earth surface. Then it tends to move in the downward direction and comes back to earth. It is clear from Eq. (12.13) that the smaller is the value of i, the smaller is n, and the larger would be the value of N_1. Thus, for smaller values of i, the point of reflection would be higher, as shown in Fig. (12.5), since the electron density increase with height. If the angle of incidence is so small that Eq. (12.13) cannot be satisfied with the largest value of N_1 in the layer, the ray passes through the layer and does not come back to earth. This is the case for the ray marked 5 in Fig. (12.5).

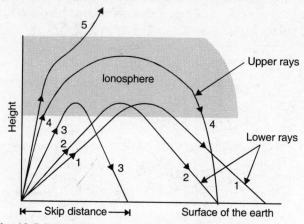

Fig. 12.5. Rays Incident at different angles on the Ionosphere.

12.4.2 CRITICAL FREQUENCY

For vertical incidence, $i = 0$; and so $n = 0$ at the reflection point. In the case of vertical incidence the maximum frequency that will be reflected-back to the earth from an ionospheric layer is called the **critical frequency** of the layer, and is denoted by f_c. Critical frequency is different for different layers.

As we know $n = \dfrac{\sin i}{\sin r} = \sqrt{\dfrac{\varepsilon}{\varepsilon_0}} = \left(1 - \dfrac{81N}{f^2}\right)^{1/2}$

By definition $i = 0$, $N = N_{\max}$ and $f = f_c$

Then the highest frequency that can be reflected back by the ionosphere is one for which refractive index 'n' becomes zero. So

$$n = \frac{\sin 0^\circ}{\sin r} = \sqrt{\frac{1 - 81 \cdot N_m}{f_c^2}} = 0$$

$$1 = \frac{81 N_m}{f_c^2}$$

Or
$$\boxed{\begin{array}{c} f_c = \sqrt{81\, N_m} \\ f_c = 9\sqrt{N_m} \end{array}} \qquad \ldots 12.14$$

where, is expressed MHz and is per cubic meter. Below EM wave will reflect back to earth by the ionized layers. The frequency is determined by N, the ionic density.

12.4.3 OBLIQUE INCIDENCE AND MAXIMUM USABLE FREQUENCY (MUF)

Equation (14) shows that a frequency greater than the critical frequency can be reflected back by the layer at oblique incidence. This frequency is known as maximum usable frequency (MUF).

For a sky wave to return to earth, angle of reflection, i.e., $i_c < r = 90°$

$$n = \frac{\sin i}{\sin r} = \sqrt{1 - \frac{81 . N_m}{f_{muf}^2}}$$

or
$$n = \frac{\sin i}{\sin 90°} = \sqrt{1 - \frac{81 . N_m}{f_{muf}^2}}$$

or
$$\sin^2 i = 1 - \frac{81 . N_m}{f_{muf}^2}$$

but $f_c^2 = 81 N_m$

or
$$1 - \frac{f_c^2}{f_{muf}^2} = \sin^2 i$$

or
$$\frac{f_c^2}{f_{muf}^2} = 1 - \sin^2 i$$

or
$$\frac{f_c^2}{f_{muf}^2} = \cos^2 i$$

or
$$f_{muf}^2 = \frac{f_c^2}{\cos^2 i} = f_c^2 \sec^2 i$$

or
$$\boxed{f_{muf} = f_c \sec i} \qquad \ldots 12.15$$

This equation, known as the **secant law**, gives the relationship between the maximum usable frequency of the reflected signal for an angle of incidence i with the critical frequency f_c for a given value of electron density. The maximum value of i (i.e., i_{max}) for a given layer is attained when the ray leaves the earth at grazing angle (see Fig.12. 6). We get from Fig. (12.6),

$$\sin i_{max} = \frac{a}{a+H} = \left(1 + \frac{H}{a}\right)^{-1} \qquad \ldots 12.16$$

where a is the radius of the earth and H is the height of the ionized layer.

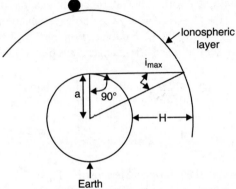

Fig. 12.6. Ray leaving earth at grazing angle.

Using typical values $N_m = 6 \times 10^{10} \, m^{-3}$ for E layer, and $N_m = 10^{12} \, m^{-3}$ for F layer, the critical frequencies for the two layers turn out to be 2.2 MHz and 9 MHz respectively. Taking $a = 6400 \, km$ the and E layer height to be 110 km, we obtain for this layer $\sin i_{max} = \dfrac{6400}{6500} = 0.983$. Hence $i_{max} = 79°$. The maximum frequency f_{max} that is reflected from this layer is found from Eq. 15 to be $2.2 \times \sec 79° = 11.5$ MHz. Taking F-layer height to be 250 km, the corresponding value $i_{max} = 74°$ and $f_{max} = 9 \times \sec 74° = 32.6$ MHz. Note that these values are illustrative only since the height and the peak electron density of a layer undergo diurnal, seasonal and sun-spot cycle variations. It may be mentioned that since the peak electron density of a layer at night is lower than its day-time value, the critical frequency of a layer at night is less than its day-time value. Hence the carrier frequencies of sky waves at night are chosen smaller than the day-time carrier frequencies. For transmission over distances of about 200 km over the surface of the earth, E-layer reflections are used. For transmission round the globe, F-layer reflections are employed. Note also that communication by sky waves is not possible for carrier frequencies in the region of 50 MHz or higher since these frequencies are not reflected by the ionosphere.

10.4.4 USABLE FREQUENCIES: LUF AND OUF

The maximum usable frequency (MUF) is the highest frequency that can be used for transmission between two points via sky wave-that is, the highest frequency at which a signal radiated into the ionosphere will be refracted back to earth with usable strength. The MUF varies between 8 MHz and 30 MHz with time of day, distance, direction, season, and solar activity as explained in section 10.4-3.

A frequency lower than the MUF can actually be used, because it will also be refracted. However, as the signal frequency decreases, the signal energy absorbed by the ionosphere increases, drastically reducing the usable signal strength. Below a certain frequency, sometimes called the **lowest usable frequency (LUF)**, the RF signal will be totally absorbed by the ionosphere. Therefore, operating at the MUF will produce the maximum received signal strength.

The (LUF) depend upon the following three factors where as the MUF depends only upon the state of the ionosphere and the distance between transmitting and receiving points.

(i) The effective radiated power,
(ii) The absorption characterization of the ionosphere for the paths between transmitter and receiver and
(iii) The required field strength, which in turn depends upon radio noise at the receiving location, i.e., signal to noise ratio (S/N),

The MUF changes constantly with atmospheric conditions, and the LUF is border line for refraction. Therefore, to obtain the most reliable sky wave propagation, the **optimum usable frequency (OUF)** is used. The OUF is well above the LUF but far enough below the MUF that the signal is not appreciably affected by the constant atmospheric changes, thus providing relatively reliable sky wave communications.

Because the usable frequencies are constantly changing, charts are available that predict them for every hour of the day, over any path on the earth, and for any month of the year. The predictions on these harts are based on extensive solar observations and can be used to optimize sky wave communications.

12.4.5 THE VIRTUAL HEIGHT

The refraction of a wave by the ionosphere is shown in Fig. (12.7). As the wave is refracted, it bends downwards gradually rather than sharply. Below the ionized layer, the incident and refracted rays follows paths that are exactly the same as they would have been if refraction had taken place

SKY WAVE PROPAGATION – THE IONOSPHERE WAVES

from a surface located at a greater height. This is called the **virtual height** of the layer, and knowing this; it is possible to calculate the angle of incidence required for the wave to return to ground at a selected spot.

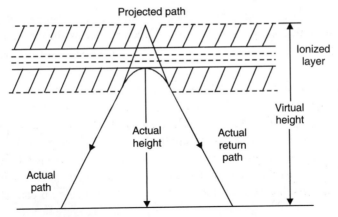

Fig. 12.7. Actual and virtual height of an ionized layer.

Measurement of virtual height is normally carried out by means of an instrument known as an IONOSONDE. The basic method is to transmit vertically upward a pulse-modulated radio wave with pulse duration of about 150 micro-seconds. The reflected signal is received close to the transmission point, and the time and round trip is measured. The virtual height (H_v) is then given by $H_v = \dfrac{cT}{2}$, where c = velocity of light is m/s = 3×10^8 m/s.

The ionosonde will have facilities for sweeping over the radio frequency range; typically, it will sweep from 1 MHz to 20 MHz in 3 minutes. It will also have facilities of automatic plotting of virtual height against frequency. The resultant graph is known as an IONOGRAM shown in Fig. (Height). This is called the **virtual height** of the layer, and knowing this; it is possible to calculate the angle of incidence required for the wave to return to ground at a selected spot.

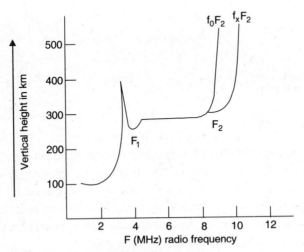

Fig. 12.8. Ionogram.

The ionogram shows two critical frequencies f_oF_2 and f_xF_2. These two critical frequencies are for **ordinary ray** denoted by f_oF_2 and for the **extra-ordinary ray** denoted by f_xF_2. The development of two components of the relative permittivity of the ionized layer is due to earth's magnetic field. The occurrence of two critical frequencies is only for F_2 layer.

12.4.6 THE SKIP DISTANCE

It is the shortest distance from a transmitter, measured along the surface of the earth at which the sky wave of a fixed frequency will be returned to the earth. The presence of a minimum distance can be explained if the behavior of the sky wave is considered with the help of a diagram (see Fig. 12. 5).

When the angle of incidence is made quite large (ray 1) the sky wave returns to the ground at a long distance from the transmitter. As the angle is reduced, the wave returns closer and closer to the transmitter (rays 2 and 3). If the angle is smaller than that for the ray 3, the ray is too close to the vertical to be returned to the earth. It is only bent but the bending is insufficient to return the wave, unless the frequency used for communication is less than the critical frequency. Finally if the angle of incidence is slightly less than that for ray 3, the wave will return to a point farther from ray 3 as it has to spend a longer time in space before being bent because the ion density is changing very slowly at this angle.

The smallest distance (at which the ray 3 is returned) between the transmitting antenna and the point at which the wave returns to earth is called the *skip distance*. The frequency corresponding to this distance is the f_{muf}. Right near the transmitter, the signal is heard via the ground wave surface wave. After leaving the ground wave range we encounter a zone of silence, called **skip zone** (Fig. 12.9), where no signal at all can be picked up. If the sky wave signal is sufficiently strong after returning to the earth it may be reflected by the earth and again by the ionosphere, thus returning to earth in a series of hops separated by the skip distance.

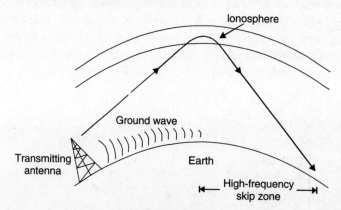

Fig. 12.9. Skip zone.

The signal may be transmitted in this way one or more times around the world (1/7th sec for one round). As the frequency increases, the ionosphere becomes progressively less effective and lower and lower angle or radiation must be used to reflect the signal back to earth. At frequencies above about 30 MHz the signal is no longer returned even for the smallest vertical angle, and the ionosphere has little effect on it. The exact frequency depends upon the density of ionization which is affected by sun spot cycles, daily and seasonal variations.

In the night the vertical height of the ionospheric layer increases than in the day time and so the skip distance two increases as shown in Fig. (12.10).

SKY WAVE PROPAGATION – THE IONOSPHERE WAVES

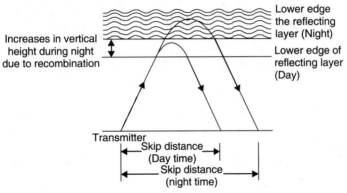

Fig. 12.10. Increase in skip distance in night due to increase in vertical height.

The transmission path is limited to the skip distance at one end and the curvature of the earth at the other. The longest signal hop distance is obtained when the ray is transmitted tangentially to the surface of the earth. For the F_2 layer, this corresponds to a maximum practical distance of about 4000 km because of the fact that the semi-circumferences of the earth is just over 20,000 km.

12.4.7 RELATION BETWEEN SKIP DISTANCE AND MAXIMUM USABLE FREQUENCY f_{muf}:

For a given frequency of propagation $f = f_{muf}$, the relation between skip distance D and f_{muf} can be obtained as

$$D_{skip} = 2H \sqrt{\left(\frac{f_{muf}}{f_c}\right)^2 - 1} \qquad \ldots 12.17$$

assuming flat earth and ionized layer to be thin with sharp ionization density gradient. These assumptions gives mirror like reflection of radio waves as shown in Fig. (12.11).

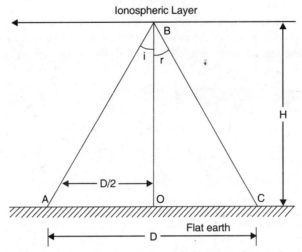

Fig. 12.11. Reflection from a thin layer on flat earth.

From the Fig. 12.11

$$\cos i = \frac{BO}{AB} = \frac{H}{\sqrt{H^2 + \frac{D^2}{4}}} = \frac{2H}{\sqrt{4H^2 + D^2}} \qquad \ldots 12.18$$

where H = height of layer and D = propagation distance AC (Fig. 12.11). We also know for f_{muf}

$$n = \sin i = \sqrt{1 - \frac{81 N_m}{f_{muf}^2}} \quad \text{or} \quad \sin^2 i = 1 - \frac{f_c^2}{f_{muf}^2} \qquad \ldots 12.19$$

or

$$\frac{f_c^2}{f_{muf}^2} = \cos^2 i = \frac{4H}{\left(4H^2 + D^2\right)} \qquad \ldots 12.20$$

$$f_{muf}^2 = \frac{4H^2 + D^2}{4H^2}$$

$$\frac{f_{muf}}{f_c} = \sqrt{1 + \frac{D^2}{4H^2}}$$

or

$$f_{muf} = f_c \sqrt{1 + \left(\frac{D}{2H}\right)^2} \qquad \ldots 12.21$$

Rearranging we get

$$D_{skip} = 2H \sqrt{\frac{f_{muf}^2}{f_c^2} - 1} \qquad \ldots 12.22$$

When earth is curved Eq. (22) is modified to

$$D_{skip} = 2\left(H + \frac{D^2}{8a}\right) \sqrt{\left(\frac{f_{muf}}{f_c}\right)^2 - 1} \qquad \ldots 12.23$$

where a is earth's radius.

In case when we take earth as curved, then reflecting region is consider to be concentric with earth as shown in Fig. (12.12). In this figure transmitting waves leaves the transmitter tangentially to be earth. Let us assume that 2θ be the angle subtended by the transmission distance D at the centre of the earth 0 then

$$\text{Angle} = \frac{\text{Arc}}{\text{Radius}}, \quad \therefore \quad 2\theta = \frac{D}{a}$$

Therefore, $D = 2a\theta$. Here $AT = a \sin\theta$, $OT = a \sin\theta$

$$BT = OE + EB - OT = H + a - a \sin\theta$$

and

$$AB = \sqrt{AT^2 + BT^2} = \sqrt{(a\sin\theta)^2 + (H + a - a\cos\theta)^2}$$

Therefore, $\cos^2 i = \dfrac{BT}{AB} = \dfrac{H + a - a\cos\theta}{\sqrt{(a\sin\theta)^2 + (H + a - a\cos\theta)^2}}$

or

$$\cos^2 i = \frac{(H + a - a\cos\theta)}{(a\sin\theta)^2 + (H + a - a\cos\theta)^2}$$

As we know that

$$\frac{f_c^2}{f_{muf}^2} = \cos^2 i = \frac{H + a - a\cos\theta}{(a\sin\theta)^2 + (H + a - a\cos\theta)^2} \qquad \ldots 12.24$$

SKY WAVE PROPAGATION – THE IONOSPHERE WAVES

Since the curvature of earth limits both the MUF and skip distance D and hence the limit is obtained when waves leave the T_X at a grazing angle $\angle OAB = 90°$. Therefore D is maximum, when θ is maximum and it may be given as

$$\cos\theta = \frac{OA}{OB} = \frac{a}{a+H},$$

However, the actual value of θ is very small. So this relation can be extended as

$$\cos\theta = \frac{a}{a\left(1+\dfrac{H}{a}\right)} = \left(1+\frac{H}{a}\right)^{-1}, \quad \cos\theta = \left[1+\frac{H}{a}+\ldots\right] \quad \text{because} \quad \frac{H}{a} \ll 1$$

or $\quad 1 = 1+\dfrac{\theta^2}{2} = 1-\dfrac{H}{a} \quad$ or $\quad \theta^2 = \dfrac{2H}{a}$

We have $\quad D^2 = 4a^2\theta^2 = 4a^2 \times \dfrac{2H}{a} = 8Ha$

$$H = \frac{D^2}{8a} \qquad \ldots 12.25$$

Here $\quad \cos\theta = \left[1+\dfrac{D^2}{8a^2}\right]$ and as θ is small

$$\sin\theta = \theta = \frac{D}{2a}$$

By putting the value in eqn. (24), we get

$$\frac{f_c^2}{\left(f_{muf}\right)^2_{max}} = \frac{\left\{H+a-a\left(1-\dfrac{D^2}{8a}\right)\right\}^2}{\left\{a^2 \cdot \dfrac{D^2}{4a^2} + \left(H+a-a\left[1-\dfrac{D^2}{8a}\right]\right)^2\right\}} = \frac{\left[H+\dfrac{D^2}{8a}\right]^2}{\dfrac{D^2}{4}+\left\{H+\dfrac{D^2}{8a}\right\}^2}$$

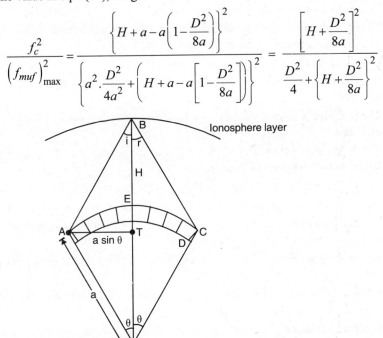

Fig. 12.12. Reflection from a thin Ionospheric layer in case when earth curvature is considered.

or
$$\frac{(f_{muf})_{max}}{f_c} = \frac{\sqrt{\frac{D^2}{4} + \left\{H + \frac{D^2}{8a}\right\}^2}}{\sqrt{\left(H + \frac{D^2}{8a}\right)^2}}$$

$$(f_{muf})_{max} = f_c \frac{\sqrt{\frac{D^2}{4} + \left\{H + \frac{D^2}{8a}\right\}^2}}{\sqrt{\left(H + \frac{D^2}{8a}\right)^2}}$$

From eqn. (25)
$$(D_{skip}) = \sqrt{8Ha}$$
Since D is skip distance
$$D = 2\left[H + \frac{D^2}{8a}\right]\sqrt{\left[\frac{f_{mud}}{c}\right]^2 - 1} \qquad \ldots 12.26$$

SUMMARY

- The various layers of the ionosphere have different effects on the propagation of radio waves.
- The ionosphere is the upper portion of the atmosphere, which continually absorbs large quantities of radiant energy from the sun, gets heated and ionized.
- Four main layers D, EF_1 and F_2 in ascending order.
- The **D layer** is the lowest layer existing at height of 50 to 90 km and an average thickness of 10 km.
- The next higher layer is the **E layer** which appears at the height of 100 km and has a thickness of 25 km.
- The **E_2 layer** is a thin layer of very high ionization density. It is also called the **sporadic E layer**, and often persists also at night.
- The **F_1 layer** exists at a height of 180 km in the day time and has a thickness of 20 km.
- The F_2 layer is the most important reflecting medium for high-frequency radio waves.
- **Diurnal variations** : the hour-to-hour changes in the various ionospheric layers caused by the rotation of the earth around its axis.
- **Seasonal variations** : caused by the constantly changing position of any point on earth relative to the sun as the earth orbits the sun.
- **Geographical variations** : caused by the varying intensity of the solar radiation striking the ionosphere at different latitudes.
- **Cyclical variations** : caused by sunspot activity over an 11-years cycle.
- The presence of electrons and ions in the ionosphere affects the propagation of a radio wave passing through it.
- The phase velocity of a plane wave in a medium of permittivity ε and permeability μ is given by

SKY WAVE PROPAGATION – THE IONOSPHERE WAVES

- $v_p = \dfrac{1}{\sqrt{\varepsilon\mu}}$

- The refractive index of an ionized medium with respect to free space is given by

- $n = \dfrac{c}{v_p} = \sqrt{\dfrac{\varepsilon}{\varepsilon_0}}$

- In the case of vertical incidence the maximum frequency that will be reflected-back to the earth from an ionospheric layer is called the **critical frequency** of the layer, and is denoted by f_c

- $f_c = 9\sqrt{N_m}$

- Frequency greater than the critical frequency can be reflected back by the layer at oblique incidence. This frequency is known as maximum usable frequency (MUF).

- $f_{muf} = f_c \sec i$

- **Skip Distance**
  It is the shortest distance from a transmitter, measured along the surface of the earth at which the sky wave of a fixed frequency will be returned to the earth.

IMPORTANT FORMULAS

$\Rightarrow \varepsilon_r = \dfrac{\varepsilon}{\varepsilon_0} = \left(1 - \dfrac{81N}{f^2}\right)$

$\Rightarrow v_p = \dfrac{1}{\sqrt{\varepsilon\mu}}$

$\Rightarrow n = \dfrac{\sin i}{\sin r} = \sqrt{\dfrac{\varepsilon}{\varepsilon_0}} = \left(1 - \dfrac{81N}{f^2}\right)^{1/2}$

$\Rightarrow f_c = 9\sqrt{N_m}$

$\Rightarrow f_{muf} = f_c \sec i$

$\Rightarrow D_{skip} = 2H\sqrt{\dfrac{f_{muf}^2}{f_c^2} - 1}$

$\Rightarrow D_{skip} = 2\left(H + \dfrac{D^2}{8a}\right)\sqrt{\left(\dfrac{f_{muf}}{f_c}\right)^2 - 1}$

SOLVED PROBLEMS

Problem 1. *Assume that reflection takes place at a height of 500 km and that the maximum density in the ionosphere corresponds to a 0.8 refractive index at 10 MHz. Determine the range (assume earth is flat) for which the MUF is 10 MHz.*

Solution. Given, $H = 500$ km; $n = 0.8$ m; $f_{muf} = 10$ MHz; $f = 10$ MHz

We know that, $n = \sqrt{1 - \dfrac{81N}{f^2}}$

And $\quad D_{skip} = 2H\sqrt{\left(\dfrac{f_{muf}}{f_c}\right)^2 - 1}$

Putting these values in above equation, we get

$$0.8 = \sqrt{1 - \dfrac{81 N_{max}}{f^2}}$$

$$0.64 = 1 - \dfrac{81\, N_{max}}{\left(10 \times 10^6\right)^2}$$

$$\dfrac{81 N_{max}}{\left(10 \times 10^6\right)^2} = 1 - 0.64 = 0.36$$

$$81\, N_{max} = 0.36 \times 10^{14}$$

$$N_{max} = \dfrac{0.36 \times 10^{14}}{81} = 0.44 \times 10^{12}\, Hz$$

$$N_{max} = 440 \times 10^9\, Hz \text{ or } 440\, GHz$$

therefore, $\quad f_c = 9\sqrt{N_{max}} = 9\sqrt{440 \times 10^9} = 5.96 \times 10^6\, Hz$

$$D_{skip} = 2h\sqrt{\left[\dfrac{f_{muf}}{f_c}\right]^2 - 1}$$

$$D_{skip} = 2 \times 500 \sqrt{\left[\dfrac{10 \times 10^6}{5.96 \times 10^6}\right]^2 - 1} = 1000\sqrt{\left[\dfrac{10}{5.96}\right]^2 - 1}$$

$$= 1000 \times \sqrt{1.82} = 1349.07\, km$$

$$D_{skip} = 1349.07\, km \quad \textbf{Ans.}$$

Problem 2. *In the ionospheric propagation, consider that the reflection takes place at a height of 500 km and that the maximum density in the ionosphere corresponds to a refractive index of 0.8 at a frequency of 10 MHz. Determine the ground range (assume earth is curved) for which the MUF is 10 MHz. (Refer problem 1)*

Solution. When the earth's curvature is taken into consideration, then

$R = 6370$ km (Radius of the earth)

$H =$ height of reflecting layer form the earth

$D = 1349.07$ km (Refer problem 1)

$$\dfrac{f_{muf}}{f_c} = \sqrt{\dfrac{D^2}{4\left(H + \dfrac{D^2}{8a}\right)} + 1}$$

$$\dfrac{D^2}{4\left[H + \dfrac{D^2}{8a}\right]^2} = \left[\dfrac{f_{muf}}{f_c}\right]^2 - 1$$

SKY WAVE PROPAGATION – THE IONOSPHERE WAVES

or
$$D = 2\left[H + \frac{D^2}{8a}\right]\sqrt{\left(\frac{f_{muf}}{f_c}\right)^2 - 1}$$

$$D = 2\left[500 + \frac{(1349.07)^2}{8 \times 6370}\right] \cdot \sqrt{1.82}$$

$$D = 2[500 + 35.71] \cdot [1.34] = 1435.70 \text{ km} \quad \textbf{Ans.}$$

Problem 3. *What is the critical frequency for reflection at vertical incidence if the maximum value of electron density is $2.48 \times 10^6 \, cm^{-3}$.*

Solution. We know that $f_c = 9\sqrt{(N_{max})}$

where f_c = Critical frequency in MHz
N_{max} = Maximum electron density in m^{-3}

$$= 2.48 \times 10^6 \times 10^{-6} \, m^{-3}$$

$$f_c = 9 \times \sqrt{2.48} \, MHz = 9 \times 1.574 \, MHz$$

$$f_c = 14.17 \, MHz \quad \textbf{Ans.}$$

Problem 4. *A high frequency radio link has to be established between two points of the earth 4000 km way. If the reflection region of the ionosphere is at height of 200 km and has a critical frequency of 5 MHz, calculate the MUF for the given path.*

Solution. We know that, $f_{muf} = f_c \sqrt{1 + \left(\frac{D}{2H}\right)^2}$

Given, $D = 4000$ km, $H = 200$ km, $f_c = 5$ MHz

$$f_{muf} = 5\sqrt{1 + \left(\frac{4000}{2 \times 200}\right)^2} = 5\sqrt{101}$$

$$f_{muf} = 50.25 \, MHz \quad \textbf{Ans.}$$

Problem 5. *Calculate the frequencies for F_1, F_2 and E for which the maximum ionic densities are 2.3×10^6, 3.5×10^6 and 1.7×10^6 electrons per cubic cm respectively.*

Solution. Since, for F_1 layer

$$N_{max} = 2.3 \times 10^6 \, cm^{-3}$$

$$= 2.3 \times 10^6 \times 10^{-6} \, m^{-3} = 2.3 \, m^{-3}$$

$$f_c = 9\sqrt{N_{max}} = 9\sqrt{2.3} = 13.65 \, MHz \quad \textbf{Ans.}$$

For F_2- layer,

$$N_{max} = 2.3 \times 10^6 \, cm^{-3} = 3.5 \, m^{-3}$$

$$f_c = 9\sqrt{N_{max}} = 9\sqrt{3.5} = 16.84 \, MHz \quad \textbf{Ans.}$$

For E layer,

$$N_{max} = 1.7 \times 10^6 \, cm^{-3} = 1.6 \, m^{-3} \quad \textbf{Ans.}$$

Problem 6. *At what frequency a wave must propagate or the D-region to have an index of refraction 0.7? Given $N = 400 \text{ cm}^{-3}$.*

Solution. Since $n = \sqrt{1 - \dfrac{81N}{f^2}}$

$$(0.7)^2 = \sqrt{1 - \dfrac{81 \times 400}{f^2}}$$

$$\dfrac{81 \times 400}{f^2} = 1 - 0.49 = 0.51$$

$$f^2 = \dfrac{81 \times 400}{0.51}$$

$$f = \sqrt{\dfrac{81 \times 400}{0.51}} = \sqrt{63529.41} = 252.05 \text{ MHz} \quad \textbf{Ans.}$$

Problem 7. *A TV transmitter has a height of 169 m and the receiving antenna has a height of 20 m. what is the maximum distance through which the TV signal could be received by space wave propagation? What is the radio horizon?*

Solution. Given $H_T = 169 \text{ m}, \ H_R = 20 \text{ m}$

$$d = 4.12 \left[\sqrt{H_T} + \sqrt{H_R} \right]$$

$$d = 4.12 \left[\sqrt{169} + \sqrt{20} \right] = 71.98 \text{ km}$$

Radio horizon $= \sqrt{2r' H_T}$

$$= \sqrt{2 \times \dfrac{4}{3} \times \dfrac{6370}{1000} \times H_T}$$

$$= \sqrt{16.9866 \times H_T} = \sqrt{16.9866 \times 169}$$

Radio Horizon $= 53.57$ km **Ans.**

Problem 8. *Calculate the transmission-path distance for an ionospheric transmission that utilizes a layer of height 200 km. The angle of elevation of the antenna beam is 20°. The earth's radius can be assumed 6370 km.*

Solution. $D = \dfrac{2H}{\tan \beta} = \dfrac{2 \times 200}{\tan 20°} = \dfrac{400}{0.3640}$

$D = 1098.9 \text{ km}$

$$D = 2a \left[\left(90° - \beta \right) - \sin^{-1} \left(\dfrac{a \cos \beta}{a + H} \right) \right]$$

$$D = 2a \left[\dfrac{\pi}{2} - \beta \right] - \sin^{-1} \left[\dfrac{a \cos \beta}{a + H} \right]$$

SKY WAVE PROPAGATION – THE IONOSPHERE WAVES

$$= 2 \times 6370 \left[\frac{3.14}{2} - 20° \times \frac{\pi}{180°} \right] - \sin^{-1}\left[\frac{6370 \cos 20°}{6370 + 200} \right]$$

$$= 12740 \left[\left(1.57 - \frac{3.14}{9}\right) - \sin^{-1}\left[\frac{6370 \times 0.9397}{6570} \right] \right]$$

$$= 12740 \left[1.2212 - \sin^{-1}(0.9110) \right] = 12740 \left[1.2212 - 65.6° \right]$$

$$= 12740 \left[1.2212 - 65.6 \times \frac{\pi}{180} \right] = 12740 \left[1.2212 - \frac{65.6 \times 3.14}{180} \right]$$

$$= 12740 [1.2212 - 1.1443] = 958.943 \; km$$

$D = 959 \; km$ **Ans.**

Problem 9. *Calculate the maximum range obtainable in single hop transmission utilizing F_2 layer for which assumed earth's radius as 6370 km. Considering height of F_2 layer is 400 km.*

Solution. Maximum range in a single hop transmission is given by

$$d = 2R \cdot Q = 2R \cos^{-1} \frac{R}{R + h_m}$$

where R = Radius of earth
 h_m = Height of the ionospheric layer

$$d = 2 \times 6370 \times \cos^{-1}\left[\frac{6370}{6370 + 400} \right]$$

$$= 12740 \cos^{-1}(0.9409)$$

$$= 12740 \times 19.8° \times \frac{\pi}{180}$$

$$= 12740 \times \frac{19.8 \times 3.14}{180}$$

$$= 12740 \times \frac{62.172}{180} = 4400.396 \; km$$

$d = 4400.396 \; km$ **Ans.**

Problem 10. *Calculate the value of frequency at which an EM wave must be propagated through the E-region of refraction of 0.6 and electron density of 4.23×10^4 electrons/m^3.*

Solution. Given, $n = 0.6$, $N = 4.23 \times 10^4$ electron per m^{-3}

We know that

$$n = \sqrt{1 - \frac{81 \; N}{f^2}} = 0.6 = \sqrt{1 - \frac{81 \times 4.23 \times 10^4}{f^2}}$$

$$0.36 = 1 - \frac{81 \times 4.23 \times 10^4}{f^2}$$

$$\frac{342.63 \times 10^4}{f^2} = 0.64$$

$$f^2 = \frac{342.63 \times 10^4}{0.64} = 5353593.8$$

$$f = 2313.7834 \ Hz \ \text{or} \ 2.313 \ KHz \ \textbf{Ans.}$$

Problem 11. *At what frequency a wave must propagate for the D region to have an index of refraction 0.8? Given $N = 500$ electron per cc for D-region.*

Solution.
$$n = \sqrt{1 - \frac{81 \ N}{f^2}}$$

$$0.8 = \sqrt{1 - \frac{81 \ N}{f^2}}$$

$$0.64 = 1 - \frac{81 N}{f^2}$$

$$\frac{81 \ N}{f^2} = 1 - 0.64 = 0.36$$

$$\frac{81 \times 500}{0.36} = f^2$$

$$f^2 = 112500$$

$$f = 335.410 \ kHz \ \textbf{Ans.}$$

REVIEW QUESTIONS

1. Explain the terms MUF and skip distance. Shown that on flat earth the skip distance for a given frequency is given by

$$D_{skip} = 2H\sqrt{\frac{f^2}{f_c^2} - 1}$$

   where H is the average height of the reflecting layer and f_c is the critical frequency for the layer? How does the curvature of earth effect the propagation of radio wave?

2. Using ray treatment, show that the refractive index of the ionosphere, in absence of magnetic field and collision is given by

$$\mu = \sqrt{1 - \frac{81 \ N}{f^2}}$$

   where N is the number of electrons per cm^3 and f is the frequency in kHz.

3. Explain the term 'virtual height', critical frequency and skip distance.
4. Discuss in detail the sky-wave propagation.
5. Describe the ionosphere and ionospheric propagation.
6. Discuss the concept of skip-distance and MUF in ionospheric propagation.
7. Define skip distance and show how it is related to maximum usable freguency.
8. What is ionosphere? Show how the electron density in the ionosphere varies with height.
9. Show the typical variations of electron density with the height of atmosphere. Explain how these variations affect the radio wave reflection from the ionosphere.
10. Prove that the phase velocity of a plane EM wave in an ionized medium exceeds the velocity of light in free space.

SKY WAVE PROPAGATION – THE IONOSPHERE WAVES

11. Define critical frequency of an ionosphere layer. Explain why the carrier frequency of a programme transmitted through sky waves at night is less than its day-time value.
12. "For the same transmitter power, the intensity of the sky wave signal received at night is greater than that received during day time". Explain.
13. Describe the mechanism of reflection of EM wave in ionospheric layers. What do you mean by skip distance and how does it vary with the hour of the day and the frequency of the wave?
14. Write short notes on :
    (i) Maximum usable frequency
    (ii) Critical frequency
    (iii) Ionospheric fading
    (iv) Virtual height
15. What is the role of the ionosphere in propagation? How do refraction and reflection occur?
16. What is the critical frequency for reflection at vertical incidence if the maximum value of electron density is 1.24×10^6 cm^{-3}?

    [**Hint:** $f_c = \sqrt{N_{max}}$ MHz]   (**Ans.** 10.026 MHz)

17. Radio communication is desired between the places separated by a distance of 1000 km of the earth's surface via an ionospheric layer whose height is 400 km and critical frequency is 11 MHz. Determine the maximum value of transmitter frequency for effective communication.

    [**Hint:** $f_0 = f_c \sqrt{1 + \left(\dfrac{D}{2H}\right)^2}$]   (**Ans.** 17.6 MHz)

18. A high frequency radio link has to be established between two points on the earth 2000 km away. If the reflection region of the ionosphere is at height of 200 km and has a critical frequency of 6 MHz, determine the MUF for the given path.

    [**Hint:** $f_{muf} = f_c \sqrt{1 + \left[\dfrac{D}{2H}\right]^2}$]   (**Ans.** $f_{muf} = 30.6$ MHz)

19. What is the radio horizon of a television antenna placed at a height of 166 m? If the signal is to be received at a distance of 66 km, what should be the height of the receiving antenna?

    [**Hint:** $d = 4.12 \left[\sqrt{H_T} + \sqrt{H_R}\right]$ km]   (**Ans.** $H_R = 9.858$ m)

20. Calculate the value of frequency at which an EM wave must be propagated through the D-region with an index of refraction of 0.5 and an electron density of 3.24×10^4 electron/m^3.

    [**Hint:** $\mu = \sqrt{1 - \dfrac{81 \, N}{f^2}}$]   (**Ans.** 1.87 kHz)

21. A TV transmitter antenna has a height of 256 m and the receiving antenna has a height of 25 m. What is the maximum distance through which the TV signal could be received by space wave propagation? What is the radio horizon in this case?

    [**Hint:** $d = 4.12 \left(\sqrt{H_T} + \sqrt{H_R}\right)$, Radio horizon = $\sqrt{2r' H_T}$]

    (**Ans.** Radio horizon = 65.92 km)

22. In the ionospheric propagation, consider that the reflection takes place at a height of 400 km and that the maximum density in the ionosphere corresponds to refractive index of 0.9 at

frequency of 10 MHz. Determine the ground range for which this frequency is the MUF. Take the earth's curvature into consideration.

[**Hint:** $\mu = \sqrt{1 - \dfrac{81\, N_{max}}{f^2}}$, $f_c = 9\sqrt{N_{max}}$, $D = 2\left\{H + \dfrac{D^2}{8\,a}\right\}\sqrt{\left[\dfrac{f_{muf}}{f_c}\right]^2 - 1}$]

(**Ans.** $N_{max} = 23.456 \times 10^{10}\, m^{-3}$, $f_c = 4.3588 \times 10^6\, Hz$, $D = 1673.86\, km$)

23. In VHF mobile radio system, the base station transmits 100 W at 150 MHz and the antenna is 20 m above ground. The transmitting antenna is a dipole for which the gain is 1.64. Calculate the field strength at a receiving antenna of height 2 m at a distance of 40 km.

[**Hint:** $E_R = \dfrac{88\sqrt{P H_T \cdot H_R}}{\lambda\, d^2}$] (**Ans.** $E_R = 11\, mV/m$)

24. Calculate the maximum range obtainable in a single hop transmission utilizing E layer for which assume earth radius as 6370 km.

[**Hint:** $d = 2R \cdot \theta = 2R \cos^{-1} \dfrac{R}{h_m + R}$; heighyt of E layer = 140 km]

(**Ans.** 2644.68 km)

25. Calculate the maximum range obtainable in a single hop transmission utilizing D-layer for which assume earth radius 6370 km and height of D layer is 90 km.

[**Hint:** $d = 2R \cdot \cos^{-1}\left[\dfrac{R}{h_m + R}\right]$] (**Ans.** $d = 219.14\, km$)

26. A pulse of a given frequency transmitted upward is received back after a period of 5 ms. Find the virtual height of the reflected layer.

[**Hint:** $h = \dfrac{cT}{2}\, m$, $c = 3 \times 10^8\, m$] (**Ans.** $h = 750$ km)

27. Two aeroplanes are flying at height of 400 m and 7000 m respectively. What is the maximum distance along the surface of the earth over which they can have effective point to point microwave communication? Assume radius of earth = $6.37 \times 10^6\, m$.

[**Hint:** $d = 4.12\left(\sqrt{H_T} + \sqrt{H_R}\right)$] (**Ans.** $d = 605.273\, km$)

28. Assume that reflection take place at a height of 350 km and that the maximum density in the ionosphere corresponds to a 0.8 refractive index at 15 MHz. What will be the range (assume earth is flat) for which the MUF is 20 MHz?

[**Hint:** $n = \sqrt{1 - \dfrac{81\, N_{max}}{f^2}}$, $D_{skip} = 2H\sqrt{\left[\dfrac{f_{muf}}{f_c}\right]^2 - 1}$, $f_c = 2\sqrt{N_{max}}$]

(**Ans.** $N_{max} = 1 \times 10^{12}\, m^{-3}$, $f_c = 9 \times 10^6\, Hz$, $D_{skip} = 1389.10\, km$)

29. A high frequency radio link has to be established between two points on the earth 400 km away. The reflection region of the ionosphere is at a height of 300 km and has a f_c of 5 MHz. Calculate the MUF for the given path.

[**Hint:** $f_{muf} = f_c \sqrt{1 + \left[\dfrac{D}{2h}\right]^2}$] (**Ans.** = 6 MHz)

SKY WAVE PROPAGATION – THE IONOSPHERE WAVES

30. In the ionospheric propagation, consider that the reflection takes place at a height 300 km and that the maximum density in the ionosphere corresponds to a refraction index of 0.8 at a frequency of 15 MHz. Determine the ground range for which this frequency is the MUF. Take the earth's curvature into consideration.

    [Hint: $\mu = \sqrt{1 - \dfrac{81\, N_{max}}{f^2}}$, $f_c = 9\sqrt{N_{max}}$, $D = 2\left[H + \dfrac{D^2}{8\,a}\right]\sqrt{\dfrac{f_{muf}}{f_c} - 1}\,$]

    (**Ans.** $N = 1 \times 10^{12}\, m^{-3}$, $f_c = 9 \times 10^6$ Hz, $D = 19188.17$ km)

31. The electron concentration at a height of 300 km is $10^{11}\, m^{-3}$. What is the maximum angle of incidence ψi; that can be used at a frequency of 10 MHz? What is the horizontal skip distance if the maximum angle ψi is used?

    [Hint: $n = \sin i = \sqrt{1 - \dfrac{81\, N}{f^2}}$, $D_{skip} = 2\, H\sqrt{\left[\dfrac{f_{muf}}{f_c}\right]^2 - 1}$, $f_c = 9\sqrt{N_{max}}\,$]

    (**Ans.** $D_{skip} = 6638$ km)

32. What are space waves? Explain their propagation and use in communication.
33. What are ordinary and extra ordinary waves? Explain their propagation?
34. Show that the maximum usable frequency is given by, where is the angle of incidence and f_c is the critical frequency.
35. Discuss the sky wave propagation. Show that for an ionospheric propagation to take place.

    $$\varepsilon_r = \left[\dfrac{1 - 81\, N}{f^2}\right]^{1/2}$$

    where N is the electron density of layer and/is the frequency of transmissions and is refractive index.

36. Derive the required formula and calculate the value of the factor by which the horizon distance of a transmitter will be modified if the gradient of refractive index of air near ground is $-0.05 \times 10^6\, m^{-1}$. The radius of earth is given as 6370 km.

    (**Ans.** The horizon distance of a transmitter can be modified by replacing r by r', i.e., 25/17 r)

37. Deduce an expression for critical frequency of an ionized region in terms of maximum ionization density.

    [Hint: $f_c = 9\sqrt{N_m}\,$]

38. What is tropospheric propagation? What are the various causes of attenuation and fading in tropospheric propagation?
39. What do you mean by fading? How it can be minimized?
40. Explain MUF and give any method for calculating the same.
41. Describe a method for determining the equivalent height of the region of ionosphere responsible for reflection of the radio waves.
42. Explain the bending of EM waves by ionized volumes of atmosphere. Derive the expressions for critical frequency of a layer.

CHAPTER 13

RADAR

OBJECTIVES

- Introduction
- Principle of RADAR
- Radar Block Diagram
- Range Determination
- Common Parameters of Radar Pulse
- Applications of Radar
- Radar Frequencies
- Classification of Radars
- Basic Pulse Radar System
- Simple form of Radar Equation
- Effect of Pulse Width on Range
- Pulse Repetition Frequency (PRF) and Range Ambiguities
- Minimum Detectable Signal (MDS)
- Radar Receiver Noise and Signal to Noise Ratio
- Continuous Wave (CW) Radar
- Frequency Modulated CW Radar
- Moving Target Indicator (MTI) and Doppler Radar
- Receiver Noise
- Low Noise Front Ends
- Duplexer
- Radar Display
- Radar Antennas

13.1 INTRODUCTION

Radar is an acronym for **Radio Detection and Ranging**. It is a system used to detect the object and determine the range (determine the distance), azimuth (bearing) and height of the airborne object. The term "radio" refers to the use of electromagnetic waves with wavelengths in the so-called radio wave portion of the spectrum, which covers a wide range from 10^4 km to 1 cm. Strong radio waves are transmitted by the antenna and a receiveing antenna listens for reflected

RADAR

echoes. By analysing the reflected signal, the reflector can be located, and sometimes identified. It can operate in darkness, haze, fog rain and snow .It has ability to measure distance with high accuracy in all weather conditions.

13.2 PRINCIPLE OF RADAR

The basic principle of operation of primary radar is simple to understand. However, the theory can be quite complex. An understanding of the theory is essential in order to be able to specify and operate primary radar systems correctly. The implementation and operation of primary radars systems involve a wide range of disciplines such as building works, heavy mechanical and electrical engineering, high power microwave engineering, and advanced high speed signal and data processing techniques. Some laws of nature have a greater importance here.

Basic Principle of Operation

Radar measurement of range, or distance, is made possible because of the properties of radiated electromagnetic energy:

- **This energy normally travels through space in a straight line, at a constant speed,** and will vary only slightly because of atmospheric and weather conditions.
- **Electromagnetic energy travels through air at approximately the speed of light,**

    300,000 kilometers per second

    or

    186,000 statute miles per second

    or

    162,000 nautical miles per second.

- **Reflection of electromagnetic waves**

The electromagnetic waves are reflected if they meet an electrically leading surface. If these reflected waves are received again at the place of their origin, then that means an obstacle is in the propagation direction.

The principle on which radar operates is very similar to the principle of sound-wave reflection. If you shout in the direction of a sound-reflecting object, you will hear an echo. If you know the speed of sound in air, you can then estimate the distance and general direction of the object. The time required for a return echo can be roughly converted to distance if the speed of sound is known. Radar uses electromagnetic energy pulses in the same way, as shown in figure 13.1. The radio-frequency (rf) energy is transmitted to and reflects from the reflecting object. A small portion of the energy is reflected and returns to the radar set. This returned energy is called an **ECHO**, just as it is in sound terminology. Radar sets use the echo to determine the direction and distance of the reflecting object.

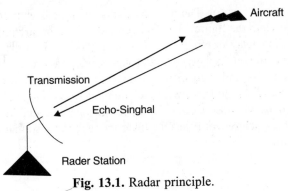

Fig. 13.1. Radar principle.

13.3 RADAR BLOCK DIAGRAM

A practical radar system requires five basic components as illustrated below:

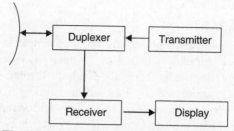

Fig. 13.2. A Simple Radar Block Diagram.

1. Transmitter

The transmitter creates the radio wave to be sent and modulates it to form the pulse train. The transmitter must also amplify the signal to a high power level to provide adequate range. The source of the carrier wave could be a Klystron, Traveling Wave Tube (TWT) or Magnetron. Each has its own characteristics and limitations.

2. Receiver

The receiver is sensitive to the range of frequencies being transmitted and provides amplification of the returned signal. In order to provide the greatest range, the receiver must be very sensitive without introducing excessive noise. The ability to discern a received signal from background noise depends on the signal-to-noise ratio (S/N).

3. Duplexer

This is a switch that alternately connects the transmitter or receiver to the antenna. Its purpose is to protect the receiver from the high power output of the transmitter. During the transmission of an outgoing pulse, the duplexer will be aligned to the transmitter for the duration of the pulse, PW. After the pulse has been sent, the duplexer will align the antenna to the receiver. When the next pulse is sent, the duplexer will shift back to the transmitter. A duplexer is not required if the transmitted power is low.

4. Antenna

A common antenna is used to transmit the EM energy in to the space and receives the reflected signals from the target. Furthermore, the antenna must focus the energy into a well-defined beam, which increases the power and permits a determination of the direction of the target.

5. Display

The display unit may take a variety of forms but in general is designed to present the received information to an operator. The most basic display type is called an **A-scan** (amplitude vs. Time delay) as shown in the figure 13.3. The vertical axis is the strength of the return and the horizontal axis is the time delay, or range. The A-scan provides no information about the direction of the target. The most common display is the PPI (plan position indicator). The A-scan information is converted into brightness and then displayed in the same relative direction as the antenna orientation. The result is a top-down view of the situation where range is the distance from the origin. The PPI is perhaps the most natural display for the operator and therefore the most widely used. In both cases, the synchronizer resets the trace for each pulse so that the range information will begin at the origin.

RADAR

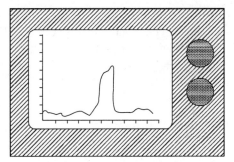

Fig. 13.3. "A"- Scope.

13.4 RANGE DETERMINATION

The detection and ranging part of the radar is accomplished by timing the delay between transmission of a pulse of radio energy and its subsequent return. If the time delay is Δt, then the range may be determined by the simple formula:

$$R = c \times (\Delta t/2) \qquad \qquad ...13.1$$

where $c = 3 \times 10^8$ m/s, the speed of light at which all electromagnetic waves propagate. The factor of two in the formula comes from the observation that the radar pulse must travel to the target and back before detection, or twice the range.

A radar pulse train is a type of amplitude modulation of the radar frequency carrier wave, similar to how carrier waves are modulated in communication systems. In this case, the information signal is quite simple, a single pulse repeated at regular intervals. The common radar carrier modulation, known as the pulse train is shown below. The common parameters of radar may be defined by Figure 13.4.

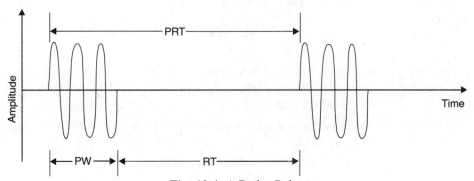

Fig. 13.4. A Radar Pulse.

13.5 COMMON PARAMETERS OF RADAR PULSE

(i) Pulse Width (PW)

PW has units of time and is commonly expressed in µs. PW is the duration of the pulse.

(ii) Pulse Repetition time (PRT)

PRT has units of time and is commonly expressed in ms. PRT is the interval between the start of one pulse and the start of another. PRT is also equal to the sum of pulse width and rest time, it may be written as

$$PRT = PW + RT.$$

(iii) Pulse Repetition Frequency (PRF)

PRF has units of time^{-1} and is commonly expressed in Hz (1 Hz = 1/s) or as pulses per second (pps). PRF is the number of pulses transmitted per second and is equal to the inverse of PRT.

$$PRF = 1/PRP$$

(iv) Rest Time (RT)

RT is the interval between pulses. It is measured in μs.

(v) Peak Power (Pt)

The power **Pt** in the radar equation is called by the radar engineer, the Peak Power. The peak power as used in the radar equation is not the instantaneous peak power of a sine wave. It is defined as the power averaged over that carrier frequency cycle which occurs at the maximum of the pulse power. Peak power is usually equal to one-half of the maximum instantaneous power.

(vi) Average Power (Pav)

The average power (Pav) is defined as the average transmitted power over the pulse repetition period. If the transmitted waveform is a train of rectangular pulses of width (T) and Pulse Repetition Period or Time (PRP or Tr).

$$PRP = 1/PRF \qquad \text{...13.2}$$
$$Pav \ (\text{Average Power}) = Pt \times (T/Tr) = Pt \times T \times PRF \qquad \text{...13.3}$$

(Average Power = Peak Power × Pulse Width /PRP =Peak Power × Pulse width × PRF)

(vii) Duty Cycle

The ratio of average power to the Peak power or pulse width to the PRP or Pulse width multiplied by PRF, is called the **Duty Cycle** of the radar. It may be written as

$$\text{Duty Cycle} = PW \times PRF \qquad 13.4$$

Example

The fraction of time the radar transmitter (*TX*) is generating pulse power is termed the Duty-Cycle. Thus, in any one second, radar producing 2 μs pulses at a rate of 1000 pulse per sec is "on".

The duty cycle may be calculated by using equation (13.4),

Duty cycle = PW × PRF
$2 \times 10^{-6} \times 1000 = 0.002$ s

The duty cycle is then 0.002s, and the average power is given by the product of duty cycle and peak power; thus if the peak power is 500 kW, the average power is than

Average Power (Pav) = Pt × Duty Cycle
Pav = $5000 \times 10^3 \times 0.001$
Pav = 1000 W

13.6 APPLICATIONS OF RADAR

Radar has been employed to detect target on the ground on the sea, in the air in space and even below ground. The major areas of radar applications are briefly described below:

1. **Military Application**. In military radar have wide applications. Radar is used by the military especially for surveillance purpose of the enemy aircraft. The surveillance radar detect ,locate and identified the enemy targets. It is also being used for providing navigational aids to military or as well as civil aircraft. It is used for tracking the target and controlling of weapons..

2. **Targeting radars** use the same principle but scan a much smaller area far more often, usually several times a second or more, where a search radar might scan a few times per minute.

RADAR

Some targeting radars have a range gate that can track a target, to eliminate clutter and electronic counter-measures.

3. **Navigational radars** resemble search radar, but use very short waves that reflect from earth and stone. They are common on commercial ships and long-distance commercial aircraft.

4. **Weather radars** can resemble search radars. These radar uses radio waves with horizontal, dual (horizontal and vertical), or circular polarization. The frequency selection of weather radar is a performance compromise between precipitation reflectivity and attenuation due to atmospheric water vapour. Some weather radar uses doppler to measure wind speeds

5. **General purpose radars** are increasingly being substituted for pure navigational radars. These generally use navigational radar frequencies, but modulate the pulse so the receiver can determine the type of surface of the reflector. The best general-purpose radars distinguish the rain of heavy storms, as well as land and vehicles. Some can superimpose sonar and map data from GPS position.

6. **Radar Proximity Fuses** are attached to anti-aircraft artillery shells or other explosive devices, and detonate the device when it approaches a large object. They use a small rapidly pulsing omnidirectional radar, usually with a powerful battery that has a long storage life, and a very short operational life. The fuses used in anti-aircraft artillery have to be mechanically designed to accept fifty thousand gravities of acceleration, yet still be cheap enough to throw away.

7. **Radar altimeters** measure an aircraft's true height above ground.

8. **Air Traffic Control** uses Primary and Secondary Radars to control and guide the aircraft.

9. **Law Enforcement and Highway safety**: The radar speed meter, familiar to many is used by police for enforcing speed limit.

13.7 RADAR FREQUENCIES

Conventional radar generally operates in what is called the microwave region (a term not rigidly defined). Microwaves are since they are defined in terms of their wavelength (λ) in the sense that micro refers to tinyness. In short a microwave is a signal that has a wavelength of 1 foot or less i.e. $\lambda \leq 30.5$ cms. So all the frequencies above approximately 1000 Mhz (1 Ghz) to about 1000 Ghz are microwave frequencies. IEEE standard Radar Frequencies letter –Band nomenclature given in Table 13.1.

Table 13.1. Ieee Standard Radar Frequencies

Band Designation	Nominal Frequency range	Specific radar frequency range Based on ITU
HF	3- 30 Mhz	138-144 Mhz
VHF	30-300 Mhz	216-225 Mhz
UHF	300-1000 Mhz	420 –450 MhZ
L	1- 2 Ghz	850-942 Mhz
S	2-4 Ghz	1215-1400 MhZ
C	4-8 Ghz	2300-2500 Mhz
X	8 –12 Ghz	2700-3700 Mhz
KU	12-18 GhZ	5250 –5925 Mhz
K	18-27 Ghz	8500-10680 Mhz
Ka	27-40 Ghz	13.4-14 Ghz
V	40-75 Ghz	15.7-17.7 Ghz
W	75-110 GhZ	24.05-24.25 Ghz
		33.4-36 Ghz

		59-64 Ghz
		76-81 Ghz
		92-100 Ghz
		126-142 Ghz
		144-149 Ghz
mm	110-300 Ghz	231-235 Ghz
		238-248 Ghz

13.8 CLASSIFICATION OF RADAR

Radar systems may be divided into types based on the designed use. This section presents the general characteristics of several commonly used radar systems:

- Air-Defense Radars
- Air Traffic Control Radars
- Fire-control radar or tracking radar
- Speed gauges
- Airborne radar
- Mortar locating radar
- Radar Satellites
- Weather Radar
- Ground Penetrating radar

Depending upon the desired information, the radar units must have different qualities and technologies.

According to the techniques radar units are classified as follows:

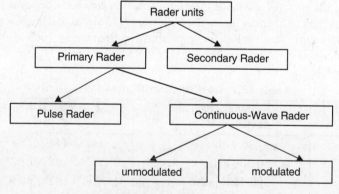

Fig. 13.5. Radar Classifications.

Primary Radar

Primary radar transmits high-frequency signals which are reflected at targets. The arisen echoes are received and evaluated. This means, unlike secondary radar units a primary radar unit receive its own emitted signals as an echo again.

Secondary Radar

At these radar units the airplane must have a transponder (transmitting responder) on board and receives an encoded signal of the secondary radar unit. An active also encoded response signal, which is returned to the radar unit then is generated in the transponder. In this response can be contained much more information, as a primary radar unit is able to acquire (E.g. an altitude, an identification code or also any technical problems on board such as a radio contact loss...) Example of secondary radar is an Identification of Friend and Foe (IFF).

Pulse Radar

Pulse radar units transmit a high-frequency impulsive signal of high power. After this a longer break in which the echoes can be received follows before a new transmitted signal is sent out. Direction, distance and sometimes if necessary the altitude of the target can be determined from the measured antenna position and propagation time of the pulse-signal.

Continuous- Wave Radar (CW Radar)

CW radar units transmit a high-frequency signal continuously. The echo signal permanently is received and processed. The receiver needn't be mounted at the same place as the transmitter absolutely. Every firm civil radio transmitter can work as a radar transmitter at the same time, if a remote receiver compares the propagation times the direct one with the reflected signal. Tests are known that the correct location of an airplane can be calculated from the evaluation of the signals by three different television stations.

Unmodulated CW- Radar

The transmitted signal of these equipments is constant in amplitude and frequency. These equipments are specialized in speed measuring. Distances cannot be measured. E.g. these equipments are used as speed gauges of the police.

Modulated CW- Radar

The transmitted signal is constant in the amplitude but modulated in the frequency. This one gets possible after the principle of the propagation time measurement with that again. It is advantage of these equipments that an evaluation is carried out without reception break and the measurement result is therefore continuously available. These radar units are used everywhere there where the measuring distance isn't too large and it's necessary a continuous measuring (e.g. an altitude measuring in airplanes or as weather radar/wind profiler).

Radar units whose transmitting impulse is too long to get a well distance resolution also use a similar principle of obtaining. Often these equipments modulate its transmitting pulse additional, to obtain a distance resolution also within the transmitting pulse with help of the pulse-compression.

13.9 BASIC PULSE RADAR SYSTEM

A very basic radar block diagram of pulsed radar system was shown in figure 13.2. A more detailed block diagram of high –power pulsed radar set is shown in figure 13.6.

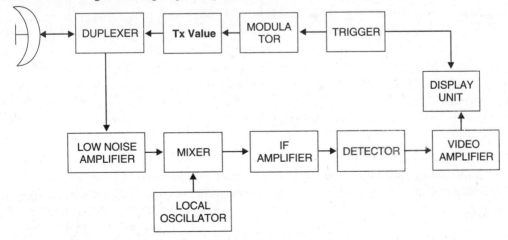

Fig. 13.6. Pulsed Radar Block Diagram.

Duplexer

This is a switch that alternately connects the transmitter or receiver to the antenna. Its purpose is to protect the receiver from the high power output of the transmitter. During the transmission of an outgoing pulse, the duplexer will be aligned to the transmitter for the duration of the pulse, PW. After the pulse has been sent, the duplexer will align the antenna to the receiver. When the next pulse is sent, the duplexer will shift back to the transmitter. A duplexer is not required if the transmitted power is low.

Antenna

The antenna takes the radar pulse from the transmitter and puts it into the air. Furthermore, the antenna must focus the energy into a well-defined beam which increases the power and permits a determination of the direction of the target. The antenna must keep track of its own orientation, which can be accomplished by a synchro-transmitter. There are also antenna systems, which do not physically move but are steered electronically.

Transmitter Valve

The transmitter creates the radio wave to be sent and modulates it to form the pulse train. The transmitter must also amplify the signal to a high power level to provide adequate range. The source of the carrier wave could be a Klystron, Traveling Wave Tube (TWT) or Magnetron. Each has its own characteristics and limitations.

RADAR Modulator

The radar modulator is a device, which provides the high power to the transmitter tube to transmit during transmission period. It makes the transmitting tube ON and OFF to generate the desired waveform. Modulator allows storing the energy in a capacitor bank during the rest time. The stored than can be put in to the pulse when transmitted. It provides rectangular voltage pulses which acts as the supply voltage to the output tube, thus switching it ON and OFF as required. Commonly thyratron valve is being used as switching device, which is a gas filled valve.

Trigger Source (Synchronizer)

The synchronizer coordinates the timing for range determination. It regulates that rate at which pulses are sent (i.e. sets PRF) and resets the timing clock for range determination for each pulse. Signals from the synchronizer are sent simultaneously to the transmitter, which sends a new pulse, and to the display, which resets the return sweep.

Receiver

Receiver is usually of superhetrodyne type whose function is to detect the desired echo signal in the presence of noise, interference and clutter. The receiver in pulsed radar consists of the Low noise RF Amplifier, mixer, local oscillator, IF amplifier, detector, video amplifier and radar display.

Low Noise RF Amplifier is the first stage of the receiver. It is low noise transistor amplifier or a parametric amplifier or a TWT amplifier. Silicon bipolar transistor is used at lower radar frequencies (below L-band 1215 to 1400 Mhz) and the GaAs FET is preferred at higher frequencies. It amplifies the received weak echo signal.

Mixer and Local Oscillator

These convert RF signal output from RF amplifier to comparatively lower frequency levels called Intermediate Frequency (IF). The typical value for pulse radar is 30 MHz or 60 MHz.

I F Amplifier

It consists of a cascade of tuned amplifiers, these can be synchronous, that is, all tuned to the same frequency and having identical band pass characteristics. If a really large bandwidth is

RADAR

needed, the individual IF amplifiers may be stagger tuned.). The typical value for pulse radar is 30 Mhz or 60 Mhz.

Detector

Detector is often a Schottky-barrier diode which extracts the pulse modulation from the IF amplifier output. The detector output is then amplified by the Video Amplifier to a level where it can be properly displayed usually on CRT (Cathode Rays Tubes) directly or via digital signal processors. Synchronizing pulses are applied by the trigger source to the display unit.

Display Unit

The received video signals are displays on the CRT (Cathode Rays Tube) for observations and action.

13.10 SIMPLE FORM OF RADAR EQUATION

The radar equation relates the range of radar to the characteristics of the Tx (transmitter), Rx (receiver), Ae (antenna), target and the environment. It is useful to determine the maximum range at which particular radar can detect a target and its makes to understand the factor affecting radar performance. In this section, the simple form of equation is derived.

The radar equation is an important tool for following aspect:

1. Assessing the performance of radar.
2. Designing of new radar systems.
3. Assessing the technical requirement for new radar procurement.

If the power of radar Tx is denoted by **Pt** and if an isotropic antenna (which will radiate energy uniformly in all direction) then the power density at a distance "**R**" from the radar is equal to the radiated power divided by the surface area $4\pi R^2$ of an imaginary sphere of radius R, or ie ,power density at range R from an isotropic antenna

$$= Pt/ 4\pi R^2 \text{ watt / m}^2 \qquad \ldots 13.5$$

However, we employ directive antenna (with narrow beam width) to concentrate the radiated power Pt in a particular direction.

The gain of an antenna is a measure of the increased power density radiated in some direction as compared to the power density that would appear in that direction from an isotropic antenna. The maximum gain "G" of an antenna may be defined as

G = Maximum power density radiated by a directive antenna / power density radiated by loss less isotropic antenna

Power density at a distance R from directive antenna of power Gain "G",

$$= Pt\ G\ /\ 4\pi R^2 \text{ watt / m}^2 \qquad \ldots 13.6$$

The target intercepts a portion of the incident energy and re radiates it in various directions. It is only the power density re radiated in the direction of the radar (echo signal) that is of interest. The radar cross section of the target determines the power density returned to the radar for a particular power density incident on the target. It is denoted by σ and is often called, for short, target cross section or radar cross section. The radar cross section is defined by the following:

$$\text{Reradiated power density back at the radar} = \frac{Pt.G\sigma}{4\pi R^2\ 4\pi R^2} \qquad \ldots 13.7$$

The radar cross section has units of area, but it can be misleading to associate the radar cross section directly with the targets physical size. Radar cross-section is more dependent on the targets shape than on its physical size.

The radar antenna received a portion of the echo power. If the effective area of receiving antenna is denoted Ae, the power Pr received by the radar is,

$$Pr = \frac{Pt.G.\sigma.Ae}{4\pi R^2 \cdot 4\pi R^2} \qquad ...(13.8)$$

The maximum range of radar Rmax is the distance beyond which the target can not detected. It occur when the received signal power Pr just equals the minimum detectable signal (mds) or (Smin).

Substituting, Smin = Pr in equation (13.8)

$$S\min = \frac{Pt.G.\sigma.Ae}{4\pi R^2 \cdot 4\pi R^2} \text{ watts} \qquad ...13.9$$

ie, $$R^4 \max \frac{Pt.G.\sigma.Ae}{4(\pi)^2 (Sm \times in)} \qquad ...13.10$$

$$R^4 \max \frac{Pt.G.\sigma.Ae}{4(\pi)^2 (Sm \times in)} \qquad ...13.11$$

$$\boxed{R_{\max} \sqrt[4]{\frac{P_t.G.\sigma.A_e}{4(\pi)^2 S\min}}} \qquad ...13.12$$

Where,
Pt = Transmitter Power,
G = Maximum Gain of Antenna
Ae = Aperture area of receiving Antenna
σ = Cross section area of the target
Smin = Minimum detectable signal (mds)
Rmax = Maximum range of radar

This is the fundamental form of radar range equation. The important parameters are the transmitting gain and the receiving effective area. The transmitter power Pt has not been specified as either the average or the peak power.

If the same antenna is used for both transmitting and receiving, as it usually is in radar, antenna theory gives the relationship between transmit gain G and the receive effective area Ae as

$$G = \frac{4\pi.Ae}{\lambda^2} \qquad ...13.13$$

Where λ = wavelength (λ = c/f where c = velocity of propagation and f = frequency) equation (13.12) can be substituted, first for Ae and the G, to give two other forms of the radar equation,

$$\boxed{R\max = \sqrt[4]{\frac{Pt.G^2.\sigma.\lambda^2}{(4\pi)^3 S\min}}} \qquad ...13.14$$

Or

$$\boxed{R\max = \sqrt[4]{\frac{Pt.Ac^2.\sigma.}{4\pi\lambda^2 S\min}}} \qquad ...13.15$$

These three forms of the radar equations (13.12, 13.14 & 13.15) are basically the same.

In the radar range is to be doubled, we have to increase the transmitter power 16 times since Rmax $\infty \frac{1}{4}$ root of peak power.

RADAR

13.11 EFFECT OF PULSE WIDTH (PW) ON RANGE

The duration of the pulse and the length of the target along the radial direction determine the duration of the returned pulse. In most cases the length of the return is usually very similar to the transmitted pulse. In the display unit, the pulse (in time) will be converted into a pulse in distance. The range of values from the leading edge to the trailing edge will create some uncertainty in the range to the target. Taken at face value, the ability to accurately measure range is determined by the pulse width.

If we designate the uncertainty in measured range as the **range resolution**, R_{RES}, then it must be equal to the range equivalent of the pulse width, namely:

$$\boxed{R_{RES} = cPW/2} \qquad \text{...13.16}$$

Now, you may wonder why not just take the leading edge of the pulse as the range which can be determined with much finer accuracy. The problem is that it is virtually impossible to create the perfect leading edge. In practice, the ideal pulse will really appear like:

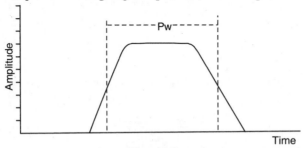

Fig. 13.7

To create a perfectly formed pulse with a vertical leading edge would require an infinite bandwidth. In fact you may equate the **Bandwidth, β,** of the transmitter to the minimum pulse width, PW by:

$$\boxed{PW = 1/\beta} \qquad \text{...13.17}$$

Given this insight, it is quite reasonable to say that the range can be determined no more accurately than c PW / 2 or equivalently

$$\boxed{R_{RES} = c/2\beta} \qquad \text{...13.18}$$

In fact, high-resolution radar is often referred to as wide-band radar, which you now see as equivalent statements. One term is referring to the time domain and the other the frequency domain. The duration of the pulse also affects the minimum range at which the radar system can detect. The outgoing pulse must physically clear the antenna before the return can be processed. Since this lasts for a time interval equal to the pulse width, PW, the minimum displayed range is then:

$$\boxed{R_{MIN} = c \times PW/2} \qquad \text{...13.19}$$

The minimum range effect can be seen on a PPI display as a saturated or blank area around the origin.

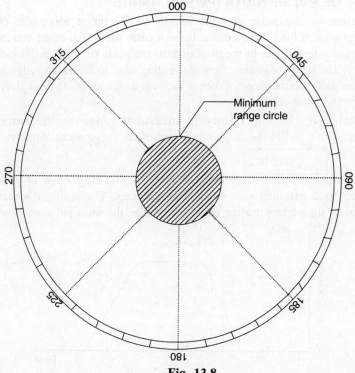

Fig. 13.8

Increasing the pulse width while maintaining the other parameters, the same will also affect the duty cycle and therefore the average power. For many systems, it is desirable to keep the average power fixed. Then the PRF must be simultaneously changed with PW in order to keep the product PW x PRF the same. For example, if the pulse width is reduced by a factor of ½ in order to improve the resolution, then the PRF is usually doubled.

13.12 PULSE REPETITION FREQUENCY (PRF) AND RANGE AMBIGUITIES

The frequency of pulse transmission affects the maximum range that can be displayed. Recall that the synchronizer resets the timing clock as each new pulse is transmitted. Returns from distant targets that do not reach the receiver until after the next pulse has been sent will not be displayed correctly. Since the timing clock has been reset, they will be displayed as if the range where less than actual. If this were possible, then the range information would be considered ambiguous. An operator would not know whether the range were the actual range or some greater value.

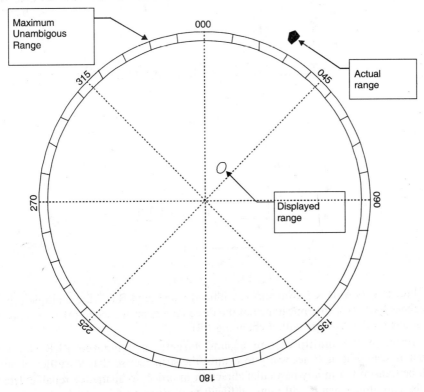

Fig. 13.9

The maximum actual range that can be detected and displayed without ambiguity, or the **maximum unambiguous range**, is just the range corresponding to a time interval equal to the pulse repetition time, PRT. Therefore, the maximum unambiguous range,

$$R_{UNAMB} = c\ PRT/2 = c/(2PRF) \qquad \ldots 13.20$$

The PRF is determined primarily by the maximum range at which targets are expected. If the PRF is made too high, the likelihood of obtaining target echoes from the wrong pulse transmission is increased. Echoes signals received after an interval exceeding the PRP (Pulse Repetitation Period) are called **multiple times around echoes.**

They can result in erroneous or confusing range measurement .The nature of some multiple-time- around echoes may cause them to be labeled as ghost or angel targets or even flying saucers.

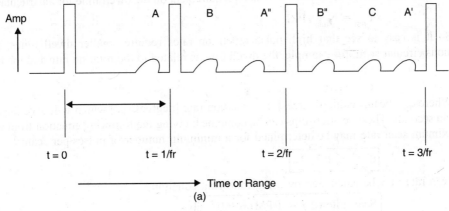

(a)

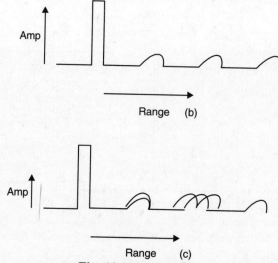

Fig. 13.10. (a,b and c).

(a) Three targets A, B & C where A is within Runamb and B &C are multiple time around echoes (targets); (b) appearance of the three targets on A –scope (c)appearance of three targets on the A-scope with a changing PRF

To minimize the multiple time around targets, a staggered PRF can be used. When radar is scanning, it is necessary to control the scan rate so that a sufficient number of pulses will be transmitted in any particular direction in order to guarantee reliable detection. If too few pulses are used, then it will more difficult to distinguish false targets from actual ones. False targets may be present in one or two pulses but certainly not in ten or twenty in a row. Therefore to maintain a low false detection rate, the number of pulses transmitted in each direction should be kept high, usually above ten.

For systems with high pulse repetition rates (frequencies), the radar beam can be repositioned more rapidly and therefore scan more quickly. Conversely, if the PRF is lowered the scan rate needs to be reduced. For simple scans it is easy to quantify the number of pulses that will be returned from any particular target. Let τ represent the **dwell time**, which is the duration that the target remains in the radar's beam during each scan. The number of pulses, N that the target will be exposed to during the dwell time is:

$$\boxed{N = \tau \text{ PRF}} \qquad \ldots 13.21$$

We may rearrange this equation to make a requirement on the dwell time for a particular scan

$$\tau_{min} = N_{min}/\text{PRF} \qquad \ldots 13.22$$

So it is easy to see that high pulse repetition rates require smaller dwell times. For a continuous circular scan, for example, the dwell time is related to the rotation rate and the beam-width.

$$\tau = \theta_B/\theta_s \qquad \ldots 13.23$$

Where q_B =beam-width [degrees] θ_s = rotation rate [degrees/sec] which will give the dwell time (t)in seconds. These relationships can be combined; giving the following equation from which the maximum scan rate may be determined for a minimum number of pulses per scan:

$$\boxed{\theta_s = \theta_B \text{ PRF}/N} \qquad \ldots 13.24$$

Scan rate may be calculated by using following equation,

$$\boxed{\text{Scan rate } (\theta_s) = (\text{RPM} \times 360°)/60 \text{ s}} \qquad \ldots 13.25$$

RADAR

13.13 MINIMUM DETECTABLE SIGNAL

The receiver is sensitive to the range of frequencies being transmitted and provides amplification of the returned signal. In order to provide the greatest range, the receiver must be very sensitive without introducing excessive noise. The ability to discern a received signal from background noise depends on the signal-to-noise ratio (S/N).

The background noise is specified by an average value, called the **noise-equivalent-power (NEP)**. This directly equates the noise to a detected power level so that it may be compared to the return. Using these definitions, the criterion for successful detection of a target is

$$P_r > (S/N) \text{ NEP},$$

where P_r is the power of the return signal. Since this is a significant quantity in determining radar system performance, it is given a unique designation, S_{min}, and is called the **Minimum Signal for Detection**.

$$S_{min} = (S/N) \text{ NEP} \qquad \ldots 13.26$$

Since S_{min}, expressed in Watts, is usually a small number, it has proven useful to define the decibel equivalent, MDS, which stands for **Minimum Discernible Signal**.

$$\text{MDS} = 10 \text{ Log } (S_{min}/1 \text{ mW}) \qquad \ldots 13.27$$

When using decibels, the quantity inside the brackets of the logarithm must be a number without units. In the definition of MDS, this number is the fraction $S_{min}/1$ mW. As a reminder, we use the special notation dBm for the units of MDS, where the "m" stands for 1 mW. This is shorthand for decibels referenced to 1 mW, which is sometimes written as dB/mW.

In the receiver, S/N sets a threshold for detection, which determines what will be displayed and what will not. In theory, if S/N = 1, then only returns with power equal to or greater than the background noise will be displayed.

Detection is based on establishing a threshold level at the output of the receiver

13.14 RADAR RECEIVER NOISE AND SIGNAL-TO-NOISE RATIO

"**Noise** is unwanted E M energy which interferes with the ability of the receiver to detect the wanted signal". It may generate within receiver it self or it may enter through the receiving antenna along with the desired signal.

The noise generated by thermal motion of the conduction electron in receiver input stages is called the **Thermal Noise or Johnson Noise**. Its magnitude is directly proportional to the bandwidth and the absolute temperature of the resistive potion of the input circuit. The available thermal noise power (watt) generated at input of a radar receiver of bandwidth Bn (hertz) at a temperature T (degree Kelvin) is

$$\text{Available thermal noise power} = K \cdot T \cdot Bn \qquad \ldots 13.28$$

Where,

$$K = \text{Boltzman's Constant} = 1.38 \times 10^{-23} \text{ Joules/degree}$$

The bandwidth of a superhetrodyne receiver is taken as at I F Amplifier (matched filter)
The bandwidth noise (Bn) is defined as

$$Bn = \frac{\infty |H(f)|^2 \, dt}{|H(fo)|^2} \qquad \ldots 13.213$$

Where

H (f) = frequency response function of the IF amplifier
fo = frequency of the maximum response (usually occurs at mid band)

The noise power in practical receiver is normally more than the thermal noise. The measure of the noise out of a real receiver to that noise out from the ideal receiver with only thermal noise is called the Noise Figure (Fn) and is defined as

$$Fn = \frac{\text{Noise out of practical receiver}}{\text{Noise out of ideal receiver at standard temp } T_0}$$

$$Fn = N_{out} / k . T_0 . B_n . G_a \qquad \ldots 13.30$$

Where,

No = noise output from receiver
Ga = available gain

The temperature To is to be taken 2130 degree Kelvin (according to the Institute of Radio Engineers)

The noise No is measured over the linear portion of the receiver input-output characteristics, usually at the o/p of the IF amplifier.

The receiver bandwidth Bn is that of the IF amplifier in most radar receiver. The

Available gain Ga is the ratio of the signal out (So) to the signal in (Si), and kToBn is the input noise (Ni) in ideal receiver.

$$Fn = \frac{Si/Ni}{So/No} \qquad \ldots 13.31$$

$$= (Si / So) \times (No / Ni)$$
$$= (Si / G . Si) \times [G \times (Ni + Nri) / Ni]$$
$$Fn = [1 + Nri / Ni]$$

Thus, $(Nri / Ni) = (Fn - 1)$

$Nri = Ni (Fn-1)$

Or $\quad Nri = k . T_0 . B (F-1) = S_{(min)}$

$k . T_0 . B (F-1) = S_{(min)}$

Where $\quad Nri$ = Noise power generated at the receiver input.

Substitute the value of $S_{(min)}$ in equation (13.12),

$$R_{max} = 4\sqrt{\frac{P_t . A_e^2 \sigma}{4\pi\lambda^2 k . T_0 . B(f-1)}} \qquad \ldots 13.22$$

13.15 CW RADAR

It is well known in the fields of optics and acoustic that if either the source of oscillation or the observer of the oscillation is in motion, an apparent shift in frequency will result. This is the **Doppler Effect** and is the basis of **Continuous wave (CW) radar.**

If the range to the target is R, than the total number of wavelength λ in the two-way path from radar to target and return is 2 R /λ. Each wavelength corresponds to a phase change of 2π radian. The total phase change in two-way propagation path is then

$$\Phi = 2 \pi \times 2R/ \lambda \text{ rad}$$
$$\Phi = 4 \pi R / \lambda \text{ rad} \qquad \ldots 13.33$$

If the target is in motion relative to the radar, R is changing and so will phase (Φ), differenting equation (13.33) with respect to time gives the rate of change of phase, which called the angular frequency,

$$\omega_d = 2 \pi f_d = d\Phi /dt$$
$$= (4 \pi / \lambda).(dR/dt)$$

RADAR

Therefore,
$$= 4\pi V_r / \lambda$$

Where $V_r = dR/dt$ is the radial velocity (meter/sec) or rate of change of range with time.

In figure (13.11) the angle between the target velocity vector and the radar line of sight to the target is θ, the $V_r = V\cos\theta$, where v is the speed or magnitude of vector velocity. The rate of change of Φ with time is the angular frequency
$$\omega d = 2\pi f d$$

Where fd is Doppler frequency shift.
$$2\pi f d = 4\pi V_r / \lambda$$
$$f_d = 2 V_r / \lambda \qquad \ldots 13.34$$
$$f_d = 2 f V_r / c \text{ (since } \lambda = c/f\text{)} \qquad \ldots 13.35$$

The radar frequency is $f = c/\lambda$ and the velocity of propagation $c = 3 \times 10^8$ m/s. If the Doppler frequency in hertz, the radial in knots, and the radar wave length in meters, we can write
$$\boxed{f_d = 1.03\ V_r\ (kt) / \lambda(m)} \qquad \ldots 13.36$$
$$V_r\ (kt) / \lambda(m) \qquad \ldots 13.37$$

The Doppler frequency in hertz can also be approximately expressed as 3.43 V_r f,
Where f is the radar frequency in GHz and V_r is in Knots.

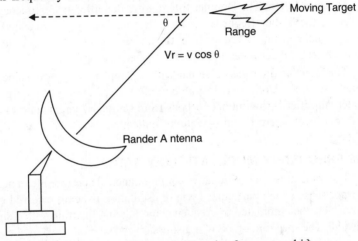

Fig. 13.11. Geometry of Radar and Target in Doppler frequency shift.

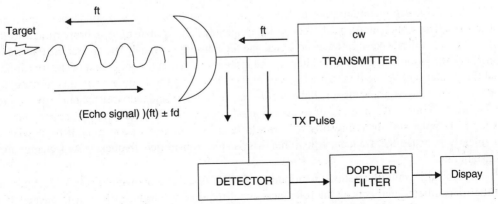

Fig. 13.12. The block diagram of simple CW Radar.

The block diagram of simple CW Radar is shown in Fig. 13.12. It utilized the Doppler frequency shift to detect moving target. A CW radar transmits while it receives without the Doppler shift produced by the movement of the target, the weak signals (echo) would not be detected in the presences of much stronger signal from the transmitter. Filtering in the frequency domain is used to separate the weak Doppler shifted echo signal from the strong transmitter signal in a CW radar.

Transmitter: The transmitter generates a continuous (unmodulated) sinusoidal oscillation at frequency ft, which is than radiated by the antenna. A portion of this radiated energy intercepted by the target and the re radiated energy is collected by the receiver antenna. If the target in motion with a velocity Vr relative to the radar the received signal will be shifted in frequency from the transmitted frequency by an amount ± fd. The plus sign applies when the distance between radar and target is decreasing. Thus the echo signal from a closing target has a higher frequency than that which was transmitted. The minus sign applies when the distance is increasing.

Detector (Mixer): To utilize the Doppler frequency shift radar must be able to recognize that the received echo signal has a different frequency from that which was transmitted. This is the function of that portion of the transmitter signal that finds its way (or leaks) in to the receiver. The TX leakage signal act as a reference to determine that frequency change has taken place. The detector or mixer multiplies the echo signal at a frequency ft ± fd with the TX leakage signal (ft).

Doppler Filter: The Doppler filter allows the difference frequency from the detector to pass and rejects the higher frequencies. The filter has a lower frequency cut off to remove the TX leakage signal and clutter echoes from the receiver output. The maximum radial velocity expected of moving target determines the upper frequency cut off.

The Doppler filter passes signal with a Doppler frequency fd located within its pass band, but the sign of the Doppler is lost along with the direction of the target motion. The purpose of Doppler amplifier is to eliminate echoes from stationary targets and to amplify the Doppler echo signal to a level where it can operate an indicating device such as frequency meter or a pair of earphone.

13.16 FREQUENCY MODULATED CW RADAR

The transmitter in CW radar is not modulated. Therefore it can neither provide range of the target nor sense which particular cycle of oscillation is being received at any instant. This major drawback can be eliminated by frequency modulating the transmitted signal even it increases the bandwidth. The spectrum of a CW transmission can be broadened by the application of modulation, amplitude, frequency, or phase. An example of an amplitude Modulation is the pulse radar.

Theory of Operation

A widely used technique to broaden the spectrum of CW radar is to frequency –modulate the carrier. The timing mark is the changing frequency. The transit time is proportional to the difference in frequency between the echo signal and the transmitter signal. The greater the transmitter frequency deviation in a given time interval, the more accurate the measurement of the transit time and the greater will be the transmitted spectrum. By measuring the frequency of the return signal, the time delay between transmission and reception can be measure and therefore the range determined as before. Of course, the amount of frequency modulation must be significantly greater than the expected Doppler shift or the results will be affected. The simplest way to modulate the wave is to linearly increase the frequency. In other words, the transmitted frequency will change at a constant rate.

A block diagram illustrating the principle of FM-CW radar is shown in Fig. 13.13. A portion of the transmitter signal acts as the reference signal required to produce the beat frequency. It is introduced directly into the receiver via a cable. Ideally, the isolation between transmitting and receiving antenna is made sufficiently large so as to reduce to a negligible level the transmitter leakage, which arrives at the receiver via the coupling between the antennas. The beat frequency is amplified and limited to remove any amplitude fluctuations. The frequency of amplitude limited beat note is usually measured with a cycle –counting frequency meter calibrated in distance.

FM CW radar application:

One of the major applications of the FM-CW radar principle has been as an altimeter on board aircraft to measure height above the ground .The large target cross section and the relatively short range required of altimeters permit low transmitter power and the low antenna gain. Since the relative motion, and hence the Doppler velocity between the aircraft and ground, is small the effect of the Doppler frequency shift may usually be neglected.

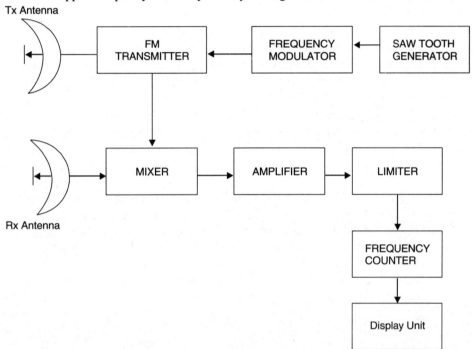

Fig. 13.13. Block Diagram of Fm Cw Radar.

13.17 MOVING TARGET INDICATOR (MTI) AND DOPPLER RADAR

The radars discussed till now were required to detect targets in the presence of noise. But in practical radars have to deal with more than receiver noise when detecting targets while they can also receives echoes from the natural environment such as land, sea and weather. These echoes are called **clutter**, since they tend to clutter the radar display with unwanted information's. Clutter echoes signal has greater magnitude then echo signal receives from the aircraft. When an aircraft echo and a clutter echo appear in the same radar resolution cell, the aircraft might not be detected. But the Doppler effect permits the radar to distinguish moving targets in the presence of fixed target even the echo signal from fixed targets has comparatively greater magnitude than the moving targets such as aircraft.

Echo signals from fixed targets are not shifted in frequency, but the echo from a target moving with relative velocity Vr will be shifted in frequency by an given by the Doppler formula, $f_d = 2v_r/\lambda$, when λ is wave length of the transmitted signal.

Pulse radar, which makes use of the Doppler shift for detecting moving targets, is either, MTI radar or a pulse Doppler radar. The MTI radar has a PRF low enough to not have any range ambiguities $R_{un} = c/f_p$,(such as multiples around echo). It does however have many ambiguities in the Doppler domain (such as blind speed).

The pulses Doppler radar on the other hand have a PRF large enough to avoid Doppler ambiguities but it can have numerous range ambiguities. These are also a medium PRF pulse Doppler that accepts both range and Doppler ambiguities.

13.17.1 MTI Radar (Using Power Amplifier as Transmitter)

Typically, MTI radar can extract the moving target echo from the clutter echo even if the clutter echo is 20-30db greater than the moving target echo. A simple block diagram of MTI radar that uses a power amplifier as the transmitter is shown in figure 13.14. In figure 13.14, an oscillator called the Coho, which stands for coherent oscillator, supplies the coherent reference.

The Coherent Oscillator (Coho) is a stable oscillator when frequency is the same as the IF used in the receiver. In addition to providing references signal, the output of the Coho fc is also mixed with the local oscillator frequency fl. The local oscillator must be a stable oscillator and is called **stable local oscillator (STALO).**

The local oscillator of an MTI radar's superhetrodyne receiver must be more stable than the local oscillator for a radar that does not employing Doppler .If the phase of the local oscillator were to change significantly between pulses, an uncancelled clutter residue can result at the output of the delay line canceller which might be mistaken for a moving target even though only clutter were present.

This IF stage is designed as a matched filter instead of an amplitude detector, there is a phase detector following the IF stage. This is a mixer like device that combines the received signal (at IF) and the reference signal from the Coho so as to produce the difference between the received signal and the reference signal frequencies. This difference is the Doppler frequencies. Coherency with the transmitted signal is obtained by using the sum of the Coho and the stable signal as the input signal to the power amplifier. Thus the transmitter frequency is the sum of the STALO frequency f_1 and the Coho frequency fc. This is accomplished in the mixer (adjacent to PA) shown in figure 13.14.

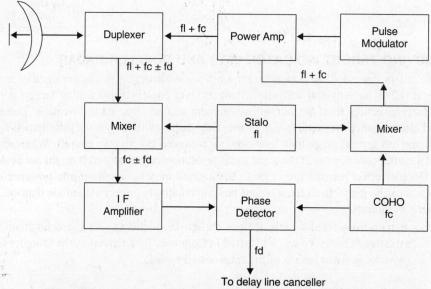

Fig. 13.14. Block Diagram of MTI Radar (Using Power Amplifier as Transmitter).

RADAR

The output of the phase detector is the input to the delay line canceller and acts as a high pass filter to separate the Doppler shifted echo signal of moving targets from the unwanted echoes of stationary clutters. The Doppler filter might a signal delay line canceller. The video portion of the receiver is divided into two channels. One is a normal video channel. In the other, the video signal experiences a time delay equal to one PRP equal to the reciprocal of the PRF. The outputs from the two channels are subtracting from one another. The fixed targets with unchanging amplitude from pulse to pulse are canceled on subtraction. However the amplitude of the moving target echoes are not constant from pulse to pulse and subtraction results in an uncancelled residue. The output of the subtraction circuit is bipolar video, just as was the input. The bipolar video must convert to unipolar video by a full wave rectifier before presenting on PPI display.

The power amplifier is a good transmitter for MTI radar, since it can have high stability and is capable of high power. The pulse modulator turns the amplifier ON and OFF to generate the radar pulses. The klystron and traveling wave tube (TWT) are usually preferred type of vacuum tube amplifier. Solid-state transistors are being used at lower frequencies. It has the advantage of good stability and does not need pulse modulator.

13.17.2 Blind Speed

Blind speed is a very serious problem in MTI radar. When the Doppler shift equals an integer multiple of PRF (Pulse Repetition Frequency), the delay line canceller not only eliminates the DC components caused by clutter but it also rejects moving targets. In other words, blind speed can be explained as the relative velocities at which the MTI response is zero called **blind speeds**.

The response from the single delay line canceller will be zero, when

$$f_d = n/T = n f_r$$

Where n=0, 1,2,3, ———— T = PRT and f_r is the PRF.

The nth blind speed V_n can be expressed as

$$V_n = n\lambda /2 T = n\lambda f_r /2$$

Where, n = 1,2,3, —

If λ is in meters, f_r in hz and the relative velocity in knots,

The blind speeds are written as

$$V_n = n\lambda f_r / 1.02 = n\lambda f_r \qquad ...13.38$$

This limitation of blind speed can be reduced up to some extent by employing following steps:

1. Operate the radar at long wavelength (means low frequency),
2. Operate with a higher PRF,
3. Operate with more than one PRF,
4. Operate with more then one RF frequency.

Limitations to MTI performance

An improvement in the signal/clutter ratio of an MTI is affected by several factors other than the design of the Doppler signal processor, equipment instabilities, internal fluctuation of clutter, scanning modulation and limiting in the receiver can all detract from the performance of MTI radar.

1. Equipment instabilities:
2. Antenna scanning modulation
3. Internal fluctuation of clutter

13.18 RECEIVER NOISE

13.18.1 Noise

The signal to noise ratio is a very important factor in a radar receiver since it determines the maximum detection range of a radar system. A weak signal from a distant target is received and displayed on an 'A' scope along with noise if the noise amplitude is more than the signal amplitude can be reduced, the signal merges in the noise and the target remains undetected. If the noise amplitude can be reduced considerably, the same weak signal is detectable. Thus, by increasing the signal to noise ratio the maximum detection range of radar set can be increased.

If there were no noise present in the receiver, it would be possible to detect any weak signal, by providing the receiver with sufficient amplification. Practically it cannot be achieved since noise is always present, and unlimited amplification of weak signals amplifies noise as well. **Noise is unwanted EM energy, which interfere with the ability of receiver to detect the wanted signal.** It may originate within the receiver itself, or it may enter via the receiving antenna along with the desired signal. There are various types of noise generated in side the receiver, which are as follows

1. Thermal noise
2. Shot noise
3. Partition noise
4. Induced grid noise
5. Crystal noise
6. Flicker noise
7. Transistor noise.

13.18.2 Noise Figure

An ideal receiver adds no noise of its own to the signal being amplified. All practical receiver, however generate noise to some extent. A **noise figure** or noise factor can be explained as a measure of the noise produced by a practical receiver compared with the noise of an ideal receiver. Thus, the noise figure Fn of a linear system may be defined as

$$F_n = S_{in}/N_{in} \times S_{out}/N_{out} \qquad ...13.39$$

Or $\qquad F_n = S_{in} \times N_{out} / N_{in} \times S_{out}$

We know, $\qquad G = S_{out}/S_{in} \qquad ...13.40$

And N_{in} is function of the standard temperature (T) and the band width (B_n), which is directly proportional to T and B_n

$$N_{in} \propto T B_n$$
$$N_{in} = K T B_n \qquad ...13.41$$

By substituting the value of N_{in} & G, It can be written as

$$F_n = N_{out} / K T B_n G \qquad ...13.42$$

Where,

F_n = Noise figure

S_{in} = Available input signal power

N_{in} = Available input noise power (KT B_n)

S_{out} = Available output signal power

N_{out} = Available output noise power

K = Boltzmen constant = 1.38×10^{-23} J/° K

T = Standard temperature (in Kelvin = 2130° K)

B_n = Band width

G = Gain

RADAR

If we express the above equation, it may be considered as the degradation of signal to noise ratio by the receiver, or it may be interpreted as the ratio of the actual available output noise power (N_{out}) to noise power, which would be available if the receiver merely amplified the thermal noise of the source. The noise figure is commonly expressed in decibels (db), that is, $10 \log_{10} F_n$. Some time, it may be called as **noise factor**.

The noise figure may be also written as

$$F_n = (KTB_n G + \Delta N) / KTB_n G \qquad \ldots 13.45$$

Or $\qquad F_n = 1 \pm \Delta N / KTB_n G \qquad \ldots 13.46$

Where ΔN is additional noise added by the receiver itself.

In order to standardize the definition, the institute of Radio Engineers specifies that $T = 290°$ k or $63°$ F be employed for the calculation of noise figure. It may be convenient to write in the factor of K To is approximately 4×10^{-21} watt/cps.

13.19 LOW NOISE FRONT ENDS

Early microwave superhetrodyne receivers did not use RF amplifiers as the first stage or front end, since the RF amplifiers at that time had a greater noise figure than when the mixer alone was employed as receiver input stage. There are now a number of RF amplifiers that can provide a suitable noise figure. Noise figure is a function of frequency for several receiver front ends used in radar applications. The parametric amplifier has the lowest noise figure of the devices described here, especially at the higher microwave frequency. However, it is generally more complex and expensive compared to the other front ends.

The transistor amplifier can be applied over most of the entire range of the frequency of interest to radar. The silicon bi-polar transistor has been used at the lower radar frequency (Below 'L' band) and the gallium arsenide field effect (Ga As FET) is preferred at the higher frequencies. The transistor is generally used in a multistage configuration with a typical gain per stage decreasing from 12 db at VHF to 6 db at Ku band. In the GaAs FET, the thermal noise contribution is greater than the short noise. Cooling the device will therefore improve the noise figure.

The tunnel diode amplifier has been considered in the past as allow noise front end, with noise figure from 4 to 7 db over the range 2 to 25GHz. It has been supplemented by the improvement made in the transistor amplifier. The traveling wave tube amplifier has also been considered as low noise front end, other devices have over taken it. Cryo – genic. Parametric amplifiers and masers produce the lowest noise figures, but the added complexity of operating at low temperatures has tempered their use in radar.

The noise figure of the ordinary Broadband mixer whose image frequency is terminated in a matched load would lie about 2db higher than the noise figure for the image – recovery mixer. There are other factors beside the noise figure, which can influence the selection of a receiver front end; cost, burnout, and dynamic range must also be considered. The selection of a particular type of receiver front end might also be influenced by its instantaneous bandwidth, tuning range, phase and amplitude stability, and special requirement for cooling. The image – recovery mixer represents a practical compromise, which tends to balance its slightly greater noise figure by its lower cost, greater ruggedness and greater dynamic range.

Utility of low noise front ends

The lower the noise figure of the radar receiver, the less need be the transmitter power and or the antenna aperture. Reduction in the size of the transmitter and the antenna always desirable if there is no reduction in performance. A few decibels improvement in receiver noise figure can be obtained at a relatively low cost as compared to the cost and the complexity of adding the same few decibels to a higher power transmitter.

There are however, limitations to the use of a low noise front end in some radar applications. As mentioned earlier the cost, burn out and dynamic range of low noise devices might not be acceptable in some applications. Even if the low noise device itself is of large dynamic range, there can be reduction of the dynamic range of the receiver as compared to other receivers with a mixer as its front end. Dynamic range is usually defined as the ratio of the maximum signal that can be handled to the smallest signal capable of being detected. The smallest signal is the minimum detectable signal as determined by receiver noise and the maximum signal is that which causes a specified degree of intermodulation or a specified deviation from linearity (usually 1db) of the output Vs- input curve.

13.20 DUPLEXER

Whenever a single antenna is used for both transmitting and receiving, as in a radar system, problems arise. Switching the antenna between the transmit and receive modes presents one problem; ensuring that maximum use is made of the available energy is another. The simplest solution is to use a switch to transfer the antenna connection from the receiver to the transmitter during the transmitted pulse and back to the receiver during the return (echo) pulse. No practical mechanical switches are available that can open and close in a few microseconds. Therefore, ELECTRONIC switches must be used. Switching systems of this type are called **duplexer.** A simple block diagram of Radar System using Duplexer is shown in figure 13.15.

The duplexer enables a radar system to transmit powerful signals and still receive very weak radar echoes. The duplexer acts as a gate between the antenna and the receiver and transmitter. It keeps the intense signals from the transmitter from passing to the receiver and overloading it, and also ensures that weak signals coming in from the antenna go to the receiver. A pulse radar duplexer connects the transmitter to the antenna only when a pulse is being emitted. Between pulses, the duplexer disconnects the transmitter and connects the receiver to the antenna. If the receiver were connected to the antenna while the pulse was being transmitted, the high power level of the pulse would damage the receiver's sensitive circuits. In continuous-wave radar the receivers and transmitters operate at the same time. These systems have no duplexer. In this case, the receiver separates the signals by frequency alone. Because the receiver must listen for weak signals at the same time that the transmitter is operating, high power continuous-wave radar systems use separate transmitting and receiving antennas.

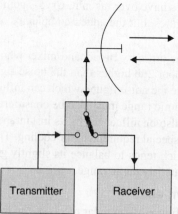

Fig. 13.15 A simple block diagram of Radar System using Duplexer

13.20.1 BASIC DUPLEXER OPERATION

In selecting a switch for this task, you must remember that protection of the receiver input circuit is as important as are output power considerations. At frequencies where amplifiers may be

RADAR

used, amplifier tubes can be chosen to withstand large input powers without damage. However, the input circuit of the receiver is easily damaged by large applied signals and must be carefully protected.

An effective radar duplexing system must meet the following four requirements:

1. During the period of transmission, the switch must connect the antenna to the transmitter and disconnect it from the receiver
2. The receiver must be thoroughly isolated from the transmitter during the transmission of the high- power pulse to avoid damage to sensitive receiver components.
3. After transmission, the switch must rapidly disconnect the transmitter and connect the receiver to the antenna. For targets close to the radar to be seen, the action of the switch must be extremely rapid.
4. The switch should absorb an absolute minimum of power both during transmission and reception. Therefore, a radar duplexer is the microwave equivalent of a fast, low-loss, single-pole, double- throw switch. The devices developed for this purpose are similar to spark gaps in which high-current microwave discharges furnish low-impedance paths. A duplexer usually contains two switching tubes.

Types of duplexer

There are different types of duplexer have been used, brief explanation of these duplexer are given below:

1. Branch type duplexer
2. Balanced duplexer
3. Ferrite duplexer

13.21 RADAR DISPLAY

The device used to display radar information is known as a radar indicator or radar display. Since indicators can be located at a point away from the other radar equipment, they are frequently referred to as remote indicators. Remote indicators are sometimes referred to as repeaters. The present-day remote indicator can operate with any of the search radars are in use.

The most common types of displays (indicators) are described in next section.

A- Scope

The A- Scope display, shown in **figure 13.16,** presents only the range to the target and the relative strength of the echo. Such a display is normally used in weapons control radar systems. The bearing and elevation angles are presented as dial or digital readouts that correspond to the actual physical position of the antenna. The A-scope normally uses an electrostatic-deflection CRT.

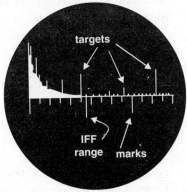

Fig. 13.16 "A"-Scope of the (Russian) P-18

PPI Scope

The **PPI Scopes** shown in this figures 13.17, are by far the most used radar display. It is a polar coordinate display of the area surrounding the radar platform. Own position is represented as the origin of the sweep, which is normally located in the center of the scope, but may be offset from the center on some sets. The PPI uses a radial sweep pivoting about the center of the presentation. This results in a map-like picture of the area covered by the radar beam. A long-persistence screen is used so that the display remains visible until the sweep passes again. Bearing to the target is indicated by the target's angular position in relation to an imaginary line extending vertically from the sweep origin to the top of the scope.

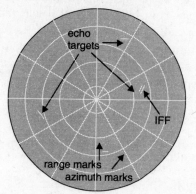

Fig. 13.17. Plan Position Indicator (PPI).

Some more radar displays

There are some more radar displays, which have not being given later designations appended below:

1. **B-Scope**: An intensity – modulated rectangular display with azimuth angle indicated by the horizontal co-ordinate and range by the vertical co-ordinate.
2. **C-Scope** : An intensity – modulated rectangular display with azimuth angle indicated by the horizontal co-ordinate and elevation by the vertical co-ordinate.
3. **D-Scope** : A C-Scope in which the blips are extended vertically to give a rough estimate of distance.
4. **E-Scope** : An intensity – modulated rectangular display with distance indicated by the horizontal co-ordinate and elevation angle by the vertical co-ordinate.
5. **F- Scope**: A rectangular display in which target appears as centralized blip when the radar antenna is aimed at it. Horizontal and vertical error are respectively indicated by the horizontal and vertical displacement of the blip.
6. **G – Scope**: A rectangular display in which target appears as a laterally centralized blip when the radar antenna is aimed at it in azimuth and wings appears to grow on the pip as the distance to the target is diminished. Horizontal and vertical aiming errors are respectively indicated by horizontal vertical displacement of the blip.

13.22 RADARS ANTENNAS

An antenna has either to receive energy from an electromagnetic field or to radiate electromagnetic waves produced by a high frequency generator. The types of antenna used in the radar system mainly depend upon the application of the radar. For example long-range detection radar (surveillance radar) needs large aperture of the antenna. The Antenna used normally for radar applications difference from the antenna used in communication system. Radar antenna must have becomes with shaped directive pattern, which can be scanned either mechanically or electronically.

RADAR

In general an antenna is a transmission device, or transducer, between a guided wave (transmission line) and a free space wave or vice versa. The basic parameter of an antenna will be discussed in the brief in the following section.

The radar antenna acts as a transducer, which converts electrical pulses from the transmitter to the free space in the form of EM waves and receives the reflected EM signals from the target in free space and converts it in to electrical signals.

In the microwave frequencies region, the most popular radar antenna is, the parabolic reflector with its optics characteristics. Some other types of antenna are also found application in radar such as, microwave lenses, surface wave antennas (for airborne radars).

Mostly radar antennas are described in detail in chapter –9.

The important parameters of radar antennas are as follows
1. Directivity or Directive gain.
2. Power gain.
3. Effective receiving aperture or total scattering cross section.
4. Polarizations.
5. Side lobes.
6. Front-to-Back Ratio

Directivity or Directive gain: The ability of radar antenna to concentrate energy in a particular direction is known as **gain.** There are two related terms of antenna gain are the power gain and the directive gain. Both the terms are of interest to the radar system engineer. The directive gain or directivity is related to radiation pattern of the antennas whereas the power gains is more suitable related to the radar range equations.

The power radiated from an antenna per unit solid angle is called the **radiation intensity** u (watts per steradian or per square degree). A radiation intensity pattern may be plot between radiation intensity and the angular co-ordinate, this plot is known as a **radiation-intensity pattern**.

Power gain

The power gain, which is denoted by (G), includes the effect of the antenna losses and any other related losses, which reduces the antenna efficiency. In the directive gain definition we have not considered the losses such as ohmic (resistance) heating, RF heating or mismatched antenna, we have considered only the radiation pattern. But these losses are taken into account while calculating the power gain. It may be defined as the ratio of maximum radiation intensity from particular antenna to the radiation intensity from (loss less) isotropic source with some power input

Effective aperture or Total scattering cross section

One of the important parameter related to the gain is the effective receiving aperture or total scattering cross section or effective area. It may be considered as a measure of the effective area presented by the antenna to the incident wave.

Polarization

The important property of an EM wave is its polarization. The direction of the electric field vector decides the direction of polarization of an antenna. The polarization may be linear, elliptical circular. Mostly radar antennas are linearly polarized that mean the direction the electrical vector either may be horizontally or vertically. In some cases the circular polarized antennas are used, in the circular polarization the amplitude of the two waves are equal but they are 130° out of phase to each other. The linear and circular polarizations are the special cases of elliptical polarization.

Major and Side (Minor) Lobes

The radiation pattern is concentrated in several lobes. The radiation intensity in one lobe is considerably stronger than in the other. The strongest lobe is called a **major lobe**; the other is a **side** (minor) **lobes**.

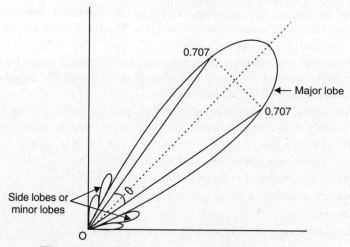

Fig. 13.18 Radiation Pattern of Parabolic antenna

Different Types of Radar Antennas

1. Half-wave Antenna
2. Yagi antennas
3. Parabolic antennas
4. Antenna with cosecant squared pattern
5. Cassegrain Antennas
6. Phased Array Antennas

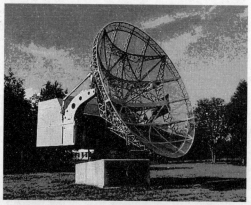

Fig. 13.19 "Würzburg Riese", a World War II radar produced in 1940 by Telefunken (Germany)

13.23 HISTORICAL OVERVIEW

Neither a single nation nor a single person is able to say, that he (or it) is the inventor of the radar method. One must look at the "Radar" than an accumulation of many developments and improvements earlier, which scientists of several nations parallel made share. There are nevertheless some milestones with the discovery of important basic knowledge and important inventions:

1865 The English physicist **James Clerk Maxwell** developed his electro-magnetic light theory (Description of the electro-magnetic waves and her propagation).

1886 The German physicist **Heinrich Rudolf Hertz** discovers the electro-magnetic waves and prove the theory of Maxwell with that.

1904 The German high frequency engineer **Christian Hülsmeyer** invents the „Telemobiloskop" to the traffic supervision on the water. He measures the running time of electromagnetic waves to a metal object (ship) and back. A calculation of the distance is thus possible. This is the first practical radar test. Hulsmeyer registers his invention to the patent in Germany and in the United Kingdom.

1917 The French engineer **Lucien Lévy** invents the super-heterodyne receiver. He uses as first the denomination *"Intermediate Frequency"* and eludes the possibility of double heterodyning.

1921 The invention of the Magnetron as an efficient transmitting tube by the US American physicist **Albert Wallace Hull.**

1922 The American electrical engineers **Albert H. Taylor** and **Leo C. Young** of the Naval Research Laboratory (USA) locates a wooden ship for the first time.

1930 Lawrence A. Hyland (also of the Naval Research Laboratory), locates an aircraft for the first time.

1931 A ship is equipped with radar. As antennae are used parabolic dishes with horn radiators.

1936 The development of the Klystron by the technicians **George F. Metcalf** and **William C. Hahn**, both General Electric. This will be an important component in radar units as an amplifier or an oscillator tube.

1940 Different radar equipments are developed in the USA, Russia, Germany, France and Japan.

SOLVED PROBLEMS

EXAMPLE 1. Calculate the range of a target if the time taken by the radar signal to travel to the target and back is 100 micro second.

Solution

$$\text{Range } (R) = C \Delta t / 2$$
$$C = 3 \times 10^8 \text{ m /sec}$$
$$\Delta t = 100 \text{ micro sec}$$
$$= 3 \times 10^8 \times 100 \times 10^{-6} / 2$$
$$\boxed{R = 15 \text{ km}}$$

EXAMPLE 2. If the transmitted peak power of radar is 100kw, pulse repetition frequency is 1000pps and the pulse width is 1 micro second than calculate the average power in dbs.

Solution

We have
$$P_t = 100 \text{kw}$$
$$PRF = 1000 \text{pps}$$
$$P_w = 1 \text{ microsecond}$$

We know that Duty Cycle = PW × PRF

By putting the value in the above equation,

We get,
$$\text{Duty cycle} = 10^{-6} \times 1000$$
$$\boxed{\text{Duty cycle} = 0.001}$$

we know, The average power = P_t × Duty cycle
$$= 100 \times 10^3 \times 0.001$$
$$= 100 \text{w}$$

or we can write in $\boxed{\text{dbs} = 20 \text{ dbs}}$

EXAMPLE 3. A typical pulse waveform of radar is shown below. In which some parameters of radar is shown. Calculate the (a) Average Power (b) Duty Cycle and (c) Maximum range of radar.

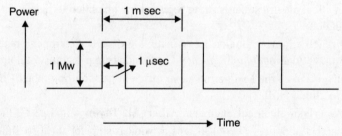

Solution

We have,

Peak power (Pt) = 1 Mw
Pulse Width (T) = 1 μsec
Pulse Repetition Period (Tp) = 1ms

We know,

$$PRf\ (fp) = 1/Tp$$
$$= 1/1000 \times 10^{-6}$$
$$\boxed{PRF = 1000Hz}$$

(a) $Pav = Pt \times T \times fp$
$= 10^6 \times 1 \times 10^{-6} \times 1000$
$= 1000w = \boxed{1kw}$

(b) Duty Cycle $= T \times fp$
$= 10^{-6} \times 1000$
$= \boxed{0.001}$

(c) Maximum Unambiguous Range (Rmax) $= c/2\ fp$
$= 3 \times 10^8 / 2 \times 1000$
$= (3/2) \times 10^5$
$= \boxed{150\ Km}$ or

Rmax = 81 nmi (1 nmi = 1852 m)

EXAMPLE 4. The following table lists the characteristics of the components of pulse-echo type surface search radar. Using the concepts presented in this chapter, complete this table.

frequency, f	5600 MHz
wave length, λ	_____
pulse width, PW	1.3μ sec
pulse repetition frequency, PRF	_____
pulse repetition time, PRT	_____
peak power	_____
average power	_____
duty cycle	8.3×10^{-4}
Antenna rotation rate, θ	16 RPM
Horizontal beamwidth, θ_B	_____
vertical beam width, ϕ_B	4°
effective aperture, A_e	0.9 m²

RADAR

power gain, G	3940
directive gain, G_D	—
number of returns per sweep, N_B	9.9
minimum discernible signal, MDS	-83dBm
receiver sensitivity, S_{min}	5.012×10^{-12}
maximum unambiguous range, R_{unamb}	—
maximum theoretical range, R_{max}	50 km
minimum range, R_{min}	—
range resolution, R_{res}	—
radar cross-section, σ	5 m²

Solution

(a) Wavelength (λ)

$$\lambda = C/f$$
$$\lambda = 3.0 \times 10^8 \text{ m/sec} / 5600 \times 10^6 \text{ hz}$$
$$\boxed{\lambda = 5.36 \text{ cm}}$$

(b) Pulse Repetitions Time (PRT)

$$PRT = PW / \text{Duty cycle}$$
$$= 1.3 \text{ } \mu\text{ sec} / 8.3 \times 10^{-4}$$
$$PRT = 1.57 \times 10^3 \text{ } \mu\text{ sec}$$

(c) Pulse repetition Frequency (PRF)

$$PRF = 1/PRT$$
$$PRF = 1/(1.57 \times 10^{-3})$$
$$\boxed{PRF = 638.5 \text{ hz}}$$

(d) Peak Power (Pt)

We know $S_{min} = P_t \cdot G.A_e \cdot \sigma / (4 \pi R^2)^2$

$$P_t = (S_{min})(4 \pi R^2)^2 / G.A_e \cdot \sigma$$
$$= (5.012 \times 10^{-12}) [(4\pi)(50 \times 10^3)^2]^2 / (3940)(0.9)(5.0)$$
$$\boxed{P_t = 279 \text{ kw}}$$

(e) Average Power (Pa)

Average Power (Pa) $= P_t \times$ Duty Cycle
$$= (279 \times 10^3)(8.3 \times 10^{-4})$$
$$\boxed{\text{Average Power} = 231.6 \text{ W}}$$

(f) Directive Gain (G_D)

Scan Rate = RPM × 360° /60 Sec
$$= 16 \times 360 / 60$$
$$\boxed{\text{Scan rate} = 96°/\text{sec}}$$

We know,
$$\theta_B = (N_B \times \text{scan rate})/PRF$$
$$= 9.9 \times 96 / 638.5$$
$$\theta_B = 1.49°$$
$$G_D = 41253 / \text{horizontal beamwidth}(\theta_B) \times \text{vertical beam width}(\Phi_B)$$
$$= 41253 / (1.49 \times 4)$$
$$G_D = 6921$$

(g) Range unambgious (R_{unamb})

$$R_{unamb} = (c \times t)/2$$
$$= [3 \times 10^8 \times 1.57 \times 10^{-3}]/2$$
$$= 4.71 \times 10^5/2$$
$$\boxed{R_{unamb} = 235.5 \text{ km}}$$

(h)

$$R \min = R_{res}$$
$$R \min = (c \times PW)/2$$
$$= (3 \times 10^8)(1.3 \times 10^{-6})/2$$
$$\boxed{R \min = 195 \text{ m}}$$

EXAMPLE 5. Calculate the minimum receivable signal in a radar receiver which has an IF BW of 1 MHz and noise figure of 6db.

Solution

We have,

$$BW = 1 Mhz$$
$$Fn = 6db \ (F = \text{Antilog } 6/10 = 4 \text{ Ratio})$$

Minimum detectable noise power $(P_{min}) = k \cdot T_r \cdot B_n$
Noise temperature of receiver $(T_r) = (F-1) \cdot T$

Therefore,

$$P_{min} = k(F-1) \cdot T \cdot B_n$$
$$k = 1.38 \times 10^{-23}$$
$$T = 290°$$
$$= 1.38 \times 10^{-23} \times (4-1) \times 290 \times 10^6$$
$$= 4 \times 10^{-21} \times 3 \times 10^6$$
$$\boxed{P_{min} = 12 \times 10^{-15} \text{ w}}$$

EXAMPLE 6. A radar is operating at 100 GHz, if the antenna diameter is 2 m, calculate the beam width of the antenna.

Solution

We know that

$$\theta = 70 \times \lambda / D \text{ Degree}$$
$$\lambda = c/f$$
$$= 3 \times 10^8 / 100 \times 10^9$$
$$\lambda = 0.003 \text{ m}$$

By putting the value of λ in above formula,

$$\theta = 70 \times (.003)/2$$
$$\boxed{\theta = 0.105 \text{ degree}}$$

EXAMPLE 7. What is the peak power of radar whose average power is 200w, pulse width (PW) is 1 μ sec and has PRF of 1000hz? Also calculate the range of this ground based air surveillance radar if it has to detect a target with a radar cross section of 2m² when it operates at a frequency of 2.9 GHz with a rectangular – shaped antenna that is 5 m wide, 2.7 m height, antenna aperture efficiency of 0.6 and mds is 10^{-12} w.

Solution

We have following parameters,

$P_{av} = 200w$, $Pw = 1 \times 10^{-6}$ sec, $PRF = 1000hz$, $\tau = 2 \text{ m}^2$, $S_{min} = 10^{-12}$ w.

Calculation of Peak power(Pt)

RADAR

$$Pt = Pav / PW \times PRF$$
$$Pt = 200 / (1 \times 10^{-6})(1000)$$
$$Pt = 200/ 0.001 = 200 \text{ kw}$$

Calculation of effective antenna area (Ae)
physical Area of antenna (A) = 5m × 2.7 m
$$= 13.5 \text{ m}^2$$

By antenna theory, we know that
$$Ae = \rho_a \cdot A$$

Where, Ae = Effective area of antenna
P_a = Antenna aperature efficiency

$$Ae = 0.6 \times 13.5 = 8.10 \text{ m}^2 \text{ (8.0 m}^2 \text{ App)}$$

Calculation for wavelength (λ)
$$\lambda = c/f$$
$$= 3 \times 10^8 / 2.9 \times 10^9 = 3/ 2.9 \times 10$$
$$\lambda = 0.103 \text{ m}$$

We know maximum range of radar is given by following equation,
$$R^4 = [Pt \times (Ae)^2 \times \sigma] / 4\pi \times \lambda^2 \times S_{min}$$
$$R^4 = [200 \times 10^3 \times 8 \times 2] / 4 \times 3.14 \times (0.103)^2 \times 10^{-12}$$
$$= (7.51879)$$
$$R = \sqrt[4]{7.51879}$$

$\boxed{R = 29.44 \text{km or } 15.89 \text{ nmi} \quad (1 \text{ nmi} = 1852 \text{ m})}$

EXAMPLE 8. Calculate the maximum range of a radar system which operates at 3cm wavelength with a peak power of 500 kw, if its P_{min} is 10^{-12} w., the capture area of its antenna is 5 m square and radar cross section area of target is 20m square.

Solution
We have,
$$\lambda = 3 \text{ cm} = 0.03 \text{ m}$$
$$Pt = 500 \text{ kW}$$
$$S_{min} = 10^{-12} \text{ W}$$
$$Ae = 5 \text{ m}^2$$
$$\sigma = 20 \text{ m}^2$$
$$R^4 = [Pt \times Ae^2 \times \sigma] / 4\pi \times \lambda^2 \times S_{min}$$

By putting the values,
$$R^4 = [(500 \times 10^3) \times (5)^2 \times (20)] / [(4 \times 3.14) \times (0.03)^2 \times (10^{-12})]$$

Solve the equation for R, We get
$$R^4 = [(500 \times 10^3) \times 25 \times 20] / [12.56 \times 0.0009 \times (10^{-12})]$$
$$R = 96990 \text{ m}$$
$\boxed{R = 96.99 \text{ km or 97 km}}$

UNSOLVED PROBLEMS

1. Determine the maximum unambiguous range and range resolution of a pulse radar whose pulse duration is 10 µsec at a PRF 1000 Hz. **(Ans. 150km, 1.5 km)**
2. A radar receiver having BW of 3 MHz, calculate the highest resolution in range. **(Ans. 50 m)**

3. A pulse radar measure the round trip propagation time with an error of 2 μsec. What will be the corresponding error in the measurement of range? **(Ans. 300 m)**
4. Calculate the duty cycle of radar which transmits a 1.5 μs pulse at a PRF of 8 kHz. If the peak power of this radar is 500 kilowatts, what is the average power? **(Ans. 12×10^{-3}, 6kw)**
5. Pulsed radar has a duty cycle of 0.016. If the resting time is 380 μs, what is the pulse width? What is the PRF? What is the minimum range, in meters, of this radar?

   (Ans. 6.18 μsec)
6. If the radar is designed for operation at 5 Ghz with an antenna of diameter 4 m, calculate the peak power required to have a maximum range of 500km with a target of cross-sectional area 20m2 and Smin is 10-12 W. **(Ans. 8.94 MW)**
7. What is the peak power of a radar whose average power 400W, pulse width of 1μs and pulse repetition frequency of 1000hz. It has to detect the target of cross-section 4 m2 when it operates at a frequency of 3 Ghz with a antenna diameter of 2 m and mds is 10-10 W. Calculate the maximum range in nmi. **(Ans. 7.212 nmi)**
8. Calculate the maximum range of a radar system which operates 2 mm wavelength with a peak power of 1000kW, if it's S_{min} is 10-12 W, the capture area of its antenna is 5m2 and radar cross-section area of target is 10m². **(Ans. 1257 km)**
9. A radar has IF bandwidth of 3 MHz, calculate the minimum detectable power, if the noise figure of that receiver is 6db. **(Ans. 3.6×10^{-16} W)**
10. If the operating frequency of surveillance radar is 10 GHz, the diameter of the radar antenna is 2m. The maximum range of the radar is 1000km with a target of cross-sectional area 20m2 and S_{min} of that receiver is 3.6 x 10-16 W. Calculate the peak power of the transmitter.

    (Ans. 2000 kW)
11. A radar transmitter has a peak power of 400kW, A PRF of 1500PPS and a PW of 0.8 μs, calculate
    a. Maximum unambiguous range,
    b. The duty cycle,
    c. The average transmitted power,
    d. Suitable bandwidth

REVIEW QUESTIONS

1. What do you understand by terms **RADAR** and explain the **echo pulse** principle.
2. What are the basic functions of radar? In indicating the position of target, what is the difference between azimuth and elevation?
3. Explain the working of pulse radar with the help of block diagram.
4. Explain the various applications of radar.
5. What are the radar frequencies, explain in brief.
6. What is the maximum unambiguous range? How it is related with pulse repetition rate?
7. Explain the limitations and applications of radars.
8. What is the difference between the pulse interval and the PRF?
9. What are the factor affecting the radar range?
10. Explain PRF and Range ambiguities.
11. What is search radar equation?
12. Derive the radar range equation.
13. What do you understand by terms Doppler Effect?

14. How do you find out the relative velocity of a target using CW radar? How does flicker noise affects its working?
15. Explain the working of CW radar with the help of block diagram.
16. Explain with neat diagram, how target velocity can be obtained by CW radar.
17. Describe the working of non-coherent MTI radar with the help of block diagram.
18. What are the advantages of digital MTI over analog MTI?
19. Describe the working of coherent MTI radar with neat diagram.
20. Derive the simple form of radar range equation and relates the Pt (Transmitter peak power) to the maximum range of the radar.
21. What are the factors affecting the radar range?
22. What do you understand by prediction of range performance?
23. What is mds to noise ratio? Why it replaces minimum detectable signal in radar system?
24. What is the maximum unambiguous range?
25. How noise affects the detection process of radar?
26. What are the different types of system losses? Explain them briefly.
27. Define the following terms in brief
    - (i) Pulse Width
    - (ii) Pulse Repetition Time
    - (iii) Average Power
    - (iv) Duty Cycle
    - (v) Range resolution
    - (vi) Multiple around echoes
    - (vii) Staggered PRF
    - (viii) Noise and clutter
    - (ix) Blind speed
    - (x) MDS
    - (xi) Propagation effect
    - (xii) Probability of detection
    - (xiii) Antenna loss

CHAPTER 14

MICROWAVE COMMUNICATION SYSTEMS

> **OBJECTIVES**
> - Introduction
> - Propagation of Microwave frequencies
> - Applications of Microwave Frequencies
> - Satellite communications
> - Microwave propagation
> - LOS propagation

14.1 INTRODUCTION

Microwave links are high bandwidth line-of-sight links. Satellite links use geostationary satellites to provide inter-continental high-bandwidth communications. Optical links (i.e. the transmission through free-space of light waves) can be used for short, high bandwidth transmission. Microwave frequencies are normally used by Radar, satellite, and mobile communication

14.2 PROPAGATION OF MICROWAVE FREQUENCIES

Microwave frequencies are usually completely blocked by obstructions. They need **line-of-sight** geometries. In addition, the wavelengths are small enough to become effected by water vapor and rain. Above 10GHz these effects become important. Above 25GHz, the effects of individual molecules become important. Water and oxygen are the most important gases. These have resonant absorption lines at 23, 69 and 120GHz shown in figure—29. Above these frequencies; absorption is increasingly affected by the Earth's *black-body* spectrum. These peaks leave "windows" that may be used for communication, notably at 38GHz and 98GHz.

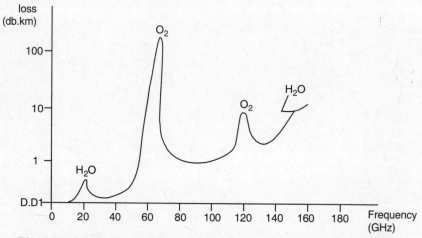

Fig. 14.1 Graph of attenuation due to water and oxygen for microwave

MICROWAVE COMMUNICATION SYSTEMS

The use of the atmosphere as a transmission medium means that control over the noise environment is not possible. There are a number of sources of noise. Natural noise is classified by its source. **Atmospheric noise** dominates the low frequencies, up to 2MHz. The primary cause of this noise is electrical discharges in the atmosphere: lightning. So-called **galactic noise** is radio noise from outside the solar system. It is important up to 200GHz. The sun also generates a small noise contribution at these frequencies.

14.3 APPLICATIONS OF MICROWAVE FREQUENCIES IN COMMUNICATION

Few microwave communication systems are discussed here in brief, which are as follows:
 (*i*) Microwave Link Communication
 (*ii*) Troposcatter communication
 (*iii*) Line-of-Sight (LOS) Communication
 (*iv*) Satellite Communication
 (*v*) Mobile Communication

14.3.1. Microwave link communication

Microwave transmission is line of sight transmission. The Transmit station must be in visible contact with the receive station (refer figure-14.2). This sets a limit on the distance between stations depending on the local geography. Typically the line of sight due to the Earth's curvature is only 50 km to the horizon. Repeater stations must be placed so the data signal can hop, skip and jump across the country.

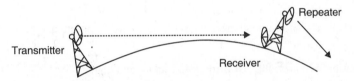

Fig. 14.2. Microwave link communication.

Microwaves operate at high operating frequencies of 3 to 10 GHz. This allows them to carry large quantities of data due to the large bandwidth.

The maturity of radio frequency (RF) technology has permitted the use of microwave links as the major trunk channel for long distance communication. The use of microwave links has major advantages over cabling systems:

- **Freedom from land acquisition rights.** The acquisition of rights to lay cabling, repair cabling, and have permanent access to repeater stations is a major cost in the provision of cable communications. The use of radio links, that require only the acquisition of the transmitter/receiver station, removes this requirement. It also simplifies the maintenance and repair of the link.
- **Ease of communication over difficult terrain.** Some terrains make cable laying extremely difficult and expensive, even if the land acquisition cost is negligible.

The use of microwave links has a number of disadvantages, which mainly arise from the use of free-space communication:

- **Bandwidth allocation is extremely limited.** The competition for RF bandwidth from various competing users leads to very strict allocations of bandwidth. Unlike cabling systems, that can increase bandwidth by laying more cables, the radio frequency (RF) bandwidth allocation is finite and limited. In practice, bandwidth allocations of 50MHz in the carrier range 300MHz to 1GHz are typical.
- **Atmospheric effects.** The use of free-space communication results in susceptibility to weather effects, particularly rain.

- **Transmission path needs to be clear.** Microwave communication requires line-of-sight, point-to-point communication. The frequency of repeater stations is determined by the terrain. Care must be taken in the system design to ensure freedom from obstacles. In addition, links must be kept free of future constructions that could obstruct the link.
- **Interference.** The microwave system is open to RF interference.
- **Costs.** The cost of design, implementation and maintenance of microwave links is high.
- The modern urban environment presents a particular challenge, in that bandwidth allocation, RF interference, link obstruction and atmospheric pollution place maximum constraints on the system simultaneously. However, urban environments also have the highest land acquisition values too. Many modern cities have found it cost effective to build a single, very high tower to house an entire city's trunk communication microwave dishes. These towers are now a common feature of the modern urban landscape.

As the demand for bandwidth increases, microwave links will become increasingly unable to deliver. The use of increased carrier frequencies in the millimeter wave region would be advantageous. However, for technical reasons, no efficient methods of producing large quantities of millimetre power have been found. This is a necessity, given the increase in atmospheric attenuation at millimeter wave frequencies.

Advantages:

1. They require no right of way acquisition between towers.
2. They can carry high quantities of information due to their high operating frequencies.
3. Low cost land purchase: each tower occupies small area.
4. High frequency/short wavelength signals require small antenna.

Disadvantages:

1. Attenuation by solid objects: birds, rain, snow and fog.
2. Reflected from flat surfaces like water and metal.
3. Diffracted (split) around solid objects
4. Refracted by atmosphere, thus causing beam to be projected away from receiver.

Figure-14.3 shows a simplex relay and duplex relay system. A simplex relay system as shown in fig-14.3 (a) provides one-way communication by using a transmitter station -1, repeater station and terminal receiver station-2. A duplex relay system as shown in fig-14.3 (b) provides two-way communications by using two simplex systems, one transmitting in one direction and the other transmitting in the opposite direction. The duplex system is further refined by using a single antenna for transmitting and receiving. This is done by using different transmitting and receiving frequencies and by using a duplexer in the transmission line. The RF equipment in terminal and repeater stations is basically the same. Terminal equipment can be converted to repeater equipment and vice-versa. Let us discuss a typical microwave transmitter and receiver.

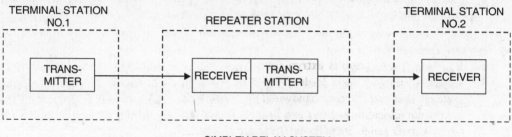

SIMPLEX RELAY SYSTEM

MICROWAVE COMMUNICATION SYSTEMS

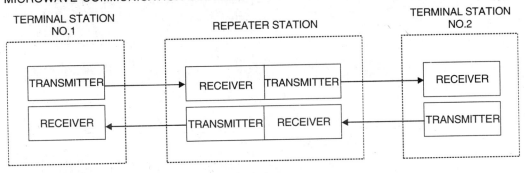

DUPLEX RELAY SYSTEM

Fig. 14.3 (*a* & *b*) Microwave Relay Systems

Microwave Transmitter: A typical microwave transmitters shown in figure-14.4. In operation, the output of a telephone multiplex terminal which consists of a frequency multiplexed AM carrier signal is applied to the terminal transmitter. This input signal (baseband signal) also could be television signal or any other form of signal to be transmitted. A pre-emphasis network accentuates the high frequencies, relative to the low, to improve the signal-to-noise ratio. The insertion amplifier accepts portion of the output power back to the klystron to the signal amplifies it and then applies the signal to the compensate for its nonlinearity. This technique allows klystron oscillator. With this method, the input signal for optimum performance with modulation densities as directly modulates the carrier frequency, resulting in a high as 1200 channels.

Microwave Receiver: A typical microwave receiver is shown in figure-14.5. During system operation, the signal from the antenna passes through a waveguide pre-selector that eliminates interference from adjacent RF channels. The signal then enters a waveguide filter tuned to its frequency which rejects all other unwanted frequencies. Next the signal passes through an isolator that minimizes intermodulation noise and holds the VSWR below 1.2:1. The signal is then mixed with the local oscillator output to produce the standard 70 MHz intermediate frequency (IF). The IF output is amplitude limited and applied to an automatic frequency control (AFC) discriminator, which controls the frequency of the local oscillator. The signal is also applied to an discriminator, a de-emphasis circuit, and a squelch circuit that disconnects the baseband amplifier and demultiplexing equipment, if noise increases above a preset level. After the squelch circuit the signal passes through a baseband amplifier and then to demultiplexing equipment, where the original intelligence is retrieved. Microwave communication system operating in the SHF (super high frequency) portion of the frequency spectrum; use the principle that propagation approaches an optical straight-line path. Propagation takes place in the lower atmosphere and is affected by meteorological factors. Communications in this medium are usually either line-of-sight or tropospheric scatter.

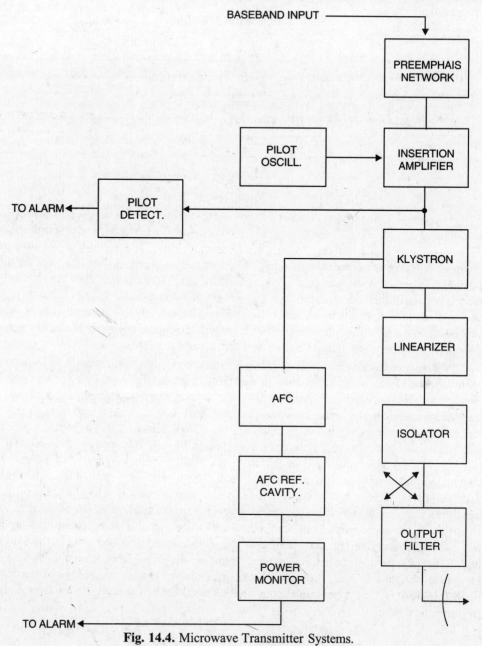

Fig. 14.4. Microwave Transmitter Systems.

MICROWAVE COMMUNICATION SYSTEMS

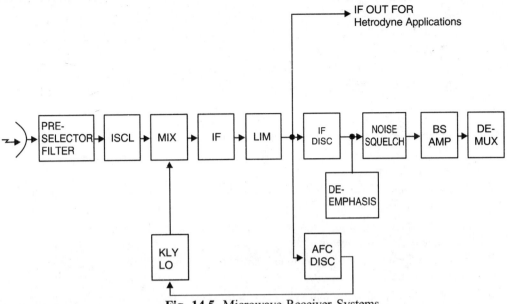

Fig. 14.5. Microwave Receiver Systems.

14.3.2 Line-of-Sight (LOS)

A lne-of-sight microwave system consists of one or more point-to-point hops. Each hop is designed to be integrated into a wordwide communication network. LOS system characteristics are as follows:

1. Propagation- Free space as affected by the troposphere.
2. Communications Capacity/ Bandwidth- Up to 600 voice channels each of 4kHz; wideband, can accept TV.
3. Range-Usually 50 to 150 km (this depends upon antenna height, earth curvture, and intervening terrain.
4. RF Power-Usually less than 10 watts.
5. Antenna-Both transmitting and receiving antennas are horn- driven paraboloids, providing high gain and narrow beamwidths. In some applications, plane reflectors are used with the paraboloids.
6. Reliability-Designed to be operational more than 99% of the time, including the periods of poor propagation.
7. Countermeasures-Because of antenna directivity, the system is difficult to jam. Additionally, the system should not be susceptible to nuclear disturbance of the ionosphere.

Applications

Because of the bandwidth capability and minimum site requirements, LOS is well adapted to

1. moderate distance point-to-point multichannel communications (with repeaters),
2. transmission of closed circuit TV,
3. transmission of radar informations from outlying sites,
4. communication relay between locations in congested areas.

14.3.3 Tropospheric Scatter System

At microwave frequencies, the atmosphere has a scattering effect on electromagnetic fields that allows for over-the-horizon communications. This type of communication is called tropospheric scatter, or troposcatter for short. Troposcatter takes place mostly at low altitudes, but some effect take place at altitudes of up to 10 miles. Under the right conditions, troposcatter can take place over hundreds of miles.

A tropospheric scatter microwave system consists of one or more point-to-point hops. Each hop is designed so it can be integrated into the worldwide communication network of the DEFENCE Communications system (DCS). Troposcatter links have the following characteristics:

1. Propagation- Free space as affected by the troposphere.
2. Communications Capacity/ Bandwidth- Up to 600 voice channels each of 4kHz; wideband, can accept TV.
3. Range-Usually up to 800 km.
4. RF Power-High; Usually upto 75 kw depending upon bandwidth, quality and range.
5. Coverage-Point-to-point only.
6. Antenna-Both transmitting and receiving antennas are horn- driven paraboloids, providing high gain and narrow beamwidths.
7. Reliability-Designed to be operational more than 99% of the time, including the periods of poor propagation.
8. Countermeasures-Because of antenna directivity, the system is extremely difficult to jam. Additionally, the system should not be susceptible to nuclear disturbance of the ionosphere.

Application

Meets the communications requirements between HF sites within its minimum skywave one-hop distance of about 400 miles and line of site of about 30 miles. It is especially useful where condition prevent the use of line of sight communications or if adverse propagation conditions interfere with other transmission methods.

14.3.4 Satellite communication

In the last two decades satellites have become the most dominant vehicles of long distant communications. Modern satellite communication has been made possible by combining the skills and knowledge of space technology with those of microelectronics industry. This has been allowed increasingly heavier, more sophisticated satellites to be launched into the orbit for ever-lower cost. The evolution of satellite is a perfect example of how the boundary between the computer and communication industries is quickly dissolving. Satellite communication is one of the most rapidly growing and evolving technologies bringing with it multitude of business opportunities in the decades to come.

The concept of satellite communication was first conceived by Arthur .C. Clarke groundbreaking article in October 1945. Signals are beamed into the space by an uplink antenna, received by an orbiting satellite electronically processed, broadcast back to the earth by a downlink antenna and received by an earth station located anywhere in the satellites footprint.

Most of the communication satellites are parked into the "Clarke's Belt" or the "geosynchronous "arc at 22,247 miles directly above the equator. The circle above the earth is unique because in this orbit the velocity of satellite matches that of the surface of the earth below. Each satellite thus appears to be in a fixed orbital position in the sky. This allows the stationary antenna to be permanently aimed towards any chosen geo synchronous satellite.

A **satellite** is an object that orbits another object. Gravity pulls the satellite closer to the primary object it orbits, but the satellite is moving perpendicular to that pull so quickly that the satellite continually avoids colliding with the primary object. All masses that are part of the solar system, including the Earth, are satellites of the Sun, or satellites of those objects, such as the Moon.

MICROWAVE COMMUNICATION SYSTEMS

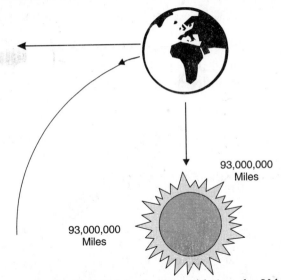

Fig. 14.6. The Earth Orbits the Sun with Angular Velocity.

In common usage, the term is usually used to describe an **artificial satellite.** A gravitational force serves as the centripetal force needed to make the object circle the primary object. The motion of the satellite around its primary gravitational source is known as *freefall*.

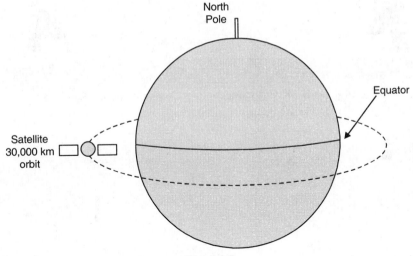

Fig. 14.7

Communication Satellite System

Satellite is a Radio Frequency (RF) repeater in orbit. Most communication satellites are in Geostationary Orbit. The primary purpose of a communication satellite is to relay the modulated microwaves between widely separated earth stations. This is accomplished by picking up the modulated microwaves on a certain frequency and retransmitting them on a different frequency.

Fig-14.8 illustrates the use of a satellite to transmit the signals from earth station (A) to earth station (B). The signal propagates by means of direct line-of-sight microwave links between each earth station and the satellite. Signals from a ground station (A) are sent to the satellite on one frequency. The satellite **Transponder** shifts the signal from the **Uplink frequency** to a different **Downlink frequency**. The path from an earth station to satellite is known as **Uplink** and the returning path from a satellite to earth station is known as **Downlink**. **Uplink and Downlink Frequencies**

To avoid influence and any filtering of signal we use in which all the frequencies are in GHz. Uplink frequency is always greater than the Downlink Frequency. The frequency range for Uplink frequency is 5.9 GHz - 6.4GHz and for downlink is 3.7GHz – 7.2GHz.

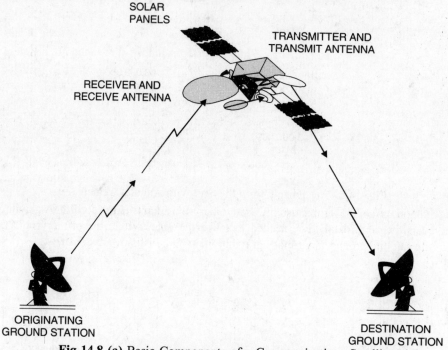

Fig 14.8 (a) Basic Components of a Communications Satellite Link

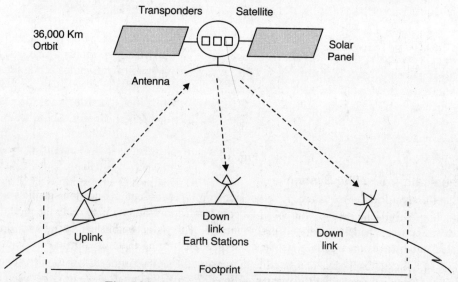

Fig. 14.8 (b) Communication satellite system.

Satellite communication became a possibility when it was realized (by Sir Arthur C. Clarke) that a satellite orbiting at a distance of 36000Km from the Earth would be **Geostationary**, i.e. would have an angular orbital velocity equal to the Earth's own orbital velocity. It would thus

MICROWAVE COMMUNICATION SYSTEMS

appear to remain stationary relative to the Earth if placed in an equatorial orbit. This is a consequence of Kepler's law that the period of rotation T of a satellite around the Earth was given by:

$$T = \frac{2\pi r^{3/2}}{\sqrt{g_i R^2}}$$

Where **r** is the orbit radius, **R** is the Earth's radius and $g = 9.8 \text{ms}^{-2}$ is the acceleration due to gravity at the Earth's surface. As the orbit increases in radius, the angular velocity reduces, until it is coincident with the Earth's at a radius of 36000Km.

In principle, three geostationary satellites correctly placed (120 degree apart) can provide complete coverage of the Earth's surface shown in figure—14.9.

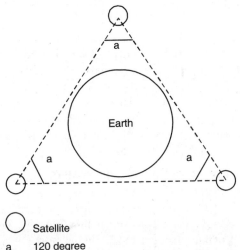

○ Satellite
a   120 degree
Fig. 14.9 Geostationary satellites providing global coverage

For intercontinental communication, satellite radio links become a commercially attractive proposition. Satellite communication has a number of advantages:

- Inherent broadcast capability
- Inherent capability of by passing the whole terrestrial networks. The laying and maintenance of intercontinental cable is difficult and expensive.
- The heavy usage of intercontinental traffic makes the satellite commercially attractive.
- Satellites can cover large areas of the Earth.
- High speed network access.
- Easy deployment and configuration.

Satellite communication is limited by four factors:

- Technological limitations preventing the deployment of large, high gain antennas on the satellite platform.
- Over-crowding of available bandwidths due to low antenna gains.
- The high investment and maintenance cost.
- High atmospheric losses above 30GHz limit carrier frequencies.

Frequency bands reserved for satellite communications are as follows shown in Table 14.1:

Table 14.1. Frequency bands reserved for satellite

Band	Downlink (GHz)	Uplink (GHz)
L-Band	(1.53, 1.559)	(1.6265, 1.6605)
S-Band	(2.5, 2.6655)	(2.6655, 2.69)
C-Band	(3.4, 4.2) + (4.5, 4.8)	(5.725, 7.075)
X-Band	(7.25, 7.75)	(7.9, 8.4)
Ku-Band	(10.7, 12.75)	(12.75, 13.25) + (14.14.8)
Ka-Band	(18.1, 21.2)	(17.3, 18.1) + (27.31)
V-Band	(37.5, 40.5)	(42.5, 43.5) + (47.2, 50.2) + (50.4, 51.5)

Passive and Active communication satellites

Communications through satellites are either passive or active. The first communications satellites were passive. **Passive satellites** reflect signals they receive to earth or other satellites. They have no transmitter or energy signals. Signals from Earth were merely reflected from the orbiting metallic sphere. Later types of satellites are active. **Active communication satellites** receive signals from Earth, electronically strengthen the signals, and transmit the signals to Earth. Active satellites can collect the data they need and transmit their own radio signals to earth (Satellites). Some active satellites have an electronic device called a transponder that receives, amplifies, and rebroadcasts signals (Types and Uses).

Current and future space communications systems

The purpose of this topic is to provide a general overview of current and future space communications systems. Satellites are used for a variety of purposes including sensor and data collection (e.g., Landsat, ARGOS, and Defense Satellite Program), weather (e.g., GOES, Defense Meteorological Satellite program), navigation and timing (e.g., GPS), weapons (e.g., "Star Wars" concepts), reconnaissance, and communications (e.g., INTELSAT, INMARSAT, Galaxy, GStar). While every satellite system employs some form of communications to accomplish its object, here we will address those satellite systems whose primary mission is communications. The focus will be on how the design of communications satellite system architectures has been evolving over the past two decades. The driving features of recently proposed satellite communications system architectures will be presented along with a discussion of the significant trends that might be expected in the near future.

INTELSAT

In August 1964, INTELSAT was formed as an international organization with the goals of production, ownership, management, and use of a global communications satellite system. INTELSAT is comprised of over 130 member nations, with the Communications Satellite Corporation (COMSAT)) acting as the US signatory. The INTELSAT system currently employs a network of 22 geostationary satellites located in orbital positions over the Atlantic, Indian, and Pacific Oceans. INTELSAT supports direct communications links among 200 countries, territories, and dependencies using more than 1,300 antennas at over 800 earth stations.

INTELSAT provides a wide range of international services including telephone, data transfer, facsimile, television broadcasting, and teleconferencing at data rates in the Mbps range. Access to the INTELSAT satellites is through a wide array of terminal types and designs with antenna sizes ranging from .5 m to 30 m and operating frequencies in both the C-band (6 GHz up/4 GHz down; nominal) and Ku-band (14 GHz up/11 or 12 GHz down; nominal). INTELSAT leases space segment capacity in increments of 9, 18, 36, 54, or 72 MHz.

INTELSAT I (or Early Bird) was the first of a long series of satellite designs developed by INTELSAT. A single INTELSAT I satellite was launched into geostationary orbit on 6 April 1965. The INTELSAT I satellite was spin-stabilized, weighed 85 lbs in orbit (at beginning of life [BOL]), and was 28 inches in diameter with a height of 23 in. The total capacity was only 240 two-way voice circuits or one television (TV) circuit. The oper-ational frequency was at C-band with a transmit power of 6 watts (W). The orbit was nominally geostationary, which is true for all of the subsequent INTELSAT satellite designs as well.

The current INTELSAT constellation represents four generations of spacecraft including INTELSAT V/V-A, VI, VII/VII-A, and VIII. Because of an ever increasing traffic projection and utilization for INTELSAT resources, every generation of INTELSAT satellites through INTELSAT VI had been designed to accommodate growth resulting in larger satellites that provided increasing levels of throughput capacity.

The trend of increasing capacity and size changed with the design of INTELSAT VII. The in orbit weight of INTELSAT VII is approximately 4,200 lbs and the orbit is geostationary. INTELSAT VII continues to support both C-band and Ku-band. A number of transmit powers are available with a maximum output transmit power of 50W. The total capacity is 18,000 two-way voice circuits plus 3 TV circuits.

As of April 1995, INTELSAT had 6 INTELSAT VIII series spacecraft on order from Martin Marietta Astro Space, the prime contractor. The first launch is scheduled for October 1996. The INTELSAT VIII satellites are being designed with improved C-band coverage and service and will include the highest C-band power level ever for an INTELSAT satellite.

INTELSAT System Satellite Characteristics are specified in Table 14.2

Table 14.2: INTELSAT System Satellite Characteristics

INTELSAT	Frequency	Power (W)	Weight (lbs)	Capacity (2-way voice channels)
I	C Band	6	85	240
V	C and Ku Bands	Up to 8.5	2260	12,000
VI	C and Ku Bands	Up to 40	4600	24,000
VII	C and Ku Bands	Up to 50	4200	18,000
VIII	C and Ku Bands	Up to ~44	3370	22,500

In summary, the history of INTELSAT is representative of the communications satellite industry up to the present time in that it employs a limited number of large satellites in geostationary orbit that provide high data rate services to a select group of users. The services are relatively expensive (i.e., not designed for personal use), require directional antennas up to 18 m in diameter, and support point-to-point communications (i.e., bent-pipe with no on-board processing). The satellites, while providing a complicated suite of services in the way of transponders and antennas, are based upon proven technology devoid of the satellite complexity that accompanies the implementation of on-board processing techniques and satellite cross links.

INMARSAT

In 1972, the Intergovernmental Maritime Organization (IMO) began to study the development of an international maritime satellite system. In April 1975, the IMO convened an international conference to begin establishing the system with 48 nations represented. It was unanimously agreed that such a system was necessary and that a new organization, INMARSAT, should be formed to operate the system. In 1976, the INMARSAT Convention and Operating Agreements were developed and were entered into force in July 1979. INMARSAT began operations in 1982 and is headquartered in London.

The initial membership of INMARSAT included 26 nations, increasing to 67 by December 1992. The INMARSAT system currently employs a network of 11 satellites in geostationary orbit located over the Atlantic, Indian, and Pacific Oceans. Approximately 25,000 mobile earth stations (MES) access these satellites, permitting worldwide communications with ships at sea, offshore oil rigs and drilling platforms through approximately 30 shore stations.

The purpose of INMARSAT, as stated in the original INMARSAT Convention, was to provide **maritime satellite communications**. However, in October 1989 the Convention was amended to permit the provision of aeronautical satellite communications. Furthermore, in January 1989 INMARSAT's Assembly of Nations authorized expansion of the organization's charter to include land-mobile satellite services.

INMARSAT services include telephone (point-to-point, conference, and group calls), record services (data transfer up to 9.6 kbps), a high speed 56 kbps ship-to-shore data service, and private line voice and data services. Land-mobile services were also introduced in 1989. This capability was primarily developed to support long-service trucking companies, but the service is not limited in this respect and is expected to support a wider range of users in the future. Finally, INMARSAT is also actively working on its INMARSAT-P initiative in which it hopes to provide a world-wide pocket-sized telephone service by the end of the decade.

Having secured an adequate initial operating capability via the use of leased satellites, INMARSAT was then prepared to sponsor the development of their own satellite design designated INMARSAT II. A contract was awarded to British Aerospace in 1985 to design and develop 3 INMARSAT II satellites with options to purchase up to 6 additional satellites. One of these options was converted into a firm order in 1988 making the total acquisition 4 INMARSAT II satellites. All 4 of these INMARSAT II satellites have been deployed and are currently the primary user support satellites in the INMARSAT space segment.

The INMARSAT II satellite body is rectangular with a deployed solar array span of approximately 50 ft. The in orbit weight is approximately 1,500 lbs at BOL. The frequency plan is unusual in that the communications payload has 1 channel for shore-to-ship transmissions and 4 channels for ship-to-shore transmissions. Since the INMARSAT satellite uses L-band (1.5-1.6 GHz - nominal) for communications with ships and C-band for communications with shore stations, the result is a single uplink C-band channel (for use by the shore-to-ship communications) and 4 L-band channels for ship-to-shore communications. The downlink is just the opposite with a single L-band channel to communicate from shore-to-ship and 4 C-band channels to communicate from ship-to-shore.

The INMARSAT II satellite transmit power is 30W and the antenna system includes a 61-element array along with two 7-element arrays; all 3 providing earth coverage. The design life for the INMARSAT II satellite is 10 years and the capacity is 250 two-way voice channels. The first 2 launches occurred on 30 October 1990 and 8 March 1991, respectively, followed by the third and fourth launches in 1992.

The INMARSAT III series will address the need for increased capacity and power. Since the spectrum for mobile satellite communications is limited, the capacity increase will come from the use of five spot beams having increased gain relative to INMARSAT II. The result will be an effective radiated spot beam power approximately twenty times greater than that of the INMARSAT II global beam. The INMARSAT III capacity will be 2,000 two-way voice circuits.

In summary, INMARSAT is a system that supports maritime voice and low data rate (i.e., 600 bps) mobile satellite communications users via a constellation of relatively large geostationary satellites using intermediate sized earth terminals with minimum dimensions on the order of a suitcase. While, relative to INTELSAT, INMARSAT took a significant step forward in supporting PCS-like services, it still cannot support user connectivity using inexpensive (i.e., < $1000), handheld terminals which is considered by many to be a must if satellite based PCS systems are going to become viable in the future marketplace.

IRIDIUM and Globalstar

The IRIDIUM and Globalstar systems represent the primary LEO contenders in providing mobile telephony in the next 2-3 years. In the phraseology of the regulatory community, these systems are referred to as Big LEOs since they would provide the full range of mobile satellite services (MSS) including voice and data and operate in the 1 to 3 GHz band.

Some of the key architecture characteristics of the Globalstar and IRIDIUM systems are shown in Table 14.3.

Globalstar includes 48 LEO satellites at an altitude of 1401 km and equally divided into 8 orbital planes. The orbits are circular with an inclination angle of 52 degrees.

The IRIDIUM system includes 66 LEO satellites at an altitude of 785 km and equally divided into 6 orbital planes.

The IRIDIUM satellite enjoys the largest capacity at 3840 full duplex (FDX) circuits/satellite followed by Globalstar with 2800 circuits/satellite. It should be noted that the "per satellite" capacity is not cumulative due to self interference and beam overlap considerations.

Each satellite is 3-axis stabilized with a mission life of between 5 and 7.5 years. The Globalstar system includes traditional bent-pipe transponders whereas the IRIDIUM satellite will employ on-board processing techniques. This is a major design feature of the IRIDIUM system and is essential to support the satellite-to-satellite crosslinks which will circumvent the need to downlink voice and data traffic to intervening hub stations. Four crosslinks would exist on each satellite; one forward within a plane, one backward within a plane, and two cross-plane links. The satellite crosslinks will operate at 25 Mbps in the 22.55 GHz to 23.55 GHz frequency range. The onboard processing feature, together with the satellite crosslink capability, provides increased flexibility in message routing at the expense of system design complexity. Motorola is aiming to be the first vendor to utilize these techniques in a commercial satellite system.

The dry masses of the satellites are currently estimated to be 704 lbs for Globalstar and 1100 lbs for IRIDIUM. The IRIDIUM satellite is heavier than Globalstar primarily due to the additional crosslink communications payload together with the on-board processing equipment.

Both IRIDIUM and Globalstar were awarded an operational license by the Federal Communication Commission (FCC) in January 1995. IRIDIUM's first launch is scheduled for September 1996 while Globalstar is planning a first launch in July 1996. The IRIDIUM system will employ a time division multiple access (TDMA) scheme to support the user traffic whereas Globalstar will employ a code division multiple access (CDMA) scheme. The schemes are incompatible with each other resulting in a unique spectrum sharing solution generated by the FCC.

Table 14.3: Comparison of System Characteristics for the Globalstar, IRIDIUM, Orbcomm, and Teledesic Systems

		Globalstar	IRIDIUM	ORBCOMM	Teledesic
Services Provided		Mobile Telophony & Low Rate Data (9.6 Kbps)	Mobile Telephony & Low Rate Data (2.4 Kbps)	Store-n-Forward Messaging (up to 300 throughput)	High Rate Fixed Satellite Services (upto 1.244 Gbps)
Const-ellation	No. of Satellites	48	66	36	840
	Inclination (°)	52	86.4	45	98.16° Sun
	Altitude (Km)	1401	785	775	695 to 705
Satellite	Transponder	Bent Pipe	Processing	Bent Pipe	Processing
	Mission Life	7.5 Yrs	5 Yrs	4 yrs	10 yrs
	Dry Mass (lbs)	704	1100	85	747
	Crosslinks	No	Yes; 4 crosslinks @ 25 Mbps; 22.55 to 23.55 GHz	No	Yes; 8 crosslinks @ 155.52 Mbps; 59 to 64 MHz

ORBCOMM

Recent technological advancements in the areas of antenna design, signal reception in fading channels, and unit miniaturization have resulted in the feasibility and manufacture of mobile communications systems supporting transmission in the 100 to 300 bps range. One such system is ORBCOMM.

ORBCOMM is designed to provide full-time global two-way digital communications services capable of supporting messaging, emergency alert functions, position determination, and remote data collection.

The space segment will be comprised of 36 satellites with 4 satellites in near-polar orbit and the remaining 32 satellites at a 45° inclination. The ORBCOMM satellites will be launched on the Pegasus launch vehicle developed by OSC.

The satellite design life is 4 years and the weight is approximately 85 lbs. Space/ground communications is at VHF with 148-150.05 MHz used for the uplink and 137-138 MHz used for the downlink. The user segment is comprised of a handheld unit operating at a transmission rate of 2400 bps to the satellite and receiving data at a rate of 4800 bps from the satellite. The effective throughput will be in the 300 bps range.

ORBCOMM can basically be viewed as a store-and-forward mail box in the sky. The source user messages will be sent via the ORBCOMM space segment to gateway earth stations that will then either forward the message directly to the destination users via leased lines or will act as a central repository to be accessed by external users upon demand. The satellite design is very simple, small, and easily deployed. The throughput is low (i.e., approximately 300 bps); however, the services being provided are global and are in demand from a variety of users. The first two satellites were launched in April 95. Limited services are currently being offered.

Teledesic

The newest contender for a piece of the future satellite communications pie is Teledesic. Teledesic in under development by the Teledesic Corporation; principal shareholders include the

Chief Executive Officer (CEO) of McCaw Cellular Communications Inc. (Mr. McCaw) and the CEO of Microsoft Corporation (Mr. Gates). The services to be provided include domestic and international fixed satellite service. The most interesting feature of Teledesic is that it is not catering to the mobile user market but, rather, it is posturing itself in the same way as AT&T was configured before the breakup. That is, Teledesic will be a wholesaler of communications capacity and offer bulk network capability to retail telecommunications providers such as U.S. West, NYNEX, etc.

The current Teledesic architecture is shown in figure 14.10. The space segment will be comprised of 840 satellites employing onboard processing techniques and supporting packet switched asynchronous transfer mode (ATM) communications.

The altitude is approximately 700 km and the orbit is sun-synchronous. Each satellite will have 8 cross links supporting a nominal data rate of 155.2 Mbps with a maximum supportable data rate of 1.244 Gbps. The crosslink frequency band is 59-64 GHz.

Connectivity with the ground is via Ka-band with 30 GHz (nominal) used for the uplink and 20 GHz (nominal) used for the downlink. The maximum achievable data rate supported by Teledesic is 1.244 Gbps to both user ground terminals and Teledesic gateways or "GigaLink" terminals. Only limited data is available in the public domain concerning the details of the ground segment except that it will be compatible with ATM/SONET technology and protocols under development and it will have the capability to interface with existing public and private networks.

Teledesic is the greatest example of showing where future satellite communications appears to be going. Other than IRIDIUM, each of the other future systems identified in this section are examples of systems that are applying proven technology to new applications (i.e., mobile voice and data communications and messaging).

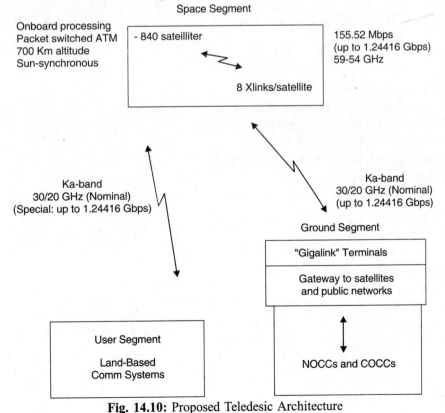

Fig. 14.10: Proposed Teledesic Architecture

Table 14.4: Comparison of Current and Future System Characteristics for Satellite Communications Systems

System Characteristic	Current	Future
Platform size	Large/Heavy	Smaller/Light Weight
Orbit	Geostationary	Low Earth Orbit (LEO) and Medium Earth Orbit (MEO)
Constellation	One to Several Satellites	10s to 100s of Satellites
Sat/Sat Connectivity	None (except for TDRSS); Bent-Pipe Operations	Cross Links; On-Board Processing
Complexity	Primarily on Ground	Primarily on Satellite
Protocol layers	Only Layers 1 and 2	All 7 Layers

Various Uses of Satellite Communications

Traditional Telecommunications

Since the beginnings of the long distance telephone network, there has been a need to connect the telecommunications networks of one country to another. This has been accomplished in several ways. Submarine cables have been used most frequently. However, there are many occasions where a large long distance carrier will choose to establish a satellite based link to connect to transoceanic points, geographically remote areas or poor countries that have little communications infrastructure. Groups like the international satellite consortium Intelsat have fulfilled much of the world's need for this type of service.

Although the initial task of satellite communications was the transmission of telephone and television signals, its mission has been extended to cover other applications. Its applications include:

1. Fixed satellite communication systems
2. Broadcast satellite communications systems
3. Land mobile communications system
4. Maritime satellite communications systems
5. Aeronautical satellite communications systems
6. Radio determination satellite systems
7. Satellite personal communications systems
8. Satellite internet access systems

The services offered by the satellite can be divided into the following categories:

Voice services

1. Telephony
2. Audio broadcasting – DAB
3. Voice conferencing

Video and image services

1. Facsimile
2. Graphics
3. Freeze-frame video
4. Full-motion video
5. Broadcast quality video
6. Teletext/videotext
7. HDTV

MICROWAVE COMMUNICATION SYSTEMS

Data services
1. Electronic mail
2. Database access
3. File transfer
4. Remote data monitoring
5. Short message transmission
6. Paging
7. Electronic funds transfer

Cellular

Various schemes have been devised to allow satellites to increase the bandwidth available to ground based cellular networks. Every cell in a cellular network divides up a fixed range of channels which consist of either frequencies, as in the case of FDMA systems, or time slots, as in the case of TDMA. Since a particular cell can only operate within those channels allocated to it, overloading can occur. By using satellites which operate at a frequency outside those of the cell, we can provide extra satellite channels on demand to an overloaded cell. These extra channels can just as easily be, once free, used by any other overloaded cell in the network, and are not bound by bandwidth restrictions like those used by the cell. In other words, a satellite that provides service for a network of cells can allow its own bandwidth to be used by any cell that needs it without being bound by terrestrial bandwidth and location restrictions.

Television Signals

Satellites have been used for since the 1960's to transmit broadcast television signals between the network hubs of television companies and their network affiliates. In some cases, an entire series of programming is transmitted at once and recorded at the affiliate, with each segment then being broadcast at appropriate times to the local viewing populace. In the 1970's, it became possible for private individuals to download the same signal that the networks and cable companies were transmitting, using c-band reception dishes. This free viewing of corporate content by individuals led to scrambling and subsequent resale of the descrambling codes to individual customers, which started the direct-to-home industry. The direct-to-home industry has gathered even greater momentum since the introduction of digital direct broadcast service.

C-band

C-Band (3.7 - 4.2 GHz) - Satellites operating in this band can be spaced as close as two degrees apart in space, and normally carry 24 transponders operating at 10 to 17 watts each. Typical receive antennas are 6 to 7.5 feet in diameter. More than 250 channels of video and 75 audio services are available today from more than 20 C-Band satellites over North America. Virtually every cable programming service is delivered via C-Band.

Ku-Band

Fixed Satellite Service (FSS)

Ku Band (11.7 - 12.2 GHz) - Satellites operating in this band can be spaced as closely as two degrees apart in space, and carry from 12 to 24 transponders that operate at a wide range of powers from 20 to 120 watts each. Typical receive antennas are three to six feet in diameter. More than 20 FSS Ku-Band satellites are in operation over North America today, including several "hybrid" satellites which carry both C-Band and Ku-Band transponders. Prime Star currently operates off Satcom K-2, an FSS or so-called "medium-power" Ku-Band satellite. Alpha Star also uses an FSS-Ku Band satellite, Telestar 402-R.

Broadcasting Satellite Service (BSS)

Ku-Band (12.2 - 12.7 GHz) - Satellites operating in this band are spaced nine degrees apart in space, and normally carry 16 transponders that operate at powers in excess of 100 watts. Typical receive antennas are 18 inches in diameter. The United States has been allocated eight BSS orbital positions, of which three (101, 110 and 119 degrees) are the so-called prime "CONUS" slots from which a DBS provider can service the entire 48 contiguous states with one satellite. A total of 32 DBS "channels" are available at each orbital position, which allows for delivery of some 250 video signals when digital compression technology is employed.

DBS

DBS (Direct Broadcast Satellite) -The transmission of audio and video signals via satellite direct to the end user. More than four million households in the United States enjoy C-Band DBS.

Medium-power Ku-Band DBS surfaced in the late 1990s with high power Ku-Band DBS launched in 1994.

Marine Communications

In the maritime community, satellite communication systems such as Inmarsat provide good communication links to ships at sea. These links use a VSAT type device to connect to geosynchronous satellites, which in turn link the ship to a land based point of presence to the respective nation's telecommunications system.

Space borne Land Mobile

Along the same lines as the marine based service, there are VSAT devices which can be used to establish communication links even from the world's most remote regions. These devices can be hand-held, or fit into a briefcase. Digital data at 64K ISDN is available with some (Inmarsat).

Satellite Messaging for Commercial Jets

Another service provided by geosynchronous satellites is the ability for a passenger on an airborne aircraft to connect directly to a land based telecom network.

Global Positioning Services

Another VSAT oriented service, in which a small apparatus containing the ability to determine navigational coordinates by calculating a triangulating of the signals from multiple geosynchronous.

MICROWAVE COMMUNICATION SYSTEMS

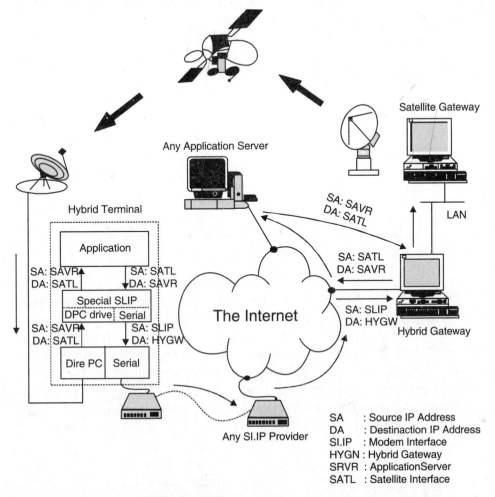

Fig 14.11. Various Satellite Applications

Why Use Microwaves for satcoms?

Frequencies above about 30 MHz can pass through the ionosphere and so are available for communicating with satellites and other extra-terrestrial sources. Frequencies below 30MHz are liable to be reflected by the ionosphere at certain stages of the sunspot cycle. The ionosphere consists of several layers of ionized gas that alter in height during the 24-hour daylight cycle. The ionosphere has an effect on satellite communications even if it does not completely prevent them.

Frequencies from about 100MHz to 2 GHz are used for communicating with low earth orbit satellites (LEOs). Since the range from ground station to satellite is only a few hundreds of km, it is not necessary to use high gain ground based antennas. Of course, there is a direct link between the beam divergence angle of an antenna and its directivity and its gain.

Microwave Frequency Bands are defined below:

Letter Designation	Frequency Range
L Band	1 to 2 GHz
S Band	2 to 4 GHz
C Band	4 to 8 GHz
X Band	8 to 12 GHz
K_u	12 to 18 GHz
K Band	18 to 26 GHz
K_a	26 to 40 GHz
Q Band	30 to 50 GHz
U Band	40 to 60 GHz
V Band	46 to 56 GHz
W Band	56 to 100 GHz

"A highly directional antenna concentrates most of the radiated power along the antenna "boresight". There is no functional difference between receive and transmit modes of an antenna except that power flow is directed inwards to the receive antenna and outwards from the transmit antenna. An antenna has the same efficiency, directivity, and polarization characteristics in receive and transmit modes. This property is called "**reciprocity**" and occurs because of the symmetry of the electromagnetic equations when the direction of time is reversed."

A high gain antenna is therefore very directional and has to be pointed with correspondingly high precision. It is an advantage if it does not have to be moved in azimuth and elevation. This restricts the use of high gain antennas largely to geostationary satellite applications.

Typically at 12 GHz the pointing accuracy needed for a 1 meter diameter dish is of the order of a degree or two of arc. Sophisticated tracking control apparatus would be required to keep such a dish pointing at a LEO satellite, particularly for mobile applications.

One way around the steerability problem is to use an electronically steerable "phased array" antenna. These have intrinsically rather less potential gain than a dish antenna, and require a high degree of electronic microwave complexity.

Frequencies from 1-30 GHz are usually called "**microwave**". From 30 GHz, to say 300 GHz the frequencies are referred to as "**millimeter wave**". Above 300 GHz optical techniques take over, these frequencies are known as "far infrared" or "quasi optical". Guided wave techniques are only used up to about 100GHz, higher frequencies use optical bench techniques and free space propagation to get the energy from one part of a circuit to another. This may occur in "**overmoded waveguide**", where the microwave energy is concentrated on the axis of the waveguide and falls off at the guide walls.

Above about 30 GHz the attenuation in the atmosphere due to cloud, rain, hydro-meteors, sand, and dust makes a ground to satellite link unreliable. Such frequencies may still be used for satellite-to-satellite links in space. It is arguable that for such applications optical technology is better than microwave carrier technology, particularly in view of the extensive development in fiber optics in recent years.

MICROWAVE COMMUNICATION SYSTEMS

6-24 GHz are useful frequencies for geostationary satcoms, since fractional bandwidths of a few percent give useful real Hz bandwidths. The **fractional bandwidth** is defined as the bandwidth in Hz divided by the carrier frequency in Hz.

Bandwidth is in short supply because there are only a limited number of geostationary satellite positional slots around the equator.

A common frequency band for satcoms is 10-14 GHz. Use is also made of 4-6 GHz but the capacity in this band is only half as much.

14.3.5 Mobile communications

The use of mobile radio-telephones has seen an enormous boost in the 1980s and 1990s.

Mobile communications are usually allocated bands in the 50MHz to 1GHz band. At these frequencies the effects of scattering and shadowing are significant. Lower frequencies would improve this performance, but HF bandwidth is not available for this purpose. The primary problems associated with mobile communication at these frequencies are:

- Maintaining transmission in the fading circumstances of mobile communication.
- The extensive investigation of propagation characteristics required prior to installation.

The use of digital mobile telephones has a number of advantages:

- Access to national and international telephone system.
- Privacy of communication.
- Data independent transmission.
- An infinitely extendable number of channels.

Mobile communication works by limiting transmitter powers. This restricts the range of communication to a small region. Outside this region, other transmitters can be operated independently. Each region is termed a **cell**. These cells are often represented in diagrams as hexagons. In practice the cell shape is determined by local propagation characteristics. Together the cells will completely cover the area supplied with mobile communication coverage (Figures - 14.12 and 14.13).

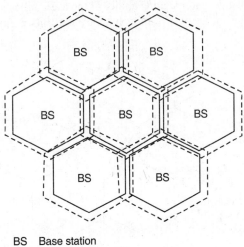

BS   Base station
—— Ideal all coverage
- - - Actual radio coverage

Fig. 14.12. Use of cells to provide geographical coverage for mobile phone service.

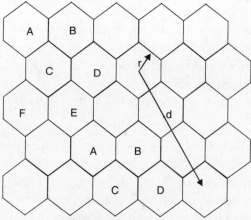

BS Base station
— Ideal call coverage
— Actual radio coverage

Fig. 14.13. Frequency re-use in cells.

Within each cell, the user communicates with a transmitter within the cell. As the mobile approaches a cell boundary, the signal strength fades, and the user is passed on to a transmitter from the new cell. Each cell is equipped with cell-site(s) that transmit/receive to/from the mobile within the cell. Within a single cell, a number of channels are available. These channels are (usually) separated by frequency. Then a mobile initiates a call, it is assigned an idle channel within the current cell by the **mobile-services switching centre (MSC)**. User uses the channel within the cell until he reaches the boundary. He is then allocated a new idle channel within the next cell.

The heart of the mobile telephone network is the MSC. Its task is to acknowledge the paging of the user, assign user a channel, broadcast user dialed request, return the call. In addition it automatically monitors the signal strength of both transmitter and receiver, and allocates new channels as required. This latter process, known as **hand-off**, is completely hidden to the user, although is a major technical problem. In addition, the MSC is responsible for charging the call, shown in figure—14.14. The decision making ability of the MSC relies to a great extent on modern digital technology. It is the maturity of this technology that has permitted the rapid growth of mobile communications.

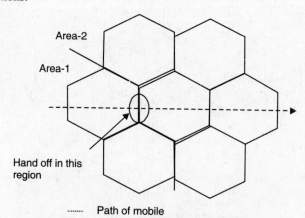

Fig. 14.14. Hand-off between cells.

The principle problem with mobile communication is the variation in signal strength as the communicating parties move. This variation is due to the varying interference of scattered radiation

MICROWAVE COMMUNICATION SYSTEMS

— **fading**. Fading causes rapid variation in signal strength. The normal solution to fading, increasing the transmitter power, is not available in mobile communication where transmitter power is limited.

The presence of fading has a severe effect on the BER (Bit-Error Rate). It can reduce the BER by several orders of magnitude over the non-fading case. In addition, some modulation schemes cope better than others. Coherent AM modulation is better than coherent FM in the fading environment. In any event BERs of 10^{-3} are typical, in comparison with BERs of 10^{-5} for non-fading channels.

The installation of a mobile telephone system requires a large initial effort in determining the propagation behavior in the area covered by the network. Propagation planning, by a mixture of observation and computer simulation, is necessary if the system is to work properly. At UHF and VHF frequencies, the effects of obstructions are significant. Some of the effects that need to be considered are:

- **Free space loss.**
- **Effect of street orientation.** Streets have a significant waveguide effect. Variations of up to 20dB have been measured as a result of street direction.
- **Effects of foliage.** Propagation in rural areas is significantly affected by the presence of leaves. Variations of **18dB** between summer and winter have been observed in forested areas.
- **Effect of tunnels.** Tunnels can introduce signal attenuation of up to 30dB according to the tunnel length and frequency.

Frequencies for Mobile communication

VHF / UHF ranges for mobile radio
- Simple and small antenna for car
- Deterministic propagation characteristics and reliable connection

SHF and higher for directed radio links
- Simple and small antenna
- Large bandwidth

Wireless LANs uses frequencies in UHF to SHF range
- Some system planned to used up to EHF
- Weather depend fading and signal loss caused due to heavy rainfall
- Limitations due to absorption by water and molecules.

First Generation Cellular Systems and Services

1970s	Developments of radio and computer technologies for 800/900 MHz mobile communications
1976	WARC (World Administrative Radio Conference) allocates spectrum for cellular radio
1979	NTT (Nippon Telephone & Telegraph) introduces the first cellular system in Japan
1981	NMT (Nordic Mobile Telephone) 900 system introduced by Ericsson Radio System AB and deployed in Scandinavia
1984	AMPS (Advanced Mobile Phone Service) introduced by AT&T in North America

Second Generation Cellular Systems and Services

1982	CEPT (Conference Europeenne des Post et Telecommunications) established GSM to define future Pan-European Cellular Radio Standards
1990	Interim Standard IS-54 (USDC) adopted by TIA (Telecommunications Industry Association)
1990	Interim Standard IS-19B (NAMPS) adopted by TIA
1991	Japanese PDC (Personal Digital Cellular) system standardized by the MPT (Ministry of Posts and Telecommunications)
1992	Phase I GSM system is operational
1993	Interim Standard IS-95 (CDMA) adopted by TIA
1994	Interim Standard IS-136 adopted by TIA
1995	PCS Licenses issued in North America
1996	Phase II GSM operational
1997	North American PCS deploys GSM, IS-54, IS-95
1999	IS-54: North America IS-95: North America, Hong Kong, Israel, Japan, China, etc GSM: 110 countries

The aims of the GSM system were:
- Good speech quality
- Low terminal cost
- Low service cost
- International roaming
- Ability to support hand-held portables
- A range of new services and facilities (ISDN)

Third Generation Cellular Systems and Services

- IMT-2000 (International Mobile Telecommunications-2000):
- Fulfill one's dream of anywhere, anytime communications a reality.

Key Features of IMT-2000 include:
- High degree of commonality of design worldwide;
- Compatibility of services within IMT-2000 and with the fixed networks;
- High quality;
- Small terminal for worldwide use;
- Worldwide roaming capability;
- Capability for multimedia applications, and a wide range of services and terminals.

SUMMARY

- Microwave frequencies are normally used by Radar, satellite, and mobile communication.
- Few microwave communication systems are discussed here in brief, which are as follows:
    - (*i*) Microwave Link Communication
    - (*ii*) Troposcatter communication
    - (*iii*) Line-of-Sight (LOS) Communication
    - (*iv*) Satellite Communication
    - (*v*) Mobile Communication
- Microwave transmission is line of sight transmission.

MICROWAVE COMMUNICATION SYSTEMS

- **Advantages of microwave transmission:**
1. They require no right of way acquisition between towers.
2. They can carry high quantities of information due to their high operating frequencies.
3. Low cost land purchase: each tower occupies small area.
4. High frequency/short wavelength signals require small antenna.
- **Disadvantages of microwave transmission:**
1. Attenuation by solid objects: birds, rain, snow and fog.
2. Reflected from flat surfaces like water and metal.
3. Diffracted (split) around solid objects
4. Refracted by atmosphere, thus causing beam to be projected away from receiver.
- A lne-of-sight microwave system consists of one or more point-to-point hops. Each hop is designed to be integrated into a wordwide communication network.
- At microwave frequencies, the atmosphere has a scattering effect on electromagnetic fields that allows for over-the-horizon communications. This type of communication is called tropospheric scatter, or troposcatter for short.
- The concept of satellite communication was first conceived by Arthur .C. Clarke groundbreaking article in October 1945.
- A **satellite** is an object that orbits another object.
- Satellite is a Radio Frequency (RF) repeater in orbit. Most communication satellites are in Geostationary Orbit. The primary purpose of a communication satellite is to relay the modulated microwaves between widely separated earth stations. This is accomplished by picking up the modulated microwaves on a certain frequency and retransmitting them on a different frequency.
- The satellite **Transponder** shifts the signal from the **Uplink frequency** to a different **Downlink frequency**. The path from an earth station to satellite is known as **Uplink** and the returning path from a satellite to earth station is known as **Downlink**. **Uplink and Downlink Frequencies**
- Communications through satellites are either passive or active. The first communications satellites were passive. **Passive satellites** reflect signals they receive to earth or other satellites. They have no transmitter or energy signals.
- **Active communication satellites** receive signals from Earth, electronically strengthen the signals, and transmit the signals to Earth. Active satellites can collect the data they need and transmit their own radio signals to earth (Satellites).
- Satellite's applications include:
1. Fixed satellite communication systems
2. Broadcast satellite communications systems
3. Land mobile communications system
4. Maritime satellite communications systems
5. Aeronautical satellite communications systems
6. Radio determination satellite systems
7. Satellite personal communications systems
8. Satellite internet access systems
- Mobile communications are usually allocated bands in the 50MHz to 1GHz band. At these frequencies the effects of scattering and shadowing are significant.

- The principle problem with mobile communication is the variation in signal strength as the communicating parties move. This variation is due to the varying interference of scattered radiation – **fading.**

REVIEW QUESTIONS

1. Explain microwave communication systems in briefs. Mention the various applications in communication.
2. Explain microwave link communication with the help of block diagrams.
3. Discuss LOS in brief.
4. Write down the various applications of satellite communication in brief.
5. Write short notes on following:
    (*i*) Active and passive satellite
    (*ii*) Advantages of satellite communication
    (*iii*) Mobile communication
    (*iv*) Tropospheric scattering communication
    (*v*) Advantages of LOS